Temperate Fruits and Nuts

A Way Forward for Enhancing Productivity and Quality

The Editors

Dr K.L. Chadha (Padma Shri Awardee) is currently President of The Horticultural Society of India as also Adjunct Professor (Hort.) at IARI. He received his B.Sc. (Agri.) and M.Sc. (Hort.) degrees from Panjab University, Chandigarh and Ph.D. from Indian Agricultural Research Institute (IARI), New Delhi. He has served in various capacities at IARI, Panjab Agricultural University, Ludhiana and Central Mango Research Station, Lucknow. He was the Director, IIHR, Bangalore, Horticulture Commissioner and Executive Director, National Horticulture Board, Govt. of India; Deputy Director General (Hort.), ICAR and ICAR-National Professor (Hort.) at IARI.

Dr Nazeer Ahmed, Former Director, ICAR-Central Institute of Temperate Horticulture, Srinagar, is currently Vice Chancellor, Sher-e-Kashmir University of Agricultural Sciences and Technology of Kashmir, J&K. Dr Ahmed served in various capacities as Scientist, Senior Scientist, Associate Director Research, Professor and Head, Dean Agriculture and also as Director Resident Instruction cum Dean Post Graduate Studies at SKUAST-K. During his 30 years of service, he has contributed significantly as a teacher, researcher, extension worker and administrator. Dr Ahmed is recipient of several prestigious awards and honours including Rajbhasha Gaurav Award, Dr R.S. Paroda Award, Dr M.H. Marigowda National Endowment Award, Dr Kirti Singh Gold Medal and Vijay Shree Award.

Dr S.K. Singh is presently Principal Scientist & Professor at the Division of Fruits and Horticultural Technology, IARI, New Delhi. He received M.Sc. and Ph.D. degrees in Horticulture from Banaras Hindu University, Varanasi and joined Agricultural Research Service of ICAR in 1994. He is serving IARI for the last 20 years and is involved in research & teaching and has developed improved varieties of grape. He underwent post-doctoral training of Tropical Fruit Crops Biotechnology at the University Florida, USA under DBT Overseas Fellowship programme.

Dr P. Kalia is presently the Coordinator of School of Horticultural Sciences and Head Division of Vegetable Science at Indian Agricultural Research Institute, New Delhi. He has significantly contributed in the field of vegetable breeding and developed 28 varieties/hybrids of different vegetable crops. He pioneered popularising exotic vegetable crops *e.g.*, broccoli and leek by developing indigenous varieties for diversification and nutrition.

Dr Kalia is recipient of N. N. Mohan Memorial Gold Medal, Commonwealth scholarship, Dr Kirti Singh Gold Medal and Himachal Gaurav Awards. He is also fellow of Horticultural Society of India, Indian Society of Vegetable Science and National Academy of Agricultural Sciences (NAAS).

Temperate Fruits and Nuts

A Way Forward for Enhancing Productivity and Quality

Editors

K.L. Chadha

Nazeer Ahmed

S.K. Singh

&

P. Kalia

Estd. 1942

The Horticultural Society of India

New Delhi

2016

Daya Publishing House®

A Division of

Astral International Pvt. Ltd.

New Delhi – 110 002

ISBN 978-93-5124-742-5 (Hardbound)
ISBN 978-93-5130-946-8 (International Edition)

Publisher's Note:

Every possible effort has been made to ensure that the information contained in this book is accurate at the time of going to press and the publisher and author cannot accept responsibility for any errors or omissions, however caused. No responsibility for loss or damage occasioned to any person acting, or refraining from action, as a result of the material in this publication can be accepted by the editor, the publisher or the author. The Publisher is not associated with any product or vendor mentioned in the book. The contents of this work are intended to further general scientific research, understanding and discussion only. Readers should consult with a specialist where appropriate.

Every effort has been made to trace the owners of copyright material used in this book, if any. The author and the publisher will be grateful for any omission brought to their notice for acknowledgement in the future editions of the book.

Published by : **Daya Publishing House®**
A Division of
Astral International Pvt. Ltd.
– ISO 9001:2015 Certified Company –
4736/23, Ansari Road, Darya Ganj
New Delhi-110 002
Ph. 011-43549197, 23278134
E-mail: info@astralint.com
Website: www.astralint.com

Laser Typesetting : **Classic Computer Services,** Delhi - 110 035

Printed at : **Replika Press Pvt. Ltd.**

Preface

The ***National Conference on Temperate Fruits and Nuts: A Way Forward for Enhancing Productivity and Quality*** is being held at Srinagar, Jammu & Kashmir 6th to 9th November, 2015 after the request received from the ICAR-Central Institute for Temperate Horticulture, Srinagar. The Horticultural Society of India, New Delhi has successfully organized several national and international events in the past. Since, 2004 six congresses have been organized at New Delhi, Shillong, Bhubaneswar, New Delhi, Ludhiana and Coimbatore during 2004, 2007, 2008, 2010, 2012 and 2014. These scientific gatherings have generated a number of successful recommendations and have been projected in the appropriate fora by consideration of relevant agencies involved in implementing of programmes and policies.

In 2014-2015, India has emerged as a major producer of horticultural crops (280 million tonnes surpassing food grain production of 251 million tonnes) and is placed second after China in both fruit and vegetable production. Also horticulture in India with its higher annual growth rate has become a major contributor to growth of Indian agriculture. It is, therefore, important that horticultural crop production is given more emphasis so that it could sustain the desired growth rate in agriculture sector across the diverse agro-climatic conditions and provide skill development and job opportunities to the emerging youth force.

Horticulture growth and development needs to be inclusive to benefit various stakeholders in the system. Hence, it is essential to reassess our programmes and devise overall strategies to make this sector more inclusive under the rapidly changing consumer and market driven economy and rapidly depleting environmental conditions especially in Temperate Himalayan region. Today, the challenges are many namely, nutritional security, threat of climate change, need for new plant genotypes for diversified end uses, production deficit and gluts, lack of infrastructure in commodity value-chain, retail chains and organized market systems, farmer producer companies, and export promotion. It is, therefore, necessary that Temperate

Horticulture sector gears to meet new challenges and to bring an overall changes in horticulture per se and also in the allied sectors leading to redefining the short and long term goals in research, production system, education and human resource development, public and private sector collaboration, market intelligence and auction system along with diligent policies to boost production, productivity, processing and export of fresh and processed horticultural produce and products.

The present national conference is quite focused on Temperate Fruits and Nuts is timely as no such scientific meeting was held for almost a decade in the past. During this conference, there are 31 lead subject matter experts, besides 45 oral papers from young scientists from different institutions and about 284 poster papers from R & D workers across the country in 8 broad theme areas. This publication contains the details of all lead papers presented in the conference. I am glad to share that this conference has generated overwhelming response from the entire horticulture fraternity of the country. The issues covered are quite relevant, i.e. need to reassess our strategies to meet the present and future challenges confronting temperate horticulture, namely, status of temperate fruit and nut industry, genetic resource management, crop improvement, innovative technologies, efficient crop production technologies, propagation and rootstocks, enhanced input use efficiency, plant protection, postharvest management, domestic trade and export promotion.

On behalf of Horticultural Society of India, I take this opportunity to profusely thank all the members of National Advisory Committee, Local Organizing Committee, Office Bearers and Executive Council of the society, Vice-Chancellor, SKUAST-K and former Director of the ICAR-CITH, Srinagar and the scientific team and all the staff of the ICAR-CITH and SKUAST-K, Srinagar, J&K who have shouldered the responsibilities in organizing this national conference in most efficient and a grand way.

I also take the opportunity to thank all the subject matter specialists & contributors from different Institutes/ Universities from India, who have responded to our request to share their research results and thus enriching the quality of this book. I am extremely thankful to my co-editors Drs Nazeer Ahmed, Sanjay Kumar Singh and Pritam Kalia for their untiring help in timely bringing out this compilation.

It is hoped that the information given in this publication will be of academic and practical interest to all engaged in Horticulture R&D and act as a ready reference for the above emergent and future issues.

(K.L. Chadha)
President, Horticultural Society of India
Chairman, National Organizing Committee

Contents

List of Contributors

Ahmed, N.
Former Director, ICAR-Central Institute of Temperate Horticulture(CITH),Vice-Chancellor, Sher-e-Kashmir University of Agricultural Sciences and Technology of Kashmir, Srinagar-180 009 J&K
E-mail: dnak59@rediffmail.com

Ahmed, Sheikh Bilal
Professor, Division of Entomology, Sher-e-Kashmir University of Agricultural Sciences and Technology of Kashmir Srinagar - 190 025, J&K
E-mail: sheikhbilala@gmail.com

Banday, F.A.
Head, Division of Fruit Science, Sher-e-Kashmir University of Agricultural Sciences and Technology of Kashmir, Srinagar - 190 025, J&K
E-mail: headfruitscience@gmail.com

Chandel, J.S.
Principal Scientist , Department of Fruit Science, Dr Y.S. Parmar University of Horticulture and Forestry, Solan - 173 230, H.P.
E-mail chandeljs@yahoo.co.in

Chaudhury, Rekha
Professor, Tissue Culture and Cryopreservation Unit, ICAR- National Bureau of Plant Genetic Resources, Pusa Campus, New Delhi - 110 012
E-mail: rscientist58@gmail.com; rekha@nbpgr.ernet.in

Dhaliwal, H.S.

Dean, College of Agriculture, Punjab Agricultural University, Ludhiana - 141 004, Punjab
E-mail: dhaliwal_mandhali@pau.edu

Handa, Anil

Professor, Department of Plant Pathology, Dr Y.S. Parmar University of Horticulture and Forestry, Solan - 173 230, H.P.
E-mail: anilhanda@msn.com

Jindal, K.K.

Former Director Research, Dr Y.S. Parmar University of Horticulture and Forestry, Solan - 173 230, H.P. and CAU, Imphal - 795 001, Manipur
E-mail: ecofriendlyhorticulture@gmail.com

Kachroo, Dileep

Chief Scientist and Head, FSR Centre and Head, Division of Agronomy, Sher-e-Kashmir University of Agricultural Sciences and Technology of Jammu, Jammu - 180 009, J&K
E-mail: dileepkachroo@rediffmail.com

Kaur, Rajinder

Professor, Department of Biotechnology, Dr Y.S. Parmar University of Horticulture and Forestry, Solan - 173 230, H.P.
E-mail: rkaur_uhf@rediffmail.com

Kumar, Jayant

Associate Director, Regional Horticultural Research and Training Station, Dr Y.S. Parmar University of Horticulture and Forestry, Bajaura, Kullu - 175 125, Himachal Pradesh
E-mail: jayantkumar_in@yahoo.com

Kumar, K.

Professor, Department of Fruit Science, Dr Y.S. Parmar University of Horticulture and Forestry, Solan - 173 230, H.P.
E-mail: drkrishankumar@gmail.com

Kumar, N.K. Krishna

Deputy Director General (HS), ICAR, KAB-II, New Delhi - 110 012
E-mail: daghort@icar.org.in; ddghort@gmail.com

Kumar, Rohitashw

Associate Professor, Division of Agricultural Engineering, Sher-e-Kashmir University of Agricultural Sciences and Technology of Kashmir Srinagar - 190 025, J&K
E-mail: rohituhf@rediffmail.com

Mankotia, M.S.

Principal Scientist , Dr Y.S. Parmar University of Horticulture and Forestry, Regional Horticultural Research and Training Station, Mashobra, Shimla - 171 007, H.P.
E-mail: msmankotia2006@yahoo.co.in

Mukhtar, Malik

Principal Scientist, Research Centre for Residue and Quality Analysis, Sher-e-Kashmir University of Agricultural Sciences and Technology of Kashmir, Srinagar - 190 025, J&K
E-mail: drmalikmukhtar@yahoo.com

Qazi, Nissar A.

Professor & Head, Division of Plant Pathology, Sher-e-Kashmir University of Agricultural Sciences and Technology of Kashmir Srinagar - 190 025, J&K
E-mail: qazinissar11@gmail.com

Rana, J.C.

Head, Division of Germplasm Evaluation, ICAR-NBPGR, Pusa Campus, New Delhi - 110 012
E-mail: ranajc2003@yahoo.com

Shah, G.H.

Member, J&K Muslim Wakaf Board, Srinagar - 190 010, J&K
E-mail: drshahgh7@yahoo.co.in

Sharma, N.

Professor & Head, Department of Fruit Science, Dr Y.S. Parmar University of Horticulture and Forestry, Solan - 173 230, H.P.
E-mail: nsharma.sol@gmail.com

Sharma, R.C.

Director of Research, Dr Y.S. Parmar University of Horticulture and Forestry, Solan - 173 230, H.P.
E-mail: dresuhf@yahoo.com

Sharma, Satish Kumar

Joint Director Research (Horticulture), Dr Y.S. Parmar University of Horticulture and Forestry, Solan - 173 230, H.P.
E-mail: satishsharma_27@yahoo.com

Singh, Brahma

Former Director, DIHAR and Directorate of Life Science, DRDO and Founder President ISPC, CPCT, IARI, New Delhi - 110 054
E-mail: brahma88@gmail.com

Singh, D.B.

Principal Scientist, ICAR-Central Institute of Temperate Horticulture, Srinagar - 190 007, J&K
E-mail: deshbsingh@yahoo.co.in

Thakur, Vijay Singh

Vice-Chancellor, Dr Y.S. Parmar University of Horticulture and Forestry, Solan - 173 230, H.P.
E-mail: thakurvs1606@gmail.com

Vaidya, Devina

Principal Scientist, Department of Food Science and Technology, Dr Y.S. Parmar University of Horticulture and Forestry, Solan - 173 230, H.P.
E-mail: devinavaidya@yahoo.com

Verma, K.S.

Dean, College of Forestry, Dr Y.S. Parmar University of Horticulture and Forestry, Solan - 173 230, H.P.
E-mail: ksvblp16@gmail.com

Verma, M.K.

Principal Scientist, Division of Fruits and Horticultural Technology, ICAR-Indian Agricultural Research Institute, New Delhi - 110 012
E-mail: mahenicar10@gmail.com

Wani, S.A.

Professor & Head, Division of Agricultural Economics and Marketing, Sher-e-Kashmir University of Agricultural Sciences and Technology of Kashmir, Srinagar - 190 025, J&K
E-mail: dr.shabirwani@rediffmail.com

Chapter 1

Potential and Strategies for Development of Temperate Fruits and Nuts in North Western Region

Vijay Singh Thakur

Dr Y.S. Parmar University of Horticulture and Forestry, Solan – 173 230, H.P.

E-mail: thakurvs1606@gmail.com

1. INTRODUCTION

The geographical features and agro-climatic conditions prevailing in entire north- western and north-eastern hilly states of India are conducive for the cultivation of temperate fruit crops. The major temperate fruit growing states are Jammu and Kashmir, Himachal Pradesh and Uttarakhand falling in North-West Himalayas. The north–eastern states of the country contributing in temperate fruit production to a limited extent include Arunachal Pradesh, Nagaland, Meghalaya, Manipur and Sikkim. At present temperate fruits constitutes about 7.56 per cent of total area under fruits in the country yet their production constitutes only 3.65 per cent of the total production (Anon, 2014). Apple is the most important which constitute 57.31 per cent share in area and 76.92 per cent in production among the temperate fruits. Although, a lot of area has been brought under cultivation of pome and stone fruits in the country yet there is a great potential for further expansion of area under fruit crops in both north-west and north-east Himalayan regions of India. Due to introduction and adaptation of low chill varieties the potential for the cultivation of temperate fruits has extended to lower elevations which were considered marginal for temperate fruit cultivation. In Himachal Pradesh, the total area under fruits is 2, 20,706 ha, out of which of pome fruits are cultivated in 1, 79,886 ha and stone fruit in 17,741 ha. Similarly, out of total fruit production of 7,51,938 MT, pome fruits production is 6,45,419 MT and stone fruit production is 19,981 MT. Among nuts, walnut and pecan nut are mainly cultivated in the state in an area of 4,539 and 855 ha, respectively. The production of walnut is 1,568 MT and pecan nut is 124 MT (Anon, 2014). Apple is the major temperate fruit accounting for 48.8 per cent of total area under fruits in Himachal Pradesh and to the extent of 65.02 per cent of area under temperate fruits. During the last five decades, the productivity has not kept pace with the rapid increase in area under pome and stone fruits due to biotic and abiotic problems. The average productivity of apple has been fluctuating between 2-9 MT/ha and stone fruits 1.5–2 MT/ha, which is far below the average productivity of advanced countries like China, New Zealand, USA and Italy. Decreased productivity and quality of pome and stone fruits due to adverse climate, monoculture of varieties, poor quality planting material, use of seedling rootstocks, low density plantation, poor soil nutrients and moisture conditions and steep topography, rain-fed orcharding, senile orchards and in-efficient insect pests and disease forecasting and management has become a serious concern. Various strategies like introduction and adoption of spur type and high yielding varieties, use of clonal rootstocks and disease-free quality planting material, high density plantations, proper canopy management, integrated water and nutrient management, application of good horticultural practices and integrated pest and disease management, can lead to improved productivity and quality of pome and stone fruits. In apple, productivity of 75.9 MT/ha in Vance Delicious variety on M9 rootstock is achievable under high density plantation in experimental blocks in Himachal Pradesh. Similarly in peaches, with the use of high yielding varieties and proper training systems productivity of 9 MT/ha with better quality fruits has been obtained under high density planting.

2. CAUSES OF LOW PRODUCTIVITY AND POOR FRUITS QUALITY

(a) Monoculture of Standard Delicious Apple

Genetically standard Delicious apple varieties have poor fruit bearing potential and tendency towards alternate bearing. There is considerably low proportion of spur type Delicious and non-Delicious apple varieties in the state. However, importance of these varieties has been realized by stakeholders and progressive farmers are opting for varietal diversification. The state department of horticulture has also taken initiative in importing plants of spur type Delicious and non Delicious apple varieties on clonal rootstocks from advanced countries of the world. Similarly in nut crops, the low productivity and poor fruit quality is associated with the terminal bearing cultivars and the varieties of seedling origin.

(b) Changing Climatic Scenario

The occurrence of spring frost and low temperature at the time of flowering and fruit setting adversely affects production of fruits. The fluctuating fruit crop production statistics for the past three decades suggest the major role of climate in determining the success or failure of the fruit crops. The most vulnerable zone to the influence of low temperature is located between 5,000 to 6,500 ft a.m.s.l. where good spring *season with* adequate sunshine promote apple flowering during mid March to mid April. But fluctuating temperature coinciding flowering period particularly rains accompanied by low temperature inhibits the transfer of pollen because of restricted bees activity and washing off of pollen as well as restricts the pollen tube growth pre-requisite for proper fruit set. The flowers are damaged at temperatures below -2.2°C and bee activity is completely stopped below 4.4°C. Thus for successful apple cultivation, the areas vulnerable to fluctuating climatic conditions during flowering and fruit set should be avoided. It is also on observation that in areas as well as particular years, when winter precipitation in the form of snowfall is less, consequently the winters are comparatively warm. In such situations chilling requirements of the plants is not met leading to bud break is delayed as well as flowering percentage is hampered leading to poor production especially sensitive crop like apple (Awasthi *et al.*, 1986). The drought spells during fruit growth and development also adversely affect the fruit yield particularly in stone and nut crops.

(c) Inadequate Pollinizers

The orchards in Himachal Pradesh have provision of 25 to 33 per cent pollinizers as recommended for effective pollination. Most of apple orchards have less than 10 per cent of pollinizers proportions that poor varietal diversity and pollinizers. There must be at least 4-5 pollinizing varieties for effective pollination and fruit set. The recent introduction of pollinizing as well as commercial varieties such as Gala group, Akene, Fuji and Granny Smith in addition to traditional pollinizers is step forward to improve productivity in the state. The crab apples should be planted on bunds and boundaries for effective supplementing pollinizer proportion.

(d) Allocation of Second Grade Land for Orchards

Most of old orchards have been planted on poor sloppy lands whereas the lands are earmarked for agricultural crop production in the past. Such areas have low fertility and undulating gradient which pose problems for various management operations and are unproductive.

(e) Fragmented Small Holdings

The management of fragmented and small land holdings is difficult as compared to contiguous large holdings. The diseases and pest management practices are also not effectively applicable as the disease and pest incidence has the impact of the surroundings which are not always assured of proper management practices. Small holdings are also uneconomic from the point of view of fruit harvesting and subsequent operations.

(f) Mindset of Orchardists towards Vigorous and Tall Unmanageable Trees

Most of the orchardists in the state are still hesitant for adoption of clonal rootstocks and spur type varieties due to the misconception and short stature and low yield per plant compared to standard trees on seedling rootstocks. On the contrary the fact is that the high density planting on size controlling rootstocks are more efficient with respect to quality and productivity of apple per unit area but needs efficient management. The management practices are convenient in HDP as well.

(g) Faulty Pruning

Most of orchardists are dependent on hired unskilled pruners particularly after intensification of apple cultivation in the state who are not scientifically trained. Such pruners usually carry out indiscriminate cuts and disturb the canopy of the tree having mindset of growing large trees yielding more than ten boxes. Under such kind of pruning most of the metabolites are utilized for vegetative growth and spur development is adversely affected. The recent trend of pruning bearing apple trees emphasizes more of thinning and less of heading back cuts, which results in diversion of metabolites for spur formation. At low elevations where the vegetative growth is excessive due to warm conditions, having pruning promotes more vegetative growth and less fruitfulness apart from infection. This is also a major reason for unfruitfulness in orchards at lower elevations.

(h) Rainfed Growing Conditions

More than 90 per cent temperate fruits and nut orchards in Himachal Pradesh are rainfed. Moreover, the rainfall is not well distributed throughout the year, as most of precipitation occurs during mid June to mid August. Generally, severe drought conditions prevail during critical stages of growth and development which affects fruit growth and development adversely. Moisture conservation and water harvesting techniques need to be strengthened to provide irrigation water at critical stages of fruit development in order to increase the productivity of fruits in the state.

(i) Poor Soil Strata and Fertility

Himachal apple orchards are on sloppy terrains having poor soil strata, shallow soil depth and poor soil fertility status. Orchards which are planted on the slopes, run off losses render the soil nutritionally and structurally poor. Usually middle and top of the slopes have deep soil, whereas the bottom of the slope is made up of poor soil strata. As such the performance of the fruit trees varies from top to bottom of the slopes. Where soils are not deep and fertile, the health of the trees is not sound and the yields are proportionately reduced.

(j) Low Planting Density

The old orchards have been planted under traditional systems of planting at a spacing of 20-25 ft on seedling root stocks. Such orchards take 15-20 years to come to commercial fruiting in comparison to the dwarf orchards which come to commercial fruiting after 6-7 years. The superiority of high density planting in apple employing MM 106 rootstock has been demonstrated, where yield up to 35-40 tonnes/ha has been advanced. However, the availability of plant material on dwarfing clonal rootstock has remained a persistent problem so far in temperate fruits and nut crops.

(k) Lack of Pollinators

Honeybees are the major agents besides many other wild pollinators for effective cross-pollination. Over the years, the population of honeybees and other pollinators has declined due to indiscriminate use of pesticides. Placement of honeybees in the orchards have also not picked up as desired due to non-availability of bee hives which has also contributed to low fruit set and poor yields.

(l) Increasing Proportion of Old and Senile Orchards

As per survey conducted by the Department of Horticulture, Himachal Pradesh during the year 2002-03 an area of 46,829 ha was under old plantations, which had completed their economic life due to gradual orchard decline and need to be rejuvenated/or replaced with new plantations. Orchards more than 30 years (pome and nut crops) and 20-25 years (stone fruits) of age have shown much more unfruitfulness than the young orchards. Such orchards do not produce adequate annual extension growth because of improper management inadequate fertilization and pruning. The senile orchards needs quite replacement with modern ordering systems

3. STRATEGIES TO IMPROVE PRODUCTIVITY AND FRUIT QUALITY

(a) Varietal Diversification

(i) Apple

Apple is the most important temperate fruit and accounts for 48.79 per cent in area and 83 per cent in production in Himachal Pradesh. Although it is cultivated in sizeable area in the state yet there is further great potential for area expansion in one lakh hectare barren cultivable area and the land which are under other vegetable and cereal crops in temperate zone. Beside, new area expansion, there is also potential to replant apple trees in approximately 40,000 ha area needed to be vacated by uprooting of declined apple trees, which have outlived their economic life. Monoculture of old Delicious cultivars and their genetic degeneration over the years ever since their introduction has led to low productivity and poor quality. The new initiatives like diversification of varieties through introduction and evaluation of promising spur type and high colouring strain early maturing, low chill and late maturing cultivars have been made in recent years to improve productivity, fruit quality and to make the availability of fresh fruits in the market from June to October. Spur type varieties like Super Chief, Scartlet Spur-II, Oregon Spur-II, Schlect Spur and Red Chief, high colouring strain like Vance Delicious, Top Red, Red Velox, Jeromine, Bright –N-Early, Gale Gala and Early Red One and low chill varieties like Anna, Dorsette Golden and late maturing varieties like Red Fuji, Coe Red Fuji, Aullvial Fuji and Granny Smith have been recommended for different agro geographical situation of the state. Spur type and high colouring strains have shown excellent performance in term of yield, colour development and fruit quality in lower elevation which have otherwise become marginal for apple cultivation due to climatic change.

(ii) Apricot

Sweet kernel type apricot has great potential for commercial cultivation in high hills and dry temperate regions, whereas, bitter kernel type for mid hills areas. Apricot varieties EMA mature 7-10 days earlier than the commercially grown New Castle has promise of cultivation in mid hills. In dry temperate region CITH 1 and 2, Helman and Rakhail Karpo and exotic introductions Nari, Shakarpara have been found promising.

(iii) Cherry

Cherry has emerged as a high value fruit crop in high hills of the Himachal Pradesh mainly in Shimla and Kullu districts. The low chill cultivars Sunburst, Lapins, Duerone –Neuro II, Summit and Compact Stella are performing well in some niche areas of District Shimla.

(iv) Nut crops

Due to congenial climatic conditions there are great potential for the expansion of area and improving the productivity of nut crops in the state yet the non availability of planting material of lateral bearing varieties in walnut and high yielding varieties in other nut crops is the biggest bottleneck. In nut crops existing plantation is of seedling origin which has long gestation period and produce low yield of poor quality kernel. The cultivars like Lara, Gillet, Chandler, Serr and Tulerae in walnut; and Cheyenne, Chicksaw, Wichita, Kanza and Pawnee in pecan nut were found promising for higher productivity and nut quality.

(v) Peach

The areas which have become marginal for apple and pear cultivation due to climate change and rise in temperature has identified potential areas for peach cultivation. July Elberta is the commercially cultivated variety, but it mature late in the season and its maturity sometimes coincides with monsoon rains thus results in production of poor quality fruits due to infection of fruit flies. Some new varieties like Suncrest, Glo Haven and early maturing varieties like Red Haven and Sun Haven have shown promise in peach cultivation of Nectarine, Fuzzless peaches and its varieties like Snow Queen, Silver King and May Fire and Independence can be grown successfully in areas which are marginal for apple cultivation.

(vi) Pear

Pear has also got great potential as diversification crop in temperate region particularly in apple growing areas. Cultivars Bartlett in higher elevation and sand pear in lower elevation are mainly grown. Bartlett and its red colour strains Red Bartlett and Max Red Bartlett has poor fruits shelf life and alternate in bearing. The pear varieties having high fruit firmness, long shelf-life and late in maturity have potential due to higher economic returns. Cultivars like Conference, Packham's, Beurre Bosc, Concorde, Carmin and Abate Fetel have better fruit firmness and extended shelf-life and better fruit quality. Such cultivars need to be encouraged in high hills for higher productivity and fruit quality. In warmer sub mountainous areas varieties like Punjab Gold, Punjab Nector, Kieffer and China pear are recommended.

(vii) Plum

In stone fruits, plum is main fruit crop and cultivar Santa Rosa is predominantly grown and accepted cultivar in mid hills. Efforts have been made for diversification

Table 1: Diversification with some New High Yielding Varieties

Fruit	Altitude (ft a m s l)	Existing Variety(ies)	Improved Variety(ies)
Apple	3,000-4,000	Tropical Beauty, Michal and Schlomit	Anna and Dorsette Golden
	4,000-5,500	Mollies Delicious	Scarlet Spur, Gale Gala, Early Red One, Schlect Spur
	5,500-6,500	Starkrimson, Vance Delicious	Super Chief, Oregon Spur –II, Ace Spur, Jeromine, Red Velox, Red Lumb Gala, Gale Gala, Bright-N-Early
	7,000-8,500	Starking Deilcious, Red Delicious, Rich-a-Red	Top Red, Red Fuji, Granny Smith
Apricot	3000-4000	New Castle	EMA
	5000-7000	Royal	Shakarpara, CITH -2, Charmagz, Suffida
Cherry	6,000-8,000	Black Heart, Red Heart	Regina, Van,Dureone-Nero -2, Sunbrust, Stella, Lapins
Nectarine	4000-6000		Snow Queen, Silver King, Independence, May Fire
Pecan nut	3,000-4,000	Mahan, Nellis, Stuart, Burkett	Cheyenne, Chicksaw, Wichita, Kanza, Pawnee
Pear	< 4,000	Sand pear (Gola)	Punjab Gold, Punjab Necter, Le Conte, Kieffer
	4,000-5,000	Baggugosha	Flamish Beauty
	5,000-8,000	Bartlett	Concorde, Conference, Packham's, Beurre Bosc, Carmin, Abate Fetel, Canal Red
Peach	3,000-4,000	Shan-i-Punjab	Florida Sun, Pratap
	4,000-5,000	July Elberta	Suncrest, Glohaven, Redhaven,
Plum	3,500-5,000	Santa Rosa, Mariposa	Black Amber, Red Beaut, Fronteir, Durate
Walnut	5,000-8,000	Govind, Willson Wonder, Sholding Selection, Kashmir Seedling	Lara, Gillet, Chandler, Serr, Tulerae

of varieties in term of early and late in maturity with better fruit firmness and quality. The varieties like Red Beaut (early), Frontier and Mariposa (late), and Black Amber and Durate have been recommended. The European type plum and prunes have also shown promise in high hills and dry temperate areas.

(b) Rootstock and High Density Plantation

Rootstocks have profound influence on the growth, productivity and quality of fruits. In pome and stone fruits seedling rootstock are predominantly used and the plants on them are highly variable in growth and fruit production as well as in fruit quality. Majority of plantation of apple, pear and stone fruits is on seedling rootstocks and are planted under low density which is the major reason of low productivity. Clonal rootstocks are precocious, size controlling and essential for high density plantation. Clonal rootstocks like M 9, MM 106, M7, MM 111 and Merton 793 for apple, Quince C and BA29 for pear, RV- 4 for peach, RV-7 and RV17 for apricot and plum, Giesela 5 and Giesela 6 for cherry have been found promising for improving the productivity and fruit quality, and need to be commercialized (Ananda, 2001). In experiment trials, yield of 75.00 MT/ha has been obtained in Vance Delicious on M_9 rootstock with the planting density of 2,222 plants/ha. There is a need of adoption of clonal rootstocks and high density planting technology and should be advocated to the growers through adaptive trials.

Table 2: Suggested Planting Distance for different Varieties

Nature of Variety	*Rootstock(s)*	*Vigour of the Scion tree*	*Planting Distance (m)*	*No. of plants per ha*
Non-spur type	MM 111	Semi-vigorous	6.0	278
	MM 106 and M7	Semi-vigorous	4.5	494
	M 9	Dwarf	1.5	4444
Spur type	seedling	Semi-vigorous	5.0	400
	MM 111	Semi-dwarf	3.5	816
	MM 109			
	MM 106 and M7	Dwarf	3.0	1111
Pear	Seedling	Vigorous	6.0	278
	Quince C	Dwarf	3.0	1111

(c) Pollination Management

Most of the commercial cultivars of apple and cherry are self unfruitful, thus require pollinating varieties for cross pollination and good fruit set. Poor pollination due to inadequate proportion of pollinizers in the orchard and poor population of pollinators at the time of flowering is also one of main cause of low productivity. More than 80 per cent apple growers in the state have pollinizers varieties less than 20 per cent and population of natural pollinators has drastically declined due to fast deforestation and indiscriminate use of insecticides. The commercial used pollinizers varieties Red Gold and Golden Delicious have severe biennial bearing tendency. To improve the productivity, at least 25 to 33 per cent pollinizers varieties need to be planted in the orchards. Cultivars like Summer Queen, Spartan, Pippins, Gloster, Gale Gala and Red Fuji besides having commercial value is also good pollinizer and should be planted in the apple orchards. In cherry, Compact Stella is a found good pollinizer. In walnut and pecanut most of the varieties are self unfruitful, therefore three to four varieties should be interplanted in the orchards to overcome the pollination problem due to dichogamy. To improve the pollination 5-6 beehives colonies/ha need to be provided for effective pollination in apple and cherry orchard (Verma, 1990).

(d) Canopy Management

Modified central leader system of training has been traditionally used for training of apple and pear. In this system due to poor light interception in the interior canopy of the tree, the fruit bud formation and fruit set is less which leads to low productivity. Due to poor light interception the colour development on fruits and fruit quality are poor. Some new systems of training like vertical axis, tall spindle, super spindle, vertical axis and spindle bush should be adopted in apple (Chadha and Awasthi, 2005) and pear under high density planting.

The stone fruits are trained traditionally with open vase system but 'Tatura trellis' has been found better in peaches under high density for improving the productivity and quality of fruits.

(e) Nutrient Management

Balanced dose of fertilizers should be provided to temperate fruits and nuts using Integrated Plant Nutrient System (IPNS). In integrated plant nutrient system, soil fertility is maintained for sustaining crop productivity through optimizing all possible sources (organic, inorganic and biological) in an integrated manner appropriate for a particular farming situation.

(i) Organic manures

Various organic manures have potential for use in nutrient management of temperate fruits and nuts which include village compost (made from decomposition of crop residues, house hold waste and yard waste including cattle dung and urine), urban compost (made from town refuse), farm yard manure, sewage and sludge, vermicompost and concentrated organic manures like oil cakes, waste from slaughter house-bone meal, blood meal and fish meal. Farm yard manure and vermicomposts are recommended and being used for nutrients management in temperate fruits and nuts. Under undulating conditions, intercropping and cover cropping practices are also recommended. Intercropping with leguminous crops helps in fixation of atmospheric nitrogen through their root nodules in the orchard soil and cover crops helps in reducing the nutrient losses.

(ii) Inorganic fertilizers

The doses of all primary and some of secondary and micro nutrients have been recommended for temperate fruits and nut crops as per the requirement of a particular temperate fruit or nut crop. Different secondary nutrient and micro-nutrient mixtures like agromin, multiplex, tracel, micron, micnelf and nuspartin etc. are available and used in temperate fruits and nut crops for correction of nutritional deficiencies and improving the yield and quality of these fruit and nut crops.

(iii) Biofertilizers

Biofertilizers are carrier based preparations containing live or latent cells of specific micro-organisms which either fix atmospheric nitrogen or increase the availability of other macro and micro-nutrients by solubilising or mobilizing them in addition to rapid mineralization of organic waste. They help in building up of soil micro-flora and thereby the soil health. Use of bio-fertilizers is recommended for improving the soil fertility in orchards. The biofertilizers which can be used in temperate fruits include nitrogen fixers (*Rhizobium, Azotobacter* and *Azospirillum*), nutrient solubilizers (P, K, Zn and Mn solubilising bacteria) and nutrient mobilizing micro-organisms (Arbuscular mycorrhizal fungi).

4. FUTURE STRATEGIES

(a) Right Time of Fertilizer Application

Fertilizers should be applied as and when they are required by the plants. In temperate fruits and nuts, soil application of nitrogenous fertilizers should be given in 2-3 split doses during the critical stages, *i.e.* flowering and fruit set. However, phosphate and potash fertilizers should be applied during dormant season in January-February well in advance before the initiation of flowering. In fertigation, fertilizers should be applied in a number of split doses at weekly or fortnightly intervals throughout the active growth periods. Micro-nutrients should be applied through foliar application during active growth after petal fall for correction of deficiency.

There are three major components for increasing fertilizer use efficiency, *viz.*, fertigation, foliar applications of fertilizers and use of controlled release fertilizers.

(b) Fertigation

Fertigation refers to the application of water soluble and liquid fertilizers with the irrigation water (drip irrigation system) directly in to the root zone. The fertilizers are applied in a number of split doses at weekly or fortnightly intervals throughout the active growth season as per the requirement of the fruit crop at different stages of fruit growth and development. Using fertigation in apple, nutrients should be applied in 15 split doses right from green tip stage throughout fruit growth and developmental stages. In each fertigation, 75 g NPK mixture (19:19:19), 30 g urea and 25 g potassium chloride should be applied. In apricot, Raina *et al.* (2005) obtained highest yields of quality fruits (96.31 q/ha) with application of 75 per cent recommended dose of NPK in 10 split doses through fertigation right from flowering to fruit maturity by applying 81.5 g urea, 113.6 g mono ammonium phosphate and 116.7 g potassium chloride per application.

(c) Foliar Applications

There are some nutrients which are subjected to leaching following rains and require special attention for their management. Although such losses can for minimizing by a number of practices used for efficient fertilizer use but foliar application of nutrients have significance under steep topographical and rainfed conditions of temperate fruits and nut crops. Dry spells and prolonged drought conditions in the orchards create temporary deficiency of N, P, K, Ca, Mg and B as fertilizer application is confined to tree basin area which do not cover the active root zone of the tree. Such kind of nutritional deficiencies can be easily overcome by foliar application of fertilizers. Foliar application provide the best possibilities to supplement the nutrient requirements, when soil application of fertilizers become limiting due to different factors such as steep gradient of the soil, poor aeration, drought conditions, root and nutrient interactions. A combination of soil and foliar applications of fertilizers is considered the ideal to optimize growth, yield and plant nutrient status in temperate climate. Recommendations of some macro and micro nutrients for temperate fruits and nuts are given in Table 3.

Table 3: General Recommendations of Traditional Fertilizers and their Concentrations for Foliar Sprays in Temperate Fruits and Nuts

Nutrient	*Suitable Fertilizer(s)*	*Conc (Per cent)*
B	Borax, boric acid	0.1
Ca	Calcium chloride, calcium nitrate	0.25-0.5
Cu	Copper sulphate	0.2-0.3
Fe	Ferrous sulphate	0.4-0.5
K	Potassium sulphate, potassium nitrate,	0.5-1.0
Mg	Magnesium sulphate, magnesium nitrate	0.4-0.5
Mn	Manganese sulphate	0.2-0.3
N	Urea, calcium nitrate	0.5-1.0
P	Potassium dihydrogen orthophosphate	0.25-0.5
Zn	Zinc sulphate	0.4-0.5

5. CONCLUSION

The climatic conditions of North West and North East Himalayas are suitable for the cultivation of temperate fruit and nut crops. Although efforts has been made to increase the area under these crops during the last three to four decades, yet the productivity is very low, the strategies like introduction of high yielding varieties on clonal rootstocks, adoption of high density planting, canopy management, good horticultural practices and integrated nutrient and water management, inset pest and disease management can improve the productivity as well as quality of fruits.

6. REFERENCES

Anonymous. 2014. Indian Horticulture Database. National Horticulture Board, MoA, GOI 85, Institutional Area, Sector 18, Gurgaon.

Ananda, S.A. 2001. New rootstocks for temperate fruits and their propagation techniques. **In:** *Enhancement of Temperate Fruit Production* (Ed.) Jindal, K.K. and Gautam, D.R., Dr. Y S Parmar University of Horticulture and Forestry, Nauni-Solan, H. P. pp. 24-31.

Awasthi, R.P., Thakur, G.C. and Chauhan, P.S. 1986. Role of meteorological components on flowering time, fruit set and yield of Starking Delicious apple. **In:** *Advances in Research on Temperate Fruits* (Ed.) Chadha, T.R., Bhutani, V.P. and Kaul, J.L. Dr Y S Parmar University of Horticulture and Forestry, Nauni-Solan, H.P. pp.11-18.

Chadha, K.L. and Awasthi, R.P. 2005. *The Apple: Improvement, Production and Post Harvest Management*. Malhotra Publishing House, New Delhi, pp.182-201.

Raina, J.N., Thakur, B.C., Suman Shashi and Spehia, R.S. 2005. Effect of fertigation through drip system on nitrogen dynamics, growth, yield and quality of apricot. *Acta Hort*. **696:** 227-231.

Verma, L.R. 1990. Apiculture and Mountain crop productivity. **In:** *Beekeeping in Integrated Mountain Development Economics and Scientific Perspectives*, Oxford and IBH Publishing Co. Ltd., New Delhi, pp. 23-65.

Chapter 2

Potential and Strategies for Development of Fruits and Nuts in Cold Arid Region of India

Brahma Singh[1], Ashish Yadav[2], Tsering Stobdan[3], S.K. Dwivedi[4] and O.P. Chaurasia[5]

[1]Former Director, DIHAR and Directorate of Life Science, DRDO and Founder President ISPC, CPCT, IARI, New Delhi – 110 054
E-mail: brahma88@gmail.com
[2]Senior Scientist, ICAR Research Complex for NEH Region, Sikkim Centre, Tadong – 737 102, Sikkim
[3]Scientist 'E', Defence Institute of High Altitude Research, Leh – 901 205, Ladakh
[4]Scientist 'F', Defence Institute of Bioenergy Research, Haldwani – 263 139
[5]Scientist 'F', Defence Institute of High Altitude Research, Leh – 901 205, Ladakh

1. INTRODUCTION

A vast stretch of harsh, inaccessible and arid temperate area (94,809 sq km) designated as 'cold arid region or cold desert' lies in the North-Western Himalayan region of India. About 87.4 per cent of this area is in Ladakh region of the State of Jammu and Kashmir and the remaining 12.6 per cent, largely in the Lahaul and Spiti region of the State of Himachal Pradesh. The cold arid regions (>2,500 m amsl) are considered to be one of the most food insecure and fragile areas of the world because of their topography, poor soil texture and structure, and harsh climatic conditions.

These experience -40°C temperatures in winter to +35°C in summers, low annual precipitation (80-300 mm), relative humidity (20-40 per cent), atmospheric pressure (493 mm Hg) and high wind velocity (8-10 km/h) mainly in afternoon, long frozen winters (5-7 months depending on altitudes), very low partial pressure of oxygen and thin atmosphere resulting in heavy influx of ultra violet and infra-red radiations. Due to sub-zero temperature during winter months, plant growth and development is possible only from April to October making harvesting possible from July to October. For remaining period fruit plants remain dormant. The region remains isolated from rest parts of the country from November to May every year due to heavy snowfall enroute on Srinagar and Leh road.

2. POTENTIAL OF FRUIT CROPS IN COLD ARID REGIONS

Indigenous/introduced temperate fruit crops are being grown either as individual trees or small groups of trees in cold arid regions of India. Fruit orchards are few and of recent origin. Temperate fruits are important sources of minerals and vitamins and provide a part of the calorie requirement in the daily diet of the high altitude people. They also provide most of the food roughage/fibre, which contributes to the prevention of disorders of the human digestive system. The nutritional status of diet is on a declining trend due to low intake of fruits in these areas. The increased production and intake of fruits by the cold arid high altitude people shall help compensate debilitating nutritional deficiencies. Fruit crop diversification is an important step for sustainable economic growth of cold arid regions. As economy grows, there is a gradual movement out of subsistence food-crop production to a diversified market-oriented production system in cold arid high altitudes.

Fruit industry in cold arid region is in infancy stage but has been expanding in recent years. Apple is the most prevalent fruit crop of the cold arid regions of Himachal Pradesh, while apricot is the major fruit crop of Ladakh region. Himachal Pradesh and Ladakh both are promoting in apple and apricot production. Almond, walnut, pear, peach, plum, cherry, mulberry, strawberry, grape and sea buckthorn (*brahmaphal*) are other fruit crops of the region. There is large natural growth of sea buckthorn (*Hippophae* sp.) in both the states having dominance of *H. rhamnoides* in Ladakh and mixed growth of *H. rhamnoides* and *H. salicifolia* in Himachal Pradesh. Due to research and development efforts of DRDO - DIHAR, Leh sea buckthorn has become important fruit crop of Ladakh commercially benefitting local population of the area and human health.

Apple and apricot fruit crops grown in cold arid regions of India are prized for their quality. They look attractive and taste better. Historically, the premium quality dried apricots mainly from Ladakh are used as one of the main trading commodity with neighbouring countries. Apricot grows naturally in the form of forest fruit in valleys and lower slopes playing a vital role in the socio-economic development of cold arid regions. It is serving multipurpose role providing fresh fruits, nuts, oils, fuel and fodder to the locals. The climatic conditions of cold arid desert are highly congenial for the production of export quality apricot, owing to long days, high light intensity, high concentration of UV rays during fruit bearing season. Clear sky high day temperature followed by low night temperatures and low atmospheric humidity at the time of active fruit growth and maturity period during July and August help in quality apricot production in this region far superior to that in the other part of the country.

Apple is the second major crop of the region. It has been found growing 3,600 m msl where even apricot failed to grow successfully. Study conducted by DRDO-DIHAR, Leh has highlighted the presence of valuable indigenous apple genotypes in the region and has identified Tha, Mangol, Karkichoo, Shing, Mar, Nas, Yangma, Squirmo, Khara, Shamer Kushu (apple) genotypes, which have not been highlighted so far from Leh and Kargil districts of Ladakh. However, detailed studies on conservation and exploitation need to be initiated. The local apple varieties bear fruits profusely which are small in size, highly juicy but have poor shelf life. The edible biomass production per tree is large. Their abiotic stress tolerance and heavy bearing traits are worth exploiting. Dhiman *et al*. (2009) reported that demonstration of Hi-tech orchards of spur type and non-spur apples at RHRSS, Tabo is an example for commercialization of apple cultivation in cold deserts. Spur type cultivars of apple namely Red Chief, Super Chief, Oregon Spur-II, Red Spur, Well Spur, Silver Spur, Scarlet Spur, Brighten N- Early, Golden Spur, Stark Spur, and Golden Delicious are producing premium quality fruits on seedlings and clonal rootstock planted under high density planting techniques. The standard cultivars *viz.*, Starking Delicious, Red Delicious, Rich-a-Red and Golden Delicious are producing export quality fruits. Recently non-delicious cultivars like Red Fuji, Coe-Red Fuji, Gale Gala and Granny Smith have been introduced on clonal rootstocks at RHRSS, Tabo. Due to high diurnal temperature variations and high IR and UV rays, fruit quality of apple is better than other apple growing areas of the country. An optimum balance between vegetative and reproductive phases results in high precocity and prolific production of fruits. Apple growing in these areas has changed the mindset of local tribal farming community and they are developing fresh marginal areas for establishment of apple orchards. Tabo success story of apple cultivation will prove to be a boon in providing nutritional and economic security to the farming community in cold arid regions of India. Ladakh region has vast area with climates of Tabo region and thus has vast potential of quality apple production there. Apple plantation is now gradually picking up there.

Another potential fruit crop is strawberry. *Fragaria* species are found in nature in Ladakh. Introduction of long day or day neutral strawberry varieties has opened up avenues for its successful cultivation in cold arid regions. The long day type varieties recommended for Ladakh conditions are Addie, Confetura, Hera and day neutral types are Brighton, Fern, Douglas, Selva, Tieoga and others. Strawberry can be grown mainly in the open and also under protected structures/polyhouses or in pots for domestic consumption. Flowering, generally, occurs from May to June under open conditions while in the polyhouse it starts by the end of March to April. Fruits are harvested 5-6 weeks after flowering and generally 3-4 pickings are available at weekly intervals. This crop with improved technology such as use of plastic mulch and drip irrigation can be made available there from May to October. This crop needs attention of planners of these areas.

3. GENETIC DIVERSITY OF FRUIT CROPS

The most common type of cultivation being through seed propagation the region represent great wealth of biodiversity in native fruits of cold arid regions particularly Ladakh *viz.* apricot and sea buckthorn. Local germplasm of apple has been mentioned above.

(a) Apricot

Apricot, locally known as *Chuli*, is broadly classified in Ladakh region into two categories based on taste of the kernel. Fruits with bitter kernel are called *Khante* meaning bitter, while those with sweet kernel are called *Nyarmo* meaning sweet. Sweet kernel is mainly consumed without processing as an important source of dietary protein, oil and fibre. Analysis of bitter apricot kernel of Ladakh region by DIHAR, Leh showed that the kernel is a rich source of oil (up to 54.21 per cent), protein (17.75-22.56 per cent), carbohydrate (21.16-35.26 per cent), crude fibre (0.84-4.71 per cent) and dietary fibre (6.03-22.24 per cent). Apricot kernel oil is a rich source (94.4 per cent) of unsaturated fatty acids, rich in oleic acid (66.2 per cent) and linoleic acid (28.2 per cent). Oil extracted from the sweet kernel is used for edible purpose either in pure form or mixed with walnut oil.

Apricot is having very rich genetic diversity in the high altitudes, which can be exploited in terms of fruit morphology, quality and yield for use as table fruit, dehydrated apricot, kernels and oil. The prominent apricot cultivars growing in cold arid regions of Ladakh are Amba, Afghani, Australian, Aam, Adzo, Arghoun, Bongti, Brochuli, Badamchuli, Chulikarpo, Charmagz, Chharap, Daychuli, Germoy, Tago, Ghoun Chuli, Kaisi, Gotsos, Gilgit Sweet, Halman, Jeembo Chuli, Ltasmet, Mamur Chuli, Margulam, Moti Tilli, Mircho, Nilipachesmo, Narmo, Narmo Gusmin, Norma Ritonil, Narma Tsangya, Nari, Pechantilli, Papa Chulli, Koban, Khantey, Rakchey Karpo, Rdongmer, Rogan, Shakanda, Shambair, Sheepa-Chesmo, Ston Chulli, Suka, Tripa, Tokpopa, Tokpopa Khantey, Thul Chulli, Urgain Chulli, Yarken, Zan Chulli, Zgoaqchulli *etc.* (Dwivedi *et al.*, 2007). Attributed to this high genetic variability, apricot is growing widely in valleys and rugged terrains at different altitudes in Ladakh. Among the apricot cultivars grown in the region, Halman and Raktsey Karpo are the most important and popular cultivars. Halman is mainly used for preparing premium quality dried apricot and the kernel is sweet. Raktsey Karpo is a dessert cultivar with white seed cover in contrast to dark brown colour in most apricot genotypes and the kernel is sweet.

(b) Sea buckthorn (recently named as Brahmaphal)

Sea buckthorn (*Hippophae rhamnoides* L.) belongs to family Elaegnaceae and has unique characteristics to grow in marginally fertile soil of cold desert. The shrub can withstand extreme temperatures from -43°C to +40°C and is considered drought resistant. These two characteristics make the shrub an ideal plant species to establish in cold deserts. *Hippophae rhamnoides*, *H. salicifolia* and *H. tibetana* are the predominant sea buckthorn species in India. Of these, *H. rhamnoides* is widely distributed in the country. Seabuckthorn (*H. rhamnoides*) is naturally distributed in six valleys, namely Leh, Nubra, Changthang, Suru, Lahaul and Spiti in Trans-Himalayan cold deserts of

India. It is also found in few pockets of Nathula in Sikkim. Seabuckthorn brings many environmental benefits, including soil and water conservation, desertification control, land reclamation, erosion and water loss control, reforestation, and the establishment of wild life habitats, especially in fragile ecosystem due to its extensive root system coupled with efficient nitrogen fixation. In 10-12 years old plant, roots extend up to 537 cm horizontally and up to 127 cm vertically making it ideal for soil binding. The shrub has been found growing well in sandy, rocky, saline and ravine soils. Roots of sea buckthorn have symbiotic association with bacterium belonging to genus *Frankia*. The symbiotic association fixes atmospheric nitrogen in soil, thereby, increasing soil fertility of nutrient deficit soil in deserts. It is estimated that sea buckthorn plantation fix 180 kg nitrogen per hectare every year (Jike and Xiaoming, 1992). Every part of the plant, *viz*. fruit, leaf, twig, root and thorns has been traditionally used as medicine, nutritional supplement, fuel and fence and therefore, sea buckthorn is popularly known as 'Wonder Plant', 'Ladakh Gold', 'Golden Bush' or 'Gold Mine' of cold arid regions. Sea buckthorn berry collection from natural habitat has become an important activity in the Ladakh region after its economic and nutraceutical potential have been recognized and publicized by the scientists of DIHAR, senior author being the pioneer. The current mean annual berry harvest is 201.8±76.8 MT. Demand for sea buckthorn has increased in last few years primarily because of development and commercialization of sea buckthorn based products. Currently, Patanjali Yogpeeth an establishment of Yoga Guru Baba Ramadeo has made arrangements to purchase pulp of Brahmaphal worth over Rs. 7 crores after taking technology from DRDO to manufacture products like multivit beverage, seed oil, herbal tea etc.

Seabuckthorn is a cross-pollinated crop and, hence, there exists high level of genetic diversity. A total of 186 superior genotypes have been identified by DIHAR, Leh-Ladakh with yield potential more than five kilogram berry per plant. Extensive survey in Ladakh region has shown presence of more than 200 genotypes in the natural population with wide variation in physico-chemical properties. It has been recorded that the shrub height varied from 1.8-7.2 m. Inverse relation has been found between the shrub height and the altitude of the region in which the shrub grows. Morphological studies in natural population of sea buckthorn in Lahaul, Spiti and Kinnaur have also shown a similar trend. Plant height is greatest in Kinnaur and lowest in Spiti. Based on physico-chemical properties of the berries and the leaves, 19 morphotypes with distinct variation have being maintained at DIHAR, Leh. A set of morphological descriptors for the genus *Hippophae* has been developed by DIHAR, Leh for germplasm obtained from various parts of India. The pattern of variability in morphological characteristics associated with habitat, plant growth habit, stem, leaf, fruit, flower, seed and biochemical parameters are taken into consideration. The suitability of these parameters in characterization of *Hippophae* germplasm has been successfully demonstrated. Using the descriptors, it is found that *H. rhamnoides* and *H. salicifolia* are genetically closer than with *H. tibetana*. *H. salicifolia* which is dominant in H. P., Uttarakhand and Sikkim has not yet been reported from Ladakh cold desert. *H. tibetana* is mainly confined to Zanskar valley of Ladakh.

Sea buckthorn has been used in traditional 'Amchi' system of medicine for centuries in Ladakh. The medicinal value of sea buckthorn was recorded as early as the 8th century in the Tibetan medicinal classic *rGyud Bzi* (Four Text of Fundamental Tibetan Medicine). More than 200 sea buckthorn-based formulations are being traditionally used either in pure form or mixed with other medicinal plants. The most popular formulations are being used in treatment of lung and phlegm disorders, blood disorder, menstrual problem, throat infection, liver problem, spleen and stomach disorder, cancer and diabetics. Modern research has supported the medicinal properties of sea buckthorn. Results of analysis of sea buckthorn accessions from Leh valley of Trans-Himalaya showed the presence of high content of multivitamins including vitamin C (275-930 mg/100 g), vitamin A (432.4 IU/100 g), vitamin E (3.54 mg/100 g), Riboflavin (1.45 mg/100 g), Niacin (68.4 mg/100 g), Pantothenic acid (0.85 mcg/100 g), vitamin B-6 (1.12 mg/100 g), and vitamin B-2 (5.4 mcg/100 g). Similarly, mineral element composition revealed high amount of minerals including potassium (647.2 mg/l), calcium (176.6 mg/l), iron (30.9 mg/l), magnesium (22.5 mg/l), phosphorous (84.2 mg/l), sodium (414.2 mg/l), zinc (1.4 mg/l), copper (0.7 mg/l), manganese (1.06 mg/l) and selenium (0.53 mg/l) (Stobdan *et al.*, 2010). The berries are, thus, storehouse of vitamins in the region where availability of other vitamin rich fruits is limited. At present DIHAR is working on six selected plants out of 200 variants, which gives a minimum of 5 kg of berries every year. There are separate male and female plants of seabuckthorn. Only the female plants produce berry fruits. Thus, if planted scientifically, 90 per cent of Sea buckthorn plantation should be of female plants their occurrence 10 per cent of male plants. Now-a-days, many farmers are aware of this profitable plant and they are cultivating seabuckthorn. In fact, Ladakh has become a premium seabuckthorn producing region in the international market. The international companies prefer Ladakh's seabuckthorn over China's because of its high quality which contains no chemicals like pesticides and herbicides. The leaves of seabuckthorn have been found nutritionally at par with that of its fruits.

(c) Grape and Others

Grape (*Vitis vinifera*) in some valleys of cold desert mainly Khalste (*Kalasi*) it performs well with and without protected structures. It is under cultivation on a limited scale. Research efforts are needed to make this important fruit crop commercial in these areas.

Mulberry is growing nicely in Turtuk region of Nubra valley in Ladakh and has potential to be commercial crop with science and technology interventions.

Besides, pear, almond, cherry, pine nut, walnut, kiwi, cape gooseberry are potential fruit crops for the region. Genetic diversity of cold desert of India is untapped and unexploited.

(d) Non-conventional Potential Fruit Plants for the Area

Like seabuckthorn there are several plants which have potential of becoming commercial fruit crops in cold arid region in addition to their well-documented medicinal values. Some of them are listed below.

Podophyllum hexandrum Royle. *P. emodi* Wall ex Royle. (Berberidaceae), Himalayan Mayapple (E), *Demokusu, Papra, Ol-mose* (L), or *Ban Kakri* : Fruits are eaten by children in Zanskar areas. This is a proven anti-cancer plant. Its tubers are used for medicinal purpose. Studies are required to explore its potential as a fruit crop also.

Rosa webbiana Wall. ex Royle. (Rosaceae), *Jangali Gulab* (H), Wild Rose (E), *Siah, Madpo* (L): Shepherds its eat ripe fruits (hips). Through crop improvement studies its cultivation as fruit crop is possible. At an altitudes of 14000 ft its luxuriant natural growth laden with hips have been witnessed by senior author.

Ephedra gerardiana Wall. ex Stapf. (Ephedraceae), *Asmania* (H), Ephedra (E), *Chhapat* or *mtshe-ldum Lug-mtshe* (L): Ripe fruits are sweet and are eaten by shepherds in Ladakh. It grows luxuriantly in dry patches and bears profusely indicating its potential as a fruit crop in such regions.

Berberis ulicina or Gorse Barberry is a small hairless shrub, 1-2 m tall, much branched and densely spiny with 5-10 mm long internodes and reddish-brown stems. Gorse Barberry is found on dry rocks of Nubra valley of Ladakh. It bears edible purple fruits.

Ribes or *Rheum* or currant bears edible fruits, stems/stalks used as vegetables. It is found in nature in Ladakh.

4. PROBLEMS OF FRUIT DEVELOPMENT IN COLD ARID REGIONS

The climatic conditions prevailing in the region, inaccessibility and lack of technical know-how are the major bottlenecks in the development of the fruit production in cold arid regions. The main problems are listed below:-

- ☆ Non availability of quality planting materials through local nurseries. Planting time in these areas is different than other altitudes where fruit nurseries are located. Dormancy break down of saplings occurs by the time road open for the area.
- ☆ Lack of location specific scientific cultivation/orchard management techniques.
- ☆ Improved cultivars of fruit crops for various purposes/locations.
- ☆ Lack of irrigation facilities and modern scientific irrigation methods.
- ☆ Lack of post-harvest technology and processing methods of fruit crops.
- ☆ Poor knowledge/infrastructure and linkage for packaging, grading, storage, transportation and marketing *etc*.

(a) Positive Aspects

- ☆ Long photoperiod coupled with low temperature is good for quality fruit production including fruit size.
- ☆ Low precipitation (rains) keep the fruit crops free of pathogens and attractive.
- ☆ Off season production of fruit crops.

- ✰ Plenty of cultivable land and snowmelt water for irrigation.
- ✰ Huge local market particularly defence forces.

(b) Strategies

Following strategies can be adopted for better fruit development in the cold arid regions of India.

(i) Introduction of new fruit crops

Experiments conducted at DIHAR have exhibited the vast potential of production of fruit crops like peach, plum, cherry, mulberry, pear and grape. Fruits like kiwi should do well thereby developing suitable varieties and production technologies.

(ii) Developing mother orchards and fruit crops nurseries

Making available planting material of apricot and apple, walnut and introduced crop varieties by establishing mother orchards and fruit crop nurseries locally is most essential and key to sustainable fruit production.

(iii) Water management

Cold arid high altitude regions are characterized by high solar radiations, long day lengths, high light intensities, and dry atmosphere with very less relative humidity. These condition result the higher total radiation absorbed by leaves, increase in the diffusion coefficient of water vapour in air at reduced barometric pressure and the increased density gradient of water vapour from the leaf to the ambient air and water evaporation from soil surface. This consequence the water deficiency in temperate fruit crops growing in cold deserts and adversely affects its production and quality of fruits. Water is a scare commodity in high altitudes. Drip irrigation is helpful in water conservation by reducing evaporation and deep drainage when compared to other types of irrigation such as flood or overhead sprinklers since water can be more precisely applied to the plant roots through a network of valves, pipes, tubing, and emitters. Each dripper/emitter, orifice supplies a measured, precisely controlled uniform application of water directly into the root zone of the plant. It saves water up to 70 per cent compared to flood irrigation. The drip irrigation method offers an opportunity for precise application of water to the soil at appropriate time with desired concentration. With drip irrigation, root development of temperate fruit crops is extensive in a restricted volume of soil. To minimize irrigation water and check weed growth, use of plastic mulch has been found quite effect in these areas.

(iv) Insect pest and disease management

Incidence of insect pests and diseases in the cold arid region is low due to low temperature and very low relative humidity. However codling moth, leaf curling aphid and gummosis are major problems. Codling moth (*Cydia pomonella*) is a serious insect-pest of apple and apricot in Ladakh. The pest is widely distributed in fruit growing areas of Ladakh. To avoid and restrain its spread to other parts of the country, the Government has imposed restrictions under SRO 397 dated 8th September 1981 under the Jammu and Kashmir Plant and Disease Act 1973, on the movement of apple and apricot from Ladakh (Dwivedi *et al., 2007*). Codling moth has a remarkable ability to adapt to a wide range of climatic conditions and geographical locations. The moth passes the winter as full grown larvae in diapauses inside a thick silken cocoon. They usually spin their cocoons under loose bark on the trunks of apple trees, under shelters about the base of the trees, or on the ground nearby. They remain dormant through the winter and are able to withstand severe and freezing temperatures. In mid spring, the larvae change in to pupal stage and adult moths emerge after two to four weeks. In Ladakh, codling moth has well synchronized its life cycle as per the prevailing conditions of the area. Codling moth damages fruits by stings where larvae bore a short distance into the flash. The deep entries occur when larvae penetrate the fruit skin, bore into the core and feed on seed cavity. Larvae may enter the sides, stem end or calyx end of the fruit. Entries into the calyx are often difficult to detect without cutting into the fruit. It is reported that the infestation of codling moth in apple was 49.7 and 42.5 per cent in Kargil and Leh districts, respectively. It is estimated that about 60 to 70 per cent loss is being observed in apple industry each year (Akbar *et al.*, 2010). Management of codling moth needs a sustainable biological control approach of pest management. An integrated management schedule involves biological control such as *Trichogramma embyrophagum*; installation of pheromone trap; sanitation of the orchards that include burning of the fallen leaves, twigs and pruned wood, burying of the damaged fruits, scrapping of the bark, manual destruction of the larvae. Involving local community for implementing the programme to eradicate this noxious pest is a must.

(v) Organic fruit production

Fruit production in Ladakh is organic by default which needs to be strengthened. Presently people are more health conscious and demand healthy food products. Organic food helps in improving the immune system because it contains the toxic free necessary nutrients which the body needs to be healthy. Because of increased consumer education and awareness there is a movement towards organic produce, which are not treated with the detrimental inputs. The market for organic produce is on the increase and there are a variety of market possibilities for producers who want to grow certified organic crops. Moreover, the environmental destruction by the chemical industrial model of agriculture necessitates a shift towards agro-ecological forms of production (Poll and Smith, 2009). There are generally two types of organic temperate fruit production depending on the purpose. One is aimed at increasing efficiency of large scale production and focuses on broad based market which also include foreign markets using organic certification system. This is extremely market and profit oriented, whereas the other focuses on food safety, environment consciousness and local sustainability (Kim, 2005). Organic producers only use inputs and materials which are of organic origin. A variety of crops or plants should be planted together to diversify the production system in cold arid regions. Soil and soil health is one of the basic principles of organic cultivation. If the soil is healthy, plants are not stressed and will considerably reduce the amount of diseases, fewer infestations by pests and weeds will also be less problematic. Conservation of bio-diversity and soil health contributes to environmental friendly practices and helps to lessen the negative effect caused by climate change in the cold arid regions.

In organic production system many things should be considered when planning a new orchard establishment. Selection of temperate fruit crop, suitable variety/ cultivar, rootstock, training system and planting spacing all need to be decided well ahead of time. Factors such as weather conditions, soil type, soil pests, water quality, water availability, irrigation system and pest pressures can all influence the decision. Even management factors including labour availability, labour skill level, packing options and marketing outlets should be considered. All these factors might interact in subtle ways. Thus, a thorough evaluation to sustain the production and productivity of temperate fruit crops in cold arid regions in organic nutrient management system is required. An integrated eco-friendly organic nutrient management is needed to maintain the soil fertility and plant nutrient supply at an optimum level for sustaining the desired productivity through optimization of the benefits from all possible sources of organic and biological components. It involves proper combination of organic fertilizers, organic manures, crop residues, N_2~fixing crops and biofertilizers suitable to the system of land use and ecological, social and economic conditions. Since certified organic temperate fruits fetch premium price in the market, therefore, emphasis should be given on economically desirable crops and its intensification for increasing the production of fruits. The new concept of high density planting is highly suitable where land is limited and even small and marginal farmers can allocate certain portion of their land for cultivation of temperate fruit crops in cold arid high altitude regions.

(vi) Holistic management strategy

For overall fruit development in cold arid regions of India, integrated compact development approach needs to be followed. Need-based support should be provided for crop production with forward and backward linkages. Inputs such as supply of quality planting material of only recommended varieties, post-harvest infrastructural facilities and processing units need to be established. Available arable land need to be utilized by changing crop priority. Wasteland need to be promoted for growing fruit crops. Cost-effective polyhouses need to be promoted in the region. Skilled manpower at different levels is required. There is a need to promote integrated nutrient management through vermi-composting, biofertilizer and others. Farm mechanisation conducive to hill farming is required to reduce labour use. Human resource need to be developed and training programmes on modern aspects of fruit crop production and post-harvest management need to be organized. Setting up of large demonstration blocks should be done on priority. Development of integrated disease and pest management modules suitable for organic fruit production is required. Hi-tech horticultural production technology such as micro-irrigation, protected cultivation, micropropagation *etc.* need to be developed in the region. Farmer advisory systems to address their day-to-day queries need to be created in every blocks. Studies on economic and marketing aspects of fruit crops need to be conducted. Special emphasis should be given for the development of processing units in the region.

5. CONCLUSION

There is a need of harnessing potential of temperate fruit production in cold arid regions in of India view of climate change and other positive aspects mentioned in the text rather than ignoring the areas because of topography, harsh climate and connectivity.

6. REFERENCES

Akbar P.I, Jahangeer, B. A. Malik, A.R. and Parray, G.A. 2010. Codling moth (*Cydia pomonella*) pest in cold arid Ladakh: problems and management. http: // krishisewa.com/articles/2010/codlingmoth.html

Ballabh, B., Chaurasia, O. P.,Pande, P. C. and Ahmad, Z. 2007. Raw edible plants of cold desert Ladakh. *Ind. J. Traditional Knowledge* 6: 182-84.

Bawa, A. S. Farath Khatoon and Singh, B. 2002. Seabuckthorn - a wonder plant. Natural Products Radiance. July- Aug. 8-14.

Dhiman, K.R., Bhardawaj, D.R., Bawa, R. and Vikram , A. 2009. Producing world class apples in cold deserts for health and economic security- A case study of Regional Horticultural Research Sub-Station, Tabo in Himachal Pradesh. In *Advances in Agriculture, Environment and Health*. (Eds. S.B. Singh, O.P. Chaurasia, Ashish Yadav, A.M. Rimando and T.H. Terrill). SSPH Delhi, India, pp. 99-104.

Dwivedi, S.K, Kareem A. and Ahmed, Z. 2007. *Apricot in Ladakh*. Published by Field Research Laboratory, Leh Ladakh, India.

Dwivedi, S.K., Stobdan, T., Yadav, A. and Singh, S.B. 2009. Fruit production in cold arid regions of India: challenges and opportunities. **In:** *Advances in Agriculture, Environment and Health*. (Eds. S.B. Singh, O.P. Chaurasia, Ashish Yadav, A.M. Rimando and T.H. Terrill). SSPH Delhi, India. pp. 61-81.

Dwivedi, S. K., Parimelazhagan, T., Singh, S. B. and Ahmed, Z. 2009. Seabuckthorn : (*Hippophae* spp.): The Golden Bush : Satish Serial Publishing House, 392 p.

Jike, Z. and Xiaoming, Z. 1992. Progress of study on *Frankia* in nodules of seabuckthorn. *Hippophae*, 2: 4-10.

Kim, K.H. 2005. Two concepts of sustainability: evaluation of organic farming. Dept. of Agricultural and Resource Economics, Graduate School of Agriculture and Life Sciences, The University of Tokyo.

Poll, R.V.D. and Smith, M. 2009. Basic principles for organic cultivation and livestock production. *Manual for Small Scale Farmers to start Organic and Livestock Production*, Surplus People Project, Athlone, South Africa.

Sharma, D. P. and Chauhan, J. S. 2008. Potential of fruit growing in cold desert areas in northwestern Himalayas- problems, prospects and future strategies. *Acta Hort.* 712: 175-178.

Singh, B. Chaurasia O.P. and Ballabh Basant. 2015. Edible wild plants of Trans-Himalayan cold desert, **In:** *Plant diversity of the Himalaya*, by P.C. Pande and S.S. Samant, (Gyanodaya Prakashan, Nainital), pp. 483-512.

Singh, B. 201–4 , K.V. Peter book, proceedings HSI, Bangalore.

Stobdan T., Chaurasia O.P., Korekar, G., Mundra, S., Ali, Z, Yadav, A. and Singh, S.B. 2010. Attributes of seabuckthorn (*Hippophae rhamnoides* L.) to meet nutritional requirements in high altitude. *Defence Sci. J.*, **60:** 226-230.

Stobdan, T. Korekar, G., Chaurasia, O.P., Balaji, B., Yadav,A., Dwivedi, S. K. Konchok Targais, K., Mundra, S and Srivastava, R.B. 2015. Seabuckthorn production for greening and sustainable income generation in cold desert of India in Innovations in agro-animal technologies (Satish Serial Publishing House, Editors: Srivastava, R.B. and Selvamurthy, W., pp. 71.

Chapter 3

Potential and Strategies for Development of Temperate Fruits and Nuts in Sub-tropical Region of India

H.S. Dhaliwal, M.I.S. Gill and Anirudh Thakur

Punjab Agricultural University,
Ludhiana – 141 004, Punjab
E-mail: dhaliwal_mandhali@pau.edu

1. INTRODUCTION

India with an area of 7.21 million ha and annual production of 88.97 million tonnes is the second largest producer of fruits after China. The fruit production in the country is dominated by tropical and sub-tropical fruits, however, temperate fruits occupy 9.2 per cent area (0.66 m ha) of the total area under fruits contributing 5.57 million tonnes (6.3 per cent) in the total fruit production. The temperate fruits are mostly grown in the North Western and Eastern Hill Region. Jammu and Kashmir (44.5 per cent), Himachal Pradesh (21 per cent), Maharashtra (13.55 per cent), Uttarakhand (11.55 per cent) and Punjab (0.79 per cent) are the five major states contributing maximum area under temperate fruits in India. The major temperate fruits under cultivation in India with sizeable acreage are apple (47.4 per cent), walnut (18.45 per cent), grapes (17.98), pear (6.36 per cent), almond (3.22 per cent), peach (2.72 per cent), kiwifruit (0.71 per cent) and peacannut (0.22 per cent). The introduction and development of low chill temperate fruit varieties in the country has expanded the cultivation of temperate fruits upto the sub-tropics. Temperate fruit varieties requiring low chilling hours have a great potential for diversification in the sub-tropical regions of the country. In light of the global warming forecasts, cultivation of low-chill temperate fruits will be a viable and suitable option even in the low valley regions of North Western and Eastern India. The low-chill temperate fruits are being grown in different states of India *viz.* Punjab, Uttarakhand, western Uttar Pradesh, Himachal Pradesh, Jammu and Kashmir, Meghalaya, Tamil Nadu and Chhattisgarh. The states of Punjab (5, 240 ha), Tamil Nadu (4, 210 ha), Chhattisgarh (1000 ha) and Uttarakhand have sizeable area under low chill temperate fruits. Among the low-chill temperate fruits the major fruits under cultivation in the sub-tropical region are pear, peach and plum. Apart from this, grapes which is a temperate fruit is also cultivated in the tropics and sub-tropics. Punjab Agricultural University (PAU), Ludhiana took a lead in early 1970's in the development and introduction of low chill temperate fruit varieties in India. PAU has a vast collection of low-chill varieties of pear, peach, plum and apple. The research work carried out on low-chill temperate fruits and their potential in subtropics is discussed in this chapter.

2. PEAR

The low chill pear is being cultivated in the sub-tropical regions of Punjab, Chattisgarh, Haryana, Uttarakhand, Himachal Pradesh and Jammu and Kashmir. In Punjab, it is primarily grown in Amritsar, Tarn Taran, Jalandhar and Ludhiana districts. The area under low-chill pear in Punjab is 2,889 ha with an annual production of 65,438 tonnes with a productivity of 22.65 tonnes per hectares. Patharnakh is the major variety of pear grown in the entire sub-tropical region of the country. Patharnakh, bears heavily even at 150 chilling h and can withstand high temperature and hot winds during the summer season of subtropics. Apart from this, varieties like Le Conte, Kieffer, Smith, Gola and Baggugosha are also grown in sub-tropical region. Patharnakh is still the dominant variety of pear and contributes more than 90 per cent in the area under pear in Punjab. At PAU Ludhiana, large number of varieties have been evaluated *viz.* Patharnakh, Le Conte, Baggugosha, Smith and Kieffer. Out of these, only Patharnakh was successful due to its high production potential, hardy nature and long shelf-life, whereas, Le Conte and Baggugosha were shy in bearing. PAU Ludhiana has around 50 strains and varieties of pear and released new pear varieties, *viz.* Punjab Beauty, Punjab Nectar, Punjab Gold, Punjab Soft, Punjab Nakh and Nijisseiki to give impetus to pear cultivation in the state. Pear varieties Gola, Le Conte, Tumaria and Bagugosha are the promising cultivars in the subtropics of Uttarakhand. Apart from this, other cultivars like Kieffer, Patharnakh, Pineapple, Hood and China pear also performs well in sub-tropical *tarai* region in Uttarakhand. In Jharkhand, a local pear variety Netarhat selection is grown in the Netarhat region near Ranchi.

(a) Description of Important Varieties

(i) Patharnakh

Tree spreading and vigorous, fruit medium, round, green with prominent dots. Flesh gritty, crisp and juicy. Keeping quality good, stands transportation well. Matures in the last week of July. Average yield is 150 kg per tree.

(ii) Punjab Nakh

Tree spreading and vigorous. Fruits are ovate, light yellowish green with prominent dots having average weight of 155 g. Flesh is gritty, crisp and juicy. Keeping quality is good and can be stored for 60 days at 0-1°C temperature and 90-95 per cent relative humidity. It matures in last week of July. Average yield is 192 kg per tree.

(iii) Punjab Beauty

Fruit is medium, yellow with red blush. The flesh is white and more juicy and sweet than Le Conte and Baggugosha. Fruits mature in third week of July. Average yield is 80 kg per tree.

(iv) Punjab Gold

Fruit is large with average weight of 166 g, ground colour golden yellow and flesh is white with 13.3 percent TSS and 0.22 per cent acidity. Mature fruit is less gritty and mellow when ripe. Matures in end of July. Average yield is 80 kg per tree. It is suitable for processing and table purposes.

(v) Punjab Nectar

Its fruit is medium to large. The average fruit weight is 138 g with ground colour yellow green, white fleshed, TSS 13.6 per cent and acidity 0.21 per cent. Mature fruits are less gritty and very juicy when ripe. Fruit matures in fourth week of July. Average yield is 80 kg per tree. It is suitable for table and processing purposes.

(vi) Punjab

Fruit size medium, ground colour yellow green, flesh whitish, TSS 11.3 per cent, acidity 0.13 per cent, mature fruits are very soft, very juicy (73 per cent juice), matures in first week of July. Fruits are suitable for processing like RTS, nectar and squash and can be cold stored for 4 weeks at 0-1°C and 95 per cent RH. Average yield is 86 kg per tree.

(vii) Baggugosha

Tree upright and vigorous, fruit greenish yellow, stem end tapering. Flesh white or cream-coloured, sweet, somewhat gritty. Matures in the first week of August and stands transportation well. Average yield is 60 kg per tree.

(viii) Le Conte

Tree medium in vigour and spreading. Fruit small to medium in size, pyriform in shape and greenish yellow in colour. Flesh whitish, juicy and sweet. Matures in the first week of August. Average yield is 60 to 80 kg per tree.

(ix) Nijisseiki

Tree spreading, medium in vigour, fruit size medium, ground colour golden brown and flesh whitish. Mature fruits are very soft with TSS 12.9 per cent and acidity 0.21 per cent, fruit firm (11.17 lbf), very juicy (74.6 per cent juice), matures in end June to first week of July, fruit ripen on tree, average yield is 80.2 kg per tree. Suitable for processing into RTS, nectar and squash. Fruits have good storage potential and can be stored for 4 weeks at 0-10°C and 90-95 per cent RH.

(b) Propagation

Rootsuckers were earlier used as rootstock for pear in the sub-tropical region. The plants propagated on root suckers produce large number of suckers under the tree. Kainth rootstock has been recommended for pear because of very low suckering. Further, Kainth is vigorous rootstock and maintains higher productivity of scion variety. Kainth seeds were stratified and germinated during winters with germination of around 60-65 per cent. To enhance germination, the freshly extracted seeds from mature Kainth fruits can be sown directly in last week of September or first week of October in moist sand in wooden boxes. These seeds germinate in 10-12 days with around 94 percent germination and the seedlings can be planted in the field at two to four leaf stage. Around 40 per cent seedlings become buddable in May-June and rest can be grafted in December-January.

(c) Canopy Management

Pear variety Patharnakh is generally planted at 7.5m x 7.5m distance thus, accommodating around 72 plants per acre. It has long juvenile period of 7 to 8 years and during the initial years produces upright branches. Planting pear at 8x4m along with bending of main scaffolds and two secondaries reduced the juvenile period by two years and resulted in higher yields than conventional system. Patharnakh is a prolific bearing variety and this leads to poor fruit size. In older plants the lower spurs of tree become less fruitful and the bearing area get confined to middle and top of the tree which leads to reduction in yield. Hence, Patharnakh pear trees should be pruned annually after they attain full size to retain the plant height at 2.0 to 2.5 m. The shoots which emerge on the top of the tree should be completely removed in the month of December-January every year. The cut ends should be treated with Bordeaux paste or copper oxychloride paste. This results in improvement of fruit size, quality and market returns.

(d) Rejuvenation of Old Pear Orchards

Pear trees tend to become less productive at an age of about 20 years. Production on such trees can be restored by rejuvenating them. It can be achieved by heading back 3 to 4 main scaffolds to about 15 cm during December-January. The rest of scaffolds should be completely removed. To avoid fungal infection, paint the cut ends with Bordeaux paint. New sprouts emerge on these stubs in March. In May, retain only one to two outgoing shoots on each stub with a total of 6 to 8 shoots per tree. Rejuvenated trees start fruiting in third year and bear commercial crop in the fifth year.

(e) Nutrition

Generally, farm yard manure, phosphorus and potassium fertilizers are applied in pear during December-January. Further, half dose of nitrogen is applied during

February and the second half dose in April after the fruit set. In Punjab, pear is harvested in July-August and routinely post-harvest fertilizer application was not recommended in pear. Post-harvest application of nitrogen is very important to maintain a favourable carbon: nitrogen ratio for proper flowering and fruiting. Hence, supplemental application of 500 g urea in the month of September has been recommended to improve fruit yield and quality.

(f) Quality Improvement

There is an increase in demand of potassium during fruit development stage. The potassium nutrition has key role in producing higher yields of quality fruits. Under sub-tropical conditions of Punjab three foliar sprays of KNO_3 @ 1.5 per cent at 15 days interval starting from full bloom improved fruit size in Patharnakh.

(g) Ripening and Storage

Unlike in temperate regions, in Punjab, pear fruits are harvested in hot and humid season. The fruits are consumed without ripening and development of proper flavour and taste. Ethylene hormone promotes fruit ripening which is characterized by changes in skin colour, flesh texture, and flavour. Further, exposure of fruits to low temperature can lead to autocatalytic production of ethylene. To improve the eating quality of pear varieties, Punjab Beauty fruits should be chilled at 0°C and ripened at 20°C for 3 days, while, Patharnakh fruits should be chilled for 3 days at 0°C and ripened at 20°C for 4 days. To induce proper ripening and improving quality of pear, fruits should be treated with 1000 ppm ethephon for three to four minutes or exposed to 100 ppm ethylene gas for 24 hrs and stored at 20°C. The Patharnakh fruits ripens in 8 days while, Punjab Beauty takes 4 days to ripen. Fruits of Punjab Beauty can be stored at 0-1°C with RH of 90-95 per cent for 60 days. These fruits have post-storage shelf life of 1-2 days at room temperature and 4 days in refrigerator. However, Patharnakh fruits can be stored at 0-1°C with RH of 90-95 percent for 60 days. These fruits have post-storage shelf life of 2 days at room temperature and 4-6 days in refrigerator.

3. PEACH AND NECTARINES

Earlier, peach and nectarines were mainly cultivated in the hilly tracts of Himalayan region. In the sub-tropical climate, peach was not cultivated because of the non-availability of low chilling, early ripening and superior varieties. The area in the plains had earlier been under Noki, Sharbati and Tikki cultivars. Apart from being inferior in quality, these cultivars mature late in the season and also had to face competition with other fruits in the market of that time. Planned breeding programme for the improvement of peach for sub-tropical climate was started in 1957 at Saharanpur to develop low chilling varieties with early maturity and good quality. Saharanpur Prabhat a cross between Sharbati and Flordasun was released for cultivation. Sharbati is a chance seedling selection from Saharanpur. In the sub-tropical India, Flordasun, Sunred, 16-33, Floradared, Florda Belle, Early Amber, 15-39 varieties which were developed in Florida, USA; Bonita, Rochan and Vantura developed in California, USA were introduced during late sixties at Punjab Agricultural University, Ludhiana. Out of these, Flordasun, TA-170 (Partap) and 16-

33 (Shan-i-Punjab) were found successful and have been released for commercial cultivation in Punjab. In 1988, seven low chilling and early maturing peach varieties were introduced from Florida and Texas at Punjab Agricultural University. Two varieties, Earli Grande and Florda Prince were released for commercial cultivation in Punjab. Seven peach and three nectarine varieties were introduced from USA in 2001. A low chill nectarine, Punjab Nectarine has been released for commercial cultivation for north Indian plains by Punjab Agricultural University, Ludhiana. In Punjab, introduction of low chilling, high yielding and early ripening cultivars of superior quality has brought a positive change in peach cultivation. The characters like precocity, early maturity, better colour and quality made peach an economic proposition which led to an increase in the area under this crop. It is being planted in solid blocks as well as a filler tree in orchards of mango, litchi and pear. The introduction of low chilling varieties has extended peach cultivation in almost all the states of the North East India but, it is most important in terms of adaptability and extent in Meghalaya. Among the low chilling varieties, Flordasun, Shan-i-Punjab and Partap were found suitable for mid hill conditions of northeast. Now-a-days, low chill peach is being cultivated in Uttarakhand, Punjab, Western Uttar Pradesh, Himachal Pradesh, Jammu and Kashmir, Madhya Pradesh, Haryana and Meghalaya. The varieties like Sharbati, Shan-i-Punjab, Earli Grande, Saharanpur Prabhat, Florda Sun, Florda Red, Earli Grande and Florda Prince are the major varieties under cultivation in the entire subtropical region of India. In Punjab, the area under peach increased from 560 ha in the year 1981 to 3,137 ha in 1991. After decreasing to 1476 ha in the year 2009 the area under peach in Punjab is now increasing on a slow pace. At present, in Punjab, peach is cultivated over an area of 1,716 ha with an annual production of 30, 340 tonnes with a productivity of 17.67 tonnes per hectares. In Uttarakhand, low chill peach is cultivated in Udham Singh Nagar and Nanital; and in Meerut, Saharanpur, Muzzafarnagar and Bulandsahar regions of Uttar Pradesh. Low chill peach and nectarines are also found successful in lower hills of Himachal Pradesh and Jammu region.

(f) Description of Important Varieties

(i) Earli Grande

Tree is semi-vigorous, high yielding and fruit maturity occurs in the first week of May (4 days earlier than Shan-i-Punjab). Fruit is large (90 g) and firm fleshed, free stone at full ripe stage and of good keeping quality. Its TSS is 10.5 per cent and acidity is 0.7 per cent.

(ii) Florda Prince

Tree is vigorous and the fruits mature in the fourth week of April. Fruit size medium (65-70 g); yellow with red blush at maturity, flesh firm and free stone at full ripe stage. TSS 12 per cent and acidity 0.5 per cent. Average yield is 100 kg per tree.

(iii) Flordasun

A low chilling cultivar developed in Florida. It is an early maturing cultivar which ripens in the last week of April. Fruits medium-sized, freestone, round in shape with red blush on the surface, pulp yellow, acidic sweet taste with 11.5 per cent soluble solids.

(iv) Khurmani

Tree is medium in vigour with an upright growth habit. Fruits mature in 1st week of June. Fruit medium in size (70 g), soft-fleshed, juicy and cling stone. TSS 13 per cent and acidity 0.4 per cent.

(v) Partap

It matures a week earlier than the Flordasun and take 76 days for maturity. Its fruits mature in the third week of April. Tree is medium in vigour. Fruit weight is 65-70 g, almost round, yellow with red over colour. Flesh is firm and free stone at full ripe stage. The fruit contains TSS 12 per cent and acidity 0.7 per cent. Average yield is 70 kg per tree.

(vi) Punjab Nectarine

Tree is vigorous and spreading, fruits mature in 2nd week of May. Fruit large, weighing 90 g, round, attractive with 90-100 per cent red blush over yellow ground colour at maturity, flesh yellow, firm, melting and free stone at full ripe stage. TSS is 11.5 per cent and acidity is 0.8 per cent. Average yield is 40 kg per tree.

(vii) Prabhat

It is a variety developed at Horticulture Technology and Training Centre, Saharanpur by crossing Sharbati and Flordasun. The trees are upright having strong and thin primary limbs. Fruits are medium in size with attractive red blush. Flesh is sweet, freestone, white having excellent flavour.

(viii) Shan-i–Punjab

This is early cultivar, maturing in the first week of May. Tree is vigorous in growth. It produces large fruits of 5-5.5 cm diameter weighing about 90 g each. The fruit yellow in colour with red blush, yellow fleshed, juicy, sweet with excellent taste and completely free-stone. The fruit is quite firm in texture and can withstand transportation. In addition to its table use, this cultivar is also suitable for canning. Average yield is 70 kg per tree.

(ix) Sharbati

It is a chance seedling selected at Saharanpur. It has medium sized fruit, which is a cling stone and is round-oblong in shape; flesh is white, soft, juicy and aromatic. TSS 13 per cent and acidity 0.33 per cent. It is a heavy yielder and ripens late in second week of June. It is very popular in plains of Western Uttar Pradesh. Average yield is 100-120 kg per tree.

(b) Rootstocks

Selection of proper rootstock is an important consideration for commercial orcharding. Peach is generally propagated on seedling rootstocks in all the major stone fruit growing areas of the world. In Punjab, Sharbati and Khurmani were earlier used as rootstock for peach which are highly susceptible to root-knot nematodes. At PAU Ludhiana, rootstocks *viz.* Nemaguard, Flordaguard, Rancho, Munson, MRS 2/8, MRS 2/5, Pesco x Bessey were evaluated for their suitability under sub-tropical conditions. The rootstock 'Flordaguard' was found promising with respect to germination, compatibility with peach varieties and resistance to root knot nematodes. This new rootstock will help in increasing the productive life span of peach orchards in the region. Flordaguard trees require around 300 chill units for flowering and fruiting.

(c) Propagation

Grafting and budding are the commercial methods for the propagation of peach. In these methods, plants usually take two years for transplanting in the field. Peach plants can be clonally multiplied through rooting of hardwood cuttings or by stenting technique (simultaneous grafting and rooting) by dipping the basal portion of cutting in 2000 ppm solution for 2 minutes in January. This practice reduces the time of propagation by one year and help in mass multiplication of peach plants to meet the increasing demand of planting material.

(d) High Density Planting

Peach is recommended to be planted at a distance of 6 x 6 meter in square system. There is a scope of increasing the production by increasing the plant density. Normally, the peach trees are trained on modified leader system in the sub-tropical conditions. Different spacings *viz.* 1.5 x 3.0m, 3.0 x 3.0m, 6.0 x 1.5m and 6.0 x 6.0 m were tested for peach cv. Shan-i-Punjab. Plantation of peach trees at 6 x 1.5 m with 'Y' system of training resulted in high productivity, early maturity and better fruit quality. In 'Y' system, plants are allowed to develop only two main limbs instead of 5-6 limbs in the modified leader system. These limbs grow from a common point at an angle of about 60 degree. The fruit bearing area is developed on these primary limbs. The 'Y' trained trees produced less vegetative growth and produced large sized fruits of better quality which matured earlier than the modified leader trained tree.

(e) Quality Improvement

In the subtropics, peach is very vigorous in growth and has a tendency to overbear. This leads to poor fruit size and quality. Therefore, various techniques have been tried to improve fruit size and quality in peach. Hand thinning of fruits before pit hardening stage by keeping a distance of 10-15 cm between the fruits is very effective in improving fruit size and quality. Girdling of tree trunk or limbs is another practice which has shown promise in improving the fruit size and quality. Girdling and thinning done at full bloom; and girdling done at 28 days after full bloom increased fruit size and advanced fruit maturity by 7 and 12 days, respectively. Application of potassium nitrate (2 per cent) and CPPU (5.0 ppm) at 14 days after full bloom also improved fruit size and quality.

(f) Post-harvest Treatment

Peach is perishable in nature which is the biggest hindrance in its successful marketing. Further, high temperature during fruit maturity and lack of cold chain also leads to high post-harvest losses. Post-harvest dips of calclium chloride (2 per cent) reduced physiological weight loss and enhanced storage life of peach fruits under cold storage conditions upto 30 days. Fruits of peach varieties have been found suitable for canning and other processed products like squash, nectar and carbonated drinks.

4. GRAPE

The success of a crop in a particular region primarily depends upon the good varieties and on adoption of improved production technologies. Grapes are not an exception to this accepted norm. Grapes are well adapted to wide range of climatic conditions ranging from tropical to temperate region. Region specific varieties and production technologies play a key role in commercial production.

In recent past, the area under grapes in Punjab has declined. The major cause for decline is that most of the grape plantations in the state are old and have low productivity. Monoculture of Perlette vareity leads to gluts in the market vis-à-vis results in low returns to farmers. High cost of establishing training structures and comparably poor quality particularly with unthinned grapes also had a detrimental effect on the adoption of grapes. The further cause for decline in area under grapes is the adoption of age old cultural practices which must also be replaced with new production, post-harvest and protection technologies. The lack of marketing infrastructure, processing facilities and incentives from government like crop insurance, minimum support price, soft loans for infrastructure development have in general affected an expansion of area under fruit crops. In grapes, PAU has played a significant role in development of new varieties and improved production technologies.

Most of the plantations in the state comprise of a single variety, Perlette. This variety is seedless, comes into bearing early and is a prolific bearer. However, its bunches are compact, have high incidence of shot berries thus, require flower bud thinning coupled with GA_3 dips to improve its quality. Moreover it is very susceptible to damage by pre-monsoon rains. To achieve high returns per unit area, the new varieties especially adapted to local conditions must be grown. PAU has recently recommended varieties *viz.* Flame Seedless and Punjab MACS Purple which have equivalent yield potential, can relatively tolerate rains, thus can be adopted by grape growers.

(a) Description of Important Varieties

(i) Punjab MACS Purple

It is rich in anthocyanins, a source of antioxidants and is suitable for processing into juice, nectar and ready to serve beverage. It contains 60-65 per cent juice with 17-18 per cent total soluble solids and 0.50 per cent acidity. It has medium and loose bunches. The berry is seeded, medium in size and purple at maturity. It matures in first week of June. The technology for red wine production (5 litre scale) has also been recommend for this variety.

(ii) Flame Seedless

Bunch is medium and well filled. Berry seedless, firm, crisp, light purple at maturity. It gives good response to quality improvement treatment. Bunches have berries with 18 per cent TSS. It ripens during second week of June. To obtain uniform colour treatment of bunches with 400 ppm ethephon at the colour break stage is recommended. Average yield is 27 kg per vine.

(iii) Beauty Seedless

It performs well under South-Western districts. Bunch is medium and well filled. Berry seedless, medium, bluish black. Highly responsive to quality improvement treatment. Treated bunches have oblong berries with 16-18 per cent TSS. Keeping quality is poor. Fruit ripens in the first week of June. Average yield is 25 kg per vine.

(iv) Perlette

This variety is cultivated on more than 90 per cent of total acreage in Punjab. Bunch is medium to large and compact. Berry is seedless, firm, crisp, juicy and light amber. It is highly responsive to quality improvement treatment. Treated bunches have large round berries, with 16-18 per cent TSS It ripens during first week of June. Average yield is 25 kg per vine.

The improved production technologies play a significant role to increase the production and quality. In grapes, over the years several improved production technologies have been developed by PAU which are described as under:

(b) Improved Training System

Training determines the productive life of vines. To ensure healthy vines with long productive life, training system should be appropriate. Most of the vines in Punjab are trained on Bower system of training. The vines trained on this system are vigorous and give higher yield, however, the quality of grapes produced is poor. To improve the yield and quality of grapes a new training system, Y-trellis has been recommended which has several advantages over the bower system. In Y-trellis system, the grapes are planted at 1.5 x 4 m spacing and accommodate almost double the number of plants per acre, mature early, produce superior quality fruits and lead to ease in cultural operations.

(c) Pruning Technology

In the subtropics, contrary to the tropics the pruning of grapes is done in January-February. There should 60-80 canes per vine with pruning of each cane at 4 bud per cane. This pruning intensity ensures higher yield and quality.

(d) Quality Improvement

To improve quality of grapes cultural practices play a significant role. In grape variety Perlette, bunches are compact and if left as it is on the vines produce poor quality bunches with shot berries. To get optimum yield and quality in grapes a package of treatments is recommended. It includes thinning of flower buds one week before flowering by leaving 100-120 flower buds per panicle. Girdling of main stem by removing a 4 mm wide ring of bark and dipping of clusters in 40 ppm GA_3 when the berry size is 4 mm. One week after the first GA_3 treatment, give a second dipping in 40 ppm GA_3. Two foliar sprays of potassium sulphate (1.5 per cent) first one week after fruit set and second at colour break stage, improve the quality and colour of Perlette grapes. In Flame Seedless grapes, for colour development ethephon spray @ 400 ppm at colour break stage is recommended. Ethephon sprays coupled with a crop load of 75-80 bunches per vine trained on Bower system further improves colour and quality of Flame Seedless grapes.

(e) Rejuvenation of Senile Vines

The old Perlette vines trained on Bower system can be rejuvenated by heading back the primary arms retaining one foot stubs in the month of December- January and training the secondary arms in subsequent year. Bordeaux paste should be applied after heading back of primary arms. This technology results in regaining the productive potential with better quality grapes from the old vine within two years.

(f) Top Working to Change the Variety

In subtopics, there was monoculture of Perlette grapes and replacement of existing variety with new variety generally takes three to four years time. To shorten this time by two years a new technology - Top working of Flame Seedless on Perlette has been recommended. To achieve maximum success of top working, grafting should be done in 2nd week of February. In addition, the Flame Seedless grapes top-worked on Perlette vines matures 4-5 days earlier than own rooted Flame Seedless.

5. PLUM

Plum was introduced in India in 1870, but their cultivation remained confined to hilly areas of North. With the introduction of low-chilling cultivars, its cultivation has gained popularity in the sub-mountaneous and sub-tropical region of North West India. Plum is rich in minerals, vitamin A, sugars and organic acids and is used either as fresh dessert fruit or processed into various products like jam, jelly, squash and nectar. The local varieties which are grown in subtropics are called *Aluchas*. Japanese cultivars having chilling requirement below 300 hours can also be grown successfully in this region.

The work on germplasm collection and evaluation was started in 1964-65 at Punjab Agricultural University, Ludhiana, Regional Fruit Research Station, Kandaghat (now in Himachal Pradesh) and at Fruit Research Station, Abohar. A great variation in tree and fruit characteristics was observed in low chilling plum varieties growing in the states of Punjab, Haryana and Uttrakhand. Earlier in this region, exclusively the indigenous varieties of plum were cultivated. Some important varieties released for cultivation in Punjab and adjoining areas were Titron, Alucha Black. Alubokhara, Howe, Kala Amritsari, Peshawari Kala, Jamun, Meeruti, Damson, Alucha Yellow and Desi Alucha. Most of the cultivars tested had resemblance with *Prunus salicina* except Damson plum which is a cultivar of *P. insititia*. The fruit ripened during the month of May except Damson which ripened in June. Among the evaluated varieties, Kala Amritsari had the highest fruit yield (31.8 kg) and Titron gave the highest TSS (17.2 per cent). All the other varieties were sub-acidic (0.78-0.92 per cent acidity). Plum varieties Kala Amritsari, Kataruchak and Kala Peshawari were found to be excellent for jam and juice making.

Under the arid irrigated conditions of Abohar, Kataru Chak was the most vigorous and Kala Amritsari was the least. Kataru Chak and Titron were high yielder and earlier to mature (1st week of May) while, Lalri was shy bearer and was the latest to ripen (3rd week of May). These varieties were recommended for arid-irrigated regions of Punjab. The fruits of all these varieties were small in size and were sub-acidic in taste and as such no variety was found suitable for table purpose. Later on, promising hybrids of plum were introduced from Gainesville, Florida (USA) at Fla 1-2 was found promising and released for cultivation under subtropics as Satluj Purple.

(a) Description of Important Varieties

(i) Satluj Purple

The fruit is medium large with average weight of 25-30g, roundish, turns into crimson colour on ripening. Fruit are thick skinned with yellow orange firm flesh, sweet in taste, 13-14 per cent TSS and 0.6-0.7 per cent acidity. It is an early variety ripening in the first week of May with average yield of 40 kg per tree. This variety is suitable for table purpose. It is self-unfruitful and requires Kala Amritsari as pollinizer which should be planted as an alternate plant in alternate rows for maximum yield. Satluj Purple and Kala Amritsari plants should be planted as 85 and 25 trees per acre, respectively.

(ii) Kala Amritsari

It is self-fruitful, high yielding indigenous variety with vigorous tree. Fruits are medium sized, round oblate depressed at both ends which on ripening turn dark purple. Flesh is yellow, moderately juicy and excellent for jam-making. Fruits are somewhat acidic with 15 per cent TSS and 1.2 per cent acidity. It ripens in mid-May with average yield of 45 kg per tree.

(iii) Amritsari Alubokhara

Fruits are large sized with attractive red colour having high TSS: acid ratio. Free stone, pink coloured flesh with uniform sweetness. Free from sourness near the pit, the major drawback of Indian plum. The maturity period is also quite earlier as compared to temperate plum varieties.

(iv) Titron

Fruit medium in size, deep purple with thin skin and yellow flesh. A self-fruitful variety. The yield improves when planted with 'Alucha Early Round' as pollinizer. The fruits ripen in second week of May. Average yield is 25-30 kg per tree.

(v) Kataru Chak

A partially self-fruitful variety but, the yield improves if planted with 'Kala Amritsari'. Fruits are large, purplish with creamy flesh. It is good for jam and squash –making.

(vi) Alu Bokhara

A self-unfruitful variety and should be planted in rows alternating with 'Howe'. Fruits are large, yellow in colour sometimes having red overcolour.

(vii) Peshawari Kala

The fruits are dark purple, thick skinned and have good keeping quality. It is a shy bearing variety.

(viii) Damson Plum

The fruits are round, medium in size, yellow coloured, thick skinned and are late in maturity. The fruits ripen in first week of June with an average yield of 40 kg per tree.

(b) Propagation

Plum is generally propagated by budding or grafting on seedlings or vegetatively propagated rootstock obtained by cuttings. Tongue grafting is done from end of December to end of January. Under subtropical conditions, the rootstock, Kabul Greengage gives better rooting and grafting success with plum cv. Satluj Purple than other rootstocks which produces vigorous plants free from incompatibility sign. It is suitable for heavy and wet soil. Research work on propagation was conducted at PAU Ludhiana to reduce the time of propagation. A new propagation technique has been developed in which plum is multiplied through stenting technique *i.e.* simultaneous grafting and rooting of rootstock cuttings. The plum cultivar Satluj Purple is propagated through this technique. The cuttings of Kabul Greengage are used as rootstock. Rootstock cuttings about 20 cm in length are tongue grafted with scion wood in the first fortnight of January. The basal parts of the grafts are given 24 hours soaking treatment with 100 ppm IBA solution. The cuttings/grafts are transplanted in well prepared nursery plots at a distance of 15-20 cm in rows spaced 30 cm apart. The plants from the cutting/grafts are available for planting after one year in the month of January as compard to two years required for traditional grafted trees. Kala Amritsari is propagated by hard wood cuttings. The time of preparation of cuttings and grafting is from end of December to end of January. Hardwood cuttings of Kala Amritsari should be 15-20 cm in length with 4-5 buds on them and should be of pencil thickness. To increase the rooting success basal part (4-5 cm) is dipped in 100 ppm IBA solution for 24 hours.

(c) Nutrition

In subtropics, the soils are sandy and characterized with high pH and low organic content which result in deficiency of nutrients *viz.* nitrogen, phosphorus, potassium and micronutrients. Application of 166 g Nitrogen, 91 g phosphorus and 216 g potassium per tree has been recommended under Punjab conditions to the fully grown trees (6 year and above). The available zinc in Punjab soils fall in deficient range which leads to reduction in auxin synthesis (IAA) which is important for normal growth of the fruits. Plum trees show zinc deficiency on light soils during summers. On the shoot tips, the affected leaves appear in whorls in rosette form. It may cause die back of the twigs and the fruits become hard and undersized. Foliar spray of zinc sulphate (3 kg) and unslaked lime (1.5 kg) in 500 litres of water should be applied on appearance of zinc deficiency. Boron has a role in pollination, fertilization and fruit set. Foliar application of boron increase fruit set and decrease fruit drop immediately after fruit set. Boron deficiency also results in shoot dieback, cork spot and fruit cracking. Three sprays of boric acid @ 0.4 per cent in April, June and September or one spray of boric acid @ 4.0 per cent during first week of February increase fruit retention, fruit size and yield in Satluj Purple plum. Interveinal chlorosis occurs on fully expanded leaves of plum during April, June and August-September due to iron deficiency. The deficiency symptoms also appear on young leaves which turn yellow with green veins. Foliar sprays of ferrous sulphate (0.3 per cent) can be used to ameliorate iron deficiency. Similarly, manganese deficiency, which starts as marginal chlorosis spreads towards mid-rib during summer months. Green bands also appear on either side of mid-rib. It can be corrected by foliar sprays of manganese sulphate (0.3 per cent).

(d) High Density Planting

Plum is either planted in solid blocks at 6 x 6 m or as a filler trees in mango, litchi and pear orchards. To improve the yield of quality fruits per unit area the research work was carried out at PAU Ludhiana. The plum cv. Satluj Purple with Kala Amritsari as pollinizer can be planted at 6 x 1.5 m distance with modified leader system. It accommodates 440 plants per acre against 110 plants using conventional planting distance of 6x6 m. This planting density ensures higher yield of better quality fruits. To maintain this planting density, light annual pruning in January and pinching (10-15 cm) during summer are recommended.

(e) Quality Improvement

Plum tends to overbear which leads to production of small sized, poor quality fruits resulting in lower return to the growers. Thinning of flowers or fruits reduces crop load which enables the remaining crop to receive a greater share of the food synthesized by the leaves. Thinning results in best effect if it is done before the end of fruit cell division. Hand thinning of the fruitlets at 10-12 cm distance before pit-hardening reduces the crop load and improves fruit weight and quality in Satluj Purple plum. Application of ammonium thiosulphate @ 1.5 per cent at full bloom is effective in reducing crop load. Ethrel 100 ppm sprayed before pit-hardening stage also reduces crop load and increases the yield of superior quality fruits. Trunk or limb girdling also improves fruit size and quality by increasing the availability of photosynthates. Girdling and thinning 14 days after full bloom reduces the fruit drop, advances fruit maturation and improves the fruit size and quality of plum cv. Satluj Purple.

The plum does not ripen uniformly at the time of harvest hence, harvested in three or four pickings which leads to higher cost and marketing problems. Foliar spray of etheral (500-600 ppm) at colour break stage results in uniform ripening of fruits and reduces the number of pickings. Application of etheral also advances fruit maturity by 5-6 days. However, fruits sprayed with etheral must be harvested and marketed as soon as they are ripe otherwise fruit drops.

(f) Management of Fruit Drop

Pre-harvest fruit drop is a major constraint in the cultivation of plum in the subtropics. The cultivar Satluj Purple show maximum fruit drop just before harvest during the first week of May. Growth hormones especially, auxins play a significant role in controlling fruit drop. Two sprays of NAA 10 ppm in 2nd and 4th week of April or one spray of Ethrel 100 ppm in the fourth week of March (after pit hardening) reduce the pre harvest fruit drop in plum cv. Satluj Purple. However, some precautions need to be followed *viz.* there should be no water stress in the orchard at the time of spray of growth regulators. Further, ethrel 100 ppm should be sprayed just after pit hardening and not before or later than this, otherwise it may increase fruit drop. Trees should be irrigated frequently during April as the fruit growth is rapid. Fruit tree nutrition also has an important role in preventing pre harvest fruit drop. Apart from application of recommended dose of fertilizers, one spray of zinc sulphate should be given during the first week of April. Mulching also helps in maintaining proper moisture and help in better retention of the fruits.

(g) Storage and Value Addition

Plum is a perishable fruit with short shelf life. Low temperature storage prolongs the effective life of fruit. Fruit should be picked when 50 per cent colour develop on the skin at the correct maturity and placed immediately in commercial cold stores where temperature is 0-3°C with 85-90 per cent relative humidity. The fruit remains in good condition for 25 days. To facilitate better marketing of plum, refrigerated vans are desirable for transportation. The Satluj Purple plum fruits harvested at colour break stage followed by post-harvest treatment of calcium nitrate (2 per cent) solution for five minutes can be stored for four weeks in cold storage (0-1°C and 90-95 per cent RH) with post storage shelf life of 2 days at ambient temperature. Non-alcoholic naturally carbonated beverage can be prepared from fruits of Satluj Purple and Kala Amritsari plum which has three month shelf life without addition of any preservative.

5. ALMOND

Almond is traditionally grown in hilly areas of Jammu and Kashmir and Himachal Pradesh. Its cultivation could not spread much because of problems like spring frosts and hail storms during the bloom; self-incompatibility; and rain damage during fruit growth. Drs. K. K. Singh, J.C. Bakshi and A.S. Narang reported successful cultivation of almond by a farmer in Jalandhar District of Punjab in 1969. The farmer obtained good yield of nuts from a local variety which was actually a peach-almond hybrid. This led to explore the possibilities of growing almond under sub-tropical conditions of North India. In temperate region, planting of late blooming varieties can overcome the problem of spring frost and rains during flowering. Cultivation of almond in low chill subtropical regions can also overcome these problems.

The almond improvement programme was initiated in 1970 at the Punjab Agricultural University, Ludhiana to develop almond varieties suitable for cultivation in the subtropics of North India. During year 1970 and 1972, 38 local selections of almond were collected from Dry Fruit Research Station, Boktu and from private plantations at Spillo, Ribba, Pooh, Jeori and Roopa Valley, Himachal Pradesh. Later, around 100 selections were done from Kashmir Valley collected from locations at Hariparbat, Bandipur, Multon, Pattel Galender, Laddo, Avantipur, Ranberpura, Lollypura, Airport area, 'Badami Bagh' and Shalimar Fruit Research Station, Srinagar during 1973 to 1975. Apart from this, 38 almond varieties were introduced from the leading almond growing countries like USA, Australia, Spain, Afganistan, Russia, Italy and France.

Most of the almond varieties had flowering during the third year and in the subsequent years barring some Russian varieties. The variety H.P. Selection 7 was the earliest to ripe and harvesting was over by the 2nd week of June. The other varieties ripened in the 3rd and 4th week of June except, Jeori Selection I and Almond Hybrid-15 which ripened in the first week of July. The three year old plants of California Paper Shell and Almond Hybrid-15 produced 1.27 kg and 1.39 kg nuts per tree, respectively. Amongst other varieties, H.P. Selection 7, H.P. Selection 10, and Pethick's Wonder produced more than half kg of dry nuts per tree. The local variety Sloh produced largest nuts (2.58 g/nut) on weight basis, which was followed by Pethick's Wonder (2.52 g/nut), B.T. 102 (2.37 g/nut), Strouts Paper Shell (2.36 g/nut) and California Paper Shell (2. 19 g/nut). The nuts of Almond hybrid-4 and B.T. 106 were smallest. Out of the varietal evaluation, varieties like California Paper Shell, Pathick's Wonder and H.P. Selections 7 and 10 performed well. The local peach x almond hybrid (Sloh) had vigorous growth and heavy fruiting with hard shell like almond however, had poor kernel quality. Thus, it can be used for hybridization purposes. One of the hybrids obtained from the cross Thin Shelled x Sloh, tested under the name 'Almond Hybrid 15' has been found to be quite promising in view of its high productivity and good quality. The Hybrid 15 yielded 1.28 kg nuts/tree with an average nut weight of 1.68 g, length 3.25 cm; breadth 1.83 cm; thickness 1.3 cm; size medium-large and fairly uniform.

(a) Description of Important Varieties

(i) California Paper Shell

The thin-shelled nuts are bold with 2.1 g nut weight and average yield 3.8 kg per tree. It ripen in the last week of June.

(ii) Hybrid 15

Nut is oblong with prominent keel towards the apex. Nut weighs 1.7g, apex protruding and pointed with roundish basal-end. Pits numerous, conspicuous and shallow to medium-deep. Colour whitish. Shell medium-thick, soft, easy to crack with teeth. It ripens in the first week of July with an average yield of 3 kg per tree.

(iii) Pethick's Wonder

The nuts are thin shelled and bigger than all other varieties. On an average, a nut weighs 2.5 g with an average yield of 3.7 kg per tree.

(iv) JK 55

It is an early variety, nuts are semi-hard and small in size. Harvested in the first week of June with an average yield of 3.5 kg nuts per tree.

(b) Propagation

Almond can be propagated by budding and grafting on peach rootstock. Peach seedlings become buddable during May-June. The grafting is done from November-January. Almond trees on peach seedling rootstocks grow vigorously, come into bearing earlier and survive better than on almond rootstock. Under subtropical conditions, Kabul Greengage is a good dwarfing rootstock for almond.

The extensive research on almond was done on a large number of varieties. Keeping in view, the potentially poor yield and market returns from almond the work has been discontinued under subtropical conditions. However, the work done at Punjab Agricultural University, Ludhiana points towards the potential of almond in the subtropics may be in future with different set of low chilling varieties.

6. FUTURE THRUSTS

(*i*) The low chill temperate fruit industry in India particularly in the subtropics is solely dependent on the introduction of varieties from other countries. The imposition of restriction on the exchange of germplasm due to Intellectual property laws and commercial interests of private players of the West is hampering the growth of low chill temperate fruit industry. Though, the efforts should continue to introduce more low chill cultivars, but local breeding programme must be strengthened to develop improved cultivars having local adaptability. Hybridization in low chill peach, nectarines and mutation breeding in pear has already been initiated at PAU Ludhiana.

(*ii*) The global warming trends indicate the potential of low chill varieties in the marginally temperate regions of lower foot hills. Hence, there is a need to initiate planned breeding programmes involving crosses between low and high chilling cultivars to widen the gene pool and develop germplasm suitable for the subtropics having improved fruit size, colour and quality.

(*iii*) The time, space and funds required for establishing and evaluating germplasm necessitates the need for collaborative research. In this regard, sharing germplasm of advanced selections for multi-locational evaluation and pollen of elite germplasm for hybridization can speed up improvement programmes.

(*iv*) Substantial work is going on canopy management and training systems for higher yield and better fruit quality at different centres but, there is a need to establish farmer friendly technologies for economic upliftment of marginal farmers.

(*v*) Technology for drip irrigation and fertigation needs to be standardized for all the crops for higher water and fertilizer use efficiency.

(*vi*) The low chill temperate fruits in subtropics mature in hot summers. Lack of cold chain infrastructure is the biggest hurdle which results in poor shelf life and high post-harvest losses in such fruits. Industry partnership is required for the development of cold chain facility for marketing of low chill temperate fruits, but it also requires improved varieties which mature over a long period of time to make the business profitable.

(*vii*) To avoid monoculture, other low chill temperate fruits *viz.* apple, strawberry, walnut, apricot and kiwifruit can be evaluated in the subtropics

7. CONCLUSION

To conclude, due to global warming vis-à-vis upward shift of core temperate fruits towards higher altitudes, the temperate fruits particularly having low chilling requirement have vast potential in the sub-tropical and lower valley regions of hills in India. However, to realize the full potential, there is a need to develop improved varieties and standardize region specific production practices. Above all, economic gains cannot be realized until smart packaging and marketing facilities are created which are preferably operational at farmer level.

Chapter 4

Potential and Strategies for Development of Temperate Fruits and Nuts in North Eastern Region

K.K. Jindal

Former Director Research
Dr Y.S. Parmar University of Horticulture and Forestry,
Solan – 173 230, H.P.
and CAU, Imphal – 795 001, Manipur
E-mail: ecofriendlyhorticulture@gmail.com

1. INTORDUCTION

Among the different regions of India, the Northeast Himalayan Region (NEHR) is blessed with the bounties of nature. Apart from agriculture and forest produce, this region is ideally suited for the production of a large number of horticultural crops. The available resources of soil, the unique diversity of temperate to tropical climate within a small area and the abundance of water sources combine to make the area the most suitable region in the country for the development of horticulture based industry. In the hills of NEHR temperate fruits, such as apple, pear, peach and plum, are being grown on a commercial scale, while crops like walnut, chestnut and apricot are also grown on limited scale in states like Arunachal Pradesh. From an agroclimatic point of view, in almost all the states, except Tripura and Assam, there lies scope for expansion of temperate fruits like pear, kiwi and stone fruits, but for the expansion of apple growing potential is availble in in Arunachal Pradesh, Nagaland, Sikkim and Meghalaya.

2. AREAS SUITABLE FOR DIFFERENT CROPS IN N.E. REGION

In North East Region, parts of Arunachal Pradesh and parts of Nagaland fall under temperate zone; Assam. Meghalaya, parts of Tripura, parts of Arunachal Pradesh fall under subtropical zone while parts of Tripura and Mizoram under coastal/tropical zone. Areas suitable for temperate fruit growing in NE Region states are presented in Table 1 and area and production in Table 2.

Table 1: Areas Suitable for Temperate Fruits in different States of North-eastern India

State	*Location*	*Approx. Range of Elevation (amsl)*	*Fruit Crops that can be Successfully Tried*
Arunachal Pradesh	Kameng, Siang, Trirap and Lohit district	900-4,000 m	Apple, apricot, cherry, chestnut, peach, pear, plum, walnut.
Manipur	Sekmai, Maram, Tadubi, Mao, Ukhrul, Tengnoupal	900-2,000 m	Apricot, peach, pear, persimmon, plum.
Meghalaya	Central plateau-Khasi and Jaintia hills districts	950-1,900 m	Plum, pear, peach, apricot, persimmn.
Nagaland	Mokokchung, Wokha Tuensang, Kohima, Phek districts.	1000-3,000 m	Apricot, peach, pear, plum.
Sikkim	West, East and North districts	1,000-4,000 m	Apple, kiwi, pear.

Table 2: Area (Acres) and Production (MT) of Temperate Fruits in North-eastern States of India

Fruit	*Arunachal Pradesh (2011)*	*Manipur (2005)*	*Meghalaya (2005)*	*Mizoram (2005)*	*Nagaland*	*Sikkim (2012)*
Apple	A 12,800	–	–	–	A 750	A 50
	P 10,000	–	–	–	P 150	P 20
Pear	A 125	A 160	A 30	A 60	–	
	P 681	P 1000	P 185	P 210		
Kiwi fruit	A 190					A 240
	P 62					P 100
Plum and peach	A 430	–	–	A 36	A 327	
	P 961	–	–	P 99	P 592	

A: Area in ha; P: Production in Tonnes.

Source: NEC Database 2005, 2011 and 2012.

(a) Arunachal Pradesh

In A.P. temperate fruit cultivation has been intensified in mid (900-1,800 m) high hills (above 1,800 m). Apple cultivation started in 1956 and with active participation of State Department of Agriculture, large area has been brought under temperate fruits like apple and walnuts. At present, over 250 hectare of land is under commercial production of apple especially in Kameng district, in the altitude range of 1,600-3,500 m above msl with a rainfall of about 900 mm. This isconsidered ideal for apple cultivation. Another belt of apple is developing in Ziro-Hapoli area of Subansiri district. Other nuts like almond, chestnut, etc are being tried and if these are successful they will be a boon to the tribal farmers. Here mainly Delicious group of varieties (Red, Royal and Golden Delicious) have been distributed. In addition to Red Delicious and Royal Delicious, Red June and Granny Smith are showing promise in the State Fruit Farm Shergaon, Bomdila and Dirang of West Kemang district. Attempts have been made at ICAR Research Complex to evaluate performance of different apple varieties at different locations of Arunachal Pradesh- Dirang of Kameng district. Ziro of Subansiri district, Pfutsero in Nagaland and upper Shillong in Meghalaya. It was observed that Maharaji variety in Pfutsero area of Nagaland was very vigorous, while quality of cvs Red Delicious and Royal Delicious was quite good. In Kameng district of Arunachal Pradesh colour development was excellent in Royal Delicious and Red Delicious. The fruits of Royal Delicious and Granny Smith were of excellent quality. In Ziro area of Subansiri district, varieties like Maharaji, Lord Lambourne, Buckingham *etc.* were moderate yielder. The quality of Red Delicious and Cox's Orange Pippin was good but later plantation declined totally. In upper Shillong area of Meghalaya, apple varieties yielded poorly.

(b) Manipur

In the valley of Manipur - Sekmai, Maram, Tabudi and Mai areas, pear is grown. While in areas like Ukhrul, Mao-Maram, Tengnoupal and Manipur valley stone fruits like peach and plum are grown to some extent. In Ukhrul, some apple plantations also exist.

(c) Meghalaya

In Meghalaya state, central plateau of Khasi and Jaintia hill districts (1,000 to 2,000 m above msl) grow high quality plum, pear, peach and apricot. Apple has been introduced to a limited scale. It has also potential for growing persimmon and cherries.

Indian Council of Agricultural Research (ICAR) Research Complex for N.E. Hill Region and States Department of Agriculture are instrumental in the execution of Research and Developmental activities of this region. Department of Agriculture. Nagaland established two stations for temperate fruits, one at Pfutsero in Phek district and other in Atoizo at Mokok Chung district. The Horticulture Research Station Pfutsero is situated at an elevation of 2,000 m and large collections of improved varieties of apple, pear, peach, plum and apricot have been made at the station covering about 31 ha. The var. Maharaji has been found better than Golden Delicious, Coxs Orange Pippin, Baldwin, Jonathan, Rome Beauty, Winter Banana, *etc.* Incidence of fruit fly is reported to be high.

Department of Agriculture, Manipur established one research station for temperate fruits at Ukhrul and one Walnut Development Farm at Tangnoupal. Inferior quality hard type pear (Nashpati) is being produced in large quantity. Pear cultivars like Le Conte, China and Smith were found promising in some parts of the state. Andro Farm of Central Agriculture University, Imphal and College of Horticulture, Passighat, Arunachal Pradesh has established experimental farm including low chilling temperate fruits in 2006. Department of Horticulture, Sikkim under Horticulture Technology Mission has taken lead in Kiwi fruit and Oriental Pear production in West Sikkim and North District where apple cultivation has been threatened by apple scab.

In Meghalaya, temperate fruits were introduced about 70 year's back when a Fruit Garden was established near Shillong at Umlyngla. As back as 1913, one Fruit Experiment Station was established at Laitumkhrah, Shillong by one Mr. Holder and it was later acquired by the Department of Agriculture, Assam in 1930. Though no systematic studies were made on varietal performance of different temperate fruits in this station, however, some varietal recommendations from the observations were made as:

Apple:	James Grieve, King David, Cox's Orange Pippin William's Favorite, Charles Ross
Plum:	Elberta, Santa Rosa, Doris Satsuma, Pond's Seedling, Green Gage
Pears:	William Bon, Chretien, Fertility, Marlliat
Apricot:	New Castle
Persimmon:	Daidai Maroo
Cherry:	Black Heart, Kentish

(d) Nagaland

In Nagaland temperate fruits like peach, plum, pear and apple are grown to a limited extent. There is scope to cultivate apples in Kohima, Zunheboto and Tcnsang districts. While peach, plum, and pear can be successfully cultivated in certain parts of Mokokchung, Tensang and Wokha districts.

3. APPLE PRODUCTION IN NE

It is evident that Arunachal Pradesh is the only leading state in the NEH Region suitable for apple production. The productivity is quite low (1.76 Mt/ha), as compared to other leading apple growing states such as Himachal Pradesh (3.9 Mt/ha) and Jammu and Kashmir (8.9 Mt/ha). Therefore, the present theme paper has been exclusively devoted to overcoming the production constraints, reasons for low productivity and postharvest management problems in Arunachal Pradesh based on a case study in the West Kameng area.

(a) Low and Irregular Fruiting in Arunachal Pradesh

It may be due to complex factors such as spring frost, hailstorms, lack of flowering in cultivars and lack of proper pollination and pollinators.

(i) Spring frost

Discussion with farmers revealed that during flowering, spring frost or a sudden drop in temperature is common in the area. The spring frost is one of the important factors affecting fruit setting and yield not only in apples but almost all temperate fruits. The kind of cold injury killing flowers is due to freezing, which occurs when there is cold wind in the region. Highly fluctuating temperatures at the time of flowering can even adversely affect fruit set. There is a need to monitor the temperature fluctuation and spring frost at different locations before deciding on new plantations. Pockets lacking proper air drainage should be avoided. Existing Poor Planting System has been observed that old orchards have been planted with closer spacing, and some are across the slope which can block free air movement. This aggravates the problem of frost and diseases. The rows of trees should run down the slope to permit proper air circulation.

(ii) Poor yielding varieties

In old orchards there is no systematic record of varieties. These seem to be procured in an unorganized way from unknown pedigree. During discussion, and from earlier reports, it was noticed that varieties in old plantations are of poor quality and mostly sour in which Baldwin, Cox's Orange Pippin, Buckinghum, Lord Lambourne and Maharaji are predominant. The new apple plantations have the Delicious group. It is an established fact that Royal Delicious is susceptible to spring frost and there is also a problem of sterility. Both these factors contribute to poor fruit yield.

(iii) Hailstorms

Sometimes hailstorms at the time of flowering can adversely affect the fruit set in apples. Single hailstorm damage can reduce the crop for two years. It has been observed in Himachal Pradesh that almonds can escape the damage of hailstorms due to early flowering. Therefore, intercropping of apple orchards with almond can compensate the farmers if the apple crop is damaged by hail.

(iv) Lack of proper training and pruning practices

The old plantations of apple orchards are neither trained efficiently nor pruned regularly. Due to uncontrolled luxuriant vegetative growth of the trees very little fruiting wood (spur) is formed. This has resulted in erratic bearing.

(v) Inadequate pollination

Almost all the commercial cultivars, especially the Delicious group, are selfunfruitful. They require cross-pollination with suitable pollinizing varieties to achive fruit setting. The existing apple plantations whether old or new lack the sufficient pollinizers and their exact proportion or placement is poor. Golden Delicious is not a good pollinizer. Since it is strong alternate bearer, that is, it flowers fully in one year and does not flower the next year. This may be one of the reasons for heavy drop of young unfertilized fruits after setting.

(vi) Lack of pollinators

The role of honey bees for pollination is well known in all major apple growing areas. In the absence of honey bees fruit set does not take place. Howevers, no bee colony or wall hive was noticed in the area. It appears that the apple orchards are being pollinated scantily due to lack of honey bees. There is a strong need for encouraging beekeepers in this area to ensure better fruiting of trees and vegetable production. The possibilities for beekeeping should be explored. In Himachal Pradesh it is common practice to maintain two to three honey bee hives/ha to facilitate pollination and increase the yield in apples.

(vii) Uncontrolled tree growth due to lack of appropriate rootstocks

Poor spacing and lack of pruning have contributed towards large unmanageable trees. At present, apple trees are also grown on seedling rootstock which are vigrous in growth, and trees take eight to ten years to come into commercial yield. This long period of growth and height of the trees can be controlled if dwarf or semi-dwarf rootstock are tried especially in areas with comparatively flat land and high soil fertility.

(viii) Lack of scientific knowledge of orchard management

Due to the lack of technical knowledge and ignorance, farmers do not follow the complete package of practices to grow horticultural crops. There is no proper method being followed for orchard plantation, preparation of pits, planting distance and other orchard management practices. No fertilizer schedule is being followed by farmers, except for as application of a meagre amount of FYM and that too, not regularly. The fertilizer mixture available to farmers should be discontinued, since it does not contain NPK in the correct proportions. It is advisable to use separate fertilizers of NPK. Since the soil is quite acidic, the apple plants show symptoms of micronutrients deficiency. Lack of nutrients especially Zn, Mn, Ca, B and Fe need to be ascertained from soil and leaf analysis. The apple orchards have been found to have grossly defective pruning and training systems. Heavy pruning or the lopping of branches is the main cause of the decline of old apple orchards in this area. Cultivation of land between the trees is along the slope which can deplete the soil. In order to check soil erosion and leaching of nutrients, emphasis must be laid on correct mulching and weed control.

(ix) Prevalence of plant diseases and insect pests

Due to relatively high summer temperatures and humidity, apple scab has assumed serious proportions in this area. The areas where soils are heavy or lack proper drainage, root rot and collar rot are common. As mentioned earlier, due to incorrect training and pruning, diseases like papery bark and canker are common in apple orchards. San Jose scale, woolly aphid and shoot and root borers have also been noticed. Almost all the apple plantations are heavily infested with lichen. One small plantation of apples owned by Ms Dorjee Sunzom showed the presence of Rubbery Bark virus. The control measures for such diseases require immediate attention to ensure good yield and quality.

4. POTENTIAL OF APPLE CULTIVATION IN ARUNACHAL PRADESH

Apple is the most important fruit crop among all the temperate fruit crops. Arunachal Pradesh situated in the Eastern Sub-Himalayan ranges possesses huge areas under temperate climate. A large portion of the temperate region experiences mild weather with moderate temperature in summer months and dry winter months with frequent sudden drop of temperature during strong cold outbreaks from the Himalayas. The soil texture of this region varies from loamy to sandy loam with rich humus content having a pH range of 5.8 to 6.5. The region is blessed with abundance of water resources fed by numerous rivers and rivulets with sufficient photoperiods for healthy metabolism of biomass. These favorable agro-climatic conditions prevalent in the high hills where other crops do not succeed, provide immense potentiality for successful cultivation of apples on commercial scale. Many species of local germplasm of apples have been found growing in wild forms in the temperate region of the state. Thus the inherent potentiality of cultivation of apples on the southern side of the Eastern Himalayas in Arunachal Pradesh is undeniable.

Apple cultivation in Arunachal Pradesh was introduced from Jammu and Kashmir, Himachal Pradesh and UP hills in the early 1970. During 1976 a state Horticulture Farm for temperate fruit crops was established at Shergaon in West Kameng district. The farm situated at an ideal location proved to be an excellent centre for cultivation of quality apples. Subsequently, plantation programme under area expansion of apples was taken up in other suitable areas of the state, namely, Ziro, Tawang, Menchukha, Anini, Hayuliang, Talihaetc besides West Kameng district. Later on some exotic varieties of apples were also introduced from USA. In due course of time the productivity and quality of apples grown in the state were assessed and found to be at par with its genetic potentiality. The apples of Arunachal has now gained recognition not only in the state, region and the country but also in the neighboring South East Asian countries. This is evident from the fact that the apples of the state were awarded several first and second Prizes for their quality parameters in the All India Apple shows at Madras (Chennai) in 1982, at Nainital in 1995 and recent exhibitions in Bengaluru and New Delhi. At present Arunachal Pradesh is the leading apple growing state in the North Eastern Region of the country. Some important varieties of apple grown in Arunachal Pradesh are listed below.

(a) Established Varieties

Royal Delicious, Red Delicious, Golden Delicious, Mollies Delicious, Red Chief, Starkrimson, Rich-a-Red, Cooper-N, Vance Delicious, Ambred, Oregon Spur, WellSpur, Tydeman's Early Worcester, Red Gold, More Super Golden, Granny Smith, Jonathan, Crofton, Rusp Pippin, Top Red, Mcintosh,Maharaji, Black Ben Davis *etc.*

(i) Recently introduced varieties

Red Fuji, Ambri, Lal Ambri,Gala Mast, Gold Spur, Pink Lady, Royal Gala, Mayan, Vista bells, Spartan, Firdous, Shireen, Silver Spur, Prime Beauty, Michale, DK Delicious *etc.*

With the launching of Technology Mission for North East in the year 2001-02, ample areas have been covered under apple cultivation in the state under area expansion programme, most of which are yet to attain commercial production. Although, considerable area has been brought under cultivation of apples through state plan, National Horticulture Mission and Private entrepreneurs so far, the major potential area of the state still remains unexplored due to difficult terrains and lack of communication infrastructure earlier. But now with the advent of trans Arunachal

Highway, connectivity roads under PMGSY and construction of defence roads along the international border, the scope for tapping huge potential areas under apple cultivation has opened up. On the other hand, it may not be wrong to advocate that the scope of area expansion in other apple growing states like Jammu and Kashmir, Himachal Pradesh and Uttarakhand has already attained saturation point. Now Arunachal Pradesh is the only state of the country where cultivation of apples can be promoted in big way in a mission mode. With the increase of population and constriction of easy income sources, more and more farmers are shifting to gardening for their sustainable livelihood.

Potential area, production trend, proposed area expansion and estimated future production of apple in Arunachal Pradesh is envisaged in the Tables 4-7 (Y Tsering, Jt. Dir. Personal Comm.).

Table 4: Potential Area (in ha) of Temperate Fruits Estimated

Total Geographical Area	*Estimated Area for Temperate Climate*	*Cultivable Area for Temperate Fruits*	*Estimated Cultivable Area Under Apple*
83,74,300	21,91,000	6,97,000	1,97,000

Table 5: Apple Production Trend since 2001-02

Year	*2001-02*	*2005-06*	*2010-11*
Area (Ha)	6,852	9,016	12,308
Production (MT)	8,588	9,759	9,986
Productivity MT/Ha	4	5	6
Remarks	Area includes young non bearing and declining trees. Productivity is the average of conventional and High density plantation.		

Table 6: Proposed Area for Expansion in XII and XIII 5 Year Plan for Apple

Year	*2012-13*	*2013-14*	*2014-15*	*2015-16*	*2016-17*	*2017-18*	*2018-19*	*2019-20*	*2020-21*	*2021-22*	*Total*
Area (ha)	500	500	2000	3000	4000	5000	5000	5000	5000	5000	40000

Table 7: Estimated Future Production

Area (in ha)	*Estimated Production (in MT)*		*Remarks*
	As per Present Productivity	*As per Potentiality*	
49,986 (by 2022)	2,99,916	3,99,888	Estimated production at minimum level
1,97,000 (Total area suitable for apple)	11,82,000	15,76,000	

Personal communication with Joint Director, Temperate Fruits Bomdila, Arunachal.

Considering the potentiality, availability of land and market feasibility coupled with receptive moods of the growers, promotion of commercial. Apple cultivation in Arunachal Pradesh has been deemed to be a gainful proposition not only for sustainable economic growth of the rural masses and the state but also for contribution towards national nutritional security and gross domestic products (GDP). It may not be out of place to highlight that a Roving team of senior officials from ICAR visited the concentrated temperate region of the state to prepare a Road Map for promotion of temperate horticulture in NE region. The team consisting of very senior scientists has also advocated promotion of temperate horticulture including the apple in the states of Arunachal Pradesh and Sikkim.

5. ROAD MAP FOR TEMPERATE HORTICULTURE IN NORTH EAST HILL REGION

Secretary DARE and DG (ICAR) Vide D.O. No. 2011/1226 dated 15/08/2011 constituted a Roving Team with technical experts to visit Arunachal Pradesh and Sikkim to:

a) Asses the present status and future scopes of temperate horticultural crops, fruits and nuts in particular, in the high hills and mountains of North Eastern states.

b) Analyze climate change parameters and suggest R&D intervention areas for successful commercial cultivation of apple, nut crops, kiwi and wine grapes in suitable pockets.

c) Suggest immediate actions under MM I and MM II of Horticulture Technology Mission for hill states.

The team comprises of Dr S.P. Ghosh, Ex- DDG (Horticulture), ICAR – Chairman, K.K. Jindal, Ex-ADG (Horticulture) and Director Research, YSPUH and F. Solan, Dr A.A. Sofi, Ex-Director, CITH Srinagar, Dr Nazeer Ahmed, Director, CITH, Srinagar and Dr S.V. Ngachan, Director, ICAR Research Complex for NEH Region, Umiam, Barapani – Convener.

Abstract of Report and Recommendations

Temperate fruits, namely, apple, pear, plum and peach are grown semi-commercially in the NE hills states, particularly in Arunachal Pradesh, Sikkim, Nagaland and Meghalaya. Walnut, strawberry and kiwi also experienced area expansion. Apple in rain shadow belts in Arunachal Pradesh with high density planting system has been reported to enhance productivity in a big way. Similarly, kiwi has caught imagination of producers in Arunachal Pradesh, where not only productivity is good, but also the quality of the produce is reported to compare favorably with Australian kiwi fruits that are in good demand in the international market. The road map for the establishment of a sound temperate fruit industry in Arunachal Pradesh and Sikkim suggested below:

(a) Arunachal Pradesh

Arunachal Pradesh is the only leading state in the NEH suitable for apple production. In more recent times kiwi and walnut are showing promise in the temperate- alpine zone, particularly in the rain shadow belts. High hills of West Kameng, Tawang, Ziro, Anini, Upper reaches of Anjaw, Upper Subansiri and upper Siang are suitable areas for growing apple.

(b) Sikkim

Sikkim has a meagre area of 200 ha in apple with limited potential for expansion. The existing orchards are sick, infected with scab and cankers and have no systematic training and pruning of trees. They need rejuvenation, proper training and technological interventions for management of pests and diseases. The other temperate crops which have potential as kiwi fruit, pear and walnut. Apple has not been included as a focused crop under Horticulture Technology Mission and State Govt. to come forward with specific programme for temperate fruits, Asiatic pear and kiwi in particular.

Arunachal Pradesh

(a) Geographical Features and Climate Data

Horticulture is considered as the backbone and future of rural economy of Arunachal Pradesh. Present status of land utilization shows that of total geographical land, agriculture covers only 3.40 per cent and horticulture covers 5.20 per cent. Barren and uncultivable land is about 10 per cent of geographical area, whereas, snow clad pastoral alpine land covers another 12 per cent area. There is vast scope for expansion of horticultural crops in Arunachal.

(b) Suggested Action Plan

(i) Immediate R&D action

The post of horticulture scientists at ICAR Research Complex to be filled immediately with a Fruit Scientist stationed at Basar. A Regional. Research Station of CITH, Srinagar at Dirang for location specific research and development may be considered in the XII Plan by acquiring Regional Apple Nursery, Dirang. For creation of infrastructure for laboratories and nursery production 50 ha community land at Zimithung (5 km from Dirang) could be acquired with the assistance of Directorate of Horticulture for layout trials on varietal and rootstock evaluations and development of production and plant protection techniques. While field trials under R&D will be funded immediately under MM I, the new HDP plantations and micro- irrigation under area expansion will be covered under MM II of HTM. KVKs in temperate fruit zones (Tawang, West Kemang, upper Siang and upper Dibang and lower Subansiri districts) may be involved in technology assessment and refinement and training activities in consultation with the scientist of ICAR-RC, Umian and Basar under funding support of MM I of HTM.

(ii) Control of apple scab

Infrastructure to control the most important disease like apple scab and premature leaf fall is urgently required in all districts of Arunachal growing apple. Under all India Coordinated Research Project on apple scab with PC-cum-HOD, Mycology and Plant Pathology of UHF, Solan (H.P.) was started in VI 5 Year Plan with a sub centre of SKUAST at Srinagar and Hill Campus Ranichauri of GB Pant University. Control measure worked out by establishing apple scab monitoring and fore casting laboratories in major apple growing areas of respective states. The scheme was terminated in VIII year plan meeting the objective of effective control measures. These five laboratories are still maintained for advisory service by UHF Solan for benefit of apple growers of the state.

On the similar lines apple scab monitoring and forecasting laboratories at Govt. Horticulture farm Shergaon, Dirang, Twang, Ziro and Anjaw (Upper Dibang) need to be established under ICAR/MM-II Grants. The technical backup for establishing apple scab monitoring laboratories can be obtained from Department of Mycology and Plant Pathology of Dr Y.S. Parmar University of Horticulture and Forestry, Nauni, Solan. The equipment required for laboratory include Reuter stocks forecasting system, automatic weather station with scab monitoring and forecasting software system, besides autoclave, BOD incubator *etc.* costing about 5 lakhs for each laboratory. Total cost for 5 such labs will be approx. Rs. 25 lakhs. Funding support under ICAR's own budget or HTM of DOAC is required as All India Coordinated Project of ICAR has been closed and replaced by Network Project.

(iii) Establishment of a centre for research and development

Establishment of Regional Research and Extension Centre at Dirang for temperate horticulture under Central Institute of Temperate Horticulture Srinagar (J&K) with following proposed mandate:

- ✰ Extensive survey to delineate areas for different horticulture crops.
- ✰ Research on the effect of weather vagaries.
- ✰ Develop cultural operations to regulate flowering.
- ✰ Introduction of varieties and cultivars for higher productivity.
- ✰ Selection and plantation of suitable pollinizer cultivars.
- ✰ Selection of suitable clonal rootstock for high density plantation.
- ✰ Techniques for mass multiplication of rootstock and nursery production be standardized.
- ✰ Effect of availability of macro and micronutrients be standardized on basis of soil and leaf analysis.
- ✰ Use economical and eco-friendly methods to control major diseases and pests.
- ✰ Forecasting and control measures for chronic pests and diseases be developed.

(iv) Establishment of modern nurseries

Establishment of a Central Nursery for different temperate fruits with mother blocks and with following infrastructures to be developed either at Dirang or Shergaon:

Polyhouses, shade net houses, tissue culture labs for micropropagation for difficult to root scion and rootstock, genotypes mist propagation and hardening units, special equipment for:

- ✰ Hot callusing unit for walnut.
- ✰- Automatic grafting/budding machines.
- ✰ Packing houses.

Note: During the last five year fruit nursery at UHF Nauni has supplied about 10,000 apple, 5,000 grafted kiwi and 500 walnut to AP whereas, for Sikkim 7-8 thousand apple and 4,000 kiwi have been supplied. The University can make arrangement for superior planting material for NE provided one year advance order is placed. However, keeping in view local demand only limited supply can be arranged therefore, calling for a strong nursery production infrastructure in NE (scientists of nursery production UHF can provide complete technical know-how).

(v) Capacity building

Extensive HRD exercise be conducted by drawing resources persons from YSP UHF, Nauni, Solan and CITH, Srinagar for training of trainers (TOT). Areas of training should include:

Healthy nursery production (propagation techniques), nursery management, high density plantation, training cum pruning technology, top working for rejuvenation of senile orchards, layout and planting, Integrated water and nutrition management and insect pest and disease management.

The District horticulture officers/other field extension functionaries posted in temperate fruit producing districts and SMS (Horticulture) of KVKs will be first batch of TOT. The Trained officers/SMS will subsequently organise farmers training in respective districts. First TOT to be organized by Central Institute of Horticulture (CIH) of DOAC, MoA, Govt. of India located at Medziphema, Nagaland in consultation with local Govt. and KVK SMS (Horticulture) of KVK will be first batch of TOT (acting Director, Temperate Horticulture, A. P. and Programme Co-ordinators of KVKs).

(vi) Rejuvenation of senile orchards

Rejuvenation through canopy management and nutrient supplement as follows suggested for declined apple orchards.

- ✰ Canopy management through pruning (light and severe pruning and old spur removal) Supplementary nutrient application (NPK and micronutrients) 700:150:350 g NPK/tree of 10 years or more during winter season and a cocktail spray of urea + $ZnSO_4$ + $MnSO_4$ (0.5 per cent each) after petal fall may be beneficial
- ✰ Application of fungicides and insecticides
- ✰ Disease-control of scab, powdery mildew root/collar rot wherever prevalent.
- ✰ Pests- San Jose scale, wooly aphid, defoliating beetles, red spider might if detected need to be controlled.
- ✰ Top working with superior cultivars of higher yields.

Location: State Horticulture farm Shergaon (AP) and selected farmers orchards in concentrated apple belts.

Note: Similar rejuvenation of declined orchards in Government farm in Lachung and Leichen in North Sikkim is required except the approach should involve only organic inputs as State Policy prohibits use of chemical fertilizers and plant protection measures.

(vii) Establishment of a controlled atmosphere (CA) storage at Bomdila in West Kameng district

It is reported that in Arunachal Pradesh over 12,000 ha area already covered with apple with a production of over 10,000 tonnes. The high altitude region apple becomes ready for harvest from mid-September to mid-October after which the atmospheric temperate gradually becomes cooler. The apple of Kumaon region of Uttrakhand comes to market first followed by apple of H. P. and J.K. The Arunachal apple has distinct advantage of late market arrival coinciding with autumn early winter season. Combining traditional storage with cold storage/CA storage may give a clear edge of AP apple in realizing higher market price in off season. C.A. storage using oxygen and carbon dioxide (1 to 5 per cent concentrations) is recommended for long term storage of apple, pears, kiwi *etc.* for up to 10 months.

It is reported that CA storage technology has increased steadily in more recent years in advanced nations, contributing significantly in extending post-harvest life of temperate fruits, apple in particular. For promoting Arunachal apple and kiwi in international market CA storage facility will be very essential. Arunachal Pradesh has long borderline with China, Bhutan and Myanmar. The Bangladesh market is also not far off (both China and Bhutan are exporting apple to Bangladesh in limited quantity) and apple and Kiwi should find access if quality can be ensured and post-harvest loss can be reduced. Apples of good quality and long storage potential can be stored gainfully in CA storage for which technologies have been perfected.

(viii) New potential crops for expansion

Almond (green), olive, European pear, cherry, wine grapes suggesting for field trials in KVKs at Dirang and Tawang for performance and suitability (Table 8).

(c) Way Forward and Action Plan of the Team

The temperate zone of Arunachal Pradesh is highly suitable for development of temperate fruits (apple, walnut, kiwi, pears and others) on commercial scale and a science based sound temperate fruit industry need to be developed on Mission mode approach with full support of ICAR and DoAC.

Two ICAR institutions, namely, ICAR Research Complex for NEH, Umiam, Barapani, Meghalaya of NRM Division and CITH, Srinagar of Horticulture Division will provide technical support with proven technologies and through generation of location specific technologies through MM I of Horticulture Technology Mission and with the institutional initiatives as suggested by the Roving Team. DDG (Horticulture) ICAR will overview the programme and co-ordinate all technical issues for faster implementations of identified activities.

The Horticulture Division of DoAC under the leadership of Horticulture Commissioner will provides support to procure planting materials and enhancing

Table 8: Activity Milestones with Time Frame, Implementing Agency and Funding Sources Suggested by Roving Team

Subject	*Activities*	*Implementing Agency*	*Training*
A. Training/HRD Training of Trainers (TOT) Improvement in existing plantations in major temperate fruit (apple, pears, kiwi, walnut) Districts Training focus TOT: a) Production technology b) Rejuvenation c) Training and pruning d) Scab control **P Farmers Training (FT)** a) Training and pruning practice b) Production practices (Spacing, Contour planting nutrient and water management, plant protection)	☆ Training of district level officers of State Dept. of Horticulture and SMS of KVKs of West Kameng, Tawang. Lower Surbansiri, West Siam, Dibang Valley to develop Master Trainers. Venue for the training may be at: Regional Apple Nursery, Dirang or at the NRC for Yak, Dirang. (Details of Training indicated in item 5 of Suggested Action Plan). ☆ Farmers Training about 100 farmers from each of the 6 districts, 2 days-duration. Venue: KVKs and State Govt. facilities at Dirang, Shergaon and other.	☆ Central Institute of Horticulture (CIH), Medziphema, Nagaland by drawing Resources persons from Dr. YSPUH and F, Solan and from CIH fund/budget. ☆ State Dept. (Jt. Director, Temp. Zone Horticulture) from State/MM II fund Master-Trainers developed by CIH as above will conduct Farmers Training.	For one week/5 days each district One week each district
B. Rejuvenation of old apple orchards Focus of rejuvenation programme as detailed in item 6 of Suggested Action Plan	Under MM II of TMNE rejuvenation in citrus already promoted. Similar way, provision of inputs (plant nutrient, plant protection, irrigation, pruning tools etc.) under MM II funding support to be worked out per ha/plant population basis.	Horticulture Commissioner DOAC/Jt. Director. Temp. Zone Horticulture/DHOs	Winter months (Dormant Season) each district
C. Immediate R&D activities Field trials with new elite varieties as detailed under 'Way Forward: Priority Activities' of the report	New Non-spur and Spur varieties with pollinizers will be introduced for immediate planting through procurement of planting materials from R&D institutes in J&K (CITH) and Himachal Pradesh (YSPUHF)	Proposal with costing for funding support under MM I to be submitted by ICAR- RC, in consultation with CITH, Srinagar. Trial sites may be at KVKs Dirang, Tawang, Ziro, Upper Siang (Geku), Anjaw and Shergaon Govt. farm	December and January every district

Contd...

Table 8–*Contd...*

Subject	*Activities*	*Implementing Agency*	*Training*
D. **Apple scab monitoring and Control including.** Creation of advisory	Establishment of apple scab monitoring and forecasting labs in major apple growing	ICAR-RC, Umiam	Proposed
E. **Establishment of R&D Centre at Dirang**	A Regional Centre of CITH Srirnagar to be proposed for creation as 12th plan activity of ICAR. Among others, new crops (Almonds, Wine Olive, Cherry, Pears etc also to be included)	DG/DDG (Horticulture)/Director CITH of ICAR under DARE/ICAR funding support Proposal has been initiated by Director, CITH, Srinagar.	EFC Memo approved byICAR Land acquisition under process
F. **Area Expansion Programme**	Initial Target and Roadmap 2012-22 given in the text untie report. Technical details, planting material requirements and others are also worked out. The funding support for area expansion from Technology Mission (TMNE) of DOAC and from other possible NHB sources. Also, indent for planting materials to he placed at least one year ahead to UHF/CITH potential supply sources	Horticulture Commissioner/Department of Horticulture, A.P to work out and allocate required fund. CITH, Srinagar of ICAR may provide technical details for strengthening nursery production programme at local level and help in establishing Mother Block of elite varieties	March – April 2012 onwards
G. **Other Infrastructures** ✰ CA Storage ✰ Nursery development ✰ Pack house development	Justification and technical details worked out	DOAC (Under MM III and MM IV) of TMNE, NHB and MOFP to suitably formulate proposals for early implementation	Yet to start

the nursery production activities at local levels, area expansion, rejuvenation of old orchards and plant protection measures (particularly for scab disease control of apple) through Directorate of Horticulture, Govt. of A. P. The Horticulture Commissioner agreed to provide possible funding support under MM I and MM II of the Horticulture Technology Mission. The Central Institute of Horticulture (CIH) of DoAC, located in Nagaland will organize HRD/training activities for field extension functionaries and farmers of temperate fruits. Area expansion programme as suggested will be supported to possible extent under DoAC funding support.

The Director, ICAR-RC, UmiamBarapani and its Regional Centre in Arunachal Pradesh, Basar, in consultation with Director CITH, Srinagar will formulate research proposals for (i) Immediate R&D activities and (ii) Apple scab monitoring and control in the line as proposed by the Roving Team in its report for funding support of MM I of Technology Mission. The Director, CITH, Srinagar will submit proposal for establishment of a Regional Centre at Dirang, Arunachal Pradesh for consideration of ICAR as XII plan R&D strengthening proposal.

ICAR-RC Umiam, Meghalaya and some KVKs in the temperate zones of A. P. will be involved in implementation of various proposed activities.

Future Thrust Areas

Arunachal Pradesh

Apple

Apple is the most important temperate fruit and is universally regarded as the King of temperate fruits. The old adage "An apple a day keeps the doctor away" focuses man's attention on the importance of apple in the -daily diet owing to it's richness in zatxihydrates, proteins, minerals and vitamins besides certain medicinal properties. Apple was introduced in the state in the early seventies. Subsequently, the State Horticulture Farm at Shergaon and the Regional Apple Nursery at Dirang in West Kameng district were established in 1976 and 1977 respectively under the sponsorship of North Eastern Council for the promotion of apple cultivation in the state. In due course of time these institutions attained the status of model objects for promotion of apple industry of the state. It is felt worthwhile to mention that the apples of the state have been rewarded the several first and 2nd prizes in All India Apple Shows at Madras and Nainital in 1982 and 1995 respectively and of late the apples of the state have received accolades from scientists, traders and consumers in national exhibitions in New Delhi, Bengaluru *etc.* It may not be out of place to highlight that China on the other side of the eastern Himalayan ranges is the largest producer of apple in the world (40 per cent). These facts prove that the potentiality of apple cultivation on commercial scale in the state is undeniable.

Kiwi Fruit

The kiwi fruit is a berry, oblong in shape with brown skin similar to sapota (*chikoo*). The fruit has a refreshing, delicate flavour, pleasing aroma and high nutritive and medicinal values with low calories. The centre of origin of kiwifruit lies in China and hence is crown as Chinese gooseberry. It is grown well in the mid hills under moderate to high rainfall conditions with a wide range of adaptability having no serious pests and diseases. The fruit ripens during October/November, which is a lean period for other fruits in the market. During this period rains stop and temperature slides down providing less chance of road blockade and spoilage thereby ensuring possibility of supply for long periods without creating glut and fetching high prices.

Although the kiwifruit is of recent introduction in the state it has assumed tremendous popularity and preference among the growers as well as consumers due to its favourable properties for easy maintenance and marketing besides having high nutritional and medicinal values. The kiwifruit of the state has already attained commercial identity not only in the local markets but also in the national as well as international markets. As the kiwifruit originated in China on the northern front of the eastern Sub-Himalayan ranges, the inherent potentiality for its commercial cultivation on the southern Sub-Himalaya ranges is obviously evident. This is the future crop of the state whhich could provide sustenance to the economy of the rural masses and the state.

Walnut

Walnut is the most important nut among the nut fruits. The nuts are rich in protein, oil. vitamins and minerals. It has high commercial value in the global market. Due to its unperishable nature the nut possesses tremendous scope for export. Although the tree requires long gestation period, its maintenance cost is pretty negligible. Besides high nutritive values of the nut the tree bears high timber value when mature. Walnut grows well at slightly lower altitudes of the temperate region with long growing seasons where other major fruits like apple and kiwi do not do well. Arunachal Pradesh having a wide range of agro-climatic conditions provides tremendous potentiality for cultivation of walnut. Walnuts introduced in Dirang and Rupa areas of West Kameng district have shown excellent results. Walnut has been identified as a potent commercial crop for economic upliftment of certain section of rural masses through efficient land utility in lower Tawang area under Horticulture Mission.

Low Chill Fruit Crops

The vast sphere of temperate region of the state has been found to be suitable for successful cultivation of minor temperate fruits like pears, peaches, plums, apricot, pecan-nut, chestnut, fig, pomogranate, pistachio nut, hazelnut, persimmon, temperate grapes, strawberry *etc*. Diversification in any farming system provides sustainability to the major occupation. Cultivation of minor temperate fruit crops can impart sustainability to the promotion of temperate horticulture venture.

Sikkim

Sikkim in bestowed with a wide range of agro-climatic conditions transitionally between the sub-tropical humid climate of West Bengal on the south and temperate dry conditions on the northern front. The climate is thus, generally characterized by wet summer months (May to Sept.) in the foot hills while it is mild in the hills with moderate temperature in summer and dry winter months and frequent sudden drop of temperature during strong cold out breaks from the Himalayas. The state is blessed with rich water resources of Teesta river and its tributaries, lakes, abundance of rains and storage of snow water in the alpine ranges during winter months. Prevalence of this sphere of agro-climatic conditions in the state provides immense potentiality of growing a wide range of tropical to temperate horticulture corps.

Horticulture in Sikkim comprises of fruits mainly Sikkim mandarin, pear, kiwi, papaya and banana; Vegetable such as beans, peas, tomato, cole crops, radish, cucurbits including chayote (chow-chow); tuber crop mainly potato; spcies mainly large cardamom, ginger, turmeric and cherry pepper; and flowers like cymbidium, gerbera etc. New diversification like bee keeping, mushroom, bamboo and medicinal plants have widened the scope of horticulture sector. The significance of horticulture in improving land use, promoting crop diversification employment generation and providing food and nutritional security to people has increased many fold with more awareness and keenness of its importance. The emergence of organic farming, protected cultivation etc has added new dimensions. Floriculture especially orchids has been able to change the entire concept of horticulture, helping in entrepreneurship and tourism. Horticulture seem to be better avocation of livelihoods for sustainable growth as evidenced in following data base of area and production.

Table 9: Estimates of Area, Production and Productivity of Fruits

Fruit Crop	*Area (000 tonne)*	*Production (000 ha)*	*Productivity (kg/ha)*
Sikkim mandarin	8.320	14.390	1,730
Passion fruit	0.015	0.030	2,000
Other fruits	4.560	5.690	1,247
Total	12.895	20.110	3,089

Table 9: District-wise Physical Achievement under Fruits during 2010-11

	North	*East*	*South/west*		*Total*
Area expansion (New plantation)					
Orange	104.25	114.80	150.00	107.07	476.12
Guava	20.0	40.0	60.0	12.50	132.50
Apple	50.0	–	–	–	50.00
Kiwi	–	–	–	11.68	11.68
Area in second year plantations					
Orange (m)	100.0	314.0	200	200	814.00
Kiwi (m)	–	–	–	240.0	240.00

Source: Sikkim Horticulture Technology Data base 2010-2011.

Horticulture mission for North East and Himalayan states (HMNEH) is the back bone of horticulture development initiatives in the state. The main objectives being focus on establishing convergence of new and synergy among different programmers in Horticulture sector. The aim of the mission is to achieve horizontal and vertical integration of their programme to ensure adequate, appropriate, timely attention to all links in production, postharvest management and consumption chain to maximize economic, ecological and social benefits. Based on its performance Horticulture Technology mission of Sikkim was adjudged as best Horticulture Mission of all Himalayan states in 2012 by Government of India.

Temperate Fruits Scenario

The important temperate fruits grown in the state includes new introduction like kiwi and Asiatic pears and papaya and banana as fillers in lower and mid hills. Apple cultivation a traditional crop in North Sikkim dwindled after decline of orchards.

Lessons Learnt for Apple Farming in Sikkim

The issues include:

- ☆ Before deciding apple areas for expansion, spring frost and temperature fluctuation parameters were not monitored. High fluctuation temperature at flowering time adversely affected fruit yield in apple.
- ☆ Varieties, *viz.* Baldwin, Lord Lambourne and Co-Pippin, were of poor quality and mainly sour in taste. The fruits of Delicious apples introduced later though better in quality but verities being susceptible to spring frost often yield poorly.
- ☆ The trees of old plantations hardly received desired training and pruning as result of which excess vegetative growth was observed resulting to poor fruit wood growth.
- ☆ Pollizing varieties to achieve good fruiting were not planted in right proportion resulting lack of pollination and fruit set.
- ☆ Apple scab, canker, root and colour rot, lichen were not controlled effectively and damage had been considerable. The control measures by organic means were not standardized.

With the introduction of new cultivars, apple growing may show up swing. One has to be cautious in selection of varieties suiting to chilling requirement and as crop diversification European with high chilling pears need to be encouraged in place of apple especially in North Sikkim, while oriental pears in areas with less chilling.

Kiwi and Oriental (Asiatic) pears have vast potential in lower and upper belts of the state respectively. There fruits provide ample scope for area expansion under Horticulture Technology Mission. Keeping in view the importance of these two crops the road map for establishing of sound temperate fruit industry in Sikkim is suggested as per visit and field demonstration in identified areas of mainly West Sikkim.

Kiwi Production in Sikkim

West Sikkim is taking lead in kiwi fruit production. More and more farmers in Richenpong, Deythang and Soren areas are growing the fruit with support from State Department of Horticulture under Horticulture Technology Mission guidance. There are now many progressive kiwi fruit growers of Dehap, Singyang, Narboo, Yuksam,

Neshe, Kaluk, Uttarey, Yangum area of west Sikkim – Some of the kiwi growing areas and kiwi field were visited during the current visit between 12-18 September, 2014. It is interesting is note that wherever kiwi is grown, large cardamom is there. This means that large cardamom can be inter-planted in new kiwi orchard.

Recommendation for Sikkim State

Sikkim has meagre area of 200 ha in apple with limited potential for expansion. The existing orchards are sick, infected with scab and canker and have no systematic training and pruning of trees. They need rejuvenation, proper training and technological interventions for management of pests and diseases. The other temperate crops which have potential are kiwi fruit, pear and walnut. Suitable varieties of these crops can be evaluated and identified along with specific technologies and infrastructure for increasing area, production and productivity. It is reported that Asiatic pear and kiwi may be successful crops in mid hills. The North Sikkim particularly in Lachung area, which is otherwise suitable for temperate fruits, have been badly affected by recent earthquake. At the time of visit at the Government Temperate fruit Orchard at Lachung on 18 September, the Roving Team observed reasonably good bearing in trees of some apple varieties. It is reported that over 20 apple varieties were introduced and 4-5 varieties were found to be promising. The fruits become ready for harvest in late September to mid-October. However on 18 September, 2011 evening due to earthquake and heavy rains the farm at Lachung was devastated totally. Presently, 20 ha area is being rejuvenated with new apple varieties.

Strategies and Priority

In Sikkim kiwi fruit and pear are experiencing expansion among the temperate fruits. Both these fruit crops have specific production problems.

Major production constraints in kiwi fruit:

- ✰ Lack of genetically superior planting material.
- ✰ High cost of plantation whole establishment.
- ✰ Lack of standardized techniques of cultivation like pruning-cum training.
- ✰ Since Sikkim is an organic state, the package of practice of organic cultivation is *adhoc.*
- ✰ Lack of use of proper proportion of pollinizers, practice of hand pollination.
- ✰ Concept of hand thinning and use of biostimulants to increase fruit size is non-existent.

Major production problems in pear:

- ✰ Poor varietal status.
- ✰ Low density plantation resulting in non-optimal utilization of resources
- ✰ Though pear is considered less affected by pollinizers, still suitable cross-pollinations varieties are scant.
- ✰ Management agro-techniques like pruning-cum-training are lacking.

- ☆ Organic pear cultivation practices are *ad hoc.*
- ☆ Prevalence of diseases and pests.
- ☆ Senile orchards.

The priority areas and activities suggested for lower belt of Sikkim are as follows:

i. **Introduction, evaluation and mass multiplication of suitable varieties of pear and kiwi which can bear biotic and abiotic stress**

Pear: Although high hills of Sikkim especially north district with rain shadow areas of Lachung are suitable for European pear. The varieties suitable for such areas are Bartlett, Max Red Bartlett and Red Bartlett hold promise. The Oriental/Asiatic pears may so well in lower belt mid hill areas, where chilling is less; The Asiatic pear varieties developed in Punjab Agricultural University, Ludhiana may be tried, besides introductions from USA and South Korea suggested in Horticulture Technology Technical Report, Dept. of Horticulture, 2014

Kiwi: The orchards in West Sikkim has mainly, Monty Allison and Bruno. Hayward variety which is very less. This variety is now being introduced widely under Technology mission.

ii. **Training and pruning**

The technique of trellis on T-bar is very scientific and needs precise practice to be performed by farmers so as to have maximum fruiting wood. The farmers are to be educated practically.

iii. **Crop and quality regulation**

All cultivars of kiwi fruit except Hayward bear heavy crop every year. The heavy crop creates a severe competition between fruits for nutrients and water, which leads to production of small size fruits. Poor management practices, improper ratio of pollinizers, their placement in plantations and lack of hand-pollination also contribute to small fruits. Therefore to increase quality kiwi crop of good size hand thinning is essential. The studies conducted in thinning of fruits to 4-6 fruits per bunch immediately after fruits set and treatment with CPPU bio-stimulant/plant growth promoter standardized by Dr Y.S. Parmar University of Horticulture and Forestry Nauni, Solan need to be replicated in kiwi orchards of Sikkim.

iv. **Rejuvination of senile orchards of apple and pear**

The treatments suggested include (a) Canopy management through light and severe pruning and old spur removal, (b) Supplementary organic nutrient application of NPK and micronutrients after petal full will be beneficial, (c) Effective control of diseases and pests by recommended organic pesticides, (d) Top working with superior cultivars of high yielding traits with proper proportion of pollinizing varieties.

v. **Improvement in propagation and quality planting material in kiwi:**

Production of quality planting material of Kiwifruit through improved propagation technologies and improvement in fruit growth by plant hormones has to be green emphasis.

Potential of other Temperate Fruits

The exotic kiwifruit has shown tremendous potential in the recent years and has been assessed as a modal fruit for diversification of mid hill horticulture in Sikkim as one of the pioneer state to demonstrate cultivation in the north east. The demonstration on the improved cultivation in production will further enhance the scope of kiwifruit in mid hills of Sikkim where stone fruits have shown declining trend in the recent past due to their perishable nature and poor marketability. The state department has to make package of practice for this crop using organic inputs.

The biggest bottleneck in the popularization of marketing of quality kiwifruit is the lack of mass production of suitable planting material and production technology, the improvement in quality and production of the kiwifruit quality planting material has to be interdicted through intervention of improved technology like softwood mass multiplication, training, pruning and use of plant growth regulators, bio-stimulants.

Low Chilling Cultivars Suitable for Subtropical Conditions of North Eastern States

Some important cultivars which have been recommended for subtropical conditions are given below. These can be introduced and evaluated under subtropical conditions in north eastern region.

Almond: Hybrid-15, JKS-55, H-98, Perkin's Wonder California Paper Shell

Apple: Early Shanburry, Gallia Beauty, Tropical Beauty, Winter Banana, Tamer, Vered, Shlomit, Maayan, Michal

Apricot: Chaubattia Alankar, Chaubattia Madhu, Chaubattia Kesri, Trevatt, NJ13

Peacaunt: Mahan, Berket and Claro

Peach: Sharbati, Safeda Early Cream, Flordared, Flordasun, Flordabelle, Saharanpur Prabhat, Shan-i-Panjab, Ranjit Bagh Early, Khurmani

Pear: Kieffer, Smith, Baghugosha, Le Conte, Patharnakh, China Pear, Gola

Plum: Jamuni, Aloo Bokhara, Alucha Black, Titron Howe, Kala Amritsari

Strawberry: Chandler, Pusa Early Dwarf, Fern, Selva, Pujero

6. CONCLUSION

Arunachal Pradesh and Sikkim located in the eastern Sub-Himalayan ranges, while the Purvanchal Region comprising Nagaland and Manipur, Meghalaya Mikir regions having huge unexplored land under temperate climate has tremendous potentiality for promotion of temperate fruits namely apple, kiwi, walnut, pecanut, Asiatic pear, low chill stone fruits especially in Arunachal and Sikkim. If the basic requirements of development like appropriate self-reliant technology, infrastructure and techno-physical marketing facilities are addressed in a mission mode with long term vision, the temperate horticulture of the state could be developed into a vital industry for sustainable economy of these rural people in particular and the state and country in general for their livelihoods and export to neighbouring areas.

7. REFERENCES

Chadha K.L. 2003. *Commercial Cultivation of Temperate Zone Fruits in the Tropics and Subtropics: some reminiscences in souvenir* VII Int, Symp. on TZFTS held by ISHS UHF, Solan (Ed. K.K. Jindal *et al.*) pp. 48-49.

Chadha, K.L. 1998. Improvement in tree fruits and plantation crops. *Indian J. Hort.* **55**: 265-296.

Dilip Singh, R.K. 2004. Temperate horticulture status in NEHR: An overview. **In:** *Recent Trends in Horticulture in the Himalayas* (Eds. K.K. Jindal and R.C. Sharma). Indus Publications, New Delhi, pp. 329-333.

Ghosh, S.P, Jindal, K.K., Ahmed, Nazeer and Ngachan, S.V. 2011. *Road Map for Temperate Horticulture in North Eastern Hills*, ICAR Roving Team Report, pp. 1-21.

Ghosh S.P. 1978. Horticulture in North Eastern region. *Indian Horticulture,* **24**: 31-36.

Jindal, K.K. 2006. Promising fruits for temperate and subtemperate zones of North East states of India. *CAU Imphal Technical Bull.* pp. 1-32.

Jindal K.K.2014. Fruit growing area of Sikkim with special reference to kiwi and Asiatic pear production Technology, Technical Report, Department of Horticulture. pp. 11.

Jindal, K.K. and Dorjee, T. 2003. Apple farming in the North Eastern Himalayan Region **In:** *Apple Farming and livelihoods in Himalayas.* K.K. Jindal, Rajan Bawa and Tej Pratap (Eds.), Bishen Singh Mohinder Singh Publ. Shiva Offset Press, Dehradun, pp. 127-140.

Jindal, K.K. and Singh, L.N. 2010. Precision Farming Initiatives in rains shadow horticulture for North Easter India. *National Seminar on Precision Farming* held at Jhalawar 28-29, 2010, pp. 680.

Jindal, K.K., Chauhan, P.S. and Mankotia, M.S. 2002. Apple production in changing climate scenario: Issues and constraints. **In:** *Enhancement of Temperate Fruits Production-in-Changing Climate,* K.K. Jindal and D.R. Gautam (Eds.) ICAR UHF Publ., pp. 45-54.

Jindal, K.K. 2006. Crop diversification with special reference on nut crops in *North East States Temperate and Sub temperate Zone* CAU, Impahal Technical Bull. pp. 20.

Jindal K.K. 2015. Canopy management in West Sikkim on training and pruning of kiwi fruits. Technical Mannual under Hort. Tech. Mission Department of Horticulture, Govt. of Sikkim , pp. 1-30.

Tsering, Y. 2012. Perspective profile of temperate zone horticulture Arunachal Pradesh, Office of Joint Driector, Bomdila, AP. (Personal Comm.).

Chapter 5

Canopy Management and Plant Architectural Engineering in Temperate Fruits and Nuts

N. Ahmed, K.L. Kumawat and Lal Chand

ICAR-Central Institute of Temperate Horticulture,
Srinagar – 190 007, J&K
E-mail: dnak59@rediffmail.com

1. INTRODUCTION

In India, Horticulture is an important sector growing rapidly and contributing to economic and nutritional security. It is a major foreign earning commodity and has great potential for further commercial exploitation. As compared to other fruits the productivity of temperate fruits and nuts continues to be low because of several factors one of which is the poor management of canopy as it has direct bearing on productivity and quality.

Basically, in agricultural plants there is a transformation of solar energy in to chemical energy by the green foliage through photochemical process of photosynthesis. This forms the primary source of energy, which ensures food, fodder, fiber and fuel therefore it need to be harnessed effectively. In fruiting tree, the fruiting potential is largely governed by its architecture, canopy density and photosynthetic efficiency (Kallow *et al.*, 2005). Where, the term canopy refers to the physical composition of plant comprising of mainly stem, branches, shoot and leaves, while the canopy density is determined by considering the number and size of leaves. Its architecture is determined by the number, length and orientation of stem, branches and shoots while canopy volume is determined by spread and height of canopy

(Pathak, 2009). The criteria used to define tree architecture relate mainly to primary growth. They are usually classified into four groups namely, type of growth (rhythmic vs. continuous), branching pattern (sympodial vs. monopodial), morphological differentiation of branches (orthotropic vs. plagiotropic), and position of reproductive structures on scaffold branches (lateral vs. terminal).

Fruits plants attain tall and becomes huge if they are not managed by proper training and pruning from the initial stage. For harvesting the sun, each plant must establish its optimum canopy spread at earliest of its life cycle. Management of canopy architecture is therefore one of the predominant technologies by which huge and unmanageable trees are properly managed to make them more productive. It not only involves just pruning and tree training; but also includes the regulation of flowering and fruit growth.

2. CANOPY MANAGEMENT

(a) Essential Features of an Ideal Canopy

- ✰ Strong frame of primary branches
- ✰ Wider crotches in scaffold branches
- ✰ Healthy and well distributed secondary and tertiary branches
- ✰ Sufficient fruiting terminals in most productive areas
- ✰ Healthy foliage with high photosynthetic efficiency to maximize the solar radiation use efficiency
- ✰ Enough space for air circulation inside canopy
- ✰ Should support adequate to nurture and protect the fruits from sun burn

(b) Objective of Canopy Management

The basic objective of canopy management is to maximize light interception, to optimize light distribution within canopy and to maintain proper airflow. Canopy management enhances productivity, improves fruit quality, facilitates cultural practices and help in management of pest and disease. In new plantations initial training and pruning is given to develop strong frame work of the tree whereas in old plantation the aim of canopy management is to reduce tree height and make provision of solar radiation inside the canopy by thinning excessive bio mass. The objective of canopy management are:

- ✰ To increase yield efficiency
- ✰ To increase quality of produce
- ✰ To make high density planting successful
- ✰ To rejuvenate old orchard
- ✰ To facilitate cultural practices
- ✰ To increase input use efficiency
- ✰ To reduce pest and disease incidence
- ✰ Efficient use of pesticide and fungicide

- ☆ To facilitate mechanization
- ☆ To Reduce cost of production

(c) Principles of Canopy Management

The basic principle of canopy management of the fruit trees is to make the best use of the land and the solar radiations for increased productivity. Light is fundamental aspect of fruit production. The canopy management optimizes the plant model in such a form that it intercepts maximum light. In many fruit crops, improved production and fruit quality has come from producing more fruit from smaller trees as small trees are better at capturing and converting sunlight into fruit than larger trees. Smaller trees safer, more efficient and produce more fruits from sun exposed part of the tree and need to be considered when developing new planting system.

Fruit production involves the capture and conversion of light energy into fruit biomass (dry matter). The main controlling factors are (i) the amount of incoming radiation (ii) the percentage of that radiation which is intercepted by the tree and (iii) how efficiently the tree converts that energy in to fruits. Both climate and length of growing season determine the amount of incoming radiation. The grower can control neither of these, except the initial choice of orchard site. The length of growing season is also important in deciduous plants which shed their leaves in autumn and thus have a finite growing season. The percentage of incoming radiation intercepted by the tree is where growers have most control so decisions relate to block size and planting density is important. In small orchard blocks the surrounding shelter trees shade a greater proportion of the trees than they do in larger blocks. The situation is similar with high density plantings where shading from adjacent trees results in large volumes of shaded canopy and leaves operating well below their photosynthesis potential. The high density is thus determined more by tree size than by planting distance. In low density plantings high percentage of incoming radiation being captured by the grass growing between the trees and not by the fruit trees. Interception of light by total orchard canopy is therefore very important and is correlated to total biomass production. According to Shikhamany (2001) the basic principles of canopy management are lies in (i) maximum utilization of light (ii) avoidance of the build-up of micro-climate congenial for the disease and pest (iii) convenience in carrying out the cultural operations (iv) maximizing the productivity and quality; and (v) economy in obtaining the required canopy architecture. In tree canopy development two aspects are important firstly the interception of available light by the total orchard canopy which is correlated to total biomass production of the crop system (Monteith,1977). Secondly, penetration and/or distribution of light within a single tree canopy which must be optimized for maximum tree performance and crop development.

(d) Light Interception

Light interception is the amount of available light intercepted by the tree canopy and not striking the orchard floor. Interception then is a function of plant density, canopy shape and size, and leafarea index (LAI, m^2 leaf area/m^2 ground area) within the canopy. Biological yield has been increased in many crops due to an enhancement in the amount of seasonal total light interception, and this generally is attributable to improved cultural practices that affect leafgrowth, leaf duration and/or canopy display (Gifford and Jenkins, 1982). The total amount of light intercepted by a fruit orchard system depends mainly on orchard design factors such as planting system, tree spacing, tree shape, tree height, alley width, row orientation aswell as leaf area index (ratio between the total area of all leaves per tree to the land area allocated to the tree) and the length of the growing season (Jackson, 1980; Lakso, 1994; Palmer, 1989; Wagenmakers, 1991).

Orchard canopies, unlike those of other crops, due to the alleyways maintained between rows, a large proportion of the land area not covered by trees (Jackson, 1980). This distribution results in low values for LAI and total light interception for orchard canopies. In addition, leaf area is not randomly distributed over the land area as in annual crops but is traped in trees, branches and spurs. Light interception is low when the orchard is just planted and increases as the orchard develops in relation to total LAI. Cain (1970) introduced the idea that orchard density should be considered over the lifetime of the orchard. Jackson (1970) has shown that widely spaced bush trees, and large at maturity, intercept very little light when they are young, yet when full-sized they have excessive shading within-tree and relatively low LAIs. The realization that significant land and light resources are wasted in the early life of an orchard has encouraged the planting of higher and higher tree densities. In 'Golden Delicious' low-density hedge row plots (1100 trees ha^{-1} on M.9 and 660 trees ha^{-1} on M.2) intercepted roughly half of the incident light at maturity and yields peaked at 40 t ha^{-1} Whereas, the high-density hedge row plots (3300 trees ha^{-1} on M.9 and 2260 trees ha^{-1} on M.2, intercepted two thirds and three-quarters of available light, respectively and yielded more than 70 t ha^{-1} in their 6th and 7th years (Verheij and Verwer, 1973). Likewise, in apple cv. 'Golden Delicious' with plant densities ranging from 853 to 3746 trees ha^{-1}, yield and light interception were approximately linearly related and light interception is related to LAI, not just planting density (Palmer and Jackson, 1974). In discontinuous orchard canopies LAI is the largest single factor influencing light interception (Jackson *et al.*, 1977). Increasing tree planting density has been the most important means of increasing the early yield and early light interception of young orchards (Wertheim *et al.*, 1986). In the same way, Barritt (1989) reported that tree density was more important than training system or rootstock for improving light interception and yield of 'Granny Smith' in the 3rd year. Pruning generally reduces total light interception because of reduced growth and canopy size. Summer pruning has been shown to improve light penetration within tree canopies (Marini and Barden, 1987). Light interception varies during a season due to leaf surface area development and solar angle, and during orchard establishment due to tree size increase. Light interception during the season closely follows leaf area development (Palmer and Jackson, 1977). Total light interception by north-south rows did not vary after complete canopy formation (Jackson and Palmer, 1972). From all these finding it can be concluded that in early life of the orchard, light interception is a function of tree planting density whereas in mature orchard, light interception is a function of the tree height alley width ratio and the density of the canopy. Thus light interception in orchards can be raised by increasing the number of trees ha^{-1}, increasing the height of the trees relative to the clear alley width, increasing the density of foliage in the canopy (Corelli and Sansavini, 1989), orienting rows in north-south direction (Wunsche and Lakso, 2000) and by summer pruning.

(e) Light Distribution

In an ideal orchard, light should be evenly distributed under tree rows and alley's and should be around 60-70 per cent. Light penetration into the tree is usually determined by the size and shape of the tree.

(f) Tree Form and Light Distribution

Two approaches have been used to improve the light distribution in tree canopies. One is to use relatively natural tree forms that allow light penetration through the canopy by providing many small openings in the foliage such as in the multiple leader, central leader, vertical axis, or slender spindle forms (Wertheim, 1968; Lespinasse and Delort, 1986). A second approach is to provide fewer large, permanent openings for light penetration into canopies restricted into geometric forms. Examples are thin restricted planes of foliage such as narrow hedgerows, tree walls, and A, V, or T forms (Chalmers and van de Ende, 1975).

(i) Round crown tree

In large round crowned trees light intensity decreased rapidly with increasing depth of foliage and that of lower and centre positions of the tree received very low light intensities of 6 per cent to 30 per cent of full sunlight. (Heinicke, 1963; Looney, 1968; Verheij and Verwer, 1973). Effective penetration depth of light into unrestricted apple canopies is 1m in round crown tree (Jackson, 1970). Heinicke (1963) proposed that 30 per cent of full sun serve as a lower limit of desired light level in apple canopies. With decreased in tree size, the heavily shaded area within the tree decreased (Heinicke, 1964; Cain, 1970).

(ii) The central leader tree

Awasthi and Mehta (2004) reported that light intensity within large Delicious apple trees decreased by 58 per cent within about 2m from the top of the tree trained on central leader system. Further, Barritt *et al.* (1991) have reported on the seasonal change in interior canopy light of 'Delicious' central leader trees. Their data show a rapid decline in light exposure to <20 per cent full sun by early June for the interior parts of the canopy. He suggested summer pruning to improve fruit color. Similarly, summer pruning improved light distribution in the middle section of the canopy trained on central leader system (Robinson *et al.*, 1991). Lakso *et al.* (1989) found that use of interstock (Empire/M.9/MM.111) and greater relative distance of the 1 m distance between scaffold branches increase light distribution within canopy.

(iii) The slender spindle tree

Slender spindle trees of 'Empire'/M.9 had very good light exposure in most of the canopy at 14DAFB. However, by 44 DAFB, light exposure levels in all but the top of the tree had dropped considerably, indicating that the gaps between limbs had been closed. By 74 DAFB, there was a large, poorly illuminated area of the canopy thus, summer pruning of slender spindle trees is essential for good fruit color (Robinson *et al.*, 1991).

(iv) Thin Restricted Plane Canopy Trees

Planar canopies have been developed to overcome the problems of light penetration into thick canopies. However, Tustin *et al.* (1989) found low light transmission through the top layer of the trellis resulting in excessive shade to the lower layers. These horizontal canopy systems suffer from excessive upright shoot growth from the top side of the trellis. This increases the amount of shading, can result in complete closure of the space between tiers. Due to the flat limb orientation red apple cultivar 'Empire' shows poor color with horizontal trellises. Vertical planar canopies have good light exposure on both sides of the canopy but have vertical stacking of fruiting branches leading to loss of vigor in the bottom of the tree. Inclined planar canopies (*e.g.*, the Y-trellis) overcome the horticultural problems of the horizontal canopies and provide light exposure to both sides of the Y. In addition, they do not vertically stack the fruiting branches (Robinson *et al.*, 1991). Thus, light distribution in orchards can be raised by decreasing tree size, summer pruning, light dormant pruning, avoiding heading cuts and by spreading or tying down branches to horizontal position (Wunsche and Lakso, 2000).

(f) Air Circulation and Humidity

Air circulation and humidity within tree canopy influence micro-climate. In ill trained tree with dance canopy, the minimum air circulation result in high humidity within canopy promotes infestation of insect pest and diseases. Ideal canopy architecture both for high and low density orchards need to be designed facilitating maximum air circulation and optimum humidity.

(g) Effect of Light Interception and Distribution on Fruit Plants

The productivity and quality of fruits largely depends upon tree vigour and weather parameter particularly light interception, temperature, relative humidity and air flow. The proper sunlight exposure of foliage and fruits and prevailing temperature play vital role in vegetative growth, photosyntheyic efficiency, flower initiation, fruit set, fruit bud differentiation, ripening, fruit colour, fruit quality *etc.* while increase air flow reduce disease pressure. Fruit trees where there is a higher proportion of 'shade' to 'sun' their leaves were found photosynthetically inactive under deeper shade and act as unproductive sink. Shaded leaves are thinner and contain less palisade layers compared to control leaves. For the induction of flowering, certain level of irradiance and photosynthates accumulation are needed.

The size and colour of the fruit is associated with light level of more than 70 per cent of full radiation. Large and best coloured fruits are borne near the outer periphery of the canopy where irradiance is high, whereas smaller poor colour fruits are borne in the shady part of the tree (Awasthi and Mehta, 2004). Tree canopy receiving less than 30 per cent light produced virtually no flower. Reduction in light from 100 to 70 per cent in tree canopy had a small reduction in flowering whereas increased shading increased fruit shedding and reduced flower setting in the following season. Jackson and Palmer (1977) observed 36 per cent reduction in the number of flowering spurs when light to Cox's Orange Pippin tree was reduced to 37 per cent by shading in early June to leaf fall. The increased shading led to progressive reduction in flower bud production and further shading decreased fruit weight and produced symptom of fruit immaturity *e.g.* decrease in fruit colour, fruit dry matter, fruit soluble solids but increased fruit firmness (Lakso, 1994). The decline in fruit quality with shading within tree canopy has also been reported in peach and cherry (Flor and Kesner, 1982). Proietti and Tombesi (1996) opined that shading reduced flower by about 50 per cent and starch content of root tissues. Low irradiance reduced fruit set in open pollinated flower of persimmon (George *et al.*, 1993).

(h) Techniques for Canopy Management

Canopy management reflects an inter-relation of the physiology underlying the relationship between growth and production. Orchard architecture largely depends upon the orchard production system, which is a combination of variety, rootstock, tree spacing, training and pruning. To optimize the utilization of light for increased yield of quality fruits, canopy management deserves greater attention by exploiting the various available techniques like training, pruning (dormant, summer and root pruning), braches orientation (bending), scoring, girdling, bark inversion, selection of proper rootstock, use of plant growth regulators, appropriate use of fertilizer, deficit irrigation, use of genetically engineered plants with altered architectural characters would help in maintaining the ideal canopies of trees.

(i) Training

Basic objectives of training is to control vegetative vigor, increase light exposure for fruit/foliage, increase air flow to prevent disease infestation and facilitate pruning and harvest methods and mechanization. Whatever system is chosen, the objective of training and pruning is to achieve maximum tree life and productivity. In training we should consider (1) ground clearance (ii) shoot orientation and (iii) ventilation and light interception into canopy. In order to harvest the advantage of light and proper air circulation, training is an important practice performed in the initial years by maintaining desired shapes and strong framework. While deploying the training systems, utilization of vertical and horizontal space should also be taken into consideration. The choice of training system is part of the whole orchard management influencing planting distances, light interception and finally the success of the orchard in terms of obtained fruit quantity and quality (Palmer, 1999).

(ii) Orchard planting system

Interception and distribution of light within tree canopies be determined by various inter-related factors. One factor is genotype of cultivar or architectural pattern ranging from the non-branching, columner type to the spreading, nonspur type. Further, a given cultivar rootstock also play an important role via the canopy volume. A second factor is the planting system which included tree arrangement in orchard (planting spacing, and orientation of row) and training of canopy by pruning as well as limb orientation (height and shape of tree). Training partly defines the proportion and distributionin space of the various shoot categories, long versus short shoots and vegetative versus fruiting shoots (Robinson *et al.*, 1991). Since whole-tree photosynthesis is primarily light limited, thus manipulation of canopy architecture to enhance global light interception is a major objective of all planting systems

On the basis of tree canopy shape Robinson (2003) categorized orchard planting systems in four basic shapes: spherical canopy shapes, conic canopy shapes, flat-fan canopy shapes and Y or V canopy shapes.

(iii) Spherical-shaped canopy systems

Traditionally, in this form of canopy shape, the trees are globe shaped with 6-8 m height and 2 m trunk length. During late 1800s and early 1900s it was most common tree form in traditional apple orchards in Europe and North America (Walker, 1980). According to (Wertheim, 1981) after development of clonal rootstocks in apple in the 1930s, a smaller version of the large globe-shaped tree on semi-dwarf rootstock developed in Holland and England, named the bush tree (the bush-tree system) and on dwarf in Germany named Spindle-bush system.The main advantage of spherical shapes is that they are generally the natural shape, however, the canopies generally have a large shaded centre core that is unproductive and produces poor fruit quality (Robinson, 2003).

(iv) Conic-shaped canopy systems

As large globe shaped tree result in less light distribution within the canopy, many researchers proposed conic or pyramidal canopy. In this tree form several planting systems developed like Central-leader system, Mini-central-leader system, Palmette-leader system, Slender-spindle system, North Holland spindle, Slender-spindle multi-row or bed system, Vertical-axis system, SolAxe system, Slender-pyramid system, HYTEC system, Super-spindle system *etc.* and now it is a dominant tree form in commercial orchards in most part of the world. The primary advantage of this tree from are that it is a natural form with high light interception and distribution in initial age of orchard establishment. However, when mature, the upper branches of the tree begin to outgrow the lower branches, causing excessive shade on the lower canopy. Keeping this in view later many new systems have been developed to increase light interception and distribution in conic shape tree form.

The conic shaped systems have chosen by majority of apple growers all over the world. In the 1960s and 1970s, the central-leader system was dominant system in the world while in the 1970s and 1980s, the slender spindle and the vertical axis were the most common systems in Europe. At the end of the 20th century, the most common systems are the slender spindle and its variants and the vertical axis and its variants. Over the last decade, the height of most slender-spindle trees has increased to close to 2.75 m and thus the trees are more similar to vertical-axis trees. This has allowed slender-spindle orchards to intercept more light and achieve higher yields, while reducing the amount of pruning required to contain the tree to a short stature of 2 m. Similarly over the last decade, the planting density of vertical axis orchards has increased from 1500 to more than 2000 trees ha^{-1}, which is similar to the density of slender-spindle orchards (Robinson, 2003).

(v) Flat planar canopy systems

From very early, European apple growers have utilized trellises to confine tree canopies to a two dimensional plant and developed many geometric shape of apple trees (Huggard, 1980) and among them espaliers was most commonly grown along walls and fences. Further, in the mid-1950s Italian fruit growers developed first commercial apple orchard with two dimensional plane canopy to improve labour efficiency and named as palmette training system. After that this system widespread in Italy and further, significant technical innovation in pruning of this system lead to hasten fruit bearing and reduce pruning costs and increased planting densities to secure the benefits of earlier and greater returns (Sansavini, 1983). With the time, a number of variations of this system have been developed by using dwarf rootstocks *i.e.* 'Regular' palmette system, 'Free' palmette system, Penn State thin-wall trellis system, Lincoln canopy system, Ebro trellis system, Solen system *etc.* The major advantages of flat planar systems are that it improve labour efficiency and allow good light distribution to all parts of the canopy.

(vi) V-shaped canopy systems

V-shaped fruit trees were popularized for commercial orchards by Chalmers and Van Den Ende (1975) with the growing system known as the Tatura trellis. This system was originally developed for peaches to increase yields which allowed mechanical harvesting but was later adapted for apples (Chalmers *et al.*, 1978). With the developement of dwarf rootstock and to improve light interception and penetration to the centre of the canopy, a number of variation of the Tatura trells have been developed *i.e.* Mini-Tatura trellis system, Geneva Y-trellis system, Mikado and Drilling systems, MIA trellis system (A-shaped trellis), Mini-V-trellis system, Gutingen V slender-spindle system, *etc.* The primary advantage of V system has high light interception that result in high yield (Robinson and Lakso, 1991).

3. TRAINING SYSTEM

(a) General Rules of Training of Fruit Plants

(i) Train tree by limb positioning inspite of pruning

Limb positioning can be done by tying of young braches by using spreaders. Bending nearly vertical limbs to an aspect 45 to 60 degrees from vertical usually stimulates fruit production earlier in the life of the tree. In some training systems, tying is restricted to the first tier of branches (vertical axis, slender pyramid and HYTEC), while, in other systems, upper branches are also tied down (SolAxe).

(ii) Spreading of branches

Where the crotch angle is less than 35° the attachment will be weak because of the inclusion of bark. The tissues in narrow crotches are slower to mature in the fall and may be injured by low temperatures. Narrow crotches are usually further weakened by water, ice, rot organism and canker. While the tree is young, spread branches that make very narrow angles with the main trunk.

(iii) Retention of branches

Leave more branches in spur-type cultivars than in a tree of standard growth habit.

(iv) Removal of unwanted shoots in summer

Remove unwanted shoots when they are small and before the base of shoot has lignified.

(v) Training programme

Follow the training program consistently, as frequently as necessary, to complete training properly as soon as possible.

(b) Different Training Systems

There is single planting system to suit all situations. Factors such as soil, cultivar, rootstock, management regimes and socio-economic conditions will determine the optimal combination for each orchard. Here are some most common systems followed in India as well as across the temperate region of world.

(i) Central leader system

Central leader system is commonly used system for training apple, pear, cherry, plum and pecan trees. A central leader tree is characterized by one main, upright trunk, referred to as the leader. Typically trees have 3 to 4 tires of branches spaced along the trunk at tree height of 4-5m. The bottom tire of branches is 60 to 90cm above the ground and has 4 to 5 branches. The second tier of branches is commonly 60 cm to 1m above the first tier and has 4 branches. The third and fourth tires are spaced abut 60-80 cm apart and typically have 3 branches. Spreaders placed between the trunk and branches,to spread the branch angles to a desirable range of about 60 degrees. The shape of a properly trained central leader tree is like that of a Pyramid shape. The lowest scaffold whorl branches will be the longest and the higher scaffold whorl branches will be progressively shorter to allow maximum light penetration into the entire tree.It is a medium density system with tree density ranging from 300 to 700 trees ha^{-1}. Semi-vigorous rootstocks are suitable for this training system in apple such as MM.106, MM.111, M.7 *etc.* (Robinson, 2003).

(ii) Modified central leader system

It is also called the multi-leader system, the modified central leader system is much like the central leader, but multiple vertical branches are trained instead of a single, main leader. The system works best with 3 to 5 leaders so the tree is well balanced. Rather than remove competing vertical branches in the first two years after planting, the desired number of leaders are left on the tree and each is trained in the same way as with a central leader system. Spreaders are almost always required in order to open up the space between individual leaders. Apple, pear, plum and pecan trees are among the most common fruits trained to this system.

(iii) Open centre system

It is also known as the open head or vase form wherein the trunk is allowed to grow upto 75 cm by cutting within a year of planting. The central leader is removed, leaving three to five evenly spaced scaffold branches growing from the trunk in a vase shape. This system lack strong crotches and provide weak framework and as such is less satisfactory. However, it may be favoured in fruits like peaches, nectarines and plum for admitting more sunlight for better colouration of the inside fruits.

(iv) Mini-central leader system

It is similar to the central leader, but above the second tier of branches no permanent branches are allowed to develop, but small fruiting branches on the leader are retained and allowed to crop and bend below horizontal. In apple semi-dwarf rootstock is suitable for this training system where tree densities of 500 to 1000 trees ha^{-1} are maintained.

(v) Dwarf Pyramid

Dwarf pyramid is low-headed compact, central leader tree mostly suitable on dwarfing and semi-dwarf rootstock. The lowest branch arise from 30-40 cm from the ground level and successive branches radiating at intervals of 15-20 cm along the main leader, gradually diminishing in length from bottom to top forming a pyramidal shape. It allows 20-25 primary branches distributed at proper space in all directions. The height of the tree is restricted to 2 m with 1.5 m canopy spread.

(vi) Head and spread

It is also known as two tier training system and suitable on dwarfing and semi-dwarf rootstock. In this training system, the first tier (4-5 scaffolds) is allowed to develop at 45-60 cm above the ground level. The second tier system (5-6 scaffolds) develops after 60 cm of first tier system. The fruiting takes place in the inner periphery of canopy due to good exposure to the sunlight to develop excellent fruit colour.

(vii) Palmette-leader system

It is a modification of central-leader system and designed to improve the light distribution in the tree canopy as in central-leader system with the tree age, there is strong tendency for the upper scaffold limbs grow more vigorously than the lower limbs, due to better light exposure resulting in excessive shade in the bottom of the trees. In palmette-leader system of training the tree is developed by removing upper east and west-growing branches of a mature central-leader or mini-central-leader tree, resulting in a flat north–south-orientated palmette (*i.e.* fan-shaped) top which result in good light penetration into the tree centre throughout the season.

(viii) Spindle Bush

Spindle bush (or free spindle) is best suited to densities up to 2,000 trees ha^{-1} at 2-3 m in height and 3-4 m x 1-2m apart. Well feathered two-year old nursery trees usually should be preferred to develop spindle bush system. At planting, a number of laterals are selected to form part of the permanent scaffolds in the bottom tire of the tree. As the leader grows, more scaffolds are selected and spaced equally. Leader dominance is important and if it is lost will result in a reduced tree canopy, whereas if it becomes too strong, lateral growth and development will be reduced. This systems can be free standing, however mostly utilize some form of support (either two to three wire trellis or individual supports).

As tree densities get higher and row spacings become more restricted, variations on the spindle system such as the vertical axis, slender spindle and super spindle are favoured.

(ix) Vertical axis

In vertical-axis, the trees are made up to a single vertical trunk, which serve to support many relatively 'weak' or small-diameter fruiting branches. Tree density is between 1,000 to 2,500 trees ha^{-1} at a spacing of 4-5 m x 1-2 m and height can reach up to 3m. The most commonly used rootstocks are M.9 and M.26 in apple. Maintaining apical dominance is important in the vertical axis system particularly during early stages of development to ensure weak fruiting branches (Lespinasse and Delort, 1986), therefore at planting and during the development years the leader is not headed back. Fruiting branches are retained along the trunk from a height of 1m upward. For apple trees on M.9 a good balance between vegetative growth and fruiting is usually obtained with 12-16 fruiting branches (Lauri and Lespinasse, 2000). Branches are systematically renewed to prevent them from becoming permanent scaffolds. Support of two to three wire trellis is required.

(x) SolAxe system

SolAxe system of training is developed through modification in vertical-axis system by focusing even more on branch bending and avoiding renewal pruning in an attempt to gain a more weeping canopy.The system combines the bending of the central axis and the fruiting branches from the Solen system with the free-growing fruiting branches and the removal of competing vegetative branches from the vertical-axis system. The tree is developed at planting by strongly bending the feathers along the trunk into an arc. Branches that are below 1.2 m are removed. The leader is left unheaded and allowed to grow to the top of the support system and when it reaches the top of the support system, it is bent horizontal. Upper fruiting branches are bent below the horizontal. As with the vertical-axis system, the best balance between vegetative growth and fruiting on a mature SolAxe tree is obtained with 12–16 fruiting branches arranged spirally along the trunk. On each fruiting branch, upright shoots are removed annually and the pendant portion is retained.

(xi) Slender Spindle

This training has been designed to improve early yields and management efficiency by planting high higher tree densities and reducing tree height to allow all management to be done from the ground.The slender spindle system involves more severe pruning than the vertical axis and is suited to densities of 1,500 to 4,000 trees ha^{-1} either single-, double-, triple-or multiple-row beds. Planting spacing is 3-3.5m x 1-1.5 m and tree height is usually restricted to 2-3 m. The tree is a narrow, fully dwarf, conic-shaped with a distinct, supported, vertical central leader branch. At the base of the tree there is a permanent frame of a varying number of branches and above these exist more or less horizontal fruiting branches shorter in length than the frame branches. Leader control is important – strong leaders and lateral growth in the top of the trees should be avoided. Generally the leader is headed back to a weaker lateral when it becomes too dominant. In apple the most common rootstock used is M.9, but other stock of the same dwarfing level, such as B.9 and G.16, are also used.

(xii) Slender-pyramid system

This system of training was developed by Tustin and colleagues (Tustin, 2000). In designing this system they considered physiological principles of light interception, light distribution and pruning, plus combined features of the vertical-axis, slender-spindle systems and McKenzie central leader (McKenzie and Mouat, 1963; Wertheim, 1978; Lespinasse and Delort, 1986). This training system can be used with relatively low tree densities (600 trees ha^{-1}) with MM.106 rootstock and with moderately high tree densities (2,000 trees ha^{-1}) with M.9. This system of training has a dominant central leader, a well-developed, spreading, basal tier of four branches bearing fruiting laterals, with a narrow pyramidal canopy above the basal tier. The mid- and upper canopy is comprised of well-spaced whorls of weaker fruiting laterals arising directly off the trunk. Trees are not headed when planted and are supported using a single- or multiple-wire vertical trellis (Robinson, 2003).

(xiii) Super spindle

The super spindle system is utilized for super high-density orchards on weaker rootstocks such as Quince C in pear. Generally there is a row spacing of <3 m and a tree distance within the row of <0.8 m, giving a density of more than 4,000 trees ha^{-1}. Tree height is generally maintained at 2-3 m.This system has closely spaced compact trees with short fruiting wood or spurs evenly spaced along the central leader.

(xiv) Tall spindle planting system

The Tall Spindle planting system utilizes highly feathered nursery trees planted at a density of 2,500-3,300 trees ha^{-1} at the spacing of 3.0-3.5 m between row and 0.9-1.2 m in row. The trees are pruned minimally after planting but feathers are tied below horizontal soon after planting. The tree is grown rapidly to 3.3 m tall with no heading of the leader and little pruning for the first 4-5 years after which tree height is limited each year to 3.3 m by cutting to a side branch. The mature tree is pruned using limb renewal pruning by removing 2-3 of the largest limbs (>2 cm diameter) in the canopy each year by cutting them back to a 2-3 cm long stub.

It is a combination of the Slender Spindle (Wertheim, 1968), the Vertical Axis (Lespinasse, 1980), the Super Spindle (Nuberlin, 1993) and the SolAxe (Lespinasse, 1996) systems. This system utilizes the concept of high tree densities from the Slender Spindle system, but utilizes lower planting densities (2,500-3,300 trees ha^{-1}) than the Super Spindle. The mature trees are tall similar to the Vertical Axis but with very narrow canopies like the Super Spindle.The system is designed to achieve high light interception (70-75 per cent) by growing the trees to a height of 90-100 per cent of the between row spacing (3-3.5 m) but also to have good light distribution throughout the canopy by keeping the canopy thin (1-1.5 m) and based on simple and small diameter fruiting limbs (Robinson *et al.*, 2014).

(xv) Cone system

Cone system of training is developed by Gandev (2009). It is a new way of training and growing non-spur apple cultivar grafted on MM.106 rootstock in intensive orchards. It consists of pruning activities, which, applied consecutively, lead to the formation of a crown conical in shape and having different numbers of scaffold branches according to the age of the trees, their growth vigor and loading with fruit.

(xvi) Trellis system

Trellising is a means of supporting trees to increase their bearing surface and hence their production. The trellis consists of a series of wires supported by posts. Limbs and branches are positioned on this wire fence to encourage fruit production. Apple varieties on M.27, Bud.9, M.9, or M.26, as well as pears on Quince C rootstock, can be trained to a trellis. This system is adopted especially for the high density planting on clonal rootstocks to optimize production by harvesting 70 per cent of available sun light. It allows more fruiting wood than the structural wood and gives production at early stages of the plant/orchard life.

(xvii) Palmette system

There are three types of palmette, *i.e.* horizontal arm (espalier), oblique arm (palmette) and combination of horizontal and oblique arm.In palmette system of training the tree is trained to a 2.5-3 m tall four- to six-wire vertical trellis. The tree has a central trunk and regular tiers of branches that are trained to the two-dimensional vertical plane along the row. The branches at each tier are tied either at an oblique angle (30°), so that they grow up across the wires (oblique palmette) or horizontally along the wire at each tier (horizontal palmette or espalier). First two branches as across the wire and after that along the wire (combination of oblique and horizontal).

Espalier system of training is suitable for apples and pears varieties that bear short fruiting spurs (not tip bearing). Trees should be planted 3-6 m apart, according to their vigour. MM.106 is most suitable for small apple trees with three or four tiers but M.26 can also be used however for large apple trees or in poor soil MM.111 is best. In pear Quince A is suitable.

(xviii) V or Y shaped canopy systems

Generally there are two basic shapes of canopies- Y shaped trees which have a vertical trunk and two opposing arms of the tree trained to either side of the trellis, and V shaped trees where the whole tree is leaned to one side of the trellis while the next tree in the row is leaned to the other side.

(xix) V Hedge

The planting distance in the V Hedge is 3.5 x 1.25 m which equals 2,057 trees ha^{-1}. These systems are planted using well feathered two year old nursery trees. Four feathers are kept as fruiting branches and considered as four central leaders on one stem. Tree height is maintained at 2m with an opening of the V of 1.4 m. These systems can be planted more intensively, however light interception can be inhibited.

(xx) Open Tatura Trellis (OTT)

It is a modification of the original Tatura Trellis system. In OTT there is a narrow strip of about 50 cm that separates alternating diagonally planted trees within each row. OTT systems are planted 4-4.5 m x 0.5-1 m with 2,000 to 5,000 trees ha^{-1}. Trees can be trained in a number of different ways. Three of the most common are: single leader, double leaders and, more recently, the cordon. Single leader OTT is similar to planting a slender spindle type system, with root systems about 0.5m apart and leaders 1m apart. Double leaders involve training two leaders on each tree (about 1m apart), to establish a high density of fruiting units at a lower tree cost. OTT with cordon allows for a moderately dense orchard of around 2,000 trees ha^{-1} with about 8,000 fruiting units growing up the wires. Nursery trees (usually whips) are bent over at planting and trained to the horizontal. Fruiting units are then encouraged at regular intervals along each cordon.

(xxi) Cordon system

In cordon system strong 2-3 year old trees are used for plantings. The trees are planted at an angle of 45 degree, 5 wire trellis are made spaced at iron post of 180 cm are fixed at 8 meter interval. The top wire is few cm below the top of the post. The first trellis is stretched at 45cm from ground level and others were fixed at 30 cm interval. In cordon system, the trees are planted at 45° angle which results in movement of auxin to the basal portions of the tree thus lateral buds become productive. Cordon training is suitable for all apples and pears that bear fruit on short side shoots (spur-bearing). The term 'cordon' simply refers to a single stem with short side shoots (the fruiting spurs). In this system, vigorous cultivars are difficult to keep within bounds and make poor cordons. Likewise, cultivars that bear fruit at the tips of their branches (tip-bearing) are difficult to train as the tips need to be pruned off and so little fruit is borne. Those apples and pears that bear fruit on both tips and side shoots, known as partial tip-bearers, can be grown as cordons if some short branches are left unpruned each year.

For this training system apple rootstock M.27 (extremely dwarfing) or M.9 (very dwarfing) are most suitable where the soil is fertile otherwise we can use M.26 rootstock (dwarfing). Pears are best grafted on Quince C or, for very poor soils, Quince A.

(c) Fruit Production in different Training Systems

(i) Apple

Palmer (1980) worked out LAI for three different training systems, namely, slender spindle, palmette and full field, capable to ensure adequate distribution of intercepted light within 90 per cent of the canopy. The LAI value for three systems were 1.6, 1.6 and 1.9 respectively, and the fraction of total intercepted light was 53 per cent for both the slender spindle, palmette and 66 per cent for full field. Light sharply decreased inside the first 0.5 to 1 m from the canopy edge in Delicious apple trained on central leader (Barritt *et al.*, 1987). Ferree (1980) observed that cumulative production in the sixth year for central leader was 41 per cent of the field for the slender spindle in

Golden Delicious. Dallabetta *et al.* (2014) carried an experiment at Italy to compare Spindle training system of training, with Bi-axis, on two main cultivars *i.e.* 'Galaxy Evolution' and 'Pink Lady'. Where, 'Galaxy Evolution' orchard was established in 2,005 at a planting distance of 3.45 x 1 m for Spindle (2,900 trees per ha) and 3.45 x 1.2 m for Bi-axis (2,400 trees per ha). While, the 'Pink Lady' orchard was established in 2008 at a planting distance of 3.5 x 0.8 m (3,570 trees per ha) for both training systems. They found that "Bi-axis Galaxy Evolution fruit" showed higher size in the lower level as compared to Spindle ones. Fruit size in 'Pink Lady'was significantly higher in Spindle as compared to Bi-axis while fruit colour was not affected. In general, they concluded that fruit harvested from "Bi-axis" resulted more homogeneous than "Spindle" training system. In apple the Tall Spindle system is more profitable than either higher (3,300 trees per ha) or lower density systems (2,500 trees per ha), early yield has exceeded 150 per ha over the first 5 years (Robinson *et al.*, 2014).

Apple trees 'Royal Gala'/M.9 were trained in five ways-Slender spindle (SS), Geneva Y-trellis (GY), a modified Solen training- 'Solen Y-trellis' (SY), or V-trellis(LDV), all at the same spacing (1.2 m × 2.8 m), giving a planting density of 2,976 trees per ha. In addition, a higher density (7,143 trees per ha) version of the V-trellis (HDV). After 8 years, the two Y-shaped training systems had 11-14 per cent greater cumulative yield per ha than the SS. Relative to the LDV, the HDV yielded less per tree, but far more per ha, particularly in the first 3 years. After 8 years, the cumulative yield per ha was still 65 per cent greater than with LDV, further, the HDV intercepted more light (73 per cent) than SS (53 per cent) (Hampson *et al.*, 2002). Higher and better quality yields (160.6 t per ha) obtained from 7-year-old non spur cultivar Jonagold/ MM-106 planted at 4.5 m x 2.5 m in cone training system compared to the yields (111.6 t per ha) harvested from trees trained to the free growing spindle type bush system (Gandev, 2009). At CITH, Srinagar a productivity of 55-66 t per ha was recorded in apple with vertical axis, espalier and cordon system of training system (Table 1). High yield obtained in these training systems because of high levels of light interception and its transmission through the canopy. Open tree canopies with the fruit bearing spurs being well exposed throughout the growing season provides not only the basis for high fruit numbers and large fruit size but also good fruit color and quality.

Table 1: Cumulative Yield under different Training Systems at ICAR-CITH, Srinagar

Canopy Management System	*Number of Plants/ha (Spacing)*	*Recommended Rootstock*	*Average Yield (t/ha) (Age of plants)*
Modified Central Leader System	625 (4x4m) 1600 (2.5x2.5) 2222 (3.0x1.5)	Seedling M-106 M-9	25-40 (10-12 yrs) 40-45 (8-10yrs) 30-35 (7-8 yrs)
Spindle bush	2222 (1.5 x 3.0m)	M-9	35-45 (7-8 yrs)
Head and Spread	2222 (1.5 x 3.0m)	M-9	40-50 (7-8 yrs)
Espalier	2222 (1.5 x 3.0m)	M-9	62-66 (5 yrs)
Cordon system	8888 (1.5 x 0.75 m)	M-9	55-60 (4 yr)
Vertical Axis	8888 (1.5 x 0.75m)	M-9	58-62 (5 yr)

(ii) Apricot

Apricot, has undergone a great renewal in the orchard training system and planting density have been substantially increased. The principal training systems used are vase, palmette, spindle and V-system. The V-training system is suitable for orchards with over 1,200 trees ha^{-1} (De Salvador *et al.*, 2000). While, Stanica and Eremia (2014) suggested Parallel Trident System (4 x 2 m) for high density planting.

(iii) Kiwifruit

Commercially grown kiwifruit plants are trained to a permanent framework, either a T-bar (like a clothesline with 4-5 wires) or pergola (solid overhead canopy with wire going in both directions). However, a T-bar system has yielded well, is easier to prune and may be better suited for small-scale production. For a T-bar trellis, vines commonly are planted 4.5 m apart in rows 4.5 m apart, while, for pergola plants spaced at 4.5 to 6.0 m apart. There are other systems like Geneva double curtain, free spindle and Tatura (V-trellis)- which allow much higher densities. Costa *et al.* (1991) compared. Geneva double curtain- 1,000 vines ha^{-1}, Free spindle- 1,500 vines ha^{-1}, Tatura trellis- 1,800 vines ha^{-1} and T-bar- 700 vines ha^{-1} and they found that both GDC and T-bar produced higher fruit yields (54.9 kg tree^{-1} or 40.6 t ha^{-1}) than the free spindle and V-trellis system.

(iv) Peach and nectarine

In peach new training system two-scaffold perpendicular V system have been suggest by many researcher. Research trails indicate that the V system is easier to prune and train than traditional systems if simple pruning and training rules are followed. The primary advantage of this system are early and sustained fruit cropping, high fruit quality and ease to management. In peaches among different training system light interception was 74 per cent in Y-shape, 71 per cent in open vase and 89 per cent in slender spindle. However, sunlight distribution in Y shape canopy was 35 per cent higher than the slender spindle canopy (DeSalvador and Dejong, 1989). At CITH, Srinagar a productivity 50 tonnes per ha was obtained in peach from V shape tatura trellis system while the traditional vase trained trees planted at low densities of less than 400 trees per ha yielded only 15-20 t per ha.

(v) Pear

Duckers and Schoofs (2001) have been suggested the V-hedge system (3.5 x 1.25 m and 3.2 x 0.8 m, 2,057 and 3,516 trees per ha, respectively), Tiense hedge system (3.5 x 1.75 m, 1,632 trees per ha) and free spindle system (3.75 x 1.5 m, 1,777 trees per ha) for high density planting of pear. Vercammen (2014) compared bush spindle shape (3.5 x 1.5 m), hedge of Tienen (3.5 x 1.5 m), long pruning (3.5 x 1.5 m), candlestick system (3.5 x 1.5 m), V-system (3.2 x 1.0 m), spindle trees (3.2 x 0.50 m) and Drapeau system (3.5 x 1.75 m-tree were planted under an angle of 45°). He found that long pruning system and hedge of Tienen had the lowest total costs after 11 years of growth. The highest yields were obtained with long pruning, followed by "Drapeau" system and V-system in 'Conference' pear on quince Adams rootstock.

(vi) Plum

In plum four different single row planting systems (vertical axis, free spindle, hedgerow and Y-trellis) and three planting densities (0.5, 1.0, 1.5 x/4.0 m for the Y-trellis and hedge trees and 2 x 4 m for the central leader trees) were evaluated by Meland (2005) on semi-dwarf rootstock St. Julien. He workout that trunk cross sectional area of the trees increased with increasing tree spacing. During the first four cropping years, the yield ha^{-1}were positively correlated with tree density. After nine years, cumulative yield ha^{-1} was highest on the Y-trellisand hedge trees with the highest density (5,000 trees ha^{-1}). Cook (2007) suggested hedge row (4.0 x 1.5 m- 1,667 trees ha^{-1}) and V system (4 x 0.75 m- 3,333 trees ha^{-1}) for high density planting of plum.

(vii) Sweet Cherry

Sweet cherry performed well oncentral leader, multiple leader bush, palmette, and Y-trellis system of training (Whiting *et al.*, 2005). Training systems that incurred theleast amount of pruning (*e.g.*, a trellised palmette system, some central leader/ spindle-type systems,) generally have had the highest yields but the smallest fruit size. Training systems with a higher proportion of pruning, particularly heading-type cuts (*e.g.*, multiple leader/bush architectures), generally have had lower yields but larger, sweeter fruit (Lang and Ophardt, 2000). Lang *et al.* (2014) compared Kym Green Bush (KGB),Tall Spindle Axe (TSA), Upright Fruiting Offshoots (UFO) and Super Slender Axe (SSA). They found that KGB, TSA, SSA and UFO is likely to be most suitable for higher vigour, moderate vigour, lower vigor and moderate to higher vigour sites and/or rootstock; respectively. Further, among the different canopy system the SSA is most precocious.

4. PRUNING

Pruning is employed to improve both the fruit crop and tree appearance, but it is a practice frequently neglected and misunderstood. Pruning is defined as the judicious removal of parts to obtain good and qualitative yield. The ultimate aim of pruning is to develop a canopy, which intercepts maximum light.

Basic objectives of pruning

- ✰ To restore root/shoot ratio
- ✰ To obtain specific form of the plant
- ✰ To develop specific geometry within the plant and its overall topology
- ✰ To maintain optimum balance between vegetative and reproductive parts
- ✰ Pruning for dwarfing and promotion maximum shoots/fruiting points

General rules for pruning of fruit tress

- ✰ Prune young tree very lightly, follow rules of minimal pruning suggested by Robinson (2003) for first four year
- ✰ Prune mature tree more heavily, especially if they have shown little growth.
- ✰ Prune the top portion of the tree more heavily than the lower portion.

- ☆ Thin out more shoot towards the end of a well-pruned branch in a mature tree. This will increase fruit size and quality on the remaining shoots.
- ☆ To reduce the height of an excessively tall tree, cut whole limbs at the top, making cuts flush with the bark of a lower limb. Stubs left in the top of a tree would not heal.
- ☆ Follow 3: 1 pruning rule which is advised by Cook and Strydom (2000) the leader should be 3 times the diameter of any of the lateral branches in the upper part of the tree. They have suggested that limbs that are larger than the 3: 1 rule should be removed early on to preserve a hierarchy of branch and leader diameter.

(a) Dormant Pruning

All pruning's have a dwarfing effect, but dormant pruning produces the most new growth, it reduces trunk size and canopy volume as compared to the same tree that is not pruned. Miller and Byers (2000) reported 21 per cent reduction in canopy volume when peach trees were severely pruned. Dormant hedging or shearing has shown success in controlling tree size, but often results in a proliferation of shoots over the periphery of the tree resulting in very dense tree canopies (Forshey *et al.*, 1992). The effect of dormant pruning may seem to have stimulated growth, but the total dry matter produced by the pruned tree is less than the non-pruned tree (Myers and Saville, 1996). Therefore, to avoid localized and invigorating effects, it is recommended that pruning consist primarily of thinning cuts and few heading cuts be used (Marini *et al.*, 1993).

(b) Summer Pruning

Summer pruning in combination with dormant pruning reduce light interception by 15 per cent (Wertheim *et al.*, 1986) but summer pruning increase photosynthetic photon flux density (PPFD) within the tree canopy (Morgon *et al.*, 1984). According to Marini and Barden, (1987) summer pruning increased light penetration within the tree canopies, but may decrease levels during the following year (Taylor and Ferree; 1984). Summer pruning improves fruit colour for coloring varieties grown in climates where color development is less than ideal and also increases size and weight of fruits along with soluble solids. Trimming of apricot shoots is an interesting approach to avoid having the productive area shift to the canopy periphery. The early pruning intervention increased the number of flowers per tree from 2,212 (June 6th) to 2,524 (June 20th) compared to the un-pruned control that exhibited only 1,938 flowers. Pruning in early July (4th) induced a 0.8 kg per tree increase in yield compared with the unpruned control. Despite the same number of flowers/tree, fruit mass of the early July pruning was 10 g less than in fruits of the unpruned control trees (Musacchi *et al.*, 2014). In peach summer pruning is useful in traditional open centre trees to increase fruit size, colour and soluble solids and to reduce pre-harvest rots. However, in the perpendicular V, summer pruning is essential to maintain fruiting wood and fruit quality along the scaffolds.

(c) Root Pruning

Root pruning reduce uptake of water and nutrients and create imbalance in plant hormones which adversely affect shoot growth of fruit plant. However, pruning of root systems has been successful in some fruits but less efficient with others. The commercial use of root pruning has increased in temperate fruit crops because of development of tractor mounted root pruner and loss of daminozide. Pruning roots of young and mature apple trees reduced shoot growth, controlled tree size (Mitre *et al.*, 2012) increased light interception, yield efficiency, fruit colour, firmness and soluble solids and reduce pre-harvest fruit drop and biennial bearing pattern (Schupp *et al.*, 1992; Ferree and Rhodus, 1993; Mitre *et al.*, 2012).

(d) Rejuvenation

Rejuvenation is a technique which involve cutting of exhausted tree with the objective to restore production potential of old, unproductive and senile orchard to extent of 1-1.5 m height above ground level or from secondary branches for getting new shoots. The technology also help in maintaining tree height with open architecture and canopy of healthy shoots with outwardly growth facilitating penetration and utilization of light.

(e) Bending

Forshey *et al.* (1992) reported reduced vegetative vigour in limbs that are bent. He suggested that reduced vegetative growth is obtained from bent shoots because, bending reduces the production of gibberellins, which are also antagonistic to flowering. Thus, the formation of moderately vigorous lateral shoots and spurs creates additional sites for flower formation. These two effects result in increased flowering and earlier fruiting. Fruit quality is also improved because of improved light penetration and because more fruit hang free instead of rubbing against the branch on more upright limbs (Ferree and Schupp, 2003). Inclining thetrunk at planting 45° or 60° from the vertical reduced the rate of shoot elongation by 40 per cent and trunk circumference growth by 17 and 27 per cent, respectively in 'Golden Delicious'/M.9 (Bargioni and Baroni, 1995). In intensive training systems *i.e.* slender spindle or hybrid tree cone orchard system (HYTEC), bending is used to weaken the leader and encourage growth and fruiting in the lower scaffolds (Ferree and Schupp, 2003).

(f) Scoring, Girdling, Bark Inversion

Scoring and girdling reduced shoot and trunk growth (Wilton, 1999). These technique effectively control shoot growth in the year of the treatment (Greene and Lord, 1978), however, it may also affect growth for one or more years after the treatment (Hoying and Robinson, 1992). Girdling of mature Macspur McIntosh trees resulted in reduction of average shoot length by 18 per cent and shoot numbers by 20 per cent (Hoying and Robinson, 1992). Further, Veinbrants (1972) recorded 25 per cent less shoot length from scored Granny Smith apple trees soon after bloom but the reduction was 12 per cent only, if scoring was delayed by 2 weeks.

(g) Pinching and Notching

Removing immature, apical leaves (Pinching) increases the number of branches oncurrent-season growth (Ouellette *et al.*, 1996). Notching stimulates lateral branching from buds that would not normally grow by interrupting the downward movement of auxin from the shoot tip (Greene and Autio, 1994). In 1-year-old 'Empire' apple removal of a horizontal band of bark (2mm wide and one-third leader circumference) down to the secondary xylem directly above each lateral bud in a zone 35 to 65 cm (35–65 cm) above the graft union resulted in more branching (Ouellette *et al.*, 1996).

(h) Use of Dwarf and Semi-dwarf Rootstock

In temperate fruits species, two main factors that modulate tree architecture and overall tree size. The first one is the genotype, which affects branching density, proportion of short vs. long shoots and flowering pattern (i.e lateral vs. terminal flowering). A second factor is the root system, which has been used as an efficient although empirical means to control tree size, with variable results on flowering (Lauri *et al.*, 2006). The root system, from "dwarf" to "vigorous," strongly influences overall tree size, with own-rooted trees generally equivalent to or larger than trees of the same genotype grafted on the most vigorous clonal rootstocks (Quamme and Brownlee, 1993) or on seedling rootstocks (Zimmerman and Miller, 1991). Trunk cross-sectional area (TCSA) and branch cross-sectional area (BCSA) are the main traits used to characterize the influence of the root system on tree growth (Moore, 1978). Dwarfing rootstock reduced tree size and increase the partitioning of dry matter to fruit production and also results in precocious flowering. Rootstock control tree size through a direct effect on growth and indirectly through enhanced crop load. With the advent of more dwarfing rootstocks, the temperate fruit industry went through a revolution in training systems especially in apple. The most dwarfing, *i.e.* M.27, V.3, G *etc.*, dwarf *i.e.* M.9, P.22, P.2, B.9, B.146, B.49, Mark, MAC.39, G.16, G.65, OAR1, Bemani, 0.3, V.2, J.9, F *etc.* and semi-dwarf M.7, MM.106, B.490, G.13, G.30, 0.8, H *etc.* are available for commercial high density planting in apple. However, among different dwarf rootstocks M.9, B.9 and Mark are most popular rootstocks. Most of dwarfing and Semi-dwarfing clonal rootstocks have poor anchorage due to brittle roots and shallow root system, requiring staking or trellis wire support. Use of dwarfing interstock, provides best means of overcoming this problem. Generally dwarfing rootstocks, M.8, M.9, M.27, C.6, B.9 and Mark –are used as interstocks to control tree size. It allows the use of well-anchored and strong root system of seedings or vigorous clonal rootstock and dwarfing effect of interstock. The extent of dwarfing effect and precocious bearing depends upon length of interstock, according to Kanwar (1987) a minimum of 15 cm interstock is necessary to produce some effect. Tustin *et al.* (2001) conducted field trail in apple cv. Fuji trained on Slender Pyramid system utilizing medium vigour (MM. 106, 5.0 × 3.0 m), semi-dwarf (M.26, 4.0 × 2.0 m), or dwarf (Mark, 4.0 × 2.0 m) rootstocks. They concluded that light interception was up to 50 per cent less in the Mark dwarf tree system. Further, continued canopy growth in the systems on MM.106 and M.26 resulted in higher yields than from the Mark planting system from the fourth year onwards. Peak yields within 6 or 7 years reached 110 t ha^{-1}.

The most common dwarfing rootstocks for pear are Old Home x Farmingdale (OH x F) and quince (*Cydonia oblonga* L.) selections that result in trees about 50 to 70 per cent of a standard size pear tree (Alonso *et al.*, 2011).Trees on quince rootstocks are reduced in vigor, precocious and productive in cropping and, in most cases, bear fruits of good size and quality. Pear growers in Western Europe use trees propagated on Quince C, Adams, A or BA.29 clone of quince rootstock (Webster, 2002). On the basis of TCSA, canopy volume and pruning weight Massi *et al.* (2008) ranked different pear rootstock in order of decreasing vigour: Kirschensaller > Farold 282 = FOX 11 > Farold 87 = FOX 16 > Farold 40 > Sydo = EM A > BA 29 > Adams > EM C.

Cherry rootstock Mahaleb is about 75-90 per cent of Mazzard size and is often called "semi-dwarf", simply because it is smaller than Mazzard. The introduction of the interspecific hybrid Gisela (Gi) rootstocks, largely based on crosses between *P. canescens* and *P. cerasus*, provided new important traits of precocious flowering, reduced vigor, and high productivity (Lang,2000). Gisela rootstocks now available commercially include Gi3 (very dwarfing), Gi5 (dwarfing), Gi6 (semi-vigorous), and Gi12 (semi-vigorous) (Lang *et al.*, 2014). There are really no commercially utilized dwarf or semi-dwarf rootstocks for peach, plums, prunes and apricots.

5. USE OF PLANT GROWTH REGULATOR IN CANOPY MANAGEMENT

Phytohormones play a major role in controlling many physiological processes of plant growth and development. They regulate division, expansion and differentiation of cells thus determining plant morphological traits. It was shown, that different amounts of phytohormones are found in the shoots and leaves of apple trees depending on their canopy architecture (Watanabe *et al.*, 2008).

(a) Induction of Feathers (Lateral Branching)

The development of branched trees in the deciduous tree-fruit nursery has become more important as availability of dwarf rootstock has made higher-density plantings commercially feasible. Planting branched (feathered) apple, pear and sweet cherry trees from the nursery improves total production earlier in the life of orchard. Lateral branch development in vigorously growing nursery fruit trees is under the control of a physiological system termed "apical dominance". Apical dominance is driven by inhibition of lateral bud development by auxin, while cytokinin promotes growth of axillary buds, and these hormones influence each other (Leyser, 2009). The relative amounts of each type of hormone, and hence the relative degree of inhibition or stimulation of lateral-bud growth, can be altered by environmental factors as well as by various cultural practices, such as exogenous applications of bioregulators or pinching techniques such as leaf removal and notching (Ono *et al.*, 2002; Sazo and Robinson, 2011). Unfortunately, pinching and notching are inconsistent and most of the nursery plant requires more than one "leaf pinching" to maintain low auxin content due to rapid stem elongation. The use of plant growth regulator chemicals has been seen as a more consistent and less expensive method to induce lateral branch-ing.

Among the many bioregulator products tested over the past half century, Cytokinin especially, 6-benzyladenine (BA) and gibberellic acid especially (GA_{4+7}) and combination of these two- (a mixture of 50 per cent BA and 50 per cent GA_{4+7}) are the most common plant growth regulators that promote formation of lateral branches. They are successfully applied for different fruit crops. BA is responsiblefor overcoming apical dominance (Muller and Leyser, 2011) and enhancing feathering in apple nursery trees (Sazo and Robinson, 2011), while the main role of GA is to elongate feathers (Hrotko *et al.*, 2000). BA is also used in young trees newly planted in orchards to induce proleptic shoot formation and to weak acrotony (Buban *et al.*, 2000). Recently, Cyclanilide has been tested for growth-related effects on nursery apple, pear, sweet cherry and plum trees and found promising (Magyar and Hrotkao, 2002; Elfving, 2010; Sazo and Robinson, 2011; Robinson and Sazo, 2014).

(b) Control of Vegetative Growth through Chemical

Typically, tree growth management is performed by physical measures, particularly by dormant and summer pruning. In extreme cases, root pruning, girdling or even stem sawing is being used. Any of these methods is cost-intensive and/or bears a high risk of failure. Furthermore, part of the trees assimilates or potential assimilates are lost. Alternatively, an improved balance between vegetative growth and fruit formation may also be achieved by employing plant growth retardants. Historically, daminozide has been most successfully applied in apples for growth control. However, relatively high levels of residues with potentially critical toxicological features have led to a general ban of this compound for use in edible crops. The registration of chlormequat chloride, particularly value in pears, has recently been terminated in many countries because of non-tolerable residues resulting from too intensive use (Frank Maas, 2006). Paclobutrazol is registered only in some countries for growth control in fruittrees. The problems of this compound are particularly related to its extreme persistency and its strict acropetal movement in treated plants, which make adequate treatments relatively difficult. Residues in fruits and flattened fruit shapes are additional obstacles to a wider use of this compound (Rademacher, 2004). Ethylene, released from ethephon and compounds inhibiting the formation of growth-active gibberellins (GAs) would be candidates for shoot growth control. Ethephon is required at dosages, which, in addition to growth reduction, mayinduce fruit drop, precocious ripening or reduce alternate bearing.

Recently prohexadione-calcium (calcium 3-hydroxy-5-oxo-4-propionyl-cyclohex-3-enecarboxylate, Pro-Ca) has been introduced (trade names: *Apogee*® and *Regalis*®) in a number of countries as a plant bioregulator for pome fruit trees may offer an attractive new solution for the chemical control of excessive shoot growth. Several reports are available dealing primarily with the effect of Pro-Ca on vegetative growth and fruit formation (Byers *et al.*, 2004; Costa *et al.*, 2004; Rademacher*et al.*, 2004). It is relatively short-lived and only low, if any, residues are detected in fruits from treated trees. Pruning costs in treated trees are reduced by approximately 30 per cent. Often, higher yields are obtained as a result of increased fruit set and the availability of additional assimilates for fruit growth. Likewise, improved fruit quality and fruit storing quality are observed in many instances (Rademacher, 2004).

Surprisingly, trees treated with Pro-Ca turned out to be less infected by fire blight (*Erwinia amylovora*) (Norelli and Miller, 2004) and apple scab (*Venturia inaequalis*) (Costa *et al.*, 2004).

(c) Deficient Irrigation

Through deficient irrigation we can reduce vegetative growth of plant without affecting yield. In pear and peach irrigation at selected time during the growing season has been shown to reduce vegetative growth without affecting yield (Chalmers *et al.*, 1981; Mitchell *et al.*, 1989). Apple tree also respond to deficient irrigation but responses are not identical to peach (Sharma *et al.*, 2009).

(d) Precise Application of Fertilizer

Adequate soil fertility is crucial for successful cropping but over abundant nutrients can cause excessive and undesirable vegetative shoot growth. It can be difficult to fertilize so that only shoot growth is controlled without adversely affecting root growth or yield. Fertilization disproportionately increased root and shoot growth as a result the root to shoot ratio decreased in apple trees (Robinson *et al.*, 1992). Control on tree growth with selective application of fertilizer is further made difficult by the potential reservoir of minerals stored in the perenniating portions of the tree. Growth of peach tree in spring drew strongly from stored nitrogen, independent of nitrogen supplied in the fall (Niederholzer *et al.*, 2001). Likewise, according to Neilsen and Neilsen (2002) root uptake of nitrogen by apple trees was negligible in early spring as tree used more of the stored nitrogen. Greater nitrogen use efficiency resulted when nitrogen was applied to the soil after reserve nitrogen was mobilized from storage tissue in apple in early spring (Stephen and Thomas, 2003).Withholding fertilizer to maintain a desired tree size is apparently difficult because of spatial and temporal variation in nutrient availability and uptake in young and mature trees but carefully managed fertilization should help reduce undesirable vegetative growth.

(e) Alteration in Plant Architecture through Genetic Engineering

Engineered trees with altered architectural characteristics should enable substantial advancements in productivity, and sustainability. By altering gene in temperate fruit we can alter their growth habit and impart dwarfness. Over-expression of rol (root loci) genes in plant has demonstrated a drastic effect on the plant architecture. The *rol* genes from *Agrobacterium rhizogenes* have been popular transgenes for rootstock. It reduce plant height, shortened internode and increase rooting number and percentage in apple rootstock (Sedira *et al.*, 2001; Igarashi *et al.*, 2002) and kiwifruit (Rugini *et al.*, 1999). Cytokinin-over-expressing transgenic peach plants containing the *Agrobacterium tumefaciens* are dwarf in stature and produce more branches (Hammerschlag and Smigochi, 1998). Correspondingly, introduction of tumour-inducing (T)-DNA from *Agrobacterium rhizogenes* in Colt rootstock of cherry produced trees with significant reductions in plant height, internode length, leaf area and petiole length (Gutierrez-Pesce *et al.*, 1998). Likewise, Negri *et al.* (1998) reported that the rice *phytochrome*A (*phy*A) gene had dwarfing effect on Colt rootstock when over-expressed in transgenic.Srinivasan *et al.* (2012) introduced FT gene form *Populus trichocarpa* in European plum (*Prunus domestica* L.) cv. Blue Byrd. They reported that

FT plum lines expressed pleiotropic phenotypes that rendered them different from normal plum trees. They did not grow with a typical upright tree growth habit, but instead became highly branched resulting in a shrub-type growth habit. Further, they found that the architectural phenotypes of increased lateral branching (*i.e.* bushy) were not correlated with higher PtFT1 expression but these traits require only low levels of PtFT1 expression. Shrub-like growth habit of FT plum potentially makes them suitable for novel production systems.

6. CONCLUSION

Canopy management and architecture engineering are the most important aspect in temperate fruit cultivation. Several efficient training systems has been proposed world over, which are required to be adopted after testing to produce quality fruits and enhance productivity. Tree architectural aspects are important consideration in modern fruit cultivation hence must be adopted to reap the benefit of such intervention.

7. REFERENCES

Alonso, J.M., Gomez-Aparisi, J., Anson, J.M., Espiau, M.T. and Carrera, M. 2011. Evaluation of the OH x F selections as an alternative to quince rootstocks for pear: Agronomical performance of 'Conference' and 'Doyenné du Comice'. *Acta Horticulturae*, **903**: 451-456.

Awasthi, R.P. and Mehta, K. 2004. Strategies for Canopy Management in Tree Fruit Crops. *Proceedings of the First Indian Horticultural Congress*-2004, pp. 320-330.

Bargioni, G. and Baroni, G. 1995. Scion inclination in *Malus domestica* Borkh. and *Prunus* spp. influences root growth and distribution. *HortScience*, **30**: 517–520.

Barritt, B.H. 1989. Influence of orchard system on canopy development, light interception and production of third year 'Granny Smith' apple trees. *Acta Hort.*, **243:** 121-130.

Barritt, B.H., Rom, C.R., Guelich, K.R., Drake, S.R. and Dilley, M.A. 1987. Canopy position and light effects on spur, leaf and fruit characteristics of 'Delicious' apple. *HortScience*, **22**: 402–405.

Barritt, B.H., Rom, C.R., Konishi, B. and Dilley, M.A. 1991. Light level influences spur quality and canopy development and light interception influences fruit production in apple. *Hort. Science*, **26**: 993–999.

Buban T. 2000. The use of benzyladenine in orchard fruit growing: a mini review. *Plant Growth Reg.*, **32**: 381-390.

Byers, R.E., Carbaugh, D.H. and Combs, L.D. 2004. The influence of prohexadione calcium sprays on apple tree growth, chemical fruit thinning, and return bloom. *J. American Pom. Soc.*, **58**: 111-117.

Cain, J.C. 1970. Optimum tree density for apple orchards. *HortScience*, **5**: 232-234.

Chalmers, D.J. and van den Ende, B. 1975. The 'Tatura Trellis', a new design for high yielding orchards. *J.Agric. Victoria*, **73**: 473–476.

Chalmers, D.J., Mitchell, P.D. and van Heek, L. 1981. Control of peach tree growth and productivity by regulated water supply, tree density, and summer pruning. *J. American Soc. Hort. Sci.*, **106**: 307-312.

Chalmers, D.J., van den Ende, B. and Van Heek, L. 1978. Productivity and mechanization of the Tatura trellis orchard. *HortSci.*, **13**: 517–521.

Cook, N.C. 2007. An overview of plum training system in South Africa. *Acta Hort.*, **732**: 435-437.

Cook, N.C. and Strydom, D.K. 2000. The South African high density system. *Acta Hort.*, **513**: 321–327.

Corelli, L. and Sansavini, S. 1989. Light management and photosynthesis related to planting density and canopy management in apple. *Acta Hort.*, **243**: 159-174.

Costa, G., Biasi, R. Giuliani, R. and Succi, F. 1991. Comparison of kiwifruit training systems. *Acta Horti.*, **297**: 427-434.

Costa, G., Sabatini, E., Spinelli, F., Andreotti, C., Spada, G. and Mazzini, F. 2004. Prohexadione-Ca controls vegetative growth and cropping performance in pear. *Acta Hort.*, **653**: 127-132.

Dallabetta, N., Costa, F., Pasqualini, J., Wehrens, R., Noferini, M. and Costa, G. 2014. The Influence of training system on apple fruit quality. *Acta Hort.*, **1058**: 55-62.

De Salvador, F.R. and Dejong, T.M. 1989. Observation of sunlight interception and penetration into the canopies of peach trees in different planting densities and pruning configuration. *Acta Hort.*, **254**: 341-346.

De Salvador, F.R., Nuzzo, V. and Mattatelli, B. 2000. Gestionedell'albero e sistemid' impiantodell' albicocco. *Rivista di Frutticoltura e Ortofloricoltura*, **4**: 19-26.

Deckers, T. and Schoofs, H. 2001. Management of high density pear orchards. *The Compact Fruit Tree*, **34**: 117-120.

Elfving, D.C. 2010. Plant bioregulators in the deciduous fruit tree nursery. *Acta Hort.*, **884**: 159-166.

Ferree, D.C. 1980. Canopy development and yield efficiency of 'Golden Delicious' apple trees in fourorchard management systems. *J. American Soc. Hort. Sci.*, **105**: 376–380.

Ferree, D.C. and Rhodus, W.T. 1993. Apple tree performance with mechanical hedging or root pruning in intensive orchards. *J. American Soc. Hort. Sci.*, **118**: 707–713.

Ferree, D.C. and Schupp, J.R. 2003. Pruning and Training Physiology. In: Ferree, D.C. and Warrington, I.J. (eds.) *Apple: Botany, Production and Uses*, CAB International, Wallingford, Oxon, United Kingdom, pp. 319-344.

Flor, J.A. and Kesner, C. 1982. Orchard design for stone fruit based on light interception. *Compact Fruit Tree*, **5**: 159-165.

Forshey, C.G., Elfving, D.C. and Stebbins, R.L. 1992. *Training and Pruning Apple Trees*. American Society for Horticultural Science, Alexandria, Virginia, p.166.

Frank Maas, 2006. Carry-over Effects of CCC-applications in Pear Orchards. *Acta Hort.*, **727**: 125-132.

Gandev, S. 2009. Training and pruning of apple tree according to the new system 'Cone'. *Acta Hort.*, **825**: 465-470.

George, A.P., Nissen, R.J., Morley-Bunker, M.J. and Collins, R.J. 1993. Effects of pollination and irradiance on fruiting of persimmon (*Diospyros kaki* L.) in subtropical Australia. *J. Hort. Sci.*, **68**: 447-454.

Gifford, R.M. and Jenkins, C.L.D. 1982. **In:** Photosynthesis, Govindjee (Ed.) Prospects of applying knowledge of photosynthesis toward improving crop production. p. 419-457. Academic Press, New York.

Greene, D.W. and Autio, W.R. 1994. Notching technique increases branching of young apple trees. *J. American Soc. Hort. Sci.*, **119**: 678–682.

Greene, D.W. and Lord, W.J. 1978. Evaluation of scoring, limb spreading, and growth regulators for increasing flower bud initiation and fruit set of young 'Delicious' apple trees. *J. American Soc. Hort. Sci.*, **103**: 208–210.

Gutierrez-Pesce, P., Taylor, K., Muleo, R. and Rugini, E. 1998. Somatic embryogenesis and shoot regeneration from transgenic roots of the cherry rootstock Colt (*P. avium* x *P. pseudocerasus*) mediated by pRi 1855 T-DNA of *Agrobacterium rhizogenes*. *Plant Cell Rep.*, **17**: 574–580.

Hammerschlag, F.A. and Smigocki, A.C. 1998. Growth and *in vitro* propagation of peach plant transformed with the shoot mutant strain of *Agrobacterium tumefaciens*. *HortSci.*, **33**: 897–899.

Hampson, C.R., Quamme, H.A. and Brownlee, R.T. 2002. Canopy growth, yield, and fruitquality of 'Royal Gala' apple trees grown for eight years in five tree training systems. *HortSci.*, **37**: 627-631.

Heinicke, D.R. 1963. The micro-climate of fruit trees. II. Foliage and light distribution patterns in apple trees. *Proc.American Soc. Hort. Sci.*, **83**: 1–11.

Heinicke, D.R. 1964. The micro-climate of fruit trees. III. The effect of tree size on light penetration and leaf area in 'Red Delicious' apple trees. *Proc.American Soc. Hort. Sci.*, **85:** 33-41.

Hoying, S.A. and Robinson, T.L. 1992. Effects of chain saw girdling and root pruning of apple trees. *Acta Hort.*, **322**: 167–172.

Hrotko, K., Magyar, L. and Ronay, Z. 2000. Improved featheringon apple nursery trees by BA application. *Acta Hort.*, **514**: 113–119.

Huggard, J. 1980. High density planting in French orchards, developments and current achievements. *Acta Hort.*, **114**: 300–308.

Igarashi, M. Ogasawara, H., Hatsuyama, Y. Saito, A. and Suzuki, M. 2002. Introduction of *rol*C into Marubakaidou (*Malus prunifolia* Borkh. var. ringo Asami Mo 84-A) apple rootstock via *Agrobacterium tumefaciens*. *Plant Sci.*, **163**: 463–473.

Jackson, J.E. 1970. Aspects of light climate within apple orchards. *J. Appl. Ecol.*, **7**: 207-216.

Jackson, J.E. 1980. Light interception and utilization by orchard systems. *Hort. Rev.*, **2**: 208-267.

Jackson, J.E. and Palmer, J.W. 1972. Interception of light by model hedgerow orchards in relation to latitude, time of the year and hedgerow configuration and orientation. *J. Appl. Ecol.*, **9**: 341–358.

Jackson, J.E., Palmer, J.W. Perring, M.A. and Sharples, R.O. 1977. Effects of shade on the growth and cropping of apple trees. III Effects on fruit growth, chemical composition and quality at harvest and after storage. *J. Hort. Sci.*, **52**: 267-282.

Kallow, G., Reddy, B.M.C., Singh, G. and Lal, B. 2005. Rejuvenation of old and senile orchard. Pub. CISH, Lucknow, p. 40.

Kanwar, S.M. 1987. Apples-Production Technology and Economics. Tata McGraw Publishing Company Limited, New Delhi, pp. 88–98.

Lakso, A.N. 1994. Apple. In: B. Schaffer and P.C. Andersen (eds). Environmental physiology of fruit crops; Vol 1: Temperate crops. CPC Press, Boca Raton, FL, pp. 3–42.

Lakso, A.N., Robinson, T.L. and Pool, R.M. 1989. Canopy microclimate effects on patterns of fruiting and fruit development in apples and grapes, p. 263-274. In: C.J. Wright (Ed.). *Manipulation of Fruiting. Proc. 47th Nottingham Easter School, Butterworths*, London.

Lang, G.A. 2000. Precocious, dwarfing, and productive – how will new cherry rootstocks impact the sweet cherry industry? *HortTech.*, **10**: 719–725.

Lang, G.A. and Ophardt, D.R. 2000. Intensive cropregulation strategies in sweet cherries. *Acta Hort.*, **514**: 227–234.

Lang, G.A., Blatt, S.S., Embree, C., Grant, J., Hoying, S., Ingels, C., Neilsen, D., Neilsen, G. and Robinson, T. 2014. Developing and Evaluating Intensive Sweet Cherry Orchard Systems: The NC140 Regional Research Trial. *Acta Hort.*, **1058**: 113–120.

Lauri, P.E. and Lespinasse, J.M. 2000. The Vertical Axis and SolAxe systems in France. *Acta Hort.*, **513**: 287–296.

Lauri, P.E., Maguylo, K. and Trottier, C. 2006. Architecture and size relations: an essay on the apple (*Malus domestica*, Rosaceae) tree. *American J. Botany*, **93**: 357–368.

Lespinasse, J.M. 1980. La conduite du pommier. II.– L'axe vertical. La rénovationdevergers. INVUFLEC (Ed.), Paris. 120 p.

Lespinasse, J.M. 1996. Apple orchard management practices in France – From the vertical axis to the solaxe. *Compact Fruit Tree*, **29**: 83–88.

Lespinasse, J.M. and Delort, J.F. 1986. Apple tree management in Vertical Axis, appraisal after ten years of experiments. *Acta Hort.*, **160**: 139–155.

Leyser O. 2009. The control of shoot branching: an example of plant information processing. *Plant Cell and Env.*, **32**: 694–703.

Looney, N.E. 1968. Light regimes within standard size apple trees as determined spectrophotometrically. *Proc. American Soc. Hort. Sci.*, **93**: 1-6.

Magyar, L. and Hrotko, K. 2002. Effect of 6-benzyladenine (BA) and gibberellic acid (GA_{4+7}) application on feathering of plum cultivars in nursery. *Acta Hort.*, **577**: 345-349.

Marini, R.P. and Barden, J.A. 1987. Summer pruning of apple and peach trees. *Hort. Rev.*, **9**: 351–375.

Marini, R.P., Barden, J.A. and Sowers, D. 1993. Growth and fruiting responses of 'Redchief Delicious' apple trees to heading cut and scaffold limb removal. *J. American Soc. Hort. Sci.*, **118**: 446–449.

Massai, R., Loreti, F. and Fei, C. 2008. Growth and yield of 'Conference' pears grafted on quince and pear rootstocks. *Acta Hort.*, **800**: 617–624.

McKenzie, D.W. and Mouat, H.M. 1963. Semi-intensive planting and orchard management, a new approach to apple growing. *Supplement to the Orchardist of New Zealand*, p. 8

Meland, M. 2005. High density planting systems of European plums-the effect of growth and productivity of three cultivars after nine years. *Acta Agriculturae Scandinavica, Section B-Soil and Plant*, **55**: 51–57.

Miller, S.S. and Byers R.E. 2000. Response of winter-injured peach trees to pruning. *Hort.Tech.*, **10**: 757–765.

Mitchell, P.D., van den Ende, B., Jerie, P.H., Chalmers, D.J. 1989. Responses of 'Bartlett' pear to withholding irrigation regulated deficit irrigation, and tree spacing. *J. American Soc. Hort. Sci.*, **114**: 15–19.

Mitre, V., Mitre, I., Sestras, A.F. and Sestras, R.E. 2012. Effect of roots pruning upon the growth and fruiting of apple trees in high density orchards. *Bulletin UASVM Horticulture*, **69**: 254–259.

Monteith, J.L. 1977. Climate and the efficiency of crop production in Britain. *Philosophical Transactions of the Royal Society of London. Series B, Biological Sciences*, **281**: 277–294.

Moore, C.S. 1978. Biometrical relationships in apple trees. *J. Hort. Sci.*, **53**: 45–51.

Morgan, D.C., Stanley, C.J., Volz, R. and Warrington, I.T. 1984. Summer pruning of Gala apple: the relationship between pruning time, radiation penetration and fruit quality. *J. American Soc. Hort. Sci.*, **109**: 637–642.

Muller, D. and Leyser, O. 2011. Auxin, cytokinin and the control of shoot branching. *Ann. Bot.*, **107**: 1203–1212.

Musacchi, S., Abeti, C., Ancarani, V., Buccil, D., Gagliardi, F. and Serra, S. 2014. Summer pruning of apricot cvs. 'Bora' and 'Carmen Top'. *Acta Hort.*, **1058**: 85–93.

Myers, S.C. and Savelle, A.T. 1996. **In:** Maib K.M. (Ed.) *Tree Fruit Physiology: Growth and Development*. Good Fruit Grower, Yakima, WA.

Negri, P., Magnanini, E., Cantoni, L., Berardi, G. and Sansavini, S. 1998. Gene transfer in woody fruit species: initial trails in growth habit control. **In:** *Proceedings of AGRO.BIO.FRUIT. Cesena, Italy*, p.13.

Neilsen, D. and Neilsen, G.H. 2002. Efficient use of nitrogen and water in high density apple orchards. *HortTech.*, **12**: 19–25.

Niederholzer, F.J.A., DeJong, T.M., Saenz, J.L., Muraoka, T.T. and Weinbaum, S.A. 2001. Effectiveness of fall versus spring soil fertilization of filed grown peach trees. *J. American Soc. Hor.Sci.*, **125**: 644–648.

Norelli, J.L. and Miller, S.S. 2004. Effect of prohexadione-calcium dose level on shoot growth and fire blight in young apple trees. *Plant Dis.*, **88**: 1099–1106.

Nuberlin, F. 1993. The super spindle apple orchard system. *Compact Fruit Tree*, **26:** 17.

Ono, T., Koike, H., Tamai, H. and Sigehara, I. 2002. Effects of leaf removal and benzyladenine application on branch development of two-year-old 'Fuji' apple nursery trees grown by pruning and bud removal for one-year-old whip trees. *Horticultural Res.* (Japan), **1**: 31–34.

Ouellette, D.R., Unrath, C.R. and Young, E. 1996. Manual and chemical branch inducement in fall- and spring-planted 'Empire' apple on two tootstocks. *HortSci.*, **31**: 82–88.

Palmer, J.W. 1980. Computed effects of spacing on light interceptionand distribution within hedgerow trees in relation to productivity. *Acta Hort.*, **114**: 80–88.

Palmer, J.W. 1989. **In:** *Manipulation of Fruiting*, C.J. Wright (Ed.) Canopy manipulation for optimum utilization of light. Butterworths, London, pp. 245-262.

Palmer, J.W. 1999. Light, canopies, fruit and dollars. *Compact Fruit Tree*, **32**: 119–122.

Palmer, J.W. and Jackson, J.E. 1974. Effects of tree population and variations in spacing within and between rows of Golden Delicious on M.9. In: *Report of the East Malling Research Station for 1973*, East Malling Research Station, Maidstone, UK, pp. 66–68.

Palmer, J.W. and Jackson, J.E. 1977. Seasonal light interception and canopy development in hedgerow and bed system apple orchards. *J. Appl. Ecol.*, **14**: 539-549.

Pathak, R.K. 2009. Canopy management in minor fruit crops. **In:** *Summer School on Management of Canopy Architecture for higher Productivity in Subtropical fruits*. Pub. CIAH, Lucknow, India. pp: 43-48.

Proietti, P. and Tombesi, A. 1996. Translocation of assimilates and source-sink influences on productive characteristic of the olive tree. *Adv. Hort. Sci.*, **10**: 11–14.

Quamme, H.A. and Brownlee, R.T. 1993. Early performance of micropropagated trees of several *Malus* and *Prunus* cultivars on their own roots. *Canadian J. Plant Sci.*, **73**: 847–855.

Rademacher, W. 2004. Chemical regulation of shoot growth in fruit trees. *Acta Hort.*, **653**: 29–32.

Rademacher, W., van Saarloos, K., Garuz Porte, J. A., Riera Forcades, F., Senechal, Y., Andreotti, C., Spinelli, F., Sabatini, E. and Costa, G. 2004. Impact of prohexadione-Ca on the vegetative and reproductive performance of apple and pear trees. *European J. Hort.Sci.*, **69**: 221–228.

Robinson, T.L. 2003. Apple orchard systems. In: D.C. Ferree and I.J. Warrington (eds.) *Apples: Physiology, production and uses*. CABI Publishing, Wallingford, United Kingdom, pp. 345–407.

Robinson, T.L. and Lakso, A.N. 1991. Bases of yield and production efficiency in apple orchard systems. *J. American Soc. Hort. Sci.*, **116**: 188–194.

Robinson, T.L. and Sazo, M.M. 2014. Effect of promalin, benzyladenine and cyclanilide on lateral branching of apple trees in the nursery. *Acta Hort.*, **1042**: 293–302.

Robinson, T.L., Hoying, S.A., Miranda Sazo, M., Dominguez, L.I. and Fachinello, J.C. 2014. Yield, fruit quality and mechanization of the tall spindle apple production system. *Acta Hort.*, **1058**: 95–103.

Robinson, T.L., Lakso, A.N. and Ren, Z. 1991. Modifying apple tree canopies for improved production efficiency. *HortSc.*, **26**: 1005-1012.

Robinson, T.L., Stiles, W.C. and Lakso, A.N. 1992. Fertigation effects on apple trees growth, cropping and dry weight partitioning. *HortSci.*, **27**: 132.

Rugini, E., Muganu, M., Pilotti, M., Balestra, G.M., Varvaro, L., Magro, P., Bressan, R. and Taratufolo, C. 1999. Genetic stability, transgene hereditability and agronomic evaluation of transgenic kiwi (*Actinidia deliciosa* A. Chev.) plants for *rol*ABC, *rol*B and osmotin gene. In: *Fourth International Symposium on Kiwifruit, Santiago de Chile* (11-14 Feb., 1999). p. 26 (Abstract).

Sansavini, S. 1983. High density orchards of various crops. In: *Proceedings of the XXIst International Horticultural Congress*, Vol. 1. International Society for Horticultural Science, Leuven, Belgium, pp. 182–197.

Sazo, M. M., DeMarree, A. and Robinson, T. 2010. The platform factor-Labor positioning machines producing good results for NY apple industry. *New York Fruit Quart.*, **18**: 5–9.

Sazo, M.M. and Robinson, T.L. 2011.The Use of Plant Growth Regulators for Branching of Nursery Trees in NY State. *New York Fruit Quarterly*, **19**(2): 5–9.

Schupp, J.R., Ferree, D.C. and Warrington, I.J. 1992. Interactions of root pruning and deblossoming on growth, development, and yield of 'Golden Delicious' apple. *J.Hort. Sci.*, **67**: 465–480.

Sedira, M., Holefors, A. and Welander, M., 2001. Protocol for transformation of the apple rootstock Jork 9 with *rol*B gene and its influence on rooting. *Plant Cell Rep.*, **20:** 517-524.

Sharma, S., Rehalia, A.S. and Sharma, S.D. 2009. Vegetative growth restriction in pome and stone fruits - a review. *Agric. Rev.*, **30** : 13–23.

Shikhamany, S.D. 2001. Canopy management of the tropical and subtropical fruit crops. *Indian J. Hort.*, **58**: 28–32.

Srinivasan, C, Dardick, C, Callahan, A. and Scorza, R. 2012. Plum (*Prunus domestica*) trees transformed with poplar FT1 result in altered architecture, dormancy requirement, and continuous flowering. *PLoS ONE*, **7**(7)e 40715. doi: 10.1371/ journal.pone.0040715.

Stanica, F. and Eremia, A. 2014. Behaviour of Some New Apricot Cultivars under the Parallel Trident Planting System. *Acta Hort.*, **1058**: 129–136.

Stephen, S.M, and Thomas T. 2003. Regulating vegetative growth in deciduous fruit trees. *PGRSA Quarterly*, Plant Growth Regulation Society of America, **31**: 8–46.

Taylor, B. H. and Ferree, D. C. 1984. The influence of summer pruning and cropping on growth and fruiting of apple. *J. American Soc. Hort. Sci.*, **109**: 19–24.

Tustin, D.S. 2000. The Slender Pyramid tree management system - in pursuit of higher standards of apple fruit quality. *Acta Hort.*, **513**: 311–319.

Tustin, D.S., Cashmore, W.M. and Bensley, R.B. 2001. Pomological and physiological characteristics of Slender Pyramid central leader apple (*Malus domestica*) planting systems grown on intermediate vigour, semi-dwarfing, and dwarfing rootstocks. *New Zealand J. Crop Hort. Sci.*, **29**: 195–208.

Tustin, S., Hurst, P., Warrington, I. and Stanley, J. 1989. Light distribution and fruit quality through multi-layered trellis apple canopies. *Acta Hort.*, **243:** 209-212.

Veinbrants, N. 1972. Effect of succinic acid-2,2-dimethylhydrazide (Alar) or scoring on growth and flowering initiation of young apple trees. *Australian J. Exp. Agric. Animal Husband.*, **12**: 89–95.

Vercammen, J. 2014. Comparison of different planting systems for 'Conference'. *Acta Hort.*, **1058**: 37–43.

Verheij, E.W.M. and Verwer, F.L.J.A.W. 1973. Light studies in a spacing trial with apple on a dwarfing and a semi-dwarfing rootstock. *Scientia Hort.*, **1**: 25-42.

Wagenmakers, P.S. 1991. Planting systems for fruit trees in temperate climates. *Critical Rev.' Plant Sci.*, **10:** 369-385.

Walker, C.D. 1980. The development of intensive orchards in England, a historical note. *Acta Hort.*, **114**: 309–317.

Watanabe M., Bessho H., Suzuki A., Komori S. 2008. Seasonal changes of IAA and cytokinin in shoots of columnar type apple trees. *Acta Hort.*, **774**: 75–80.

Webster, A.D. 2002. Factors influencing the flowering, fruit set and fruit growth of European pears. *Acta Hort.*, **596**: 699–709.

Wertheim, S.J. 1968. The training of the slender spindle. Pub. No. 7, ProefstationFruitteelt, Wilhelminadorp, the Netherlands (37 p.).

Wertheim, S.J. 1978. Intensive apple orchards with Slender Spindles. *Acta Hort.*, **65**: 209–216.

Wertheim, S.J. 1981. High density planting, development and current achievements in the Netherlands, Belgium and Germany. *Acta Hort.*, **114**: 318–327.

Wertheim, S.J., de Jager, A. and Duyzens, M.J.J.P. 1986. Comparison of single row and multi-row planting system with apple, with regard to productivity, fruit size and color and light condition. *Acta Hort.e*, **160**: 243–258.

Whiting, M.D., Lang, G. and Ophardt, D. 2005. Rootstock and training system affect sweet cherry growth, yield and fruit quality. *HortSci.*, **40**: 582–586.

Wilton, J. 1999. Vigour control hints. *Orchardist*, **72**: 14–18.

Wunsche, J.N. and Lakso, A.N. 2000. Apple Tree Physiology–implications for orchard and tree management. *The Compact Fruit Tree*, **33**: 82-88.

Zimmerman, R.H. and Miller, S.S. 1991. Orchard growth and fruiting of micropropagated apple trees. *J.American Soc. Hort. Sci.*, **116**: 780–785.

Chapter 6

Biotechnological Interventions in Temperate Fruit Crops

Rajinder Kaur

Dr Y.S. Parmar University of Horticulture and Forestry,
Solan – 173 230, H.P.
E-mail: rkaur_uhf@rediffmail.com

1. INTRODUCTION

Temperate fruits are of immense importance due to their neutraceutical value and are also central to a country's economy. But there are various problems which hinder their propagation and production which include long juvenile phase and reproductive cycles with concomitantly long breeding spans. As a result great benefits can be derived from biotechnological interventions.

Conventional breeding techniques will not produce the major improvements sought without long-term selection and lengthy back-crossing procedures that will normally exceed the service life-span of the fruit breeder. The emerging trends of biotechnology hold promise of rapid selection for desirable characteristics (if they can be identified) and so would enable large numbers of mutants/recombinants to be examined within a short time and limited space. Biotechnological interventions which can be used for enhancing efficiency of production of temperate fruits:

1. Tissue Culture
   a. Micropropagation
   b. Meristem culture
   c. Anther culture
   d. Embryo rescue

e. Protoplast culture

2. *In vitro* Induction of Variation
   a. Somaclonal variations
   b. Somatic hybridization
   c. Genetic transformation
   d. TILLING
3. Germplasm conservation via cryopreservation of tissue cultures and synthetic seeds
4. Molecular breeding/Genomics: Marker assisted selection of desirable traits with identified markers
   a. Linkage mapping
   b. Association mapping

2. TISSUE CULTURE

(a) Micropropagation

Micropropagation is rapid multiplication of stock plant material to produce a large number of progeny plants using plant tissue culture methods.

Micropropagation has a number of advantages over traditional plant propagation techniques:

- ✰ Rapid clonal propagation.
- ✰ Produce disease-free plants through meristem culture.
- ✰ This is a plant tissue culture technique. It does not require highly expensive, sophisticated equipments. So it can be afforded by labs with less infrastructure.

One of the problems which hinder micropropagation of temperate fruits is the secretion of phenolic compounds in culture medium, but can be overcome by certain modifications. But still micropropagation has been achieved successfully in many temperate fruit crops , *viz.*, apple, blueberry, prunus, pear, cherry, strawberry etc.

(b) Meristem Culture

Culturing of meristem tips is known as meristem culture. A meristem is the tissue in plants containing rapidly multiplying undifferentiated cells (meristematic cells), found in zones of the plant where growth can take place. In this technique meristem and a few subtending leaf primordial are placed into a suitable growing media. A disease free plant is produced by this method since the meristem is presumed to be free of viruses *etc.* Classical example includes the production of apple free from apple mosaic virus and strawberry from strawberry motel virus.

(c) Anther Culture

Anther culture is the process of using anthers and pollens to obtain haploid plantlets.

Some of the advantages which make this a valuable method for obtaining homozygous plants:

- ✰ This is simple tissue culture technique and homozygous plants can be obtained quickly.

(d) Embryo Rescue

Fruit breeder have succeeded in obtaining distant hybrids in several temperate horticultural species. Two cultivars 'Gold Crest' peach and 'Mayfire' nectarine have resulted from embryo culture and are among the earliest of commercially grown cultivars in California. In certain cases, seeds show poor germination due to nutritional deficiencies and/or physiological reasons. Such inherently week embryos can be rescued. Interspecific hybridization in kiwi fruit, is restricted due to embryo and endosperm abortion in some combinations. Embryo culture has been used to rear hybrids between *Actinidia deliciosa* x *A. eriantha* and *A. deliciosa* x *A. arguta*. Consequently this technique results in the breeding of distantly related species, overcoming seed dormancy and shortening breeding cycle.

(e) Somatic Hybridization

Protoplast is a cell without cell wall which is obtained by wall degrading enzymes and then cultured in a suitable culture medium in controlled conditions for regeneration of plantlets. If conventional hybridization fails, hybrids can be obtained by somatic hybridization (fusion of protoplast of two different genotypes). Somatic hybridization overcomes crossing barriers.

i. Cherry plum fusions for the transfer of plum pox resistance to the hybrid rootstocks, as all cherries are resistant to the plum pox virus. A good example would be 'Pixy' (plum, *Prunus insititia*) fused with Colt (*P. pseudocerasu*) to produce a 9x *Prunus* combining plum pox resistance from cherry with the more dwarfing habit of plum;

ii. Pear quince (Cydonia) fusions for the production of rootstocks that have greater dwarfing capacity and better scion compatibility than existing stocks;

iii. Apple pear rootstock fusions for the introduction of better dwarfing and greater compatibility with pear scions.

3. *IN VITRO* INDUCTION OF VARIABILITY

(a) Somaclonal Variations

Variations occurring in the tissue cultured plants are somaclonal variations. These variations may be spontaneous occurring in the callus cultures or induced by modification of culture medium. These can be for both biotic as well as abiotic stresses. The stresses include resistance to diseases, herbicides and tolerance to environmental stresses, as well as for increased production of secondary metabolites.

(b) Genetic Engineering

Genetic engineering, also called *Genetic modification*, is the direct manipulation of an organism's genome using recombinant DNA technology. It involves the introduction of foreign DNA or synthetic genes into the genome of an organism of interest. The introduction of new DNA does not require the use of classical genetic methods, however subsequently traditional breeding methods are typically used for the propagation of recombinant organisms. An organism that is generated through the introduction of recombinant DNA is considered to be a genetically modified organism and known as GMO or transgenics. Following are the techniques of gene transfer:

(i) *Agrobacterium*-mediated gene transfer

Agrobacterium tumefaciens has proved to be efficient vehicle for the introduction of genes into plants and plant cells (Table 1). Most transgenic plants produced to date were created through the use of the *Agrobacterium* system. *Agrobacterium tumefaciens* is the etiological agent of crown gall disease and produces tumorous crown galls on infected species. The utility of this bacterium as a gene transfer system was first recognised when it was demonstrated that the crown gall were actually produced as a result of the transfer and integration of genes from the bacterium into the genome of the plant cells. Virulent strains of *Agrobacterium* contain large Ti-(tumour inducing) plasmids, which are responsible for the DNA transfer and subsequent disease symptoms.

Table 1: Some Examples of *Agrobacterium* Mediated Gene Transfer in Temperate Fruits

Crop	*Trait*	*Transferred Gene (transgene)*
Apple	Scab	*Vfa1, Vfa2, or Vfa4* (identified within the Vf locus of *Malus floribunda* 821)
	Fire blight	*fb_mr*5 originating from the crab apple accession *Malus robusta*5 (Mr5)
	Powdery mildew	*Malus hupehensis*-derived NPR1 (*Mhnpr*1)
	Fungal disease resistance	*Chitinase*
	Non-browning	'*pgas*' (a hybrid sequence carrying four gene groups *i.e.* PPO2, GPO3, APO5 and pSR7)
Apricot	Sharka (plum pox virus)	*PPV-cp* (Plum pox virus-coat protein)
Cherry	Stimulation of rooting	Rooting locus B (*rolB*) gene
Kiwifruit	Grey mould disease	*Osmotin*
	Salt stress tolerance	*Arabidopsis thaliana* Na+/H+ antiporter (*At nhx1*)
Peach	Efficient transformation protocols	Green fluorescent protein (*gfp*)
Pear	Fire blight	Exopolysaccharide (EPS) depolymerase and *ferritin*

Contd...

Table 1–*Contd...*

Crop	*Trait*	*Transferred Gene (transgene)*
	Abiotic stress tolerance	*Malus domestica* spermidine synthase (*Mdspds1*)
	Reduced juvenile period	*Malus domestica* terminal flower 1(*Mdtfl1*)
Persimmon	Abiotic stress tolerance	Codeine A (*CodA*)
	Production of dwarf plants	A *Prunus mume* ortholog of Arabidopsis terminal flower 1(*tfl1*)
Plum	Plum pox virus (PPV)	Plum pox virus coat protein (*PPV-cp*)
	Early flowering and fruiting	Transformation with flowering locus T1 (*ft1*)
Strawberry	Glyphosate tolerance	CP4.Enolpyruvylshikimate-3-phosphate synthase (CP4.EPSPS)
	Fungal disease resistance	Chitinase (*chit42*)
	Cold tolerance	CBF1 (C-repeat binding factor)
	Salt stress tolerance	*osmotin*
	Resistance to crown gall and root lesion nematodes (RLNs)	Rooting locus ABC (*rolABC*), *Pratylenchus vulnus* (*Pv*) 010 and *Pv*194

(ii) Direct gene transfer techniques

Most commonly used methods for direct gene delivery:

a. **Particle gun/Particle bombardment** – Some cells, tissues and intracellular organelles are impermeable to foreign DNA, especially plant cells. To resolve this problem in gene transfer, the gene gun machine is used for the transfer of desired gene. The gene gun is part of the gene transfer method called the biolistic method. In this method, DNA or RNA adheres to biological inert particles, *i.e.* gold or tungsten. By this method, DNA-particle complex is put on the top location of target tissue in a vacuum condition and accelerated by powerful shot to the tissue, then DNA will be effectively introduce into the target cells.

b. **Electroporation:** Involves a pulse of high voltage applied mainly to protoplasts to make transient (temporary) pores in the plasma membrane which facilitates the uptake of foreign DNA. Some of these cells will incorporate the new DNA and express the desired gene.

c. **Microinjection:** Microinjection is the direct mechanical introduction of DNA under microscopical control. A target can be a defined cell within a multicellular structure such as embryo, ovule and meristematic cells or a defined compartment of a cell. By examination with a microscope, a cell is held in place with gentle suction while being manipulated with the use of a blunt capillary. A fine pipette is then used to insert the DNA into the cytoplasm or nucleus. This technique is effective with plant protoplasts and tissues. Some examples of direct gene transfer into temperate fruits are given in Table 2.

Table 2: Some Examples of Direct Gene Transfer into Temperate Fruits

Species	*Method*	*Purpose*	*Gene Transferred*
Malus x *domestica*	PEGmediated	Method optimization	Green Fluorescent Protein (*gfp*)
Actinidia sp.	PEGmediated	Method optimization	*GusA, nptII*
Actinidia deliciosa	Particle bombardment	Method optimization	*GusA*
Vaccinim macrocarpon	Particle bombardment	Method optimization and Pest control	*nptII, gusA, Bt*
Vaccinim macrocarpon	Particle bombardment	Herbicide resistance	*bar, aphII*

(c) TILLING (Target Induced Local Lesions in Genome)

TILLING allows the rapid detection of induced point mutations in populations of physically/chemically mutagenized individuals. The seeds are treated with mutagens like EMS (ethyl methane sulphonate) and raised to harvest M_1 plants (first mutagenized population), which are consequently, self-fertilized to raise M_2 population. DNA extracted from M_2 plants is used in mutational screening. Unlike conventional mutation breeding in which the mutation frequency is unknown or estimated from mutations conveying a visible phenotype, TILLING provides a direct measure of induced mutations. An extension of TILLING is EcoTILLING, which uncover natural mutations present in the population.

4. GERMPLASM CONSERVATION

(a) Synthetic Seeds

Synthetic or artificial seeds have been defined as somatic embryos engineered for use in the commercial propagation of plants.

In case of apple (*Malus* sp.), *in vitro* conservation using synthetic seed (Synseed) has been demonstrated. This procedure involves the formation of sodium alginate beads around the explant to afford desiccation tolerance for prolonged periods.

(b) Slow Growth Cultures

Minimal growth storage is the most direct way of restricting growth and development of *in vitro* materials, and is usually applied to differentiated plantlets and developing meristem cultures.

The cultures can be induced into a state of slow growth by various means, including culture at reduced temperature in the presence of hormonal growth inhibitors (*e.g.* abscisic acid), reduction in sucrose concentrations, or the addition of osmotically active additives (*e.g.* 3 per cent *w/v* mannitol). Temperatures in the range of 0-5°C are employed with cold tolerant species, *e.g.* apple (*Malus domestica*), plum (*Prunus*) or strawberries (*Fragaria* x *ananassa*).

(c) Cryopreservation

The term 'cryopreservation' is used exclusively to cover the storage of living biological material at ultralow temperatures, normally at or near the temperature of liquid nitrogen (-196°C). At this temperature, all cellular divisions and metabolic events are stopped. The plant material can thus be stored without alteration for extended periods, with limited maintenance.

Table 3: List of Vegetativaly Propagated Plant Species Used for Cryopreservation

Species	*Technique*
Fragaria spp.	Synthetic seeds Slow freezing
Malus spp.	Synthetic seeds Droplet
Prunus domestica	Cryopreservation
Prunus dulcis	Cryopreservation
Pyrus spp.	Synthetic seeds Cryopreservation Slow freezing
Rubus spp.	Slow freezing
Strawberry	Cryopreservation

(d) Genomic Library

This technique holds promises for the conservation of genes or DNA segments from threatened species for future use. Germplasm can be conserved by developing genomic/cDNA libraries and preserving at -70° C.

5. MOLECULAR BREEDING/GENOMICS

Marker assisted selection of desirable traits with identified markers a) Linkage mapping b) Association mapping. Foremost requirement for molecular breeding is DNA based markers. Marker is DNA sequences that is readily detected and whose inheritance can easily be monitored. These are the signs of identification of any trait of an organism based on DNA variation in the genome. This DNA sequence studies by primer or probe can differentiate one plant/organism from the other. Several types of DNA based markers *viz.*, RFLP (Restriction Fragment Length Polymorphism), RAPD (Random Amplified Polymorphic DNA), SSRs (Simple Sequence Repeats), ISSR (Inter Simple Sequence Repeat), SNP (Single Nucleotide Polymorphism) etc. are used to assess genetic relationship, DNA finger printing, QTL analysis among various temperate fruits.

(a) Linkage Mapping for MAS

In MAS plants are selected for one or more (up to 8-10) alleles. Major steps involved in MAS: (i) Identification of molecular marker(s) associated with trait(s) of interest, (ii) Validation of identified marker(s) in the genetic background of the targeted genotypes to be improved.

MAS is estimated to be more precise or less time-consuming than the protocols based on available phenotypic selection. Therefore, "positive" MAS selection in favor to a specific allele can even be very informative even for easy-to-score phenotypic traits.

There are number of identified markers in apple and pear, which can be used in marker-assisted selection but MAS efficiency varies depending on the estimated genetic distance between the marker and the linked gene. Molecular-based selection in pome fruit breeding progenies can be particularly important for traits that are difficult to evaluate (*e.g.* mildew resistance in apple) or delayed in time by juvenility (*e.g.* fruit traits) but today the broadest molecular screening can only be efficiently performed for the Vf scab-resistance gene. MAS may be based on QTL identification through linkage mapping or association mapping.

QTL identification is the process of constructing linkage maps and conducting QTL analysis to identify genomic regions associated with traits is known as QTL mapping (Table 5).

Table 5: Some Examples of QTL Identification in Temperate Fruits

Crop	*Trait*
Apple	Growth and vegetative characters, fire blight, volatile compounds
Apricot	Plum pox virus
Peach	Resistance to bacterial spot and aphids
Pear	Fruit traits
Strawberry	Fruit aroma, day neutrality

(b) Association Mapping for MAS

Association mapping is considered an alternative strategy to linkage mapping for identifying marker-trait associations. It utilizes natural variation and linkage disequilibrium existed in natural population to identify the gene(s)/genomic regions associated with trait rather than the offspring of crosses and associations in natural populations are typically on a much finer scale because they reflect historical recombination events.

Association mapping has been used to identify linked markers for bacterial resistance, growth and fruit quality in pomegranate collections from India. A cultivated clone of pomegranates was identified in close proximity to Daru types of wild-type pomegranates that grow naturally near the foothills of the Himalayas by using microsatellite markers. This example holds good for future use of association mapping in temperate fruits.

6. CONCLUSION

Biotechnological tools are handy to be used along with conventional breeding approaches. These modern tools bring precision in making the disired change in traits. The need to be adopted by the breeders to reduce the time requirement and evolving scion variety and rootstock genotypes with the desired traits.

7. REFERENCES

Kaur R. Kumar K. Sharma, R. Kashyap, A. and Bhardwaj, S.V. 2012. *In vitro* propagation of some *Prunus* clonal rootstocks. *Crop Improv.* **1:** 1695-1696.

Kaur, R. Sharma, N. Kumar, K. and Sharma, D. R. 2006. Studies on embryo culture of walnut (*Juglans regia* L.). *Scientia Hort.*, **109:** 383-388.

Kaur, R. Shilpa, Vaidya, E. and Kumar, K. 2015. Development, characterization and transferability of peach genic SSRs to some Rosaceae species. *Adv.Res.* **3:** 165-180.

Kaur, R. and Vaidya, E. 2015. Analysis of frequency and functional annotation of SSRs in ESTs of cole crops. *J.Appl. Life Sci. Int.* **2:** 189-196.

Kaur, R. Kumar, K. Sharma, R,. Kashyap, A. and Bhardwaj, S.V. 2012. *In vitro* propagation of some *Prunus* clonal rootstocks. *Crop Improv.* **1:** 1695-1696.

Kaur, R. Sharma, N. Gupta, M. and Sharma, G. 2012. *In vitro* Propagation of a Commercial Strawberry Cultivar 'Selva'. *SKUAST J. Res.* 14: 1-8.

Sharma, D. Kaur, R. Kumar, K. and Bhardwaj, S.V. 2012. Genetic diversity among selected genotypes of almond *Prunus dulcis* Miller D.A. Webb assessed by random amplified polymorphic DNA (RAPD) markers. *African J. Biotech.* **11 :** 14877-14883.

Chapter 7

Rejuvenation and Replanting of Temperate Fruits in India

R.C. Sharma

Dr Y.S. Parmar University of Horticulture and Forestry, Solan – 173 230, H.P.

E-mail: dresuhf@yahoo.com

1. INTRODUCTION

Fruit trees require constant care and attention throughout their life for successful fruit production. As the fruit trees grow old, their care and attention has to be proportionately increased in order to continue to get good yields from them. In general orchards more than 30 years (pome fruits), 20 years (stone fruits) and 30-40 years (evergreen species) of age have shown much more unfruitfulness than the young orchards. Such orchards/trees do not produce adequate annual extension growth, because of inadequate fertilization, improper pruning, poor soil conditions and adverse effects of incurable insects, pests and diseases.

Decreasing productivity of fruit crops has become a matter of concern for research institutes, line departments and growers in the country. Because of limited land resources farmers are compelled to put back old orchard sites, under new plantations. It has been pointed out by various researchers that replantation of old orchard sites is also cumbersome as it leads to drastic economic loss due to uprooting of old trees to the orchardists and poor establishment of new plantations on old orchard sites vis a vis its juvenile period. Thus every effort should be made to rejuvenate old and neglected trees in order to restore their productive potential (Singh, 1977). However, it must be noted that every tree cannot be rejuvenated. For instance, it would be useless to try to rejuvenate a tree which is too old, is suffering from an inherent virus diseases or is growing poorly because it has been planted on a poor soil, hard pan or rock.

Rejuvenation produces best results on trees which have otherwise healthy root system, good vigour, but are unproductive because of defects in pruning or cultural practices. The rejuvenation strategy will depend upon the objective, such as either change of old variety or rejuvenation of old trees of existing cultivar. The techniques of rejuvenation will be different for very large apparently healthy trees, stunted trees and aged trees.

Reduced crop productivity in fruit trees because of repeatedly planted to the same or closely related species has been definitive and recognized by various names, including soil sickness, soil exhaustion, replant disorder, replant problem and replant disease (Verma and Sharma, 1999). In temperate fruits, replant problem and replant disorder have been attributed to diminished soil fertility, degraded soil structure and residual herbicide activity. Replant disease has generally been attributed to biotic factors. Replant disease symptoms are consistently observed shortly after planting (1-3 months), which represents the most susceptible physiological state of the plant. Typical symptoms include uneven growth throughout the orchard, stunting and shortened internodes. These above ground symptoms are associated with discoloured roots, root tip necrosis and reduced root biomass. If death of plant does not occur during first year, a delay in fruit production, reduced fruit yield and quality are common. The primary etiologic agents, which vary in relative dominance from site to site include multiple species of the genera *Rhizoctonia*, *Phytophthora*, *Cylindrocarpon* and *Pythium* at times acting in concert with the lesion nematode *Pratylenchus penetrans* (Mazzola and Manici, 2012). Abiotic factors may exacerbate overall symptoms, they do not appear to function as the primary cause of the disease.

2. REJUVENATION

It is a common sight to see every large tall tree in a neglected orchard or an orchard which has not been pruned with care, especially in case of temperate fruits which could be due to:

(a) Being vertical, the top most branches grown vigorously till they are out of easy reach.

(b) Some branches may start as very vigorous shoots in the lower parts of a tree and may grow very tall in search of sunshine, over reaching the rest of the tree.

3. PRUNING STRATEGIES

Although in the initial stages, the top branches bear heavy crops due to greater exposure to the sun, they also bend with the weight of the crop. This produces an abundance of vigorous, vertical shoots and water sprouts in the top of the tree. Such trees have less quantity of fruiting wood, which is generally located on the top or side periphery (Mahajan and Dhillon, 2002). The entire inner and lower parts of the tree are shaded and fruiting buds/spurs if any are weak. These trees can often be rejuvenated by restructuring them through corrective pruning.

(a) Extent of Pruning

It is best to renovate the tree over a three year period by removing 1/3rd to 1/2nd of the excess branches each year for three years. If a tree has been abandoned for a long time, cut out diseased and damaged branches before removing the 1/3rd of the live wood. In a situation where the whole top needs to be cut off, the tree will be highly stressed and may not produce fruits for a few years (Kumar *et al.*, 2005).

In oversized trees, it is best to start pruning by thinning out or completely removing the upper branches. These branches due to their vigorous growth and greater consumption of nutrients and shading prevent the lower branches from growing and fruiting (Sharma *et al.*, 1995).

Prune out any dead or diseased limbs. Then begin to remove competing branches that cross or crowd one another. Remove branches that go straight up or hang down. Branches that have a slight upward angle are best to leave. The canopy should appear thinned out, with no dense cluster of branches anywhere in the canopy (Kaundal *et al*, 2002). The following points should be kept in mind while performing pruning operations:

i) Pruning should be done in dormant season only.
ii) It is better to make few large cuts than many small cuts.
iii) Remove about a third of the excess limbs each year for three years.
iv) Start pruning in the top where it will do the most pruned progressively less as you move down in the canopy.

(b) Application of Fertilizers

Fertilizer is less important than pruning when it comes to revitalizing old senile trees. Get the soil tested to determine the need for lime, potassium (k) and magnesium (Mg). Nitrogen (N) is the one nutrient that should be applied with caution. Rejuvenation pruning is usually quite severe, and if too much nitrogen is provided, the vegetative growth can be excessive (Raj Kumar *et al, 2008*).

After resorting to serve rejuvenating pruning, there must be 12 to 18 inches (in case of apple) annual shoot growth. If there is less than 12 inches apply same as in the previous year and if more than 18 inches apply normal dose of nitrogen (Thakur, 2012). However, the dose of fertilizer will be different for different fruit crops.

(c) Top Working

Top working is mainly confined to changing the tops of the seedlings and undesirable varieties of the declined plants. The top working can be done either by developing stem builders by cutting back the top of declined trees to certain height or frame working the main scaffolds at many points to recover earlier tree volume in short span of time for early returns. The top working can be performed either by side veneer or side tongue or cleft grafting depending upon the field situation and fruit crop in question.

4. MANAGEMENT OF REPLANTED SITES

(a) Cultural Practices

In a traditional fruit orchard there is grass in between the rows and weed free area within the row. Numerous cultural practices have been applied in a preplant setting as a means to manage replant disease. Apple was found to grow well after a cover crop of oats, but plants were severely stunted when grown in the same soil after a cover crop of rye. It was because, rye was a better host for the root lesion nematode than oat (Verma and Sharma, 1999).

Cultivation of wheat prior to apple replantation, resulted in reduced replant problem. It is because of generation of soil microbial community with the capacity to suppress disease incited by *Rizoctonia solani* AG-5 (Mazzola and Gu, 2002). Development of soil suppressiveness to *Rhizoctonia* root rot was due to changes in the composition of fluorescent pseudomonas population. However, trees raised in the wheat cropped soils did not exhibit any significant reduction in root infection by *Cylindrocarpon* spp. or *Pythium* spp.

(b) Application of Organic Matter

The positive effect of increased soil organic matter on plant development is largely supported by the better control of soil borne diseases through application of composts or organic wastes as soil amendments. However, disease control achieved through these soils amendments are variable. Incorporation of fresh waste or green material may increase population of certain soil borne pathogens (*Pythium, Rhizoctonia)* over the short term. In contrast, highly degraded composts increase the organic content without any negative effect.

In soils possessing soil organic matter content higher than 2.5 per cent, independent of management system, organic orchards exhibited greater degree of soil suppressiveness which was highly correlated with density of total soil fungi. The frequency of root colonization by fungal pathogens (*R. solani, Pythium intermedium, Cylindrocarpon* spp.) was lower in organic than conventional orchards.

Residue from plants belonging to the family Brassicaceae commonly used to suppress replant disease. Members of this plant family produce glucosinolates, which upon hydrolysis yield biologically active compounds, including isothiocyanates. Oilseed cakes from *Brassica* spp. are also quite effective in controlling replant disease. The compounds selectively modify composition of various biological communites, such as protozoa, nematodes and fungi. It is also reported that active compounds other than isothiocyanates are also responsible in controlling soil borne pathogens. *Brassica* oil seed cake also provided systemic protection against soil borne pathogens of apple and disease control was abolished in response to pasteurization of the seed meal amended soil (Cohen and Mazzola, 2006). In *Brassica juncea* oil seed cake, apple root infection by *R. solani* is suppressed by the biologically active volatile allylisothiocyanate. This compound also generate significant changes in fungal community composition, which were associated with development of a disease suppressive soil. Similarly, short term suppression of *Pratylenchus penetrans* is response to *B. napus* oil seed cake was obtained. The nematophagous fungus *Arthrobotrys* sp.

was recovered from roots of apple grown in *B. napus* oilseed cake. It also helps in suppression of *P. penetrans*. Soils treated with Brassica oilseed cakes are more resilient to reinfestation by *P. penetrans* than soils fumigated with 1,3 – dichloropropene – chloropicrin.

(c) Chemical Control

Pre-plant soil fumigation is the primary measure employed for the control of replant disease world over. Generally pre-plant fumigation is more important than post plant treatments. Problems in the use of broad spectrum biocides include difficulty in application, high cost and potential hazards to human health. Soil fumigants with broad spectrum activity, *i.e.* formatin, methyl bromide, chloropicrin, 1,3-dichloropropene, ethylene dibromide and metham sodium have been in use against replant disease. Weekly or biweekly application of metalaxyl has been reported effective in certain locations. Carbendazim and Dithane Z-78 have also been proved effective in old orchard sites (Kumar *et. al.* 2005).

Effective use of individual narrow spectrum methods and chemicals yielding control of replant disease also exist in literature. In some cases these narrow spectrum chemicals fail to provide any disease control. For instance, application of difenoconazole enhanced growth of apple seedlings in 4 orchard soils but resulted in increased mortality in fifth orchard (Mazzola and Gu, 2002). This consequence was due to increased seedling root infection by *Pythium* and *Phytophthora* spp., the dominant soil borne pathogens at the fifth site. Plant mortality in the firth orchard was eliminated when difenoconazole was applied in concert with oomycete specific metalaxyl fungicide.

(d) Integration of Host Genetics to Enhance Disease Control

Host genetics influence the microbial community that resides in the rhizosphere in terms of both abundance and composition (Mazzola and Munici, 2012). This type of behaviour of host will influence the level of biological control. In the rhizosphere of rootstock CG 6210 large number of *Burkholderia cepacia* population were detected which are well known to suppress a variety of pathogens, including those such as *Pythium* spp. that are causative agents of apple replant disease. Similarly *B. cepacia* was dominant component of bacterial community recovered from soil that was suppressive to *Rhizoctonia* root rot of apple (Mazzola and Manici, 2012). Rootstocks with a greater capacity to select for such a bacterial community from resident populations could be expected to exhibit greater tolerance to replant disease. Similarly rootstocks that possess greater capacity to support populations of resident *Streptomyces* spp. have greater ability to protect apple plants from *Rhizoctonia* root rot. In case of peach, Nemaguard and Viking rootstocks are suitable for replant soils.

5. CONCLUSION

Senile and old fruit trees need to be made productive by adopting rejuvenation to restore productivity and fruit quality. This technique has been perfected and has been very effective in several fruit crops. This approach helps to realise production from unproductive orchards and thus enhance the orchard longivity. Though a tedious practice need to be adopted scientifically to avoid tree mortality or excessive demage beg insect pests and diseases. Though, standardized, the same shoud be popularized as a package for different fruit and nut crops.

6. REFERENCES

Cohen, M.F. and Mazzola, M. 2006. Effect of *Brassica napus* seed meal amendment on soil population of resident bacteria and *Naegleria americana*, and the unsuitability of arachidonic acid as a protozoan specific marker. *J. Protozol. Res.* **16**: 16-25.

Kaundal, G.S., Singh, Satinder, Kanwar, G.S. and Chanana, Y.R. 2002. Effect of pruning techniques on growth, production, quality and nutrient status of peach cv. Pratap. *J. Res.* **(PAU). 39** : 362-67.

Kumar, A., Pande, N.C. and Tripathi, V.K. 2005. Influence of pruning severity on the flowering and fruit of peach. *Farm Sci. J.* **14** : 12-13.

Mahajan, B.V.C. and Dhillon, B.S. 2002. Effect of pruning intensities on the fruit size, yield and quality of peach cv. Shan-i-Punjab. *Agric. Sci. Digest,* **22** : 281-82.

Mazzola, M. 1998. Elucidation of the microbial complex having a causal role in the development of apple replant disease in Washington. *Phytopathology,* **88:** 930-38.

Mazzola, M. and Gu, Y.D. 2002. Wheat genotype specific induction of soil microbial communities suppressive to disease incited by *Rhizoctonia solani* anastomosis group (AG)-5 and AG-8. *Phytopathology,* **92** : 1300-07.

Mazzola, M. and Manici, L.M. 2012. Apple replant disease : Role of microbial ecology and control. *Annu. Rev. Phytopathol.* **50** : 45-65.

Raj Kumar, Dimri, D.C. and Singh, S.C. 2008. Effect of nitrogen supplement on rejuvenation in apple (*Malus domestica* Borkh.) cv. Red Delicious. *Pantnagar J. Res.* **6.** : 296-98.

Sharma, N., Chauhan, J.S. and Awasthi, R.P. 1995. Regulation of growth, fruiting and fruit quality in Santa Rosa plum with pruning. *Hortic. J. 8: 53-59.*

Singh, D., Chauhan, J.S. and Kaith, N.S. 1997. Pruning in peach : A review. *Agric. Rev.* (Karnal) **18** : 147-54.

Thakur, K.K. 2012. Effect of rejuvenation pruning and nitrogen levels on growth, yield and quality of apricot (*Prunus armeniaca* L.) cv. New Castle. M.Sc. Thesis, UHF, Nauni, Solan, H.P.

Verma, L.R. and Sharma, R.C. 1999. *Diseases of Horticultural Crops : Fruits* . Indus Publishing Company, New Delhi, 724 p.

Chapter 8

Impact of Climate Change on Temperate Fruit and Nut Production

N. Sharma

Department of Fruit Science,
Dr Y.S. Parmar University of Horticulture and Forestry,
Solan – 173 230, H.P.
E-mail: nsharma.sol@gmail.com

1. INTRODUCTION

Climate has undoubtedly a dominating role among the various factors governing the distribution of fruits. Each fruit has a distinct climate requirement to grow and produce successfully and based on the temperature requirements and response to different climatic conditions, fruit crops have been classified as tropical, sub-tropical and temperate. Temperate fruits are commonly found in cold regions relishing a mild and temperature climate. These fruit trees endure cold and go to rest or dormancy by shedding off all their leaves during winter.

India can be broadly divided into four climatic zones, namely temperate, subtropical, tropical and arid. Temperate climate suitable for growing of temperate fruits can be experienced in the high altitudes of Himalayas, forming a broad continuous arc for nearly 2600 km along the northern fringes of the Indian sub-continent, from the bend of the Indus river in the northwest to the Brahmaputra river in the east. The Himalayas range, averaging 320 to 400 km in width, rises sharply from the Gangetic plain. North of this mountain belt lies the Tibetan Plateau. This zone comprises the states of Jammu and Kashmir, Himachal Pradesh and Uttrakhand in the north-west and Arunachal Pradesh, Sikkim, north-West Bengal, parts of Assam, Meghalaya, Nagaland and Manipur in the north-east. However, within the Himalayas, climate varies depending on elevation and location, and can further be divided into different zones namely sub-temperate, temperate, dry temperate/or cold dessert. Physically, the Himalayas forms three parallel zones: the Great Himalayas, the Middle Himalayas (also known as the Inner or Lesser Himalayas), and the Sub-Himalayas, which includes the Shiwalik Range and foothills and the *Terai* and *Duars* piedmont (an area of land formed or lying at the foot of a mountain or mountain range).

2. DIFFERENT HIMALAYAN ZONES

(a) Great Himalayan region

The Great Himalayan region is one of the few remaining isolated and inaccessible areas in the world, even today. There are small clustered settlements in some high valleys of this region. Extremely cold winters and a short growing season limit the farmers to grow one crop per year, most commonly potatoes or barley. Besides, the cold arid regions between 1,550-3,650 m a m s l with rainfall of 25-40 cm are suitable for dry nuts and drying types of apricots (Sharma *et al.*, 2005), prunes (Sharma *et al.*, 2011), apple and pear (Sharma *et al.*, 2003). In the past few decades, fruit cultivation has picked up in some valleys of this region, particularly in Himachal Pradesh.

(b) Middle Himalayas Range

The Middle Himalayas have a width of about 80 km and altitude between 1,830 and 3,050 m amsl. This region covers moderately populated some major valley centres such as Kashmir comprising districts of Anantnag, Badgam, Baramulla, Pulwama, Srinagar, and Kupwara; and Kangra, higher riches of Chamba, Mandi and Kullu, hills of Shimla, Mussoorie and Kathmandu (Nepal), and Darjeeling, Sikkim and Arunachal Pradesh. Among these centres, Kashmir valley is surrounded by snow-capped mountains and has the true cold temperate climate and hence has been more favourably located for growth of most temperate fruits. Apple, cherry, pear, walnut, almond and chestnut are commercially grown in entire Kashmir valley and in hills at elevation between 1,550-2,700 m amsl. In the North-Eastern hills, apple can be grown in some suitable pockets of Arunachal Pradesh and Sikkim. In the North-West, stone fruits like peach, plum and apricot are grown commercially up at elevation between 900-1,500 m with annual rainfall of 90-100 cm.

(c) Sub-Himalayas

This is the southernmost and the lowest zone, borders the plains of north India and Pakistan. It comprises the Shiwalik range and foothills as well as the narrow piedmont plain at the base of the mountains. The width of the Sub-Himalayas gradually narrows from about 48 km in the west until it nearly disappears in Bhutan and eastern India. A characteristic feature of the Sub-Himalayas is the large number of long, flat-bottomed valleys, which are usually spindle-shaped and filled with gravelly alluvium. South of the foothills lies in the *Tarai* and *Duars* plains, and is heavily cultivated. This region can grow low chilling temperate fruits such as peach, plum, apricot, pecan nut, kiwifruit *etc.* and some sub-tropical fruits like mango, citrus, litchi, guava *etc.*

(d) Other Temperate Fruit Growing Regions

The Nilgiri hills in South India having high altitudes have mild temperate climate and are suitable for growing low chilling temperate fruits species and cultivars.

3. ENVIRONMENTAL ISSUES IN TEMPERATE FRUIT PRODUCTION

(a) Climate Change

Climate change refers to a statistically significant variation in either the mean state of the climate or in its variability, persisting for an extended period (typically decades or longer). As per United Nations Framework Convention on Climate Change (UNFCCC), change of climate is attributed directly or indirectly to human activity that alters the composition of global atmosphere and which is in addition to natural climate variability observed over comparable time period.

The earth's climate, although relatively stable over the past 10,000 years or so, has always been changing, mainly due to natural causes such as volcanic activity. However, since the second half of 20th Century more rapid change has taken place, which has been attributed mainly due to human activities. Temperate region of the country, comprising mainly north-west and north east Himalayan region have witnesses a dramatic change in climate over the past 30 years, where adverse effect on temperate fruit production has been noticed.

Among different climate variables, temperature, precipitation and atmospheric CO_2 exert a pronounced effect on crop production.

(i) Temperature

Inter governmental Panel on Climate Change (IPCC) estimated that the current global mean surface temperature was about 0.42° to 0.54°C above the 1961-1990 annual average (IPCC, 2007). Himalayas are also warming and it is a matter of great concern as (i) the region has more snow and ice than any other region in the world outside the Polar caps, (ii) Himalayas are the maker of climate of much of the South Asia, (iii) to the Himalayas glaciers are connected to 9 river basins including the Gangetic basin, which alone has about 500 million population; and (iv) the Himalayas glaciers are receding, though recent study has contradicted it. Temperature increases can have both positive and negative effects on yields in different crops; however in general, it has a negative effect on temperate fruit production.

(ii) Precipitation

Uneven precipitation (*i.e.* level and timing), particularly lack of snow in winter (Vedwan and Rhoades, 2001) and prolonged drought in summer may adversely affect temperate fruit crops. However, snowfall during 2011-12 has been exceptionally high, but variable due to unpredictable western disturbances.

(iii) Atmospheric CO_2

An atmosphere with higher CO_2 concentration would result in higher net photosynthetic rates (Long *et al.*, 2006). Higher CO_2 concentrations may also reduce transpiration (*i.e.* water loss) as plants reduce their stomata apertures, the small openings in the leaves through which CO_2 and water vapour are exchanged with the atmosphere.

The net change in crop yields is determined by the balance between these negative and positive direct effects on plant growth and development, and by indirect effects that can affect production. These indirect effects have been largely ignored in the assessment of climate change effects. Indirect effects may arise from changes in the incidence and distribution of pests and pathogens (Sutherst, 1995), increased rates of soil erosion and degradation, and increased tropospheric ozone levels due to rising temperatures (Adams, 1986). Additional indirect effects may arise from changes in runoff and groundwater recharge rates, which affect water supplies, and changes in capital or technological requirements such as surface water storage and irrigation methods.

4. TEMPERATE FRUITS' RESPONSE TO CLIMATE CHANGE

In Himalayas, there has been a phenomenal increase in temperature during last 3-4 decades (Shrestha *et al.*, 1999; Liu and Chen, 2000), which is quite comparable to the global average of 0.75 per cent over last century (IPCC, 2007). Temperature increases are more during winter and autumns than during summers, and it increased with the increase in altitude (Jianchu *et al.*, 2007). For example, decadal temperature rise remains up to 0.2°C up to 2,000 m altitude, while above 2,000 m it often exceeds 0.3°C. Climate change has affected the fruit crops such as apple, European pear, cherry, walnut *etc.* more than those crops grown in the sub-temperate regions and the effect has been noticed in more than one ways.

i. Lack of winter chill hours for fruits such as apple, pear, cherry, walnut, high chilling apricots *etc.*
ii. Sudden rise in temperature during spring often result in early flowering, which may be vulnerable to spring frosts.
iii. Freezing injuries to fruit trees.
iv. Poor fruit set due to inclement weather conditions at flowering.
v. Deterioration in fruit size and quality and excessive fruit drop due to summer drought.
vi. Poor fruit colouration due to warmer summer (disturbance in diurnal temperature), shortening of growing period and excessive rain at fruit maturation.

(a) Lack of Winter Chill Hours

Most of the commercially grown temperate fruits in the Himachal Pradesh have their origin to the North Temperate Zone, which includes Europe, Asia and North America. These fruits have developed natural ability to cope up with winter by dropping their leaves in autumn and for the buds to break in spring as temperatures become warmer. One of the ways to overcome a warm spell in winter was to develop a requirement of a certain time at low temperatures before the buds would break. This stopped bud break coming too early, and the buds being destroyed when the cold weather returned. In temperate fruits, the requirement of certain hours of winter chill (below 7°C) to break dormancy in spring and then flowering is termed "chilling units or chilling hours". However, chilling requirement varies with fruit spices and also among varieties within the species. In general, temperate fruits require anywhere from 100 to 1500 chilling hours. Winter chilling requirements are high in apple (1200-1500 h), European pear (1200-1500 h), cherries (1100-1300 h), European plum (800-1100 h), pistachio, Persian walnut; intermediate in apricots (700 -1000 h), Japanese plum (700-1000 h), Kiwifruit (600-850 h), peaches (650- 850 h) and low in almond, chestnut, pecan and persimmon.

Because low temperature plays a vital role in temperate fruit production, warming of winter due to climate change in the last few decades has been adversely affecting the winter chilling requirements of temperate fruits and nut trees in their respective growing regions. As a consequence of global warming, mid hills and marginal growing regions in particular are experiencing severe reductions in available winter chill due to rise in temperature and consequently, fruit crops produced in this region of the country has been greatly influenced in different ways depending upon kind of fruit and their varieties:

- ✰ Some flowers too early risking damage from late frosts.
- ✰ Non-synchronization of flowering due to flowers' bud opening of main cultivars at different time from their pollinator(s).
- ✰ Lack of chilling may leads to the development of 'blind buds' that fail to develop in spring.
- ✰ Uneven opening of flowers over a prolonged period have lead to pollination problems.
- ✰ Poor leaf quality at flowering time may lead to poor fruit set.

The temperature ranges assumed to be appropriate for the cultivation of apple are 6-14°C on annual mean temperature and 13-21°C on mean temperature basis from April to October (Sugiura *et al.*, 2005). Chilling requirements of most apple cultivars is between 1,200- 1,500 hours to break bud rest and subsequent flower in spring. However, apple produced in India has been greatly influenced by warming over the years. Inadequate of chill makes buds to break slow in spring or not at all, particularly in the mid-hills and marginal areas. The flowering period may be 4-6 weeks rather than 1-2 weeks, with very irregular bud break all over the tree. This may mean that the cross pollinating variety's flowers are over or not yet in bloom as different varieties have different requirements.

During the second half of 20th Century, lower Kullu valley, Rajgarh in Sirmaur district, Chial in Solan district, Chindi-Churag and other mid hill areas in Mandi district of HP made a mark in apple production and farmers of these regions were greatly benefited from its cultivation, because of early arrival of fruits in the market from these regions. Thereafter, its cultivation failed due to lack of chill units in these belts as a result of rising temperature. Now, apple cultivation has almost been completely wiped out from these belts. Furthermore, most parts of districts Shimla, Kullu and Mandi once considered highly congenial for apple cultivation are now gradually and slowly becoming marginal due to warming. Climate change is also potentially threatening current production in major apple growing region of J&K and Uttrakhand. It was predicted that the favourable regions to cultivate apples will gradually move upward towards the higher altitudes of Himalayas and total area having favourable environment for its successful cultivation will squeeze by the middle of the 21st Century. Likewise, "Ecological niche" of other temperate fruits and nuts will be shifted. Already, higher reaches of north-western Himalayas such as Ladhak region of J&K; Kinnaur, Lahaul and Spiti and Pangi in Chamba district in Himachal Pradesh; and Pauri Garhwal areas of Uttrakhand are now becoming more suitable for apple cultivation. Apple produced in some of these regions, particularly those from Kinnaur and Lahaul and Spiti in Himachal Pradesh are rated best in the country. Recent studies in Himachal Pradesh indicated that higher altitude regions of Kalpa and Reckong Peo situated at 2,780 and 2,547m amsl, respectively are now better suited for apple production than any other location in Shimla district (Belsare *et al.*, 2015). Further, within Shimla district, higher altitudes of Kotkhai and Rohru had more production and better quality apples than their respective lower areas. In the past few decades, yield and quality of apple have been adversely affected due to climate change, which further may pose risk of substantial economic losses for apple industry worth an estimated Rs. 10,000 crores annually.

(b) Unseasonal and Erratic Flowering

Extreme low minimum temperatures are often associated with poor production of apricot, plum, peach and pear in many growing regions of the world. In these crops, the bud burst phase of trees is the most critical phase in relation to low temperature and frost since the most parts of the bud, especially the ovary, are very sensitive to low temperatures. Tree phenophases are becoming earlier in spring and are closely associated with rising temperatures. During the period 1948-2006, Arctic outflows were associated with spring frost injuries in these fruits world over (Quamme *et al.*, 2010). Though, this synoptic weather pattern was an infrequent event but had a great impact on production. A decrease in frequency and increase in minimum temperature of Arctic outflows appeared to be associated with the warming trends of the region during winter and early spring. In a study in south-western Cape of South Africa, the change in mean full bloom dates for three apple cultivars, namely Golden Delicious, Sayaka, and Granny Smith and one pear cultivar William Bon Chretien were correlated with temperature and precipitation changes in the same region between 1973 and 2009 (Grab and Craparo, 2011). Significant early spring (August/September) temperature increases of +0.45 degrees C/decade was found to be associated with a mean full bloom advance of 1.6 d/decade over the last 37 years. Golden Delicious apple trees have the strongest sensitivity (+4.2 d/degrees C) to climate change in the region, whilst Granny Smith apple trees have the lowest (+2.4 d/degrees C). Warming during spring results in early bud break in these fruits and therefore increasing risk of injuries to opened flowers if frost follows thereafter. In the past couple of decades, premature flowering have become a general trend in the lower and middle elevation of Himalayas, which resulted in decrease in the production of temperate fruits following the damage from exposure to frost and lower fruit set. Although, scientific data on this aspect is lacking, but certainly premature flowering pose risk to frost injuries and lower fruit set.

(c) Freezing Injuries to Fruit Trees

Freezing injuries to fruit trees may result from untimely frosts in the fall or spring as discussed above, or from exposure to extreme subfreezing temperatures in winter. Most perennial horticultural crops from the temperate regions have plant parts that freeze extra-cellularly. In these tissues, ice first forms within the extracellular spaces and water are progressively withdrawn from the cell to the extracellular sites of nucleation by diffusion along a gradient of vapour pressure (Rajashekar and Burke, 1982). Perennial plants tolerate sub-freezing temperatures by avoiding the freezing of their intracellular content through deep super cooling or by the initiation of extracellular freezing. In some plants, such as apple and peach, both responses to freezing have been observed; the water in bark tissue freezes extracellularly while water in xylem ray parenchyma, pith, and flower bud cells supercool (Pearce, 2001).

In cold desserts of J&K and Himachal Pradesh, winter temperature frequently drops to the levels that limit growing of the crops. Only few of horticultural crops can be grown in these regions that too with high risk of damage from sub-zero temperature. Apricot and apple fruit trees are being cultivated in some suitable pockets of these agro-climatic zones. However, these trees can be injured if the temperatures drop below the limit of their inherent tolerance. During 2011-12, a prolonged sub-zero winter temperatures prevailed due to frequent Western disturbances in these growing regions. In some of the areas, a temperature of -25°C or even lesser has been recorded during January 2012, which was severe enough to cause freezing injuries to apricot and apple trees. When the occurrence of –25°C temperatures is quite frequent, *i.e.* greater than 1 yr in 7, production of these crops becomes unprofitable (Quamme, 1987).

Types of freezing injuries

(*i*) **Above ground wood and bark injury:** Freezing injuries to above ground plant parts such as black heart, crotch injury, crown or collar injury, splitting of bark, killing back of young branches and twigs, injury to leaf and flower bud *etc.* has been noticed in extreme cold winters.

(*ii*) **Killing of roots:** Roots are more susceptible to winter killing in comparison to the above ground parts. Roots has lesser hardiness in autumn, however, hardiness increases as winter advances till late winter and then again decline as growth starts.

(d) Inclement Weather at Flowering

Weather plays an important role in spring-blooming fruit crops due to the combined effects on bee activity, flower opening, pollen germination, and fertilization. Weather at blossom time plays a dominant role in pollination. In apple, low temperatures (below 41 degrees F) can cause the pollen not to germinate and pollen tube growth is very slow below 51 degrees. As temperature falls to 28 degrees F., ice formation within the flower tissue can cause injury to fruit finish. At 27°C degrees and below, the styles and ovules can be killed, preventing fertilization. Wind do play significant role in cross pollinating tree fruit. However, it does have an adverse effect in insect activities that are required for cross-pollination. If cold weather prevails from green-tip to half-inch green, while the pollen is in the process of production, the quality of the pollen may be seriously reduced. Hot and dry weather will reduce the stigma receptivity. When the stigma is receptive to pollen germination, it glistens with a sugary exudate. Once it starts to dry and turn brown, it is no longer receptive to pollination. Bee flight is reduced by rain, wind, and cool temperatures, therefore, pollination may fail even though all other factors are positive.

Global warming has changed weather pattern all over the world. During the last few decades, inclement weather conditions during the flowering period of temperate fruits in spring prevailed quite frequently. Low temperature, rainfall and hailstorms just before, during, and after flowering represent a serious constraint to fruit production in more than one way:

(*i*) Irregular flower bud opening and prolonged flowering period.

(*ii*) Low temperature injury to flowers.

(*iii*) Rainfall washes away pollen or pollen become heavier.

(*iv*) Low or no bees' activity due to low temperature, rains or cloudy weather.

(*v*) Irregular flowering cause non-synchronization of flowering period of cross pollinating varieties.

(*vi*) Poor pollination due to all above factors; slow pollen germination and seed abortion due to low temperature and finally poor or no fruit set.

(e) Lack of Soil Moisture in Summer

Water stress is one of the most important factors limiting production in temperate fruit crops. Global climate change has further exacerbated the water scarcity problem. The most critical period of water requirement however, is during the summer months, when active vegetative growth and fruit development takes place. In the north-western Himalayas, winter rains and snow maintain soil moisture supply to the fruit crops at least until April and thereafter, water supply become scare due to lesser summer rains in monsoon type of climate. During the past two decades, climate change has reduced the winter rainfall and snow, and water supplies from shrinking glaciers have already been depleted, which has further aggravated the problem of water stress in summer. Water stress in summer affects the fruit trees in the following ways:

(*i*) Annual shoot growth and leaf area are reduced.

(*ii*) Early spring drought affect fruit set.

(*iii*) Fruit growth is slower, resulting in smaller fruits.

(*iv*) Mid and late summer drought trigger June as well as pre-harvest fruit drop.

(*v*) The incidence fruit cracking in fruits like cherry increase when dry spell is followed by rains.

(*vi*) Water stress leads to boron deficiency, which affect flowering and fruit development. Early boron deficiency leads to blossom blast in stone fruits and in apple corky spots appear on the fruit surface.

(*vii*) Water stressed fruits are less juicy and generally has lower TSS.

5. CLIMATE CHANGE STRATEGIES

The strategies identified for the horticulture industry are:

(a) Adaptation

One of the best action plans for fruit industry in Himalayas is to explore the possibilities to adapt fruit production to the changing climate. Farmers of different zones of Himalayas should be made aware of the adaptive horticultural production systems, which are less vulnerable to climate change and climate variability. Scientists and farmers should appropriately adjust the production system to current choices in response to future climate.

(i) Diversification

The choice of the fruit crop in a particular zone should change as per existing climate; this would avoid the risks in the existence production system and grab the opportunities of growing new crop suitable to the existing climate. In mid-Himalayan regions grow only those fruits and varieties, which have low chill requirement.

(ii) Breeding of low chill cultivars

Fruit scientists need to adapt to changing conditions by breeding tree cultivars for lower chill requirements. Farmers should select low chilling varieties in each "Ecological niche" of the temperate fruits (Table 1).

Table 1: List of Low Chilling Fruit Cultivars

Crop	*Low Chill Varieties*	*Current Potential Growing Regions*
Apple	Anna, Dorsett Golden, Gala, Michal, Tropical Beauty, Shlomit, Parlin's Beauty, Fuji	Mid hills of North-Western Himalayas, and all the sub-temperate regions of North-Eastern Himalayas. (cultivars Anna and Dorsett Golden also suitable for lower hills)
Apricot	New Castle, Ema	Mid hills of North-Western Himalayas, and all the sub-temperate regions of North-Eastern Himalayas.
Peach	Sharbati, Shan-a-Punjab, Partap, Florda Gold	Low hills and Northern plains
Pear	Kieffer, LeConte, Patharnakh, Kosui, Hosui, Shinseki etc.	Mid hills of North-Western Himalayas, and all the sub-temperate regions of North-Eastern Himalayas.
Plum	Alucha Purple, Titron	Low hills and Northern plains

Among nuts, almonds, walnuts and pecans are more suitable for dry temperate, temperate and sub temperate climates, respectively.

(iii) Grow late flowering cultivars

In the context of global warming, the general trend towards growing early flowering cultivars of many temperate tree species is likely to result in an increased risk of damage from exposure to frost. In this context, growing of late blooming fruit cultivars to escape freezing injury would be an appropriate method to adapt to such a climate change scenario. In Turkey, apple cultivars Starkingspur Golden Delicious, Granny Smith and Canada Reinette were found to be the least damaged due to late spring frost at various flower buds stages from tight cluster to first bloom (Aygun and San, 2005). Lesser damages to the buds due to spring frost has been established in late flowering cultivars of apple, *viz.*, Golden Delicious, Smoothee, Royal Gala and Ligol and pear cv. Concorde in Poland (Lipa *et al.*, 2008).

(b) Efficient Water Management

In order to ensure the supply of requisite quantity of water at critical stages of growth and development of fruit crops, efficient management of water is required. In rain fed areas, three essential components of water management are:

- *In situ* moisture conservation.
- Water harvesting.
- Efficient utilization of conserved/harvested water through improved cultural practices.

(i) *In situ* soil moisture conservation

Under rainfed conditions, the rain water and its run-off should be conserved rightly in the field for the optimum growth, development and production in fruit crops.

a. **Mulching:** Mulching helps in conserving soil moisture, be it mulching with organic or inorganic material. All the organic residues, *viz,*. straw, hay, crop residue, leaves, saw dust and organic wastes, *viz.*, farm yard manure, spent mushroom compost *etc.* and inorganic mulches, *viz.*, black polythene can be useful in conserving soil moisture.

b. ***In situ* rainwater harvesting:** In fruit plantations, soil working techniques like 'Crescent Bund with Open Catchments Pits', 'Trench Systems', 'V-ditch' *etc.* provide satisfactory *in-situ* harvest of rain water (Sharma and Singh, 2014). These techniques are found to be useful for conserving soil moisture in the root zone of trees for the dry months.

(i) Water harvesting

Two natural water resources that can be effectively harvested are rainfall and low water springs (as low as 1 to 30 l/min) and rivulets. About 10 to 14 per cent of total rainfall, depending on soil and rainfall characteristics may be lost as surface runoff. Water harvesting can be done through *in situ* rainwater harvesting as discussed above and water harvesting in farm ponds.

(ii) Efficient utilization of conserved/harvested Water

The water conserved/harvested from natural resources must be used very efficiently for fruit crops. The water should be possibly be used through pressurized irrigation system *e.g.*, drip irrigation. Drip irrigation system is ideal for fruit crops, because it enable to achieve higher water use efficiency and also help in maximizing crop production within limited water resource.

a. **Select drought tolerant rootstocks:** Lack of moisture in soil in rainfed region have adverse effect on fruit production. Rootstock plays vital role in water use efficiency and consequently on the growth and development of plants. Rootstocks that aid efficient water use by the tree will become more important in future. In rainfed regions, fruit plants should be established on drought tolerant rootstocks (Table 2).

Table 2: Drought Tolerant Rootstocks for some Temperate Fruits

Fruit Crop	*Drought Tolerant Rootstock(s)*	*Reference*
Apple	MM 111, EMLA 9	Fernandez *et al.* (1997)
Cherry	Victor, sour cherry, Giesela 5, Edadriz, Maxima 14, Mahaleb	Goncalues *et al.* (2008)
Peach	?	Mannini and Gallina (2001)
Pear	Kainth, BA 29	Sharma and Sharma (2008)
Plum	Peach-almond hybrid GF 677	Ryazanova (2001)

A survey (Rana *et al.*, 2009) conducted in three apple growing districts *viz.* Kullu, Shimla and Lahual and Spiti representing different elevation to examine the perception of farmers about for climate change revealed that the temperature in apple growing regions of Himachal Pradesh showed increasing trends whereas, precipitation showed decreasing trends in the regions. In this study, the chill units calculated (data not given) showed decreasing trends of chill units up to 2,400 m amsl from Bajaura in Kullu at 1,221 m amsl to Sarbo in Kinnaur at 2,400 m amsl. The Dhundi station situated at 2,700 m amsl showed increasing trend of chill unit at the rate of 25.0 CUs per year. The increasing trends of chill unit at 2,700 m suggested that area is becoming suitable for apple cultivation in higher altitude. There are few success stories related to the adaptation strategies which growers of Himachal Pradesh have successfully employed to manage their enterprises in an already changing climate:

(*i*) In recent years, the changes in climate has resulted in decreased productivity of fruit crops in different growing zones and farmers in marginal areas has shifted to more dependable and profitable cropping pattern. Farmers of lower Kullu valley have abandoned apple cultivation due to lack of chilling units since long back and shifted to plum in 1980's and 1990s and lately to pomegranate and vegetable cultivation. Simultaneously, farmers of higher region of the valley realized that climate of their area is more suitable to apple. This has led to the shift of apple cultivation from lower to the higher elevation.

(*ii*) Similarly, farmers of other marginal areas of HP have abandoned apple cultivation as climate change made apple farming non-profitable and switched to stone fruits and "off season" vegetable crops.

(*iii*) In most parts of Kinnaur district of Himachal Pradesh, apple cultivation has completely replaced almond and traditional agricultural crops like potato in the last few decades, which has immensely benefited the people in this tribal area, as apple is highly enumerative.

(*iv*) Lately, farmers of Lahaul and Spiti and Pangi in Chamba have successfully shifted to apple production system from the traditional potato farming.

The average land use per farm in Lahual and Spiti showed more than two percent shift towards apple cultivation but it showed reverse trend in other apple growing regions. The income of the farmers increased more than 10 percent in Lahual and Spiti whereas, it showed a decrease of more than 27 percent in Kullu and Shimla districts from fruits in recent decade compared to 1995. These examples clearly reflect that apple cultivation is expanding to higher altitude.

(c) Mitigation

(i) Rest breaking

Temperate fruit trees do not get sufficient chilling in the warmer winter. Insufficient chilling results in delayed foliation, protected bud burst and flowering, thus necessitating artificial bud break. Earlier, dinitro-o-cresol (DNOC) was in use in many warmer apple growing regions of the world receiving insufficient chilling, but its use now has been discontinued. Now hydrogen cynamide (HCN) has become the replacement for this purpose. It has been very effective in apple and other fruits. In South Africa, mineral oil at 4 per cent plus Dormax at 1 or 2 per cent were sufficient to break dormancy in "Golden Delicious" and "Royal Gala" apples (Sagredo, 2008). Thidiazuron (TDZ) has also shown the ability to break dormancy in apple. In general, all the chemicals (DNOC, HC and TDZ) enhanced final vegetative bud burst as compared to control. The treatment compressed and advanced flowering period, but this effect was not evident during warmer summer. The treatments also synchronized flowering between the trees of two cultivars. In a recent study (Kunz and Blanke, 2011), analysis and correlation of the meteorological data from the last 50 years showed two distinct climate phases in Klein-Altendorf, an earlier 30 year period (1958-1987) with a temperature of -0.42°C below the long term, 50-year average of 9.40°C, followed by a 20 year period of a +0.60°C temperature rise (1988 to date). A comparison of the phenological data of phase 2 (1988 to date) with phase 1 (1958-1987) showed 10 days earlier full bloom, 11 days earlier harvest and 4 days earlier leaf drop in the later phase, resulting in a 5 day longer growing period for cultivar 'Golden Delicious' at Klein-Altendorf. In Himachal Pradesh, the application of 1.0 per cent Dormax in combination with 4.0 per cent HMO in Delicious apple trees induced enhanced and uniform flowering under low elevation apple orchards. Time of application was found to be critical for enhancement of flowering. The application in the last week of January was found to be most effective, whereas, application after 15th February was ineffective. Significant positive correlation ($r = +0.70$) was observed between per cent bloom and yield with application of Dormax in the last week of January. Whereas, negative correlation ($r = -0.365$) resulted with late application, *i.e.* first week of March.

(ii) Delay bloom

In temperate fruits, early flowering as a result of rising temperature in early spring is a matter of great concern especially in frost affected pockets, because it directly affects fruit set and yield. In some years, severe and repeated frost can cause 100 per cent fruit loss and can turn in an "off year". Attempt to delay bloom by evaporative cooling and chemical means in some temperate fruits in frost affected areas have yielded positive results.

- ☆ **Tree misting:** Evaporative cooling of flowering buds in spring has a delaying effect on bloom in stone fruits (Saini and Sharma, 2010), which can significantly reduce the risk of freezing injury. In this practice, trees are sprayed with water during day time at frequent interval from bud swell stage until bloom. Frequency of tree misting depends upon the evaporation rate, which is done as and when water film on the aerial parts of the tree dries up. Evaporation of water from the shoots cools the buds and therefore delays their break.
- ☆ **Use of PGRs:** Use of some plant growth regulators and other chemicals has potential to increase cold hardiness and delay bloom in spring to provide freeze protection. Ethylene releasing compounds and gibberellic acid (GA_3) has been used to increase cold hardiness and delay bloom in many deciduous fruits. Autumn application of GA_3 and ethrel cause delaying of bloom in spring in apples and stone fruit, which can be a useful practice to increase fruit-set in frost affected areas. Another advantage of PGRs treatment is that it facilitates better cross-pollination among varieties having different (non-synchronous) bloom period. For example, blooming of Red Beaut plum can be delayed by 4-7 days with fall application of GA_3 at 200 ppm, and Ethrel at 200 ppm to help in better synchronization of flowering with cultivar Santa Rosa, inter-planted for cross pollination (Saini and Sharma, 2010). Concentration and time of application however, depend upon fruit crop and environment.

(iii) Fruit set improvement

Inclement weather during flowering often limit fruit set and production in temperate fruits. In this regard, results with the use of some PGRs have been encouraging in pear. Use of GA_3 at 10-20 ppm at bloom can increase fruit set in pear, particularly in those cultivars which have strong tendency towards parthenocarpy, *e.g.* cultivars Bartlett and Conference (Tromp and Wertheim, 1980). Furthermore, results are better when conditions are not ideal such as low temperature before or during bloom. In Himachal Pradesh, combined application of GA_3 at 10 ppm + BA at 5 ppm has been beneficial in improving fruit set in pear cultivars Flemish Beauty (Negi and Sharma, 2005) and Baggugosha (Sharma and Singh, 2008). Fruit set of apple can be improved with the bloom application GA_3 at 50 ppm (Singh and Chadha, 1990) however; response is erratic and unpredictable in comparison to pear.

(iv) Fruit quality improvement

Climatic factors affect shape and colour of fruits in apple. Apple elongation and calyx development are promoted by low temperatures during early fruit growth. Delicious apples are characterized by "typeyness", *i.e.*, higher length/diameter ratio and prominent calyx lobes. Because of its shape and colour, it fetch premium price in the market. However, these apples have lost its their peculiar shape due to warming and fruits produced in the warmer region are more round and flat in shape. It has been observed that with the increase in average mean temperature during April (bloom and post bloom stage) there was a decrease in L/D ratio in Delicious apples (Mankotia, 2010). Regression equation (Y = - 0.0102 x + 1.0495) showed that the variation accounted by temperature on fruit shape was 19.7 per cent. However, grouping of fruiting years into above 0.09 and below 0.90 ratio showed high co-efficient of determination in above 0.90 L/D years. Further, the regression equation (Y = - 0.0277 x + 1.3214) indicates decrease in L/D ratio with the increase in temperature and the variation accounted by temperature was 66.2 per cent. Promalin® at 60 ppm, as a single dose or 15 + 15 ppm as split dose applied at 'King Blossom' stage improved the fruit shape by developing well lobed oblong fruits (Mankotia *et al.*, 2004). Foliar application of ethrel at 1000 ppm (4 ml ethrel/litre of water) to mature fruits just one week before expected harvest is a standard practice for improving surface colour on Delicious apples in the marginal areas of Himachal Pradesh. In order to control fruit drop caused by ethrel, NAA at 10 ppm (Planofix® at 2.2 ml/10 litre) is added to the spray solution. However Ethrel treated should not be stored but only meant for fresh market.

6. CONCLUSION

Temperate region of the country has witnesses an increase in winter temperature and changes in precipitation as a result of global warming, which has impacted the fruit crops. Some of the adverse effects noticed includes, lack of winter chill hours for fruits such as apple, pear, cherry, walnut, high chilling apricots *etc.*, unseasonal and erratic flowering, poor fruit set due to inclement weather conditions at flowering, prolonged summer drought, and deterioration in fruit quality and colouration. Apple cultivation has almost been completely been wiped out from some mid hills and marginal areas of Himalayas as it is no more profitable and farmers in these regions have shifted to other crops. Its cultivation is gradually moving upward towards the higher altitudes and it is predicted that "Ecological niche" of other temperate fruits and nuts will be shifted. Crop diversification as per existing climate, breeding of fruit cultivars having lower chilling requirements and efficient water management should be the adaptation strategies to the changing climate. Application of dormax at 1 per cent in spring has been beneficial in many temperate fruit crops receiving insufficient chilling in the warmer growing regions. In the changing climate, improved fruit set and quality can be obtained by chemical means in different temperate fruits.

7. REFERENCES

Adams, R.M. 1986. Agriculture, forestry and related benefits of air pollution control. *Amer. J. Agric. Econ.* **68** 885-894.

Aygün, A. and San, B. 2005. The late spring frost hardiness of Some apple varieties at various stages of flower buds, *Tarim Bilimleri Dergisi*, **11**: 283-285.

Belsare Chaitanya, R., Sharma, N., Sharma, D.P. and Mankotia, M.S. 2015. Impact of different locations and altitudes on fruit quality parameters of apple cv. Starking Delicious in Himachal Pradesh. *Green Farming*, **6** (September-October).

Fernandez, R.T., Perry R.L. and Flore, J.A. 1997. Drought response of young apple trees on three rootstocks. II. Gas exchange, chlorophyll fluorescence, water relations, and leaf abscisic acid. *J. Amer. Soc. Hort. Sci.* **122**: 841-848.

Grab, S. and Craparo, A. 2011. Advance of apple and pear tree full bloom dates in response to climate change in the southwestern Cape, South Africa: 1973–2009. *Agricultural and Forest Meteorology,* Volume: 151, Issue: 3, Publisher: Elsevier B.V., pp. 406-413.

Goncalves, B., Correia, C.M., Silva, A.P., Bacelan, E.A., Santos, A., Ferreira, H. and Moutinlo-Pereira, J.M. 2008. Variation in xylem structure and function in root and stems of scion-rootstocks combinations of sweet cherry trees (*Prunus avium* L.). *Tree Struct. Func.* **21**: 121-130.

IPCC (2007). *Climate Change 2007: Impacts, Adaptation and Vulnerability*. Contribution of Working Group II to the Fourth Assessment Report of the Intergovernmental Panel on Climate Change, Cambridge University Press, Parry (M.L., Canziani, O.F., Palutikof, J.P., van der Linden, P.J. and Hanson, C.E., Ed.), ISBN 978-0-521-88010-7.

Jianchu X., Shrestha, A., Vaidya R., Eriksson, M. and Hewits, K. 2007. The melting Himalayas, ICIMOD Technical paper, ICIMOD, Nepal.

Kunz, A. and Blanke, M.M. 2011. Effects of global climate change on apple cv. 'Golden Delicious' phenology– based on 50 years of meteorological and phenological data in Klein-Altendorf. In: T.L. Robinson (Ed.). *Acta Horti.*, **903**: 1121- 1126.

Lipa, T., Lipecki, J., Janisz, A., and Sienkiewicz P. (2008). Effects of spring frosts in selected apple and pear orchards in the Lublin region in the years 2000, 2005 and 2007. *Acta Agrobotanica,* **61**: 131-139.

Liu, Xiaodong and Chen, Baode 2000. Climatic warming in the Tibetan Plateau during recent decades. *Int. J. Climatology,* **20** : 1729-1742.

Long, P.S., Ainsworth A.E., Leakey, D.B.A., Nosberger, J. and Ort, R.D. 2006. Food for thought: Lower-than expected crop yield stimulation with rising CO_2 concentrations. *Science,* **312** : 1918-1921.

Mankotia, M.S. 2010. *Annual Report, RHRS, Mashobra (2009-2010)*, pp. 30-31.

Mankotia, M.S., Chauhan, P.S. and Sud, A. 2004. Role of Promalin (GA 4+7 +BA) application on fruiting and quality of Delicious apples. **In:** *Temperate Horticulture: Current Scenario.* D.K. Kishore, S.K. Sharma and K.K. Pramanick (Eds), pp. 249-255.

Mannini, P. and Gallina, D. 2001. Effect of irrigation on peach varieties with different precocity [*Prunus persica* (L.) – Emilia –Romagna]. *Rivista di Frutticoltura,* **63**: 75-78.

Negi, N.D. and Sharma, N. 2005. Growth, flowering and cropping response of Flemish Beauty pear to bloom application of gibbrellic acid and benzyl adenine. *Acta Hort.* **696**: 295-98.

Pearce, R.S. 2001. Plant freezing and damage. *Ann. Bot.* **87**: 417-424.

Quamme, H.A. 1987. Low temperature stress in Canadian horticultural production - An overview. *Canadian J. Plant Sci.* **67**: 1135-1149.

Quamme, H.A., Cannon, A.J., Neilsen, D., Caprio, J.M. and Taylor, W.G. 2010. The potential impact of climate change on the occurrence of winter freeze events in six fruit crops grown in the Okanagan Valley. *Canadian J. Pl. Sci.* **90**: 85-93.

Rajashekar, C.B. and Burke, M.J. 1982. Liquid water during slow freezing based on cell water relations and limited experimental testing. **In**: Plant Cold Hardiness and Freezing Stress, Vol 2. PH Li, A Sakai, eds, Academic Press, New York, pp. 211-220.

Rana, R.S., Bhagata, R.M., Kalia, V. and Lalb, H. 2009. Impact of climate change on shift of apple belt in Himachal Pradesh. In: *ISPRS Archives XXXVIII-8/W3 Workshop Proceedings: Impact of Climate Change on Agriculture*, pp. 131-37.

Ryazanova, L. 2001. Drought resistance evaluation of myrobalan plum cultivars. Scientific works of the Lithuanian Institute of Horticulture and Lithuanian University of Agriculture. *Horticulture and Vegetable Growing- Babtai: Lithuanian Institute of Horticulture,* **20** : 58 - 62.

Saini, Pooja and Sharma, N., 2010. Effect of plant bioregulators and evaporative cooling on fruit set, yield, and quality of plum (*Prunus salicina* Lindl.) cv. Red Beauty *Hort. J.* **42**: 220-24.

Sagredo, K.X. 2008. Effect of rest-breaking and fruit thinning treatments on reproductive development in apple. Thesis [Ph.D(Agric) (Horticulture)]- University of Stellenbosch, 2008. P.143. UR I: //hdl. handle. net/10019.1/1260.

Sharma, D. P., Sharma, N., Bawa, R. and R. Kaler. 2005. Potential of apricot growing in the arid-cold dessert region of north-western Himalayas. *Acta Hort.* **696**: 61-64.

Sharma, N., Sharma, D.P. and Sharma, D.D. 2011. Evaluation of prunes in the dry temperate region of Himachal Pradesh. *Himachal J. Agric. Res.* **37**: 45-49.

Sharma, N., Gautam, D.R.; Jindal, K.K. and Sharma, D.P. 2003. The pear industry of Himachal Pradesh in India. *Acta Horti*. 596: 97-99.

Sharma, N. and Singh, K. 2008. Effect of plant growth regulators on fruit set, yield and fruit quality in pear cv. Baggugosha. *Asian J. Hort. Sci.* **3**: 352-56.

Sharma, N. and Singh, Karan 2014. Anatomical and biochemical characteristies of olives as influenced by *in-situ* moisture conservation during monsoon. *Indian J. Hort.* **71**: 35-39.

Sharma, S. and Sharma, N. 2008. Rootstocks affect growth, water relations, gas exchange and anatomy of Flemish Beauty pear under stress. *J. Hort. Sci. Biotech.* **83**: 658-62.

Shrestha, B.A., Wake, P.C., Mayewski, A,P. and Dibb, E.J. 1999. Maximum temperature trends in the himalaya and its vicinity: An analysis based on temperature records from Nepal for the period 1971–94. *J. Climate*, **12**: 2775-86.

Singh, R.P. and Chadha, T.R. 1990. Effect of bloom spray of growth regulators on flowering and productivity of Delicious apples. *Indian J. Hort.* **47**: 392-96.

Sugiura, T., Kuroda, H., Sugiura, H. and Honjo, H. 2005. Prediction of effects of global warming on apple production regions in Japan. *Phyton Horn.* **45**: 419-422.

Sutherst, R.W. 1995. The potential advance of pests in natural ecosystems under climate change: Implication for planning management. **In**: *Impacts of Climate Change on Ecosystems and Spicies, Terrestial Ecosystems,* [Pernetta, J.C., Leemans, R., Ekler, D. and Humphery, S. (eds.)]. IUCN, Gland, Switzerland, pp. 83-98.

Vedwan, N. and Rhoades, R.E. 2001. Climate change in the Western Himalayas of India: A study of local perception and response. *Climate Res.,* **19**: 109-117.

Chapter 9

Climate Change and Temperate Fruit and Nut Production

K.S. Verma

Dean, College of Forestry,
Dr Y.S. Parmar University of Horticulture and Forestry,
Solan – 173 230, H.P.
E-mail: ksvblp16@gmail.com

1. INTRODUCTION

Climatic changes indicated through increased temperatures, erratic precipitation, melting of glaciers, variation in river flows and unprecedented changes in the natural habitats of flora and fauna has now been accepted globally. All researches and tools in their hands to predict the extent of regional and global climate change in terms of physical weather elements using different climate change scenarios have predicted that temperature during the last one hundred years have risen by 0.74°C and will increase in the range of 1.5 – 3.5°C by the end of 21st Century.

General climatic requirements for temperate fruits are that the winter temperature must not be so cold that kill the trees. The winter must be cold enough to give buds adequate chilling to break winter rest, the growing season must be long enough to mature the crop and the temperature and light during the growing season must be adequate to develop fruit of good quality. Those fruits classified as temperate have one major factor in common; they need a distinct cold dormant period to crop satisfactory. In Himachal Pradesh, apple is the only commercial temperate crop.

2. CLIMATIC SCENARIO OF NORTH-WESTERN TEMPERATE HIMALAYA

Rising global temperature has triggered large-scale changes in the energy exchange processes (radiative forcings) affecting the atmospheric circulation and precipitation patterns (Fallot *et al.*, 1997; Zhai *et al.*, 1999; Beniston, 2003). The ecosystems which are most vulnerable to the specter of climate change are high mountain areas such as the Alps, the Rockies, the Andes and the Himalayas, etc; (Beniston, 2003). Among these, the Himalayan region remains poorly studied, largely because of inadequate meteorological and hydrological database. The Himalayas in the Indian Sub-continent, act as greatest mountain barrier on the earth, where polar, tropical and mediterranean influences interact and affect the monsoon systems.

The North-Western Himalaya, covering the states of Jammu and Kashmir and Himachal Pradesh with the river basins of the Jhelum, Satluj, Chenab, Ravi, Beas are bound by longitudes 72°E – 80°E and latitudes 30°N – 37°N. Altitudes in the region vary from about 350 m. in the Shivalik in the south to about 8000m in the Karakoram Himalayas in the North. The temperate climate prevails beyond the mid hill regions ranging from 1,800m to 8,000 msl.

Air temperature is generally recognized as a good indicator of the state of climate globally because of its ability to represent the energy exchange process over the earth's surface with reasonable accuracy (Thapliyal and Kulshreshta 1991). In the North Western Himalayan Region, very limited number of studies explaining the state of climate change in different mountain states *e.g.* Himachal Pradesh, J and K and Uttarakhand have been done for the want of adequate instrumental data.

3. LONGTERM TRENDS IN TEMPERATURE AND PRECIPITATION

(a) Temperature

Long-term trends in maximum, minimum and mean annual air temperatures across the north-western Himalayas (NWH) during the twentieth century (Table 1) have revealed significant rise with winters warming at a faster rate (Bhutiyani *et al.*, 2007). For the NWH as a whole, on the century scale, the annual warming rate is estimated to be about 1.6°C, with the minimum temperature rising relatively at lower pace than the maximum temperature.

Dissimilarity in physiographic characteristics of different regions in the NWH region has caused varied increase in the rate of increase of temperature (Table 2). Shimla in Hiamchal Pradesh, experienced significantly higher rate of warming in comparison with other locations of Srinagar and Leh. Regions those fall in lower-snow climatic zone are warming at higher rate.

(b) Precipitation

Precipitation in NWH occurs under the influence of westerly disturbances during the month of October to May and due to south west monsoon from July to September. Different parts of the region experience varied climate. The extreme eastern part with

Table 1: Results of Trend Analysis of Annual, Winter and Monsoon Precipitation in the NWH

Sl.No.	Station	Data Span	Trend Analysis	
			Mann-Kendall's Non-Parametric Text	Linear Regression Coefficient 'b'
a) Total annual precipitation				
(i)	Shimla	1966-2006	(–)*	(–)*
(ii)	Srinagar	1901-1989	(+)*	(+)*
(iii)	Leh	1901-1989	(–)*	(–)*
(iv)	NWH	1866-2006	(–)*	(–)*
b) Total winter precipitation				
(i)	Shimla	1966-2006	(–)	(–)
(ii)	Srinagar	1901-1989	(+)	(+)
(iii)	Leh	1901-1989	(–)	(–)
(iv)	NWH	1866-2006	(+)	(+)
c) Total monsoon precipitation				
(i)	Shimla	1966-2006	(–)*	(–)*
(ii)	Srinagar	1901-1989	(–)	(–)
(iii)	Leh	1901-1989	(–)	(–)
(iv)	NWH	1866-2006	(–)*	(–)*

(+), increasing trends; (–), decreasing trend.

* Significant at 95 per cent confidence level.

Table 2: Rates of Increase of Winter, Monsoon and Annual Air Temperatures in ºC in the Last Century, Computed by Linear Regression Slope (b)

Station	Season	Winterb X 100	Monsoonb X 100	Annualb X 100
Shimla	Mean maximum	2.6*	2.8*	2.4*
	Mean minimum	1.0*	-0.01	0.5
	Average annual	1.8*	1.5*	2.0*
Srinagar	Mean maximum	1.1*	0.2	0.5
	Mean minimum	1.2*	0.2	1.0*
	Average annual	1.1*	0.2	0.8*
Leh	Mean maximum	1.3*	1.0*	1.7*
	Mean minimum	0.4*	1.1*	1.3*
	Average annual	0.6*	1.1*	1.6*
NW-Himalaya	Mean maximum	1.7*	1.3*	1.6*
	Mean minimum	1.7*	0.4*	1.1*
	Average annual	1.7*	0.9*	1.6*

*Significant at 95 per cent confidence level.

little vegetative cover near Tibet experiences a cold dry climate with limited precipitation. The western part is endowed with comparatively moist weather with moderate precipitation, evergreen to semi-evergreen forest cover with many rivers and valleys. Annual winter snowfall is at a maximum in the Pir Panjal Range and it decreases as one goes northwards towards the Great Himalaya, Zanskar, Ladakh and Karakoram ranges. In higher altitudes, temperature remain sub-zero for the major part of the year and the winter precipitation is mostly in the form of snow. Precipitation due to the monsoon is highest in Shiwalik and the Pir Panjal ranges and it reduces northwards into Great Himalaya, Zanskar, ladakh and Karakoram ranges (Rakhecha *et. al.*, 1983).

4. FUTURE TEMPERATURE SCENARIO IN HIMACHAL PRADESH

Winter temperatures for the mid period 2021-2050 based on hadRM3 Model Scenario A1B (Table 3) reveals that the maximum temperature and minimum both will increase in all the districts. Av. maximum temperature will rise by 2.17°C and minimum by 2.77°C.

Table 3: Projected Winter (Nov.-Feb) Temperatures in Himachal Pradesh for Mid Period (2021-2050) Based on HadRM3 Model Scenario AIB

	T_{max}(°C)	$T_{min.}$(°C)	Average (°C)
	Sub-temperate – Temperate region		
Shimla	2.51(2.75)	2.54(2.38)	2.52(2.56)
Kullu	2.04(1.85)	2.78(2.53)	2.41(2.19)
Lahaul and Spiti	2.24(2.14)	2.82(2.50)	2.53(2.32)
Kinnaur	2.06(1.91)	2.71(2.53)	2.38(2.22)
Chamba	2.00(1.98)	2.99(2.37)	2.49(2.17)
Average Temp.	**2.17(2.13)**	**2.77(2.46)**	**2.47(2.29)**

* Figures in brackets are average values of Dec-Feb months only.

5. IMPACT ON AREA AND PRODUCTION OF TEMPERATE FRUITS

Horticulture in Himachal Pradesh has grown tremendously. Fruit cultivation, vegetable growing as well as floriculture are the major activities practiced by the farmers. The area under fruit crops during 1950-51 was just 1,200 ha, which has increased to about 1,97,446 hectare during 2006-07 and to 2,24,352 hectare during 2014-15.

Temperate fruit *viz.* apple, plum, peach, apricot, pear, cherry, kiwi, persimmon, pomegranate and strawberry etc. are cultivated in wet and dry temperate areas. Nuts and dry fruits *viz.* green almond, almond, walnut, picannut, hazelnut are also common but are at minor scale.

(a) Area and Production Trends

Apple cultivation in Himachal Pradesh has experienced progressive increase in land area brought under apple since 1965-70. But during the decade of 1985-95, the progress in terms of production came almost to a point of halt despite area under apple cultivation continued to increase (Fig. 1). The period thus, can be termed as a "sluggish growth period" in the history of apple cultivation in Himachal Pradesh. The land area under apple plantation continued to grow between 7000 to 9000 ha per five year period since 1970-75 to 1990-95. Experiencing a very slow increase in total production during the decade of 1985-95, a tremendous increase in acreage under apple, which was more than double over the past increase, was observed in 1995-2000. An additional area of 21,000 ha was brought under plantation. However, fruit production declined to a low of 249.09 thousand tones from the level of 274.62 thousand tonnes. During the period 2000-05, the area increased marginally compared to previous increase but the increase in production was all time high. The quantum jump of about 133.11 thousand tonnes obtained during 2000-05 over 1995-2000 was the highest since systematic cultivation of apple was started in Himachal Pradesh during 1965-70 (Fig. 1).

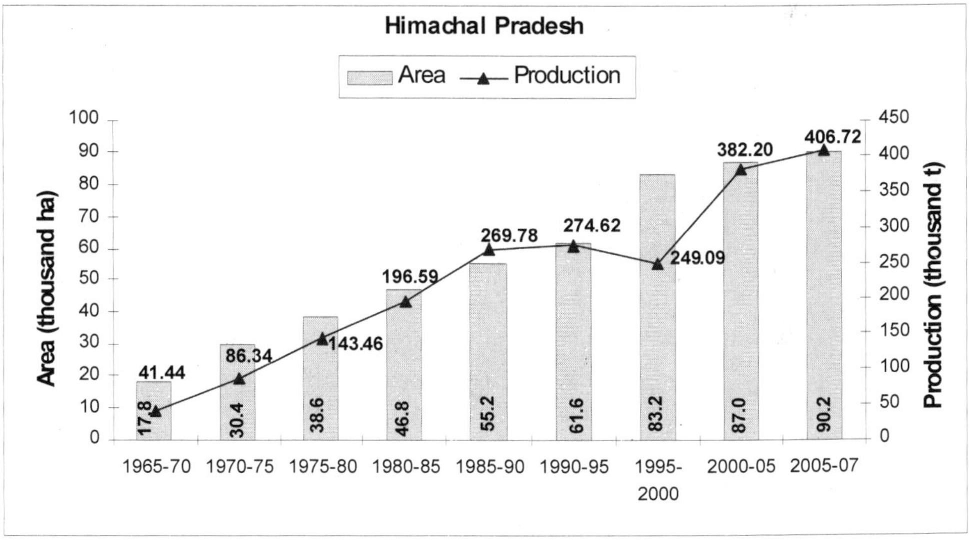

Figure 1: Land area under apple and production trends in Himachal Pradesh

Apple production in the state has continued to grow increasingly. It was 41.44 thousand tonnes in 1965-70 which rose to 406.76 thousand tonnes in 2005-07. The area under apple during the above period increased from 17.8 thousand hectare to 90.2 thousand hectares. Precisely saying, there was an increase of about 10 times in apple production and of about 5 times in area under apple plantation. Also apple cultivation experienced a "sluggish growth phase" or even negative growth in fruit production during 1990-2000. Climate studies have also revealed the decade as the hottest and top hottest years of the century were observed during this decade. Hence, it could be inferred that apple cultivation may experience adverse growth in some regions due to negative influences of global warming on the regional climate. This negative effect may enhance unless appropriate mitigation and adaptation measure are put in place, in time.

During the period 2004-05 to 2013-14, though there has been a steady increase in the area that was 107.686 thousand hectares in 2013-14 compared to 86.20 during 2004-05. However, the production remained highly unstable and reached as low as 275 thousand tonnes during 2011-12. The highest production was 738 thousand tonnes during 2013-14. The productivity, however, during this period ranged from 2.65 mt. tonnes (2011-12) to 6.8 mt. tonnes during 2013-14 (Table 4).

Other temperate fruits except pomegranate did not show any appreciable change in their area and production. Similarly, among the nut crop namely, almond, walnut, picannut and hazelnut, the situation is almost static (Table 5).

(b) Apple Productivity Trends

Apple productivity has been determined on five year basis, since 1965-70 as an average value for entire state whereas since 1981-85 for different production locations/ districts. Maximum productivity of 4.89 t ha^{-1} was achieved during 1985-90 which declined to 2.99 t ha^{-1} in 1995-2000. Later, during 2000-05, productivity regained its upward trend showing its value of 4.39 t ha^{-1} and the current state productivity was 4.51 t ha^{-1} during 2005-07. However, it is much less than the well managed apple orchards which are giving an average productivity of 8 – 9 t ha^{-1} in Shimla.

6. FUTURE IMPACTS OF CLIMATE CHANGE SCENARIO AIB ON TEMPERATE HORTICULTURE PRODUCTION IN MEDIUM TERM (2050) PERIOD

Due to the increase in winter temperature both maximum (2.00-2.50°C) and minimum (2.54-2.99°C) in the entire temperate region coupled with the rainfall pattern, temperate fruit production will have following differential impacts:

☆ Shimla i.e., wet temperate region increased temperature and less rainfall in winters especially January-June will experience following adverse influences:

i) there will be reduction in chill accumulation units.

ii) needed soil moisture availability will reduce and evapo-transpiration will increase resulting into more water stress conditions.

iii) less amount of snow but faster melting will aggravate water deficit conditions both ground and surface waters.

iv) enhanced temperature during February-March may induce early flowering and better fruit set in apple resulting into higher yields, if fruit trees are not constrained with moisture stress. Moisture conditions however shall be adverse which can result into enhanced flower and fruit dropping; increased insect-pests infestation especially at elevations around 1,800-2,000 msl.

Table 4: Area and Production of Trends of Apple during 2004-05 to 2013-14 in Himachal Pradesh

	Year									
	2004-05	*2005-06*	*2006-07*	*2007-08*	*2008-09*	*2009-10*	*2010-11*	*2011-12*	*2012-13*	*2013-14*
				Apple						
Area (thousand-ha)	86.202	88.560	91.804	94.726	97.438	99.564	101.485	103.644	106.440	107.686
Production (thousand-tonne)	527.601	540.356	268.402	592.576	510.161	280.105	892.112	275.036	412.395	738.723
Productivity mt. /ha	6.12	6.09	2.91	6.29	5.23	2.81	8.83	2.65	3.88	6.8

Table 5: Area and Production of Trends of other Temperate Fruits and Nuts during 2004-05 to 2013-14 in Himachal Pradesh

	Year									
	2004-05	*2005-06*	*2006-07*	*2007-08*	*2008-09*	*2009-10*	*2010-11*	*2011-12*	*2012-13*	*2013-14*
Plum										
Area (ha)	8335	8360	8396	8428	8420	8456	8477	8530	8546	8556
Production (Mt)	19717	11763	10546	9373	9591	10413	13717	9842	12107	15991
Peach										
Area (ha)	4966	5000	5134	5175	5170	5195	5182	5181	5159	5132
Production (Mt)	6564	7823	8173	9224	9935	5162	9527	5101	11276	6271
Apricot										
Area (ha)	2947	3034	3178	3230	3392	3444	3483	3556	3577	3600
Production (Mt)	2415	2890	2768	3311	3224	2200	3341	2437	3263	3165
Pear										
Area (ha)	7564	7596	7662	7561	7405	7382	7370	7333	7283	7220
Production (Mt)	29590	23990	12039	29573	15450	17381	32075	11760	25212	35214
Cherry										
Area (ha)	302	313	345	374	405	453	480	492	456	453
Production (Mt)	415	447	438	698	453	419	1039	433	412	1183
Kiwi										
Area (ha)	111	113	118	120	128	128	123	120	117	113
Production (Mt)	102	138	77	137	118	154	112	150	555	114
Pomegranate										
Area (ha)	585	636	759	936	1085	1263	1422	1709	1968	2196
Production (Mt)	141	223	183	196	311	475	622	749	1351	2539

Contd...

Table 5–*Contd...*

	Year									
	2004-05	2005-06	2006-07	2007-08	2008-09	2009-10	2010-11	2011-12	2012-13	2013-14
Olive										
Area (ha)	54	54	54	53	51	51	48	48	46	37
Production (Mt)	9	38	8	8	9	7	9	13	15	8
Persimmon										
Area (ha)	332	358	371	397	407	418	420	421	402	403
Production (Mt)	164	211	174	152	224	196	335	199	209	519
Strawberry										
Area (ha)	27	57	57	55	55	57	58	54	55	54
Production (Mt)	425	485	524	480	453	447	466	284	348	480
Nut crops										
Almond										
Area (ha)	5732	5766	5784	5701	5647	5610	5578	5543	5473	5412
Production (Mt)	533	1303	1211	1176	1323	1345	1815	1069	1134	929
Walnut										
Area (ha)	4729	4785	4832	4718	4672	4628	4649	4607	4557	4539
Production (Mt)	3099	1798	1555	1596	1776	1294	1632	1242	1482	2390
Piecannut										
Area (ha)	627	646	698	747	764	786	812	876	860	855
Production (Mt)	94	173	144	144	146	171	193	178	191	159
Hazelnut										
Area (ha)	12	13	14	15	13	13	13	13	12	13
Production (Mt)			2	4	3	1	0	0	1	0

☆ Kullu and Chamba will experience almost similar adverse influences as Shimla. But due to increased rainfall in each month barring June, fruit growers in this district will harvest comparatively better crop yields.

(b) Dry Temperate Regions

☆ Climate conditions in Kinnaur revealed that fruit production in low and mid altitudes may suffer adversely but at higher altitudes production, "very likely" will increase. In district Lahaul-Spiti due to increased rainfall every month and also increase in temperature during every season, more congenial climatic conditions will be created for better plant growth and yield. It will open up opportunities for cultivation of new crops and fruits besides existing vegetables and cereals in the region.

7. IMPACT ON CHILLING HOURS ACCUMULATION

Data presented in the table/revealed that the actual mean temperature of three winter months (December, January and February) for the period 1960-1990 was 6.1°C. The corresponding chilling hours calculated on the basis of regression equations were 1217 h (0-7.2°C) and 1374 (<7.2 °C). The regression equations were derived from the actual chilling hours experienced during the period 1994-95 to 2009-10. According to Scenario AIB, the mean temperature of three months for the period 1960-90 was 0.9 °C and the corresponding chillings hours were very high. In the projected mid period (2021-2050) the mean temperature was 3.4 °C and the corresponding chilling hours were 1614 h (0-7.2 °C) and 1876 (< 7.2 °C). The data further revealed that mean temperature by 2021-50 will rise by 2.50°C over the baseline.

The projected impact of the increased mean temperature over the actual temperature of 6.1 °C was thus be as follows:

Base mean temperature of 6.1 °C corresponded to 1217 h (0-7.2 °C) and 1374 h (<7.2°C). The projected increase of 2.5 °C over the base temperature will make it 8.6 °C. This will reduce the chilling hours to 849 (0-7.2 °C) and 909 (<7.2°C) (Table 6). Further, if the mean temperature increase above 10.6 °C then in such areas chilling hours accumulations will be impacted badly.

Table 6: Mean Temperatures and Corresponding Chilling Hours Accumulation Based on the Actual Data as well as Data from HadRM3 Model Scenario AIB for Temperate Wet Region of Himachal Pradesh

Actual Recorded 1960-1990			*HadRM3 Model Scenario AIB*					
			1960-1990			*2021-2050*		
T Mean (ºC)	*Chilling Hours*		*T Mean (ºC)*	*Chilling Hours*		*T Mean (ºC)*	*Chilling Hours*	
	0-7.2ºC	*< 7.2ºC*		*0-7.2ºC*	*< 7.2ºC*		*0-7.2ºC*	*< 7.2ºC*
6.1	1217	1374	0.9	2340	1981	3.4#	1614	1876

Chilling hours of the three periods were calculated on the basis of mean temperatures of winter months (DJF) using the following equations:

(i) Chilling hours (0-7.2°C) Y= -147.08X+2113.7 (R^2 = 82 per cent)

(ii) Chilling hours (>7.2°C) Y= -185.94X+2508.2 (R^2 = 89 per cent)

Increase in temperature based on AIB scenario: 2.5°C

It is, therefore, inferred that chilling hours accumulation will decrease with the increase in mean minimum temperature to 8.6 °C. But most of the apple cultivars will be able to fulfill their chilling needs owing to their optimum chilling hours requirement limits. But an increase to a temperature of 1′0.6 °C will reduce the chilling to 555 hours (0-7.2 °C) and under such situation the chilling will be limiting factors for apple growing especially cultivars with medium to high chilling requirements.

8.CONCLUSION

Glowal warming is the reality and has been created due to unchecked burning of fossil fuels. The impact of climate change of temperate horticulture is evedent and is taking huge toll on production and quality. Hence, long term studies carried by different groups need to be analysed and adaptive/mitigation measures needs to be adopted to restore production and productivity. Likewise biotic stress challenges, pollinator, pollinizers *etc.* need to be addressed and the complete plan need to be extended in vulnerable areas to sustain production and quality. Alternate crops/ new crops need to be explored. Beside, long term simulation studies need to be undertaken to suggest strategies for better preparedness for future. Leads obtained world over need to be studied and accordingly measures need to be suggest for vulnerable regions.

9. REFERENCES

Beniston, M. 2003. Climatic change in mountainous regions: A review of possible impacts. *Clim. change* **59**: 5-31.

Bhutiyani, M.R., Kale, V.S. and Pawar, N.J. 2007. Long-term trends in maximum, minimum and mean annual air temperatures across the northwestern Himalaya during the 20th century *Climatic Change*, **85**:159-177.

Fallot, J.M., Barry, R.G. and Hoogstrate, D. 1997. Variation of mean cold season temperature, precipitation and snow depths during the last 100 years in the former depths during the last 100 years in the Former Soviet Union (FSU). *Hydrol Sci. J.* **42**:301-327.

Rakhecha, P.R., Kulkarni, A.K., Mandal, B.N. and Dhar, O.N. 1983. Winter and spring precipitation over the northwestern Himalaya. *Proceedings of the First National Symposium on Seasonal Snow Cover*, New Delhi, pp. 175-181.

Thapliyal, V. and Kulshrestha, S.M. 1991. Decadal changes and trends over India. *Mausam* **42**: 333-338.

Zhai, P., Sun, A., Ren, F., Xiaonin, L., Gao, B. and Zhang, Q. 1999. Changes in climate extreme in China. *Clim. Change* **42**: 203-218.

Chapter 10

Management of Genetic Resources of Temperate Fruits and Nuts Under Emerging Intellectual Property Right Regimes in India

J.C. Rana[1], Mohar Singh[2], Rashmi Yadav[3] and Dinesh Singh[4]

[1]ICAR-NBPGR, Pusa campus, New Delhi – 110 012
E-mail: ranajc2003@yahoo.com
[2]ICAR-NBPGR Regional Station, Shimla – 171 004, H.P.
[3]ICAR-NBPGR, Pusa campus, New Delhi – 110 012
[4]Dr Y.S. Parmar University of Horticulture and Forestry, Solan – 173 230, H.P.

1. INTRODUCTION

Temperate fruits are one of the important components of the Plant Genetic Resources for Food and Agriculture (PGRFA). Major crops that constitute temperate fruits are apple, pear, peach, apricot, plum, cherry, walnut, almond, pecans, hazelnut, chestnut, currants and berries. Although their share in the overall fruits biodiversity is less as compared to tropical and sub-tropical fruits; nonetheless they contribute greatly for satisfying nutritional security and to integrate and diversify dietary and economic needs of people of the temperate regions. The cultivation of fruit crops like apple, European pear, strawberry, almond, and kiwi and also collection of naturally occurring minor fruits like pomegranate, hazelnut, seabuckthorn, pine nut *etc.* from natural growing are the source of people's livelihood in hills. This paper deals in brief on management of PGR of important temperate fruits including emerging IPR and management strategies.

2. CENTRES OF ORIGIN

One of the first scientists to become aware of the importance of plant genetic diversity was the Russian Explorer and Geneticist Nikolai I. Vavilov. His geographical survey revealed that most of the cultivated plants are linked to one of the seven basic geographical centres of origin. Almost one third of the world's species originated in south-eastern Asia and most of the major fruit and vegetable crops came from the east, west and central Asiatic centre and Mediterranean centre of origin, while roots and tubers in tropical. The primary centre of origin of apple is thought to be the region which includes Asia Minor, the Caucasus, Soviet central Asia and Himalayan India and Pakistan (Wilcox, 1962). It has existed in Europe, both in wild and cultivated form, since prehistoric times and has been well developed there at the beginning of Christian era (Shoemaker and Teskay, 1959). Records of apple cultivars are available as early as 100 BC (Magness, 1941). The genus *Pyrus* has probably originated in the mountainous regions of the western China where from it moved both in east and west directions and evolution took place in different geographical and ecological regions (Bell and Hough, 1986). Vavilov in his centres of diversity has proposed three centres of origin for pear, *i.e.* Chinese centre for *Pyrus pyrifolia, P. ussuriensis, P. betulifolia, P. calleryana*; Central Asiatic centre for *P. communis, P. salicifolia, P. regeli* and *P. pashia* and Near eastern centre for *P. coummunis, P. syriaca* and *P. caucasica*. The origin of peach is said to be China where its culture dates back at least to 3000 years (Li, 1984; De Candolle, 1967; Hedrick, 1917). Important wild species such as *P. davidiana* (an ornamental peach and a very good rootstock), *P. mira* and *P. ferganensis* grow wild in northern China, Tibet plateau and Sinkian province, respectively. *Prunus persica, P. jacquimontii, P. undulata, P. salicina* and *P.* sp. (*behmi*) occurs commonly in the interior of Himalayas (Hooker, 1879).

Plum, another important crop of prunes has been identified to be originated from five different centres, namely, Europe for *Prunus domestica*, western Asia (*P. insititia*), western and central Asia for *P. cerasifolia*, China for *P. salicina* and North America for *P. americana* (Watkins, 1976). The cultivated apricot has been originated precisely in northeast China between the town Kan-techeou and Russian border (Crossa-Raynaud and Audergon, 1987). However, Kostina in 1936 reported the distribution of apricot in temperate zone of Asia between 33°and 70°E longitude and 53° and 30°N latitude. Cherries are considered natives of south central Europe, Asia Minor and as far east as northern India and China (Westwood and Wann, 1954; Hedrick, 1915). Three species of cherry, *viz. Prunus avium* (sweet cherry), *P. fruiticosa* (ground cherry) and *P. cerasus* (sour cherry) are involved in the development of present form of cherry. The cultivated almond (*P. dulcis = P. amygdalus*) is native to central Asian mountain areas of India, Iran and Pakistan where many wild species still occur. According to Russian scientists (Popov, 1929; Vavilov, 1930) cultivated almond is a hybrid between *P. fenzlina* and *P. bucharica* because both are closely related to cultivated almond. Walnut is believed to have originated in Iran and the areas surrounding it. However wild seedlings of walnut have been found growing in vast regions right from the Carpathian mountains in eastern Europe, across Turkey, Iraq, Southern USSR and Afghanistan to north western Himalayas (Woodroof, 1979; Forde, 1975; Serr, 1969). The common walnut is considered indigenous to Persia to Kashmir (Forde, 1975).The filberts are believed to have originated in western Asia, Europe and northern America (Anon, 1959) Other important group raspberries (*Rubus* sp.) are native of the temperate region of northern hemisphere and have spread over different centres of diversity (Chinese-Japanese, Indo-Chinese, Indonesian, Hindustan, European-Siberian, South American and North American centres (Zeven and Zhukovsky, 1975).

3. GENETIC RESOURCES IN THE INDIAN GENE CENTRES

The diversity of temperate fruit plants in India is mainly confined to Indian Himalayas. The region constitutes a core area and the centre of origin and diversity for >20 major agricultural and horticultural crops (Vavilov, 1951). Of this, temperate fruits constitute an important group in which wide variation is attributed to topography, geographical position, and climatic conditions within the region. A number of species belonging to the genera *Malus, Prunus, Pyrus, Sorbus, Crataegus, Cydonia, Docynia, Rubus, Cotoneaster, Hippophae, Fragaria, Diospyros, Corylus*, and *Actinidia* occur both as cultivated and wild. The brief account of diversity among important temperate fruits is given below:

(a) Apple (*Malus* sp.)

Malus genus comprised about 30 species, which are are known to occur world wide. Out of these only two Malus baccata (popularly known as crab apple) and *M. sikkimensis* occur wild in India. Two varieties of *Malus baccata* has also been identified as *M. baccata* var. *himalaica* in north western region and Meghalaya and *M. baccata* var. *dirangensis* in Arunachal Pradesh. Besides, about seven ecotypes of *M. baccata* have also been collected from different temperate region. These morphotypes have been temporarily designated with their collection sites but they need to be studied taxonomically as they have shown very high level of morphological and genetic variation (Randhawa, 1987). About 20 species including some introgressions such as *M. prunifolia, M.* x *zumi, M. mandshurica* x *M. sieboldii, M.* x *robusta M. baccata* x *M. prunifolia* have also been introduced for use as rootstocks and pollinizers etc. Varietal diversity of apple in Indian Himalayas is primarily introduction from Europe and North American countries. However, some seedling selection and mutants, viz. Ambri and its selections, Maharaji, Kesari and Hazratbali have been selected and released for cultivation in Kashmir valley. More than 700 varieties have been introduced at various levels but only four varieties Starking (Royal) Delicious, Red Delicious, Rich–a Red Delicious and Golden were occupying major area. New cultivars, *viz.* Top Red, Vance Delicious, Hardeman, Skyline, Supreme Delicious, Red Spur, Red Chief, Oregon Spur II, Wellspur, Golden Spur, Scarlet Spur, Super Chief, Bright-n-Early and Silver Spur, Gala, Scarlet Gala, Gale Gala, Galaxy Gala, Fuzi and Red Fuzi, Coe Red Fuzi, Jonathan, Schlect Spur, Rome Beauty, Granny Smith, Compact Winter Banana, Honey Crisp, Early Red One, Washington Apple, Braeburn, Jeromine, Red Velox, Gala Red

Lum, Florina, Coop 12, Real McCoy, Sansa, Anna, Dorsett Golden *etc.* Are also being tried for their adaptability and consumer preference.

(b) Apricot (*Prunus armeniaca*)

The apricot occurs in semi-wild conditions in mid hills through high hills to cold deserts areas of north-western Himalayas. The preponderance particularly of the drier types of apricot is more in Trans-Himalayan region. Black apricot (*P. dasycarpa*) considered a natural hybrid between *P. armeniaca* and *P. cereasifera* and occurs in Kashmir (Sharma, 1993). Another *Prunus* species *P. mume* a very useful rootstock for high rainfall areas and *P. ansu* are also maintained at Regional Station, Shimla of National Bureau of Plant Genetic Resources. Genetic variability for size and shape of fruit and sweetness of pulp and kernel has been observed and collected from cold region. Two variant of dry type apricot red fruited and white fruited are widely grown in Kinnaur and Leh areas. Another local variant known as *chuli* or *zardalu* is extensively used as common rootstock for *Prunus* species and its kernel yield quality oil comparable to almond oil. Field Research Laboratory, Leh, Central Institute of Temperate Horticulture, Srinagar and NBPGR, Shimla have collected as many as 50 types of apricot from Ladakh and Spiti. Varietal diversity for both indigenous and exotic cultivars exists in temperate region and some of the promising cultivars are St. Aambroise, Moorpark, Royal, Shakarpara, Charmagaz, Suffaida, Kaisha, New Castle, Sahib, Viva Gold, Harogem, Nugget, Abricoticsourdi, Cinef-B and Nari.

(c) Cherry (*Prunus avium*)

Cherry occupy an important position among temperate fruits all over the world but the areas in India is less as compared to other stone fruit species. There are as many as five wild species occurring in India but cultivated types belong to *P. avium* (sweet cherry) and *P. cerasus* (sour cherry) which are of European, Caucasia, western Asian origin. The wild species occur in India are *P. nepaulensis*, *P. cornuta* (=*P. padus*), *P. cerasoides* (= *P. puddum*), *P. prostrata* and *P. tomentosa*. Among these *P. cerasoides* (locally known as *paaja*) has been successfully used as rootstock for sweet cherry. The species also has ornamental value in temperate region because it flowers in November when all other temperate fruits enter into dormancy. Similarly, *P. nepaulensis* shed leaves in September and produce dark green foliage in the extreme winter months when there are hardly any green leaves found on other trees. Another species *Prunus cornuta* (locally known as *Jamman*) occur in high hills and fruits of which are edible, black, and acidic with deep purple flesh and also brewed. This species, although susceptible to powdery mildew but gives very good graft compatibility when used as a rootstock for cultivated cherry (*Prunus avium*). There is not much varietal diversity and about 64 cultivars are maintained at different research stations. Some of the notable introductions, which have performed well in the Himachal and Kashmir Himalayas are Pink Ealry, Black Tarterian, Bing, White Heart, Red Heart, Black Heart, Early River, Pobeda, Early Purple, Noir, Gross, Lapins, Lambert, Sam, Sue, Sunburst, Stella, Durone Nero and Bing.

(d) Nuts

Nuts is another important category of temperate fruits. This includes walnut (*Juglans regia*), almond (*Prunus dulcis*), pecan (*Carya illinoensis*), hazelnuts (*Corylus avellana*) and chestnut (*Castanea sativa*). The genus *Juglans* includes about 20 species among which *J. regia* is the most important being commercially cultivated in many countries. The other species are grown for timber and occasionally for edible oil purpose as they contain hard shelled nuts. Some of the species, namely, *J. nigra, J. sieboldiana, J. cordiformis, J. major, J. chenovo, J. mandshurica* and *J. drianoski* have been introduced in India from various countries. Varietal diversity arose through seedlings is also very high and hundred thousands of trees occur in Jammu and Kashmir, Himachal Pradesh and Uttarakhand. These populations have tremendous variability both for fruit and tree characters. Some 20 very thin shelled clones have been selected by Central Institute of Temperate Horticulture (CITH), Srinagar, NBPGR, Shimla and Dr YS Parmar UHF, Solan from the natural populations of Kashmir, Chamba, Kinnaur and Chakrata area. Few varieties *viz.* Gobind, Roopa, Ratna, Tapan and Luxmi have been released through seedling selection. Exotic germplasm of about 50 accessions has also been introduced and tested under Indian conditions. The varieties, *viz.* Lake English, Colby, Blackmore, Hartley, Amigo, Chico, Vina, Tehama, Eureka, Turtle – 16, Turtle 31, Orth, Serr, Payne and Natra Attiro have performed better.

The cultivated almond belong to *P. dulcis* (= *P. amygdalus*) widely distributed in Himalayas. Wild population of bitter almond (*P. dulcis* var. *amara*) also occur in the interior of Kashmir, Kullu, Kinnaur and Shimla. It is considered as an important rootstock for cultivated almond. High quality papery shelled almond also grow in dry region of Kinnaur and Kashmir. Some 200 cultivars both of indigenous and exotic origin are being maintained at different places. Among these Non-Perail, Thin shelled, Drake, Ne-Plus Ultra, Tree No. 69, Tree No, 109, Telangi Selection, Merced, Thompson, Badam Jor, Spillo 2D, Kashmir Seedling, Gaura, Nikitskij, Priayanj, Spilo No. 7 and Texas White, Brandis have found promising. Pecan is an introduced crop which have adapted very well in the mid hills. Plantations in the hills generally comprised of seedlings selection and very few introduced varieties. About 10 exotic cultivars of pecan are maintained at NBPGR, Shimla and some old accessions introduced by the Punjab Government in 1937-38 are also maintained at Palampur. The cultivar introduced were Winter Nellis, Mahan, Western Schley, Halbert, Sumner, Chetopa,Major, *Cheyenne,* Choctaw, *Chickasaw,* Desirable, Giles, Cape Fear x Orth, Colby and Roentina. Among these Mahan and Western Schley have performed well in the mid-hills.

The hazelnut popularly known as *bhotia badam* or *thangee* is essentially an introduced nut in this country even though few plantations of native hazel nut *Corylus colurna* occur in Pangi, Shimla, Kinnaur, Kargil and Drass and parts of Uttaranchal as well. Localised plantations of *C. ferox* and *C. jacquimontiana* can also be encountered in the interior of Himalayas. Varietal diversity of some introduced material, *viz.* Barcelona, Alpha, White Aveline, Duchilly, Kruse, and Nibber is maintained at Chaubattia in Uttarakhand, Boktu, Bajaura and Shimla in HP. Chestnut is also an introduced plant and trees planted by Missionaries way back in the 19th Century can be still located in the cooler regions particularly near the forest bungalows. Two accessions IC 19400 and IC 19401 are also maintained at NBPGR, Shimla.

(e) Peach and Nectarine (*Prunus persica*)

Peach is widely grown in India since ancient times. It is very versatile crop as it grows from hot north Indian plains to high hills. One of the species *Prunus mira behmi* grows in extremely cold areas of Kinnaur district of Himachal Pradesh. Wild and cultivated forms of *P. persica* grow throughout the sub-Himalayan tracts. Owing to almost self-pollination, genetic variability in peach is limited. However, cross pollination through bees have led to selection of new variant arose through seedlings. Seeds harvested from bitter almond and from wild plantations are grown as seedlings and used as rootstocks. Other species like *P. jacquimontii*, *P undulata* and *P. salicina* occur in the interior of Himalayas have been tried as rootstocks but success rate is low. Peach also has high varietal diversity of both indigenous and exotic. Natural growing posses variability for fruit size, fruit colour, stone freeness, stone size, fruit ripening and also susceptibility to diseases and pests. More than 200 different varieties are being maintained in various field genebanks. Low and moderate chilling cultivars are generally preferred and some of the notable introduction and locally released varieties are Alexender, Babcock, Baifeng, Candor, Early Amber, Flordasun, July Elberta, JH Hale, Alton, Elberta, Red Haven, Shan-e-Punjab, Florda Prince, Earligrande, Pratap, Tropic Sweet, Sharbati, Stark Earliglo, Florabella, Early White Giant, CO-Smith, Paradelux, Summer Glo, Nema Guard, Stanford, Okubo, Kanto-5, Nishiki, Luna Ambri, Early Elberta, Suncrest, and Glohaven. Summer Queen, Fire Prince, Sun Red, May Fire, Silver King, Sun Rise, and Suncrest nectarine are some of the important nectarines.

(f) Pear (*Pyrus* sp.)

In India, the improved cultivars were introduced in the later part of 19th century and cultivation got momentum with the success of Bartlett and Gola in hill areas and *Patharnakh* in sub-mountainous regions (Rathore, 1992). Present day's pear cultivars can be broadly classified into two groups. First having butter fleshed fruits with inconspicuous grit cells is European pear (*P. communis*) and the second having firm and hard flesh with prominent grit cells is Oriental/Asiatic/Chinese pear (*P. pyrifolia* var. *culta* Nakai). Besides, about 15 pear species also occur wild in the Himalayan region. Some of these species and particularly *P. pashia* (popularly known as *Kainth*), *P. serotina* (zrainth), *P. pashia* var. *kumaonii*, *P. pyrifolia* and *P. jacquimontiana* have been used as resistant rootstocks for cold hardiness and disease resistance. Other wild species are *P. foliosa*, *P. polycarpa*, *P. griffthii*, *P. khasiana*, and *P. thomsoni* are confined to Indian Himalayas. There are as high as 200 cultivars being conserved in the field genebanks but the diversity under cultivation is less and few varieties, *viz.* Bartlett, Conference, Flemish Beauty, Megness, Doyenne-du-Comice, Beurre Hardy, Manning Elizabeth, Max Red, Sensation, Shimsui, Laxton's Superb and Bartlett are being cultivated. Some strains of Bartlett and Starkrimson being red in colour were also found promising. Asian pears Chojuro, Nijisseki and Shinseik, Hosui and Kosui have also been found promising in sub-mountainous areas. Another type of European pear known as *Buggugosa*, which has butter flesh and less grit cells is fully adapted to the hill conditions.

(g) Plum (*Prunus salicina* and *P. domestica*)

Genetic diversity in plum is largely distributed in sub-mountainous tracts. The cultivars grown in temperate region belong to European species *P. domestica*, whereas, cultivars grown in sub-mountainous region of India belong to *P. salicina* and its inter-specific hybrids such as Santa Rosa with American species *P. simonii* (apricot plum) and *P. americana* (wild plum). Local variety known as *Alubokhara* widely grown in the subtropics is different from *P. salicina* (Hayes, 1957). Other species, *viz. P. angustifolia, P. cerasifera, P wallichii, P. umbelleta, P. undulata, P. jenkinsii, P. prostrata, P. jacquimontii, P. tomentosa, P. punctata* and *P. rufa* are also widely distributed in Indian Himalayas. Varietal diversity of plum is generally confined in the mid-hills, however, some varities of prunes such as Italian d' agen and Sugar performed well in the cold dry region also. Although about 100 cultivars have been introduced but only few such as Santa Rosa, Frontier, Burmosa, Stanley, Greengage, Drooper, Methley, Red Beauty, Queen Ann, Au Amber, Au Rosa, Ruby Sweet, Red Plum, Red Ace, Black Plum, Black Amber, Kanto-5, Tarrol, Friar have been found promising. Some of the indigenous varieties and germplasm accessions such as Kohinoor, IC 20836, IC 20828, IC 20085 and IC 204448 are also doing well in western Himalayan region.

(h) Soft Fruits

This climate of sub-Himalayan tract is most suitable for growing soft fruits. The important fruits that come under this group are *Rubus, Ribes, Fragaria, Actinidia* and *Duchesnea*. *Rubus* is the most complex variable genus widely distributed in the Himalayas. More than 429 species have been reported the world over, of these 57 species occurs in Indian sub-continent and 33 species and four varieties (taxonomic) confined to north-western and western Himalayas. *Rubus* species that occur in the region are: *R. ellipticus, R. fruticosus, R. niveus, R. paniculatus, R. biflorus, R. macilentus, R. calycinus, R. acuminatus, R. hexagynus, R. hamiltonii, R. assamensis, R. insignis, R. ferox, R. moluccans, R. birmanicus, R. lucens, R. opulifolius, R. lasiocarpus, R. rosaefolius, R. gracilis, R. irritanus, R. pedunculosus, R. purpureus, R. saxatilis, R. pungens, R. discolor, R. antennifer, R. ulmifolius, R. almorensis, R. foliosus, R. hoffneistrianus, R. idaeus*, and *R. alpestris*. The two species *R. ellipticus* and *R. niveus* have wide range of variability for foliar and fruit characters. About 150 species of *Ribes* occur in the temperate region of the world and only seven species *R. glaciale* (= *R. acuminatum*), *R. nigrum, R. rubrum, R. orientalis, R. alpestre* (= *R. grossularia, R. emodense* (=*R. himalayense*) and. *R. takare*, have been reported to occur in India. The cultivated strawberry *Fragaria* x *ananassa*, is a result of hybridisation of two Ameriacn species, *F. chiloensis* and *F. virginiana*. About 130 strawberry cultivars are being maintained in the various genebanks of the region. The promising among them are Chandler, Selva, Douglas, Confictura, Dana, Belrubi, Gorella, Addie, Pajaro, Fern, Tioga, Phenomenal, Brighton, Etna and Sweet Charlie. Wild species such as *F. nubicola, F. daltoniana, F. nilgerrensis, F. vesca* and *Duchesnea indica* (= *F. indica*) also occur in India. The kiwi fruit (*Actinidia chinensis*) introduced in late sixties by NBPGR has been found promising in the sub

Himalayan region. Recently, two wild species although yet to be verified taxonomically have also been collected from north-eastern region.

(i) Other Minor Temperate Fruits

There are numerous other temperate fruits plant species called minor fruits such as *Punica, Sorbus, Cotoneaster, Ficus, Vitis, Viburnum, Myrica, Docynia, Elaeagnus, Hippophae, Crataegus, Pyracantha, Pinus, Diospyros, Cydonia, Rosa, Cornus* and *Ziziphus* exist in the temperate region of India. The predominant species among these genera are: *Sorbus foliolosa, S. acuparia, S. cuspidata, S. granulosa, S. lanata; Cotoneaster multiflora, C. microphylla, C. accuminata, Vitis lanata, V. parviflora, V. himalayana, V. divaricata, V. capreolata, V. vinifera, V. semicordata, V. latifolia, V. trifolia; Elaeagnus angustifolia, E. latifolia, E. umbellata, E. conferta, Cydonia oblonga, Doycinia indica, D. hookeriana, Myrica nagi, Hippophae rhamnoides, H. salicifolia; Crataegus oxycantha, C. songarica, Punica granatum* and *Ziziphus jujuba*.

3. MANAGEMENT ACTIVITIES

Keeping in view, the threats to the indigenous genetic resources of temperate fruits following activities have been taken up at various crops based research institutes and particularly at National Bureau of Plant Genetic Resources. The systematic explorations and collection activities of germplasm of temperate fruits are around two decade old. During these years extensive surveys have been conducted across the north-west and north-east region and >1500 germplasm accessions of around 30 fruit crops have been collected.

NBPGR along with institutes like Dr YS Parmar UHF, Solan, CITH, Srinagar, IARI, Regional Horticultural Station, Shimla have made rigorous efforts for evaluating the germplasm. As a result, > 3500 germplasm accession predominantly of apple, pear, peach, apricot, plum, cherry, walnut, persimmon, pecan, almond, strawberry and kiwi have been evaluated by different research institutes and nearly 45 cultivars have been identified as elite types which can be grown as commercial crops. Wild species such as *Malus baccata* (crab apple) has been found promising and produce semi to very vigorous trees when used as rootstock along with high degree of resistance to powdery mildew (Sharma *et al.*, 1998). 25 indigenous accessions of *Malus* species were also screened at IARI, Shimla and 2 accession, *viz.*, *M. baccata* from Shillong and *M. baccata* from Srinagar showed very high level of resistance to wooly aphid, scab and powdery mildew (Kishore and Randhawa, 1993). Other wild relatives such as *Pyrus pashia, P. pyrifolia* for pear, *Prunus cerasoides* for cherry, *P.* sp. (*bhemi*) for peach have been evaluated and have been recommended as very good rootstocks. New fruits like *Actinidia chinensis, Ziziphus juujba* and *Feijoa sellowiana* have also been evaluated and popularised in the region.

Fruit plants in general and temperate fruits in particular are poorly represented in the genebanks, however in recent years efforts have been made to enhance the field collection at various gene banks (Table 1).

Table 1: Germplasm Accessions* being Maintained in the Field Gene Banks at Various ICAR Institutes and SAUs in North West Himalayas

Crop	*NBPGR Shimla*	*NBPGR Bhowali*	*IARI Shimla*	*CITH Srinagar*	*YSP UHF, Solan*	*GBPUAT Ranichauri*	*SKUAST Srinagar*
Almond	21	–	25	01	83	01	80
Apple	285	29	73	16	180	50	235
Apricot	44	09	23	48	47	01	16
Cherry	12	–	13	02	28	02	12
Hazelnut	23	–	–	–	04	–	02
Pomegranate	98	–	–	–	45	–	08
Persimmon	05	03	–	–	04	–	–
Pear	104	10	–	01	126	10	35
Peach	58	12	–	–	46	12	39
Pecan nuts	50	–	–	–	06	–	08
Plum	41	06	–	–	45	01	44
Raspberry	24	21	–	–	12	–	–
Rubus/Ribes	29	30	–	–	03	–	–
Strawberry	24	82	39	–	15	–	08
Walnut	189	27	–	21	69	–	31
Wild and Minor fruits	39	20	12	05	21	04	19

* Some of the field genebanks may have duplicate accessions.

Sources: Dhillon and Rana 2004 and updated.

Germplasm in the genebanks also provide several research opportunities and also facilitate characterization and evaluation. Moreover, due to cross pollinated nature of fruit crops new variant through seeds propagation can be obtained. It is worth mentioned here that majority of the present days most popular temperate fruit varieties are seedling selections. However, field genebank, besides being vulnerable to natural calamities, insect-pests and diseases, require more space, labour and other resources, and are expensive to maintain (Jarret and Florkowski, 1990). Possibilities now exist for conserving the germplasm using in vitro methods, cryo-banking, establishment of clonal repositories etc., as alternative or complementary strategies to field genebanks have been envisaged. About 68 germpalsm accessions of 14 temperate fruit species have been conserved in cryo genebank, at NBPGR, New Delhi. NBPGR has almost completed the database on all the introductions made so far and also on the passport data collected within the country. Documentation of characterisation and evaluation data of fruit crops has also been documented in the form of catalogues and technical bulletins. Data on the field genebanks is also maintained by the concerned crop curator.

4. PROTECTING HGR IN THE IPR REGIME

Creation of new knowledge utilising living materials is now perceived as a means of generating commercial goods by converting it into useful value-added products, production process and services. The shift in the perception has been aided by the unprecedented enhancement in the last two decades in the level, scope, territorial extent and role of intellectual property rights (IPR) that were traditionally applied to mechanical inventions of one kind or another, or to artistic creations.

The complexities witnessed in the subject related to management of genetic resources have largely emanated from two global developments namely the Convention on Biological Diversity (CBD) and the Agreement on Trade-Related Aspects of Intellectual Property Rights (TRIPS) administered by the World Trade Organisation. These two major developments have confronted the plant scientists with new dimensions of obligations particularly related to access, utilisation and benefit sharing. There are now many statuary provisions and regulatory obligations that make access and utilisation of genetic resources not as unrestricted as it used to be and also provide an opportunity to establish ownership on the technologies, products and processes developed by scientific organisations. In addition, to the legal issues there are others which are political, ethical and related to environment and food safety. However, before dealing with these issues it is pertinent to first understand the various provisions and developments consequent to the adoption of these treaties which impinge upon research in genetic resources.

(a) Commission on Genetic Resources for Food and Agriculture

The genetic resources were considered as common heritage of humankind and there was no organized International effort for their management. In 1983, FAO established an intergovernmental forum: the Commission on Plant Genetic Resources (now the Commission on Genetic Resources for Food and Agriculture). The Commission has developed a global system for the Conservation and Sustainable Utilization of PGRFA, of which the Report on the State of the World's Plant Genetic Resources and the Global Plan of Action (GPA) are key elements (http://www.fao.org/ag/cgrfa/en). The GPA is intended as a framework, guide and catalyst for action at community, national, regional and international levels. It made significant contribution to the successful implementation of the CBD and seeks to create an efficient system to:

- ☆ **ensure** the conservation of plant genetic resources for food and agriculture as the basis of food security.
- ☆ **promote** sustainable use of plant genetic resources to foster development and reduce hunger and poverty.
- ☆ **promote** the fair and equitable sharing of the benefits arising from the use of plant genetic resources.
- ☆ **assist** countries and institutions to identify priorities for action.
- ☆ **strengthen** existing programmes and enhance institutional capacity.

(b) Convention on Biological Diversity

The CBD had emerged out of the global concern for conserving biodiversity. It recognised the sovereign rights of nations over their biological diversity and has three main objectives; (i) conservation of biological diversity, (ii) the sustainable use of its components, and (iii) fair and equitable sharing of the benefits arising from the use of genetic resources, including by appropriate access to genetic resources and by appropriate transfer of relevant technologies.

There are a range of articles incorporated in the treaty that reflect the concern of Parties for the issues of access to genetic resources, access to and transfer of technology, handling of biotechnology and distribution of its benefits. It can be seen from these provision that CBD reflects not only the commitment of global community for conservation and sustainable use of biodiversity, but also makes the facilitated access to genetic resources by developing countries a two-way process by effectively linking it with equitable sharing of benefits. The CBD incorporates two instruments for providing access namely "Mutually Agreed Terms" (MAT) and "Prior Informed Consent" (PIC). The concept of MAT establishes a new participatory relationship among country of origin, provider and user of genetic resources. It is a contractual arrangement, executed on a bilateral basis that provides an opportunity for the providing country to negotiate a share of the benefits derived from the use of the genetic resources. It also provides an opportunity to the provider of the genetic resource to specify the permitted or prohibited uses of the genetic resources provided, including whether or not it may be commercialised, any IPR rights which may or may not be taken over the resource or its derivatives and benefits that are to be shared. The PIC empowers the provider of genetic resource to establish an authority that can grant or refuse access following a request from the applicant as per the national legislation. The applicant is required to provide information concerning the genetic resources required, purpose for which required and any proposal for benefit sharing thereby providing the providing Party an opportunity to become equal partner and monitoring the use. As part of the PIC procedure, the authority may consult indigenous and local communities or other stakeholders concerned before granting the access.

(c) Access and Benefit Sharing

Implementation of the CBD is a dynamic process and the Conference of the Parties (CoP) is the governing body of the Convention that takes decisions on outstanding issues at its periodic meetings. In order to facilitate better implementation of the CBD obligations with respect to access and benefit-sharing Parties have undertaken several initiatives. The terms of references of these Working Groups and Expert Panel included mainly the development of a framework that would help assist Parties to develop national and regional regulations or guidelines on ABS, with special focus on evolving standards and principal elements of PIC, MAT, Material Transfer Agreements (MTA) and Monetary and Non – monetary Benefit Sharing Agreements.

(d) International Treaty on Plant Genetic Resources for Food and Agriculture

The ITPGRFA provides a legal framework that not only recognizes the need for conservation, sustainable agriculture and food security for the PGRFA but also defines a regime for ABS linking it with IPRs. It provides broad guidelines but entrusts the responsibility for realizing farmers rights to member states. The specific obligations include the need to keep an inventory of PGRFA, promote their collection, promote farmers and local communities efforts to manage and conserve on-farm their PGRFA, promote in situ conservation of wild crop relatives and wild plants for food production, and cooperate to promote the development of an efficient and sustainable system of ex situ conservation. In exercise of sovereign rights of nations over their PGRFA, the Treaty provides for a multilateral system (MS) covering a list of 35 food crops/group of crops and 29 forage crops/group of crops (provided as Annexure 1 of the ITPGRFA) selected on the criteria of food security and interdependence.

(e) The Biodiversity Act 2002

As the national response to CBD and The Biological Diversity Act, 2002 Biological was enacted on December 11, 2002 and came in force since April 15, 2003 with the notification of the Biological Diversity Rules 2003. The Act is primarily aimed to regulate, in the territorial jurisdiction of India, access to genetic resources and associated knowledge with the purpose of realizing equitable sharing of benefits arising out of the use of these resources with the local people. The implementation of the Act is coordinated by three functional bodies namely the National Biodiversity Authority (NBA), the State Biodiversity Boards (SBB), and the Biodiversity Management Committee (BMC). The NBA is the national competent authority to discharge all decisions pertaining to ABS to persons who are not Indian citizens, Indian citizens who are non-resident and a body corporate, association or organization not incorporated or registered in India or incorporated or registered in India under any law which has any nonIndian participation in its share capital or management. Also the prior approval of the NBA is mandatory before applying for IPRs based on biological resources or traditional knowledge obtained from India. Further, the Authority, while granting the approval shall ensure equitable sharing of benefits between the applicant and local bodies concerned or the benefit claimers, on mutually agreed terms and conditions. It may develop guidelines to impose and fix criteria for benefit sharing fee or royalty or both or impress conditions including the sharing of financial benefits arising out of the commercial utilization of such resources; and shall notify the specific details of benefit sharing formula in an official gazette on a case-to-case basis. The National Biodiversity Authority is also empowered to take necessary measures to oppose the grant of IPRs in any other country on any biological resource or associated knowledge derived/obtained from India.

The Indian citizens or a body corporate, association or organization, which is registered in India, shall obtain any biological resource for commercial utilization, or biosurvey and bioutilization for commercial utilization only after giving prior intimation to the concerned State Biodiversity Board. The State Biodiversity Board, on receipt of such intimation, may, in consultation with the local bodies concerned, prohibit any such activity if it is considered detrimental or contrary to the objectives of conservation and sustainable use of biodiversity or allow use on the basis of equitable sharing of benefits. Exception to this has been provided to the local people and communities of the area, including growers and cultivators of biodiversity, and *vaids* and *hakims*.

(f) Protection of Plant Varieties and Farmers' Rights Act

This is the second Indian IPR legislation which India adopted to comply TRIPS requirements. TRIPS provide option to member countries for protecting them by patents, or by an effective *sui generis* system or by a combination of patent and *sui generis* system. In the case of plant varieties, India chose not to give patent to plants and to protect plant variety by the *sui generis* system. It can also assume a wider scope to cover those aspects of IP system embodying the rights of communities, which are not protectable under the patent system. It grants an exclusive right to the innovator of a plant variety for producing, processing, and stocking, commercializing, importing or exporting the propagating material of the protected variety. This *sui generis* system of law to protect plant variety called 'Protection of Plant Varieties and Farmers Rights Act, 2001 (PPVFR Act)'. This Act provides for three rights- the Plant Breeder's Right (PBR), the Farmers' Rights (FRs) and the Researcher's Rights (RRs). The process of granting *sui generis* right or the PBR is called registration.

The FRs in the PPVFR Act includes the right on seeds, the right to equitable benefit sharing, the right to register farmers' varieties, the right for reward and recognition for conservation and improvement of plant genetic resources, the right for compensation when a registered variety does not perform to the promised level *etc.* RRs refer to the free and unrestricted right to researchers in accessing any plant material including the varieties registered under this Act for the purpose of research, including breeding new varieties. Also commercialization of such varieties bred by using a registered variety does not require prior approval from the PBR-holder of the registered variety or payment of royalty. This right is important to promote research leading to the development of more and more superior varieties, which would benefit farmers and promote the cause of national agriculture. PPVFR Act partially came into force from November 2005, with the establishment of the PPVFR Authority at New Delhi to implement this law.

A variety will become eligible for registration on qualifying for the distinctness, uniformity and stability test (DUS). Registration granted is valid for 18 years for vines and trees and 15 years for varieties of annual species. The process is on to register extant varieties (which are in commerce or public domain), new varieties (those bred by individuals including farmers and institutions and possessing novelty) and farmers' varieties (traditional varieties developed and conserved by farmer communities) in 30 crops species. Out of 102 crops for which registration has been started seven, *viz.* apple, pear, peach, apricot, cherry, almond and walnut are temperate fruits.

(g) Geographic Indications

Geographical Indications (GI) is the third important IPR law covered under the TRIPS and concerned to agriculture. Indian law on GI of Goods (Registration and Protection) Act was enacted in 1999 and enforced from 15th September 2003 with the GI Registry located at the Regional Patent Office in Chennai. Unlike other IPRs, which are created and owned by individuals or institutions, the GI is a collective IPR of many people belonging to a territory of a country or the country itself. For this reason GI is not transferable or licensable. GI refers to an essentially attributable association between a given quality, reputation or other characteristics of any goods such as agricultural, natural or manufactured, which is valued in commerce, with its either origin or production or manufacture or any act of processing or preparation in a definable territory, or a region of a country. For this exclusive relationship between the region/territory/country and the quality and distinctiveness of a given goods, it command a commercial reputation conferring premium market space in demand and price. Few examples of fruit crops in general having distinctiveness for GI are Coorg orange, Nanjanagud banana, Allahabad Surkha, Laxman Bhog mango, Khirsapati (Himsagar) Mango, Fazli Mango grown in the district of Malda, Virupakshi Hill banana, Sirumalai Hill banana, mango Malihabadi Dusseheri, According to Indian GI Act, any association of persons or producers or any organization or authority established under any law representing the interest of producers of the concerned goods are eligible to apply for the GI. Using the flexibility of this provision, efforts are underway in the country to appropriate the collective IP of farmers associated with some reputed agricultural commodities by the organized processors and traders.

5. ISSUES IN TEMPERATE FRUIT GERMPLASM CONSERVATION AND UTILISATION

Plant Genetic Resources is a high priority activity which could be achieved through judicious admixtures of traditional and modern PGR management approaches. The major issues which hamper conservation and utilisation of temperate fruit genetic resources are given below:

- Fruit crops are generally perennial in nature, produce non-orthodox seeds and propagated through vegetative means.
- The numbers of cultivars growing are mainly introductions and very less in number. Similarly, the numbers of crops being cultivated are far less than the number of edible wild and minor fruits occurring in India.
- Due to high dependence on exotic cultivars, the indigenous germplasm of cultivated, wild and minor fruits which still maintain a large genetic base has received low priority both in conservation and research.
- The temperate fruit plants in general have long gestation period hence evaluation and utilization of germplasm in the crop improvement is difficult.
- Poor networking among all the institutes working on temperate fruits. It is always difficult for a single institute to handle all the germplasm and many times same set of accessions are being maintained by several institutes.
- Lack of computerized and easily accessible data bases which may provide information on several aspects like passport data and important breeding characters *etc.*
- Field genebanks are facing paucity of land, and are subjected to various natural calamities disasters and also involve high maintenance cost.
- Wild relatives and minor fruits are under represented in the genebanks.
- The newly emerged areas of research such as value-addition traits/bio-prospecting, search and transfer of desirable genes in the wild relatives,

pre-breeding, and selection of potential rootstocks and application of biotechnology have received low priority.

- Poor knowledge of Intellectual Property Rights.

6. STRATEGIES FOR BETTER CONSERVATION AND UTILIZATION OF TEMPERATE FRUIT GERMPLASM

Before developing a strategy for conservation of utilization of PGR of temperate fruits we must buildup information on the status, occurrence, collection and conservation. For this purpose, inventories and biodiversity maps can be prepared. Once we have this information various strategies as given below can be implemented:

- Setting priorities of identifying taxa and habitats for which protection is urgently needed including wild taxa. This requires planning and conduct systematic and collaborative explorations.
- Identify the gaps, prioritize the needs and then decide what, where, when and how much is available and on what terms and conditions. On this basis, we can go for need based introductions of new varieties, new crops and rootstocks and other diverse and superior germplasm
- Mostly introduced cultivars have been evaluated only for agronomic traits with few exceptions for quality and adaptability. Hence, compile and document this available information and conclude what more is required. The evaluation programmes can be strengthened on screening of PGR for biotic and abiotic stresses, quality traits and adaptability to changing climatic conditions.
- Long gestation period in case of temperate fruit plants require detailed collaborative long term evaluation programme. Field evaluation should be taken up in the native habitats as well as in the hot spots (screening for biotic and abiotic stresses). Pre-breeding programmes can also help in reducing the long gestation period in the improvement programmes as well in the utilization of genes from wild relatives.
- Much advancement has been made now in the field of biotechnology and, therefore, use of biotechnological tools such as marker aided selection, embryo rescue, molecular characterization etc. should be made. Because of long gestation period and crossing barriers these techniques can play a leading role in the utilization of indigenous resources. The genome mapping project of European Union on "*Prunus*" and "*Malus*" are already under way and these will be very useful in tagging useful genes.
- Germplasm conserved in different genebanks may have many duplicates and, therefore, there is a need to workout the status without duplicates in seed, *in vitro*, cryo and field genebanks.
- Strengthening and maintenance of field genebanks although laborious and resource consuming but should be given top priority particularly in the absence of *in vitro* backups.

- ☆ As most of the temperate fruits produce recalcitrant or intermediate seed and are vegetatively propagated therefore, cryo preservation and *in vitro* conservation should be encouraged. For this purpose use of already standardized *in vitro* and cryo protocols can be made. The crops difficult to propagate have high economic importance, and wild germplasm should get top priority particularly under *in vitro* conservation. Establishment of DNA bank should also be considered for those species which have reached at the verge of extinction.
- ☆ Develop a crop-wise network for field genebank conservation because a single institute cannot afford to create field genebank for all the crops.
- ☆ Research efforts to develop new protocols for *in vitro* and cryo preservation should be strengthened.
- ☆ Wild relatives must get top priority in the collection as well as conservation programmes as they possess valuable genes and their habitats are being eroded. It is desirable to acquire, evaluate, maintain, pre-breed and sustainably use the wild relatives for improvement of cultivated crops. Many desirable genes, particularly for stress tolerance/resistance, are often available in wild species and unadapted germplasm and these are hardy and resilient
- ☆ *In situ* conservation and *on-farm* conservation has recently attracted the attention. It should be promoted particularly for planting wild and minor fruit crops in the protected areas, vacant community land, road side plantation etc. However, it raises question on equitable benefit sharing, farmers' needs and interests but these issues are currently undergoing scrutiny and lively debate. *In situ* conservation can be taken up at specific ecological niches, *e.g.* hazelnut grow only in Pangi, dry types of apricot in cold dry region of Kinnaur and Leh, and *chilgoja* in dry region of Kinnaur and Pangi.
- ☆ There is a need to develop a computerized information systems that can accommodate large data linking the traits for horticultural performance, nutritional quality, resistance/tolerance to stresses with passport data and adaptation to soil edaphic and climatic factors
- ☆ There is a vast potential of value additions particularly in minor fruits for multiple uses: product and processing, medicinal and ornamental, for instance, raspberries, pomegranate, wild grapes, dry apricot, sea buckthorn *etc.* Therefore, separate research programmes focusing on this valuable wealth should be developed.
- ☆ There is an urgent need to create biodiversity conservation awareness at grassroot level particularly on women, school and college curriculums and also efficient implementation of Biodiversity Act.

7. CONCLUSION

There is need for strengthening and planning and collaborative management of HGR in the Indian Himalayan Region. We also need to prepare more exhaustive and informative inventory of collections maintained at various research stations to facilitate their use and eliminate duplicates. GIS based mapping will further help to increase germplasm collection and utilization. Greater research and development realizing the importance of horticultural crops in food and nutritional security, diversification and sustainability, is required to be undertaken on areas like adaptability, breeding for specific trait to promote specific uses including export, valuation of HGR as sources of desirable traits and rootstocks, exploitation of wild relative for incorporation of resistance to biotic and abiotic stresses, nutritional quality and yield.

Since majority of HGR produce recalcitrant or intermediate seeds and are vegetative propagated therefore, cryo preservation and *in vitro* conservation activities have to be strengthened. The crops which are difficult-to-propagate, has high economic importance and wild germplasm needed to be considered at priority. The national networking on HGR management which include acquisition, conservation, evaluation and utilization of PGR, strengthening database on information and monitoring national and international regulations in policy to suitably ensure efficient PGR management is required and for this purpose greater role of NAGS for evaluation and maintenance of germplasm has been advocated in many forums. Bilateral programmes need to be developed with resource countries for the introduction of germplasm of desired traits. The threat of biopiracy and infringement of plant variety related IPR necessitates thorough characterization, including DNA fingerprinting of the elite germplasm. There is also a need to have integrated conservation of genomic resources to collect, validate and facilitate the use of useful genes and gene constructs generated in the country. The efforts are under way for creating genomic resources centre especially for those species, which have reached at the verge of extinction. Plant variety protection, geographical appellations, patents and trademarks would help to protect and utilize PGR more efficiently. It also requires a significant commitment from governments, industry sectors and the wider community to ensure a long-term balance between sustainable resources use and conservation.

8. REFERENCES

Anon, 1959. The Wealth of India, CSIR, New Delhi. pp. 298-305.

Ashmore, S.E. 1997. *Status Report on the Development and Application of In Vitro Techniques for the Conservation and Use of Plant Genetic Resources.* International Plant Genetic Resources Institute, Rome, Italy, 67 p.

Bell, R.L. and Hough, L.F. 1986. Interspecific and intergeneric hybridization of *Pyrus, Hort. Sci.* **21**: 62-64.

Chandel, K.P.S. 1995. Biodiversity conservation and use through in vitro strategies. **In:** Rana, R.S., Chandel, K.P.S., Bhat, S.R., Mandal, B.B., Karihaloo, J.L., Bhat, K.V. and Pandey, R. (Eds) *Plant Germplasm Conservation: Biotechnological Approaches.,* National Bureau of Plant Genetic Resources, New Delhi, pp. 23-35.

Chaudhary, R. 2000. Cryopreservation of seeds, embryos, embryonic axes and pollen at national cryobank of NBPGR. **In:** Engelmann F. and Takai, H. (Eds) *Cryopreservation of Tropical Plant Germplasm. Current Rresearch Progress and Application.* JIRCAS/IPGRI, Rome, Italy, pp. 457-459.

Crossa-Raynaud, P. and Audergon, J.M. 1987. **In**: Rootstocks for Fruit Crops, Rom R.C. and Carlson, R.F. (Eds.), John Wiley Sons, Inc., New York, pp. 295-320.

De Candolle, A. 1967. *Origin of Cultivated Plants*, Hafner Publishing Co. New York.

De Langhe E.A.L. 1984. The role of *in vitro* techniques in germplasm conservation. **In:** Holden J.H.W. and Williams J.T. (Eds), *Crop Genetic Resources: Conservation and Evaluation.*, George Allen and Unwin, London, pp. 163-179.

Engelmann F. 1997. *In vitro* conservation methods. **In:** Ford-Lloyd, B.V., Newbury, H.J. and Callow, J.A. (eds) Biotechnology and Plant Genetic Resources: Conservation and Use. CABI, Wellingford, pp. 119-162.

Forde, H.I. 1975. **In:** *Advances in Fruit Breeding*. Janick, J. and Moore , J.M. (Eds.), Purdue University Press, Indiana. Lafayaette, pp. 439-455.

Hayes, W.B. 1957. *Fruit Growing in India*. Kitabsuan, Allahabad, 532 p.

Hedrick, U.P. 1915. *The Cherries of New York*. Albany, NY.

Hedrick, U.P. 1917. *Report New York Experimental Station*, 1916.

Hooker, H.D. 1879. *Flora of British India.*, L. Reeve and Co. Ltd. The Oat House Brook, NR. Ashford, Kent, England II, pp. 307-388

Jarret, R.L. and Florkowski, W.J. 1990. *In vitro* vs field genebank maintenance of sweet potato germplasm: major costs and considerations. *Hort. Sci.* **25:** 141-146.

Kishore, D.K. and Randhawa, S.S. 1993. Wild germplasm of temperate fruits. **In**: *Advances in Horticulture - Fruit Crops, Vol.1.* Chadha, K.L. and Pareek O.P. (Eds) Malhotra Publishing House, New Delhi, pp. 227-241.

Li, Z. 1984. Peach germplasm and breeding in China. *Hort. Sci.*, **19:** 348-351.

Lynch, P.T. 1999. Tissue culture techniques in *in vitro* plant conservation. **In:** Benson E.E. (Ed) *Plant Conservation Biotechnology*. Taylor and Francis, London, pp. 41-62.

Magness, J.R. 1941. *USDA Farmers' Bulletin*, 1883.

Niino, T., Sakai, A., Enomoto, S. and Kato, S. 1992. *Cryopreservation of in vitro grown shoot tips of mulberry by vitrification. CryoLett.* **13:** 303-312.

Popov, M.G. 1929. *Bull. Appl. Bot.*, 22: 241-283.

Randhawa, S.S. 1987. *Wild Germplasm of Pome and Stone Fruits*. IARI Regional Station, Shimla, pp. 51.

Rathore, D.S. 1992. Some outstanding temperate fruit introductions and future strategies. **In:** Chadha, K.L., Uppal, D.K., Pal, R.N., Awasthi, R.P. and Ananda, S.A. (Eds.), *Emerging Trends in Temperate Fruit Production in India*. NHB Technical Communication, No.1, National Horticultural Board, Gurgaon, pp. 32-38.

Reed, B.M. 1992. Cold storage of strawberries *in vitro*: A comparison of three storage systems. *Fruit Var. J.* **46:** 98-102.

Reed, B.M., DeNoma, J. and Chang, Y. 2000. Application of crypreservation protocols at a clonal genebank. **In:** Engelmann, F. and Takai, H. (Eds.), *Cryopreservation of*

Tropical Plant Germplasm. Current Research Progress and Application. JIRCAS/IPGRI, Rome, Italy, pp. 246-249.

Sakai, A. 1997. Potentially valuable cryogenic procedures for cryopreservation of cultures plant meristems. **In:** Razdan, M.K. and Cocking, E.C. (Eds.), *Conservation of Plant Genetic Resources In Vitro. Vol. 1: General Aspects*. Oxford and IBH Publishing Co. Pvt. Ltd., New Delhi, pp. 53-66.

Serr, E.F. 1969. *Handbook of North American Nut Trees*, Jaynes R.A. (Ed.), The W.F. Humphery Press Inc. Geneva, New York, pp. 240-263.

Sharma, R.L. 1993. Genetic resources temperate fruit. **In:** *Advances in Horticulture - Fruit Crops*, Vol.1. Chadha, K.L. and Pareek, O.P. (Eds.), Malhotra Publishing House, New Delhi.

Shoemaker, J.S. and Teskey, B.J.E. 1959. *Tree Fruit Production*. John Wiley and Sons. Inc, pp. 1-129.

Singh, H. 1993. Apple cultivation in India. **In:** *Horticultural Development in the Hindu Kush Himalayan region*, Teotia, S.S. (Ed.), Oxford and IBH Pub. Co. Pvt. Ltd., New Delhi, pp. 49-74.

Watkins, R. 1976. **In:** *Evolution of Crop Plants*, N.W. Simmonds (Ed.) Longmans, London, pp. 243-246.

Westwood, M.N. and Wann, F.B. 1954. *Nutrition of Fruit Crops*. Childers, N.F. (Ed.), Horticultural Publications, Rutgers University, New Brunswick, pp. 158-173.

Wilcox, AmN. 1962. The apple. I. Systematics. *Handbuch der Pflanzenziichtung*. 6. Berlin, (2nd edn.) pp. 637-645.

Wilkins, C.P., Newbury, H.J. and Dodds, J.H. 1988. Tissue culture conservation of fruit trees. *FAO/IBPGR Plant Genet. Resources. Newsletter*, 73/74 : 9-20.

Woodroof, J.G. 1979. *Tree Nuts*, AVI Publishing Co. Inc. Westport, Connecticut, pp. 616-55.

Vavilov, N.I. 1930. *International Horticultural Congress*, London.

Vavilov N.I. 1951. The origin, variation, immunity and breeding of cultivated plants. *Chronica Bot.* **13:** 364.

Zeven, A.C. and Zhukovsky, P.M. 1975. *Dictionary of Cultivated Plants and their Centers of Diversity*, Center for Agri. Publ. and Doc., Wageningen.

Chapter 11

Advances in Varietal and Rootstock Improvement in Temperate Fruits and Nuts: Future Strategies

N. Ahmed, J.I. Mir and H.R. Wasim

ICAR-Central Institute of Temperate Horticulture,
Rangreth, Srinagar – 190 007, J&K
E-mail: danak59@redffmail.com

1. INTRODUCTION

In India, almost all deciduous temperate fruits such as pome (apple and pear), stone (peach, plum, apricot and cherry) and nuts (almond, walnut and pecannut) are grown in considerable quantity in the North-Western States of Jammu and Kashmir, Himachal Pradesh and Uttarakhand and on limited scale in some parts of North-Eastern Hills of Arunachal Pradesh, Nagaland, Meghalaya, Manipur and Sikkim. Some low chilling cultivars of crops like peach, plum and pear are being grown commercially in certain areas of the North Indian plains. Among all the deciduous fruits, apple is the most important in terms area and production followed by walnut and pear. The apple cultivar 'Ambri', is considered to be indigenous to Kashmir and had been grown long before Western introductions; while Britishers introduced apple in Kullu Valley of H.P as far back as 1865, followed by coloured "Delicious" cultivars in Shimla hills in 1917. Pears and other deciduous fruits were domesticated successfully in the early part of the 20th century, although some of them were reported to occur under semi-wild conditions much earlier. Apricot was found growing in the drier pockets of North-Western Himalayas and two local apricot varieties, namely 'Halman' and 'Rakshey Karpo' are reported to be indigenous to Leh-Ladakh areas of J&K State.

Sweet cherry was introduced from Europe before India's independence in 1947, while commercial cultivars of sour cherry have been brought mainly from USA in more recent years. The European and Japanese plum varieties are grown both in high and low hill areas with variety 'Santa Rosa' occupying major area. Low chilling cultivars of peach and nectarine such as Flordasum, Flordared and Sunred nectarine are successful introductions to the North-Indian plains. Some local selections of peach (Shan-e-Punjab, Sharbati), plum (Jamuni, Kala Amitsari, Alubhokhara) and sand pear (Patharnakh) are also cultivated on a commercial scale in sub-tropical - marginal chilling areas of North India. In India, temperate fruit cultivation is spread over an area of 5.03 lakh hectares with an annual production of nearly 31.14 lakh MT. Apple occupies 54 per cent of this area, followed by walnut, pear, almond, Japanese plums, peaches, apricots and cherries. In India the varietal improvement of these crops in the past has been mainly through introduction and selection of better performing cultivars in a particular climatic region. While as improvement through hybridization, some sporadic attempts have been made and few improved cultivars of apple, apricot, peach and almond were developed through hybridization but these have not gained much popularity. Through clonal selection some good number of cultivars in almond, walnut, peach, apricot and cherry were developed which are doing well. Due to lack of trait specific germplasm and ill-conceived programmes with lack of human and capital resources to continue long term breeding programmes. The cultivars with excellent and commercial importance are yet to be developed and exploited in the country. In other parts of the world especially in USA and Europe a good amount of work has been done and several varieties and rootstocks have been developed and commercially exploited. In this paper, an attempt has been made to delineate crop specific breeding objectives, gene sources and breeding strategy to evolve improved fruit and nut cultivars.

2. TEMPERATE FRUIT AND NUT IMPROVEMENT

(a) Challenges in Temperate Fruit and Nut Improvement

- ✰ Narrow genetic base and genetic erosion.
- ✰ Highly cross pollinated and heterozygous condition.
- ✰ Long juvenile period and breeding cycles.
- ✰ Sexual incompatibility and scarcity of dominant markers.
- ✰ Abiotic stress (chilling, freezing, heat, drought).
- ✰ Biotic stress (scab, fire blight, powdery mildew, canker, *Phytopthora*, red spider mite, white grub, woolly aphid, fruit fly etc.
- ✰ Incidence of viruses (Plum pox, mosaic, leaf curl).

(b) Strategies for Temperate Fruit Crop Improvement

- ✰ Augmentation of germplasm having desired source of resistance, quality, storability, yield etc.
- ✰ Identification of new source of durable resistance.
- ✰ Use of morphological and molecular markers for faster screening of various quantitative and qualitative traits.

- ☆ Reduction in the juvenile period using dwarfing rootstocks and transgenic techniques.
- ☆ Development of cross compatible and self-fertile lines.
- ☆ Development of varieties having superior yield and quality traits.
- ☆ Development of varieties and rootstocks resistant/tolerant to biotic and abiotic stress.
- ☆ Development of easy and fast propagation methods utilizing microprapagation techniques.

Traits to be Improved in different Temperate Fruits and Nuts

Fruit	*Trait to be Improved*
Apple	Uniform colour and size, sweet, crispy, white flesh, clean surface, low/medium/high chilling, early/mid/late maturing, resistant to biotic and abiotic stresses, self-fertile, regular bearing, long shelf life, processing type
Pear	Red colour, butter flesh, juicy, crispy, long shelf life
Apricot	Reddish skin, large fruit size, small stone or stone free, sweet type, drying
Plum	Large, red to deep colour, stone free, sweet
Cherry	Bright blood red colour, firm, sweet, resistant to cracking
Peach	Large, yellow with red blush, stone free, sweet and crispy, yellow to orange flush
Walnut	Lateral bearing, thin shelled, early bearer, dwarf type
Almond	Early maturing, late blooming, high yielder, paper shell, large size nut

(c) Accomplishments

There are a large number of public and private breeding programs are going on around the world and in most of the programmes common objective is to combine in new cultivars of high fruit quality with disease and pest resistance with major focus on scab (*Venturia inaequalis* (Cooke) Wint.) and powdery mildew (*Podosphaera leucotricha* (Ell. and Ev.) resistance, develop cultivars adapting to an area of extreme climatic conditions and genetic improvement of plant for tree habits that allow high productivity and regular bearing.

3. APPLE

(a) Varietal Improvement

Advances in Apple Varietal Improvement in USA and Canada

Country	*Research Institute*	*Breeding Objectives*	*Important Varieties Developed*
USA	Univ. of Arkansas	Varietal improvement	Stellar, Arkcharm
	Univ. of Minnesota	Disease resistance, cold hardy, high quality	Sweet Sixteen (1979), Honeycrisp (1991), Zester (1999)
	Cornell University, Geneva, New York	Scab resistant	Cortland, Macoun, Empire, Jonagold, Liberty (1978), Freedom (1983).

Contd...

Contd...

Country	*Research Institute*	*Breeding Objectives*	*Important Varieties Developed*
	Rutgers Univ., New Jersey	Fruit ripening variants	Disease resistance: Mollies Delicious (1966), Vista Bella (1974), Summer treat (1981)
	Washington State Univ.	Adoptability to hot, dry and sunny climate	WA 2
	Univ. of Florida	Low chilling	Tropic Sweet, Dorsett Golden, Anna
Canada	British Columbia, Summerland	Consumer preference and sensory fruit quality	Spartan (1936), Summered (1964), Scarlett Sentinel (1997), Golden Sentinel (1997)
	Quebec	Fresh market, long storage, disease and insect resistance	Belmac (1996), Primevere (1996), MacExcel (1996)
	Nova Scotia, Kentville	Scab resistance	Nova Easygro (1971), Novamac (1978), Novaspy (1986)
	Manitoba, Morden	Tree, shrubs, adaptability, landscape trade	Ornamental crab: Almey, Sundog, Garry, Selkirk, Kelsey; scion cultivars: September Ruby, Fall Red, and Red Sparkle

Apple Varietal Improvement in Australia, Brazil and New Zealand

Country	*Research Institute*	*Breeding Objectives*	*Important Varieties Developed*
Australia	Western Australia	Long storage	Cripps Pink (Pink Lady) (1985), Cripps Red (Sundowner) (1985)
	Queensland Horticultural Institute	High quality red apples, scab resistance	Applethorpe Earlidel (1988), Applethorpe Summerdel (1989)
	Heverlee	Transformation, haploid induction, molecular breeding	Merlijn (King Jonagold x Liberty)
Brazil	Empressa de Paequisa Agro. Ext. Rural Santa Catarina	Low chilling, disease resistance	Scab resistant: Princesa (1986), Primicia (1986); low chill: Catarina (1996), Imperatiz (1997), Baronesa (1997), Condessa (1997), Eva, Anabela, Caricia
New Zealand	DSIR, Havelok North	Long shelf life and fruit quality	Gala, Splendour, Braeburn Gala, Joy, Festival, Pacific Rose, Southern Snap, Sci Early, Sci Red

Apple Varietal Improvement in Asian Countries

Country	*Research Institute*	*Breeding Objectives*	*Important Varieties Developed*
China	Hebei Changli Fruit Res. Institute	High quality	Yanshanhong (Ralls Janet x Richared Delicious (1989)
	Zhengzhou Fruit Res. Institute	High quality	Huaguan (Golden Delicious x Fuji) (1988), Huashuai (Fuji x Starkrimson) (1988)

Contd...

Contd...

Country	Research Institute	Breeding Objectives	Important Varieties Developed
	Liaoning Fruit Research Institute	High quality	Golden Delicious 463 (russet resistant)
India	SKUAST(K), CITH J&K, YSPUHF, H.P. HETC, Uttarakhand CITH, R.S. Mukteshwar; IARI, RS, Shimla	Fruit quality, shelflife, high colour, early ripening and scab resistance	Lal Ambri, Sunehari, Akbar, Firdous, Shireen, Ambred, Ambrich, Ambstarking, Ambroyal, Chaubattia Anupam, Chaubattia Princess, Swarnima, CITH Lodh Apple 1, Pusa Gold, Pusa Amartara Pride
Japan	Aomori Apple Experiment Station	Self-compatibility	Megumi (1950), Mutsu (1949), Sekaiichi (1974), Hokuto (1983), Natsumidori (1983), Mellow (1990); Self-thinning: Aori 9
	NIFTS, Marioke	Fruit quality, storage, disease resistance	Fuji (1962), Akane (1970), Iwakami (1985), Sansa (1986), Kizashi (1991)

Apple Varietal Improvement in European Countries

Country	Research Institute	Breeding Objectives	Important Varieties Developed
Czech Republic	Research and Breeding Institute of Pomology, Holovouzy	Scab polygenic resistance	Angold, Julia, Produkta, Zuzana, Jarka, Julia, Resista, Selena
England		High quality, long storage, resistant to mildew, scab	Fiesta (1984), Columnar (1991): Maypole, Tuscan, Trajan, Charlotte, Obelisk
France	Institut National de la Recherche Agronomique, Angers	Resistance to pest, disease, reduction in orchard labour, improved fruit quality	Processing and Juice making: Jurella, Judaine, Judor; scab resistant: Chanteline; Others: Chantecler, Florina (1985), Baujade (1992), Initial (triploid).
Germany	Dresden-Pollnitz	High quality, regular, disease resistance	Piros, Pilot, Pingo, Pirella, Pinova, Remo, Rewena, Rebella, Reanda, Evelina
Italy	University of Bologna	Scab and mildew, resistance, compact habit, high fruit quality	Prime Red (Prima x Summerred) (1999)
The Netherlands	CPBO-DLO, Wageningen	Quality, scab, mildew resistance	Elstar, Elan, Elise, Ecolette CPRO, Santana CPRO, Bellida

Useful Horticultural Traits Carried by Apple Species

Species	Ch. No(2n)	Useful Characteristics
Malus angustifolia (Aiton) Michx.	34, 51, 68	Ornamental tree
Malus x *atrosanguinea* (hort. Ex Spath) C. K. Schneid	34	Resistance to fire blight
Malus baccata (L.) Borkh	*Malus baccata* (L.) Borkh	Monogenic (V_b, V_{bj}) resistance to apple scab. source of monogenic resistance to powdery mildew, Semi-dwarfing rootstock

Contd...

Table 5.1–*Contd...*

Species	*Ch. No(2n)*	*Useful Characteristics*
Malus coronaria (L.) Mill		Better source of precocity, ornamental tree
Malus domestica Borkh	34, 51, 68	Single dominant gene (*Co*) identified for compact columnar habit, QTL identified for malic acid gene
Malus floribunda Siebold ex Van Houtte	34	Monogenic (V_f) resistance to apple scab, semi-dwarfing rootstock
Malus fusca (*Raf.*) C. K. Schneid	34	Resistance to fire blight
Malus hupehensis (Pamp.) Rehder	51	Source of gene resistance to wolly apple aphid
Malus mandshurica (Maxim.) Kom. Ex Skvortsov	34	Semi-dwarfing rootstock
Malus × micromalus Makino	34	Monogenic (V_m) resistance to apple scab, semi-dwarfing rootstock
Malus orientalis Uglitzk	34	near dwarfing rootstock
Malus prunifolia (Willd.) Borkh	34	Better source of precocity, cold hardiness, semi-dwarfing rootstock
Malus pumila Mill.	34	Dwarf tree
Malus×robusta (Carriere) Rehder	34	Resistance to WAA (*E. lanigerum* Hausm.) source of monogenic resistance to powdery mildew, Major gene for resistance to apple leaf curling midges, Resistance to Rosy apple aphid (*Dysaphis plantaginea*), Sd3 Resistance gene identified for Rosy curling aphid (*D. devecta*), Resistance to fire blight.
Malus sargentii Rehder	51, 68, 85	Polygenic resistance to apple scab, source of monogenic resistance to powdery mildew (*Podosphaera leucotricha*), Source of gene resistance to wholly apple aphid, Semi-vigorous rootstock.
Malus sieversii (Ledeb.) M. Roem.	34	Monogenic (V_r) resistance to apple scab, Better source of precosity

(b) Apple Rootstock Breeding and Objectives

- ☆ Freestanding and free from suckering
- ☆ Precocious widely adaptable
- ☆ Good anchorage to reduce the cost of support system
- ☆ Easy to propagate
- ☆ Dwarfing
- ☆ Fully graft compatible
- ☆ Strong drought tolerance and resistance to weed competition for soil and water minerals

- ☆ Resistant to wholly apple aphid
- ☆ High resistance to severe winter cold
- ☆ Resistant to fire blight and collar rot
- ☆ Resistant to replant problem

Horticulture Research International- East Malling (formerly East Malling Research Station) is one of the pioneering centres for the breeding and development of rootstocks for temperate fruits. in 1920s Drs. Wellington and Hatton, initiated work on apple rootstocks who verified distinct rootstocks and distributed them throughout the world as Types, Type I to Type IX. Later, further types were added to the original nine. Although distributed under type names, the apple rootstocks later rapidly became known under Malling (or M.) designations and eventually the Roman numerals were replaced by the more normal Arabic ones. Only two of these original Malling selections namely M.9 and M.7, are now used extensively.

(i) Malling merton series

In collaboration with the John Innes Institute, Merton, a further series of rootstocks were produced, having resistance to the woolly apple aphid (*Eriosoma lanigerum*) (Preston, 1955), a very serious pest in many parts of the world, which attacks the root systems of apple trees. MM.106 and MM.111, two of the Malling Merton (MM.) series of rootstocks that gained resistance to woolly apple aphid from their Northern Spy parent, are still used extensively throughout the world. A further apple rootstock breeding program at East Malling, using the dwarfing M.9 as one parent, produced two more valuable dwarfing rootstock clones (Preston, 1967) namely M.26, a cross between M.16 and M.9, is semi-dwarfing and exhibits better tolerance to winter cold and drought than M.9 and M.27, a super dwarfing rootstock originating from a cross between M.13 and M.9, which is mainly used for vigorous scion cultivars planted in high density planting systems on fertile soils. Rootstock breeding was further focused on size control (dwarfing), tolerance to low temperatures (hardiness), tolerance to pathogens and pests, and adaptability to different soil conditions (Brown and Maloney 2005). With joint efforts by East Malling and Long Ashton Research Stations in England, EMLA 7, EMLA 9, EMLA 26, EMLA 27, EMLA 106 and EMLA 111 a virus-tested material of (M and MM) rootstocks was developed. Similarly in USA, the breeding programme was initiated in Cornell University who developed several selections of Cornell Geneva series (CG) resistant to fire blight (CG.16, CG.CG.30, CG.65). The other most well-known American root stock are 'Michigan Apple Clone 9', MAC9, which however performed poorly in hot dry soils (Webster and Wertheim 2003). New apple dwarfing rootstocks are also developed in other countries like Russia ('B.146' and 'B.491'), Sweden ('BM 427'), USA ('G.65' and other CG- and G-rootstocks), Japan (JM.1, JM.5, and J.M.8), Czech Republic ('J-TE-G'), UK ('M.20'), Poland (P.22,P.59,P.61,P.66), Canada (V.3), Germany (Supporter 1 to 4), and Romania (Voinesti 2).

The First Bine Paradise Malling Apple Rootstock Selections and their Renaming

Type	Common Name/Original Name	Malling Clone name	vigor
I	Broad-leavedEnglish Paradise	M.1	Vigorous
II	Doucin	M.2	Vigorous
III	No name	M.3	Semi dwarf
IV	No name	M.4	
V	Doucin Ameliore(Improved Doucin)	M.5	Vigorous
VI	Rivers' Nonsuch Paradise	M.6	Very vigorous
VII	No name	M.7	Semi-dwarf
VIII	French Paradise(Clarke Dwarf)	M.8	Dwarf
IX	Jaune de Metz	M.9	Dwarf

(ii) Sub-clones of M.9

Difficulties in propagating the original Malling selections of M.9, including the virus-free EMLA clone, prompted several nurserymen in European countries to re-select within the existing M.9 populations for subclones exhibiting improved propagation on the stoolbed. Now many improved M.9 subclones have been developed in Netherlands (NAKB 337-340 and Fleuron 56), Belgium (Nicolai 29), Germany (Burgmer) and France (Pajam 1 and 2) as well as the well-known EMLA subclone. Several of these subclones are slightly more invigorating than average (*e.g.*, EMLA, Nicolai 29 and Burgmer 984), whilst others such as NAKB 337 and Pajam 1 were slightly more dwarfing than average. One subclone not tested at East Malling, Fleuron 56, is reported by Wertheim (1997) to be the most dwarfing one tested so far.

Sub-clones of M.9

Clone Name	Country of Origin	Remarks
B.9	Begium	Slightly dwarfing than M.9 EMLA
M.9 Fleueren 56	The Netherlands	More (10 per cent) dwarfing than M.9 in Dutch trails
M.9 T337	The Netherlands	More (10 per cent) more vigorous than M.9 EMLA in UK trails
M.9 Pagam1	France	Slightly (10 per cent) more dwarfing than M.9 EMLA
M.9 Pagam2	France	Slightly (10 per cent) or slightly more invigorating than M.9 EMLA
Burgmer 984	Germany	slightly more invigorating than M.9 EMLA
M.9 Nicolai 29	Belgium	Upto (10 per cent) more invigorating than M.9 EMLA

(iii) New apple rootstock from HRI-East Malling, Kent

New rootstock selections from the HRI-East Malling apple rootstock breeding program with vigour similar to M.27 or intermediate between M.27 and M.9.

Rootstock Clone No.	Source of Origin
AR10-2-5	Source of origin M.27 x MM.106
AR69-7	AR10-2-6 open pollinated
AR360-19	M.9 x M.27
AR628-2	Ottawa 3 x MM.106
AR672-1	M.9 x 3426
AR669-1	M.7 x M.27
AR682-6	M.26 x M.793

New rootstock selections from the HRI-East Malling apple rootstock breeding program with vigour similar to M.9.

Rootstock Clone No.	Source of Origin
AR10-120-242	M.27 x MM.106
AR295-6	Robusta 5 x Ottawa 3
AR486-1	Ottawa 3 x M.27
AR680-2	M.26 x M.27
AR852-3	AR363-16 [M.27 x Ottawa 3] open pollinated

4. PEAR

(a) Breeding Programme and Objectives

- ☆ High and regular bearing, self-fertile
- ☆ Excellent fruit quality, with no grit cells
- ☆ Winter pears with storage longevity,
- ☆ Red or bi-coloured attractive fruits and
- ☆ Fire blight and virus resistance.
- ☆ Drought tolerance
- ☆ Minimizing the chilling requirement, winter frost tolerance
- ☆ Dwarfness

(b) Breeding Achievements

Many new cultivars have been developed and released during the last 20 years and the major improvements have been in the quality, storage shelf life of the early season and late season pear varieties (Bellini *et al.*, 2000). Selection for dual purpose cultivars with Bartlett type flavour and aroma has resulted in the release of Harvest Queen in Canada (Quamme and Spearman, 1983), and several clones with improved storage and post-harvest life from European programmes. The main varieties worldwide are 'Bartlett' ('Williams'), 'Conference', 'Beurre Bosc', 'Abate Fetel', 'Anjou' (USA) and 'Comice'. New varieties are needed for better resistance, better storability, longer shelf life and varieties for different climatic and soil conditions. Low chilling

Different Apple Rootstock Breeding Orogrammers in various Countries

Breeding Programme	*Series*	*Objectives*	*Rootstocks*
Geneva Apple Rootstocks	Geneva(G) series	Fireblight, Phytophthora resistant and tolerance to apple replant disease resistance cold hardiness	G.11, G.30, G.65, G.41, G.202, G.16, G.935, G.210, G.212, G.890, G.969
Polish apple rootstocks	Polish (p) series	Cold hardiness	P.1, P.2, P.22, P.14, P.16, P. 59, P.60, P 61, P.62, P.92
Russian Breeding programme	Budagovskij (B) or Bud Series	Winter cold tolerant	B-9, B.146, B.491
USA rootstock breeding		Better anchorage and Easier propagation	Alnarp' and 'Robust 5' MAC-9
Canada breeding programme		Hardiness, yield efficiency, Precocity, dwarfing ability	Ottawa 3, V.2, V.4
Germany rootstocks	Pillnitzer Supporter series	Good anchorage, early and high productivity and resistance to scab, mildew and wooly aphid	Pillnitzer Supporter 1, Pillnitzer Supporter 2, Pillnitzer Supporter 3, Pillnitzer Supporter 4
Sweden Breeding programmes	BM Series	Cold hardiness	Bemali
The Czech Republic	JTE or JOH series	–	JTE clones
Japan		–	Marubakaido' x M.9 clones
Research Institute Voinestia, Romania	Voinesti series		Voinesti 1, Voinesti 2

cultivar 'Kieffeir' cultivated extensively in the Kodaikanal area in South India (Singh, 1982) and Punjab and Haryana (Uppal *et al.*, 1993) is a cross between the French pear (*Pyrus communis*) and the Oriental pear (*Pyrus pyrifolia*) (Hayes, 1970; Uppal *et al.*, 1993). An, interspecific hybrid 'Baggugosha' reported by Dhatt and Grewal (1987), is another important variety of Kashmir valley which can also be grown successfully in sub-Montanian tract. It bears soft fruits with better quality but it shows irregularity in bearing thus requires to be improved by adding genes responsible for regular bearing (Chattopadhyay, 2009). Inter specific *P.communis* x *P. pyrifolia* hybrid adopted to low chilling have been released by University of Florida. With the exception of Garden Pearl (Bellini *et al.*, 2000) genetically dwarfed scion cultivars have not commercially introduced, although programmes in Italy (Rivalta *et al.*, 2002) and France have been breeding for reduced stature, using the short internode dwarf 'Le Nain Vert' as a source of dwarfing. Recent fire blight resistant cultivar releases include Potomac (Bell *et al.*, 1996), Blake's Pride (Bell *et al.*, 2002), Harvest Queen, Harrow Delight (Quamme and Spearman 1983), Harrow Sweet (Hunter *et al.*, 2002a) and Harrow Crisp (Hunter *et al.*, 2002b). Resistant to pear scab in European pears is now found in new cultivars from Germany (Fischer and Miildenberger, 1999) and Romania (Bellini *et al.*, 2002). Cultivars of *P. pyrifolia* such as 'Hosui' are resistant to black spot (Kantato *et al.*, 1982) and an irradiation induced, mutant of the susceptible cultivar 'Nijisseiki' has been developed (Sanad *et al.*, 1988) and released as cultivar (Kajiura, 1992). Improved resistance or tolerance to pear *Pysylla* has been reported in the Romanian cultivars 'Haydeea', 'Euras', 'Ina Estival' and 'Getica' (Branista, 2002).

Different Pear Breeding Programmers in Various Countries

Breeding Programme	*Objectives*	*Important Varieties*
Italy	Dwarf growing, more or less tolerant to fire blight and pear psylla	Tosca', Turandot', 'Norma', 'Carmen'
Canada	Fire blight resistant varieties	Harrow Delicious, Harrow Red, Harobig, Harrow Gold', Harrow Crisp'
USA	Better hot climate adapted varieties	Elliot, Gourmet, Potomac, Summercrisp', 'Blacke's Pride, Rubia, Red Satin, Red Jewel, Red Spot, Red Silk'
Russia	Frost-resistant and scab-tolerant varieties	Krasavitsa Chernenko, Bronzovaja', Svetljanca', 'Yanvarskaja', 'Smugljanca'.
Estonia, Latvia	Better adaption to climatic conditions	—
Czech Republic	—	Bohemica, Erika', 'Dicolor, Delta, 'Decora', 'Dita', 'Jana', 'Omega', 'Barbara
France	—	Angely, Delmire, Delwini, Delsavor, 'Delbuena', 'Bronstar', 'Beauroutard'
Co-operative breeding program of Switzerland and Great Britain	Less resistant to scab and mildew varieties	—

Useful Horticultural Trait Carried by Native Pear (Pyrus) Species (Bell 1991)

Scientific Name	*Useful Characteristics*
European	
P. communis L.	Resistance to fire blight, white spot, pear scab, pear decline, pear psylla; cold hardy
P. nivalis Jacq.	Resistance to pear psylla; cold hardy
P. cordata Desv.	Resistance to codling moth; adaptation to warm winters, high pH, drought
Circum–Mediterranean	
P. amgydaliformis Vill	Resistance to codling moth; adaption to warm winters, high pH, drought and clay soils
P. elaeagrifolia Pall.	Resistance to codling moth' sold hardy
P. syriaca Boiss.	Adaptation to drought
P. longipes Coss. and Dur.	Cold hardy' adaptation to wet and dry soils
P. gharbiana Trab.	Adaptation to drought
*P. manorensis*Trab.	Adaptation to low pH
Mid Asian	
P. glabra Boiss	Adaptation to drought
P. salicifolia Pall	Adaptation to drought
P. regelii Rehd.	Resistance to pear psylla
P. pashia Buch. Ham. Ex D. *Don*	Adaptation to warm winters, low pH, clay soils
East Asian	
P. pyrifolia (Burn,) Nak	Resistance to pear scab
p. pseudopashia T.T. Yu	Resistance to codling moth; winter hardy
P. ussuriensis Maxim	Resistance to fire blight, pear psylla; winter hardy
P. calleryana Decne.	Resistance to fire blight, Fabraea leaf spot, pear psylla, codling moth; adaptation to warm winters, low pH, wet, dry and clay soils
P. betulifolia Benge	Resistance to pear psylla, codling moth; adaptation to warm winters, low pH, wet, dry and clay soils
P. fauriei C.K. Scheid.	Resistance to fire blight, pear psylla, codling moth; adaption to warm and cold winters, low pH, wet and clay soils
P. hondoensis Kik. and Nak	Resistance to fire blight
P. dimorphophylla Mak.	Resistance to powdery mildew, pear psylla, codling moth; adaptation to low pH and wet soils
P. kawakamii Hayata	Resistance to codling moth; adaptation to warm winters

(c) Rootstock Breeding

The main breeding aims for pear rootstocks:

- ☆ Dwarfing and free standing of trees
- ☆ Precocity and productivity of varieties
- ☆ High yield efficiency

Genetics of pest and disease resistance in pear.

Pest and Disease	*Observations and Sources*
Fire blight (*Erwinia amylovora*)	High levels of resistance are present in *P. calleryana and P. ussuriensis*, but not immunity (Hartman, 1957),
Pear scab (*Venturia pirina*)	Resistant genotypes have been found in *P. communis, P. pyrifolia, P. ussuriensis*, and *P. novalis* (Kovalev, 1963; Westwood, 1982; Bell, 1991).
Pear decline	*Pyrus communis* is generally tolerant, *P. betulifolia* is very tolerant, while *P. pyrifolia, P. ussuriensis* and *P. calleryana* are susceptible (Lombard and Westwood, 1987).

- ✰ Positive influence to fruit quality and size
- ✰ Efficient propagation ability
- ✰ Compatibility
- ✰ Cold hardiness
- ✰ Tolerance to iron and calcium chlorosis
- ✰ Resistance to fire blight
- ✰ Resistance to pear decline

(d) Rootstock Improvement

In rootstock breeding, the European pear, (*P. communis*) and quince, (*Cydonia oblonga*) are most common while in Asian pears, other species like *P. pyrifolia, P. pashia, P. calleriana, P. ussuriensis,* and *P. betulifolia* are used. All species without *P. communisand, P. pyrifolia* used as rootstocks have problems with incompatibility (Jones and Aldwinckle, 1990). Most of the breeding programmes using cross-combinations between *P. communis* x *P. communis; P. communis* x *wild species and Cydonia* x *Cydonia*. Unlike the cultivated apple, which has a large number of clonal rootstocks offering wide range of tree vigour, pears have traditionally been hampered by a more limited number and a more limited range of tree vigour control among its rootstocks. Quinces, widely used in Europe, have given the best precocity and vigour control coupled with easy propagation via stool beds. Quince rootstocks also offer slightly greater tolerance to pear decline than *Pyrus* stocks. However their use is, often limited by their lack of cold hardiness, lack of adaptation to high pH soils and drought, and their limited graft compatibility. The latter has not been insurmountable in Northern Europe, as cultivars such as 'Conference' and 'Comice' are graft compatible with quince and the use of inter stock such as 'Beurré Hardy' has proven useful in overcoming graft incompatibility. There is also considerable genetic variation within *Cydonia oblonga,* with demonstrated variation in cold hardiness (Wertheim, 2002), susceptibility to fire blight (Bobev and Deckers, 1999), graft compatibility (Sotiropoulos, 2006) and tolerance to calcareous soils (Fideghelli and Loreti, 2009).

East Malling has developed new quince rootstocks and more recently, some *Pyrus* rootstocks. Although QR193/16 (now known as EMH) was originally comparable in vigour to QC, further work showed that this quince rootstock induces vigour between QC and QA. The quince rootstock C132 was selected from seed sent from Sukhumi and assumed to come from Transcaucasia. The 708 series of *Pyrus* rootstocks derived from a cross between 'Old Home' and BP1 received the widest testing, particularly QR 708/36, which showed good yield efficiency on a tree somewhat larger than on QA (Johnson *et al.*, 2005), although interest in the 708 series has declined because of susceptibility to pear decline. For intensive plantations, dwarfing vegetatively propagated rootstocks (clonal rootstock) is needed. In USA, the OH x F rootstocks were selected (crossings between the fire blight resistant varieties 'Old Home' x 'Farmingdale) while Bologna released the *Pyrus* rootstocks Fox 11 and Fox 16 in 1996 and Fox 9 in 2008, derived from old Italian pear cultivars. In terms of vigour, all are more vigorous than quince BA29 (Fox11 and 16 c. +30-40 per cent and Fox 9 c. +20-30 per cent). Fox 9 and 11 are tolerant of calcareous oils. Currently all these rootstocks are propagated by micropropagation. In Pisa, the breeding effort concentrated on lime-tolerant quince rootstocks and a number of selections were made prefixed by Ct.S (Cotogno Scaramuzzi). Ct.S 212 gave good vigour control (slightly below BA29) and tolerance to lime-induced chlorosis but was very susceptible to fire blight. In Germany, two pear breeding programmes ceased when the breeders retired, Helmut Jacob in Geisenheim and Manfred Fischer in Dresden-Pillnitz. Jacob's programme was based on a population derived from 'Old Home' × 'Bonne Louised' Avranches', which resulted in the release of two rootstocks 'Pyrodwarf®' (Rhenus 1) and 'Pyroplus' (Rhenus 3), both of which showed good graft compatibility with European and Asian pears, tolerance of lime-induced chloros is, ease of propagation and high frosthardiness (Jacob, 2002). Although earlier work suggested trees in 'Pyrodwarf®' were inthe size range between quince C and quince A, however more recent work indicates vigour similar or slightly lower than that with BA-29, or even greater (Kviklys and Kvikliene, 2008). In USA, root suckers have been noted as a problem in some areas. Trees on 'Pyroplus' are vigorous and is used USA, as an alternative to OHxF 87.

The programme in Dresden-Pillnitz used a wide range of inter specific crosses and selected a range of seven new *Pyrus* rootstocks, with a vigour range of "very dwarf" to "medium strong" (Fischer, 2007). Two selections went for further testing in Germany and the USA. Pi-BU 2 and Pi-BU 3, both of which included *P. longipes* as a parent. German trials have shown some tree losses with Pi-BU 3, which might indicate a graft incompatibility problem possibly linked to pear decline. Trees on Pi-BU 2 have given good tree size control, comparable with trees on QA, but have not matched the trees on quince for yield or yield efficiency (Michael Zoth, pers. commun.).Experience in the USA of pear trees on these two rootstocks has also been dis appointing in terms of precocity and these may be showing problems with pear decline. The French programme is currently the world's largest and most diverse pear rootstock breeding programme. There is not enough dwarfing, but are very important by having resistance against fire blight, pear decline and calcium chlorosis. The 'Perry' pears from France (RV 139) possess medium growing capacity; they are well suited for trees in landscape and not in intensive plantations. The 'Retuziere' rootstock series derived from the variety 'Old Home' (OH 11, 20, 33) grows like quince A and is free standing.

The BP rootstocks from South Africa and the Fox rootstocks from Italy grow vigorously. 'Pyrodwarf' from Germany (Geisenheim) and some OH rootstocks from France ('Pyriam'=OH11) have a medium fire blight resistance and they grow moderately. The Polish activities are for more frost-rèsistant quince rootstocks (Quince 'Sydo'). All others are focused on the work for better compatibility, more dwarfing, better resistance to pests, and tolerance to chlorosis and better free standing of trees (Great Britain, France, Italy), but not all attributes could be realized in a single genotype. In comparison to apple rootstocks, pear needs much more breeding and field test work. Quince has not enough resistance against frost and fire blight and not all varieties are compatible with quince. All other rootstocks have deficits in dwarfing, free standing and disease resistance. Therefore, though new rootstocks have advantages in many characteristics, but it is still a long way to achieve the 'ideal' rootstock.

Compilation of Active and Inactive Pear Rootstock Breeding Programmes

Country	*Institution*	*Genus/Species*	*Primary Trait*	*Rootstocks*
Australia	B. Morrison	*Pyrus communis*	Yield efficiency	BM 2000
Belarus	Belorussian Res. Inst. for Fruit	*Pyrus communis*	Yield efficiency	Seedling rootstock
Germany	Fruit Genebank, Dresden-Pillnitz	*Pyrus communis*	Yield efficiency	Pi-Bu series
	Geisenheim Research Institute	*Pyrus communis*	Dwarfing, precocity	Pyrodwarf, Pyroplus
Italy	CIV, Ferrera	*Pyrus communis*	Yield efficiency	Ct.214, Ct.214
	Univ. of Bologna	*Pyrus communis*	High pH soil tolerance	Fox series
Poland	Poland Institute of Pomology and Floriculture	*Pyrus communis*		Elia, Belia, Doria
South Africa	Bion Donne Research Station	*Pyrus communis*	dwarf	BP1, Bossiar series RV.138, G.54-11
	Warsaw Agricultural University	*Pyrus communis*		GK series - seedling
USA	MCAREC, Oregon State University	*Pyrus communis*	Yield efficiency	Horner series
UK	East Malling Research	*Pyrus communis, Cydonia*	Yield efficiency	708 series, C132QR I93-16
France	INRA-Angers	*Pyrus spp*	High pH soil tolerance	Pyriam, BA 29
Spain	IRTA	*Pyrus spp*	High pH soil tolerance	Interspecific crosses
USA	USDA, ARS,	*Pyrus spp.*	Yield efficiency, Fire blight resist.	OHF × US selections
China	Several institutes and universities	*Pyrus* (Asian)	Dwarfing	Several

Contd...

Contd...

Country	Institution	Genus/Species	Primary Trait	Rootstocks
Japan	Gifu University	*P. bet* and *P. call*	Dwarfing	SPRB and SPRC series
Lithuania	LIH-Babtai	*Cydonia*	Cold hardiness	K series
Netherlands	Flueren	*Cydonia*	Yield efficiency	Eline
Poland	Institute of Pomology and Floriculture	*Cydonia*	Cold hardiness	S-1
Ukraine	Ukranian Academy of Agrarian	*Cydonia*	Drought tolerance, soil pH	IS, K, and R series

5. WALNUT

(b) Breeding and Objectives

- ☆ Yield related traits: Higher yield, high number of fruitful laterals, precocity, late leafing, early harvesting, reduce tree size, self-fertile and homogamy are desired.
- ☆ Nut related traits: smooth shell texture, light colour, round to oblong in shape, paper shell, strong shell seal.
- ☆ Kernel related traits: High kernel recovery, plump, bold, easy to remove kernel halves, light in colour, sweet in taste, well filled.
- ☆ Biotic and abiotic traits: Resistant to *Xanthomonas juglandis*, tolerant to pistillate flower abscission, heat tolerant, winter hardiness etc.
- ☆ Rootstocks related traits: Rootstocks are also essential part of the composite tree. Therefore, the important objectives include compatibility between stock and scion, easy to propagate through vegetative means, tolerant to blackline, Phytopthora rot etc.

(b) Breeding Programme for Scion Varieties and Rootstocks

There has been an active walnut, breeding program at UC Davis since 1948. Who released cultivars Chandler, Howard, Vina, Tulare and Serr. Recently, three more new cultivars (Sexton, Gilletand, Forde) were released (patent pending) through controlled pollinations, which are characterized by high early yields, low blight scores, large light-coloured kernels and harvest dates before Chandler. Several cultivar selections for hypersensitivity to the cherry leaf roll virus (causal agent of black line disease) have also been made. These are derived from two to three generations of backcross breeding from Paradox (*Juglans hindsii* × *J. regia*) with *J. regia* as the recurrent parent. Superior clonal Paradox rootstock selections have been identified and are in retesting or field trials. These include an extremely vigorous individual that appears to tolerate nematode (*Protylenchus vulnus*) infestation, but is not having resistant per se. five selections of a vigorous black line-tolerant hybrid (*J. hindsii* × *J. regia*) × *J. regia*, three clones that are less damaged than others by *Phytophthora citricola* in greenhouse tests, and one clone that exhibits limited gall formation when challenged by

(*Agrobacterium tumefaciens*) are in preliminary tests. Genetic engineering through gene silencing has also resulted in clones that do not exhibit crown galls after inoculation in vitro.

Walnut Improvements in the World

Country	*Research Institute*	*Varieties Developed*
USA	University of California, Davis, USDA; Oregon; Idaho; Napa Valley; Washigton.	Waterloo, Midland, Pioneer, Carmello, Adams 10, Howard, Payne, Sunland, Vina, Ashley, Serr, Lompoc, Chico, Tehama, Eureka, Gustine, Placentia, Marchetti, Cisco, Chandler, Amigo, Hartley, Tulare, Early Ehrhardt, Conway Mayette, Pedro, Trinta, Poe, Weeper, Cascade, Sexton, Gillet, Forde,Gillet, Ivanhoe, Solano, Robert Livermore.
France	INRA(l'institut national de la researche agronomique), France	Mayette, Lozeronne, Laciniata 93-5, Meylan, Franquette
Spain	–	Concha
China	–	Chinese 85-8, Manregian, Sharkey
India	Central Institute of Temperate Horticulture, Kashmir, Srinagar	CITH Walnut-1, CITH Walnut-2, CITH Walnut-3, CITH Walnut-4, CITH Walnut-5, CITH Walnut-6, CITH Walnut-7, CITH Walnut-8, CITH Walnut-9, CITH Walnut-10
	SKUAST-Shalimar, Kashmir, Srinagar	Hamdan, Sulaiman
	Hort. Expt. and Training Centre, Dir. of Hort., Chaubattia, Uttarakhand	Hanaul Kagzi, Siri Kagzi, Kuna Kagzi,Dabla Kagzi, Nada Kagzi, Dabra Kagzi, BagiKagzi, Kandoi Kagzi, Chakrata Lehman Kagzi
Iran	Collaborative	Jamal (protandrous),Damavad (protogynous)

6. ALMOND

(a) Breeding Objectives

- ☆ Resistant to spring frost and greater winter hardiness
- ☆ Adaptable to mild climate.
- ☆ Resistant to fungual disease –gummosis, stagmina blight,
- ☆ High consistent yield-no. of nuts/tree and nut size
- ☆ Ease of nut harvest, separation of hulls from shell, and shell cracking without kernel damage.
- ☆ Moderate to high shelling percentage with sealed shell.
- ☆ Uniform kernels in shape, thickness, few or no double kernels, none or medium pubescence, relatively smooth surface, well filled-out for thickness, no sharp or irregular tips, no defects *i.e.*, gum, blank, callus etc., good flavour.
- ☆ Cultivars interact in relation to bloom and harvest for more efficient production.
- ☆ Cultivar that bloom together but harvest separately

- ☆ Increased protein content and vitamins
- ☆ Tree structure that require little or no pruning

(b) Achievements

Ten cultivars from California USA and nine from Yalta region of Russia were released between 1932 to 1996 for commercial production (Kester and Gradziel, 1996) but among them 'Nonpareil' and 'Mission' are still the dominant cultivars of California USA. In the early 1900 s, formal plant breeding programs were established in most major production areas to accelerate this selective process through controlled crosses and related genetic manipulations with varied breeding goals suiting to different environments, disease, and pest problems. At the same time, the globalization of the almond market imposed more stringent limits on acceptable kernel and shell characteristics. Despite inherent obstacles to rapid genetic improvement, including large plant size and the long seed-to-seed generation period of 4 years or more, many commercially successful cultivars have resulted from such controlled crossing programs in the last decades. Examples include the cultivars Ferragnes, Ferraduel, and Ferrastar from France; Butte, Ruby, Sonora, Padre, and Winters from California; and Guara from Spain. An almond scion breeding programme using controlled crosses was started at IRTA Mas de Bover spain in 1975. The main aim was to obtain new cultivars with late blooming (to prevent spring frost damage), self-fertility (to reduce pollination drawbacks), high production capacity, kernel quality, easy training and pruning, sufficient vigor, and tolerance to adverse conditions (frosts, diseases, and drought). Recently, four new cultivars, 'Vairo', 'Marinada', 'Constantí', and 'Tarraco', have been released.

Almond Breeding in the World and the Important Varieties Developed

Country/Institution	*Varieties*
California breeding programs	Non Pareil, Ne Plus Ultra, I.X.L, La Prima, Peerless, Texas, Drake, Harriott, Jordanolo, Harparel
Australian breeding programs	Chellastan
French breeding programs	Sultana, Ferragnes, Ferraduel and Ferrastar
Spain breeding programs	Marcona, Planeta, Pascuala, Patanzela, Guara, Vairo, Marinda, Constanti and Tarraco
SKUAST, India	Waris, Makhdoom, Shalimar

(c) Rootstock Breeding and Objectives

- ☆ Should impart high to medium vigour to support maximum nut yield.
- ☆ Possess adoption to diverse soil types and have good anchorage.
- ☆ Resistant to diseases and pests.

(d) Achievements

Almond seedlings have been the traditional rootstocks used under non irrigated and well-drained soil conditions. Advantages include easy propagation from seed, excellent compatibility with almond cultivars, deep rooting ability, and high tolerance to drought and calcareous soils. However, almond rootstocks perform poorly on excessively wet soils during active growth. The most significant commercial plum rootstocks for almond are the Marianna hybrids, a group of clones arising from a breeding line believed to be *Prunus myrobalan* x *P. hortulana*. of this group, 'Marianna 2624' is an important rootstock for almond in California for use in finely textured soils with poor drainage and where oak root fungus has occurred. 'Marianna 2624' is also nematode-resistant. Some almond cultivars including the major cultivar 'Nonpareil' can be incompatible on 'Marianna 2624' and related clones; however, almond x peach hybrids show outstanding characteristics as rootstock for almond (Felipe *et al.*, 1998). All self-fruitful almond tested have grown vigorously and produced consistently higher yields when grafted on the peach x almond hybrid rootstock GF 677.

7. PLUM

(a) Breeding and Objectives

Modern plum-breeding activities aim at the development of varieties that are

- ☆ Adapted to different climates
- ☆ Attractive fruits with good quality for profitable marketing.
- ☆ Winter hardiness and lower chilling requirements
- ☆ Higher productivity and resistance, shipping ability and late-ripening.
- ☆ Long storage ability.

(b) Achievements of European Plum (*Prunus domestica*)

The first European plum cultivar obtained by controlled hybridization namely 'The Czar', developed by Thomas Rivers in England by crossing Prince Engelbert' and 'Rivers Early Pacific' in 1843 and introduced in 1873. This variety was named in honour of the visit of the Czar of Russia in the same year while 'President' variety of commercial interest was found as chance seedling by Thomas Rivers. After the Second World War, plum breeding was intensified and a sophisticated programme in a scientific level and with clear objectives was started in Romania in 1950 by Constantinescu and was continued by Cociu (Cociu *et al.*, 1978). From 1967 to 1991, 26 new varieties were developed in Romania through open pollination, simple crossing, double crossing, pyramidic crossing and also from the induction of mutagenesis using x-rays (Botu and Botu, 2007). At Germany, plum breeding of *P. domestica* was carried out for resistance against PPV, using the hypersensitivity found in 'Jojo' and from sister variety, 'Ortenauer Stanley 34' (Hartmann and Petruschke, 2000).

From 1990 to 2006, 13 new varieties were released, some of them are of commercial interest. Some of the new varieties are tolerant to PPV like Haroma. The first absolutely Sharka-resistant European plum cultivar 'Jojo' was introduced in 1999. Since 1996, hybrids between 'Jojo' and 18 different varieties and with a lot of other genotypes not yet released as varieties have been made (Hartmann 2007). About 2000 seedlings derived from 'Jojo' and its sister is waiting for selection. A successful breeding programme was also set up by Jacob in Geisenheim since 1985 (Jacob 1998). From 1993 to 2005, 14 varieties have been developed just by selection using the existing plant material. In 1986, a breeding programme for European plum was started in Davis (California, USA).

Important Sources of Germplasm in Plum Breeding

Species	*Chromosome Number*	*Useful Characters*
P. allegehaniensis Porter	16	Resistance to crown gall
P. americana Marsh.	16	Tough skin; very winter hardy
P. angustifolia Marsh.	16	Resistance to bacterial leaf spot; limited tolerance to plum leaf scald
P. besseyi Bailey	16	Late bloom; high heat threshold; very winter hardy; resistant to crown gall
P. cerasifera Ehrh.	16 (24,32,48)	Earliness; nematode resistance
P. domestica L.	48	High flavor and fruit quality
P. hortulana Bailey	16	Resistance to bacterial leaf spot
P. maritime Marsh.	16	Late bloom; high heat threshold
P. mesicana S. Wats.	16	Large tree; low suckering
P. munsoniana Wight and Hedr.	16	Good fruit; productive
P. nigra Ait.	16	Very winter hardy
P. salicina Lindl	16 (32)	Good size, color and attractiveness; exceptional firmness and keeping quality at high temperatures; very winter hardy
P. simonii Carr.	16	Firmness; upright tree
P. spinosa L.	16	Disease resistance
P. subcordata Benth.	16	Drought tolerance; high chill requirement
P. umbellate Ell.	16	Resistance to crown gall

(c) Japanese Plum

Active plum-breeding programmes were initiated in California, Georgia, Florida and Texas in cooperation with the United States Department of Agriculture (USDA), in 1932 and released some commercially important prunes for drying. The University of California at Davis (UCD) began plum breeding in 1932 jointly with the USDA and released cultivars 'Burmosa' (1950), 'Redheart' (1950), 'Laroda' (1954), 'Redroy' (1954) and 'Queen Ann' (1954). 'Durado' was released by UCD alone in 1976. All except 'Redheart' and 'Redroy' have been important commercially. Private breeders and growers in California have bred or selected (mostly) many important commercial plums. Fred Anderson released 'Red Beaut', 'Black Beaut', and 'Grand Rosa', in addition to 'Amazon' (1960), 'Ebony' (1961), 'Grandoro' (1964), 'Andy's Pride' (1965), and 'Royal Garnet' (1981), and newer plums include 'Black Noble', 'Early Beaut', 'June Beaut', 'Purple Majesty', and 'Red Noble'. They also developed plumcots 'Red

Velvet' and 'Royal Velvet'.John Garabedian selected 'Angeleno', which is still increasing in importance. Selections from open-pollinated populations include 'Black Jewel' (1987), 'Gar Jumbo' (1987), 'Santa Rosa-2' (1987), 'Rancho 9-Golden' (1987), 'Gar Red' (1989), 'Rancho Cinco' (1989), and 'Rancho Ocho' (1989).'Rosemary' (1975), 'Rose Zee' (1984), 'Dolly' (1987), 'Royal Zee' (1985), 'Mid Red' (1986), 'Blue Giant' (1989), 'Golden Globe' (1991) and other complex inter specific hybrid rootstocks, such as 'Citation', 'Viking', and 'Atlas'. They also have been very active in the development of plum–apricot hybrids and back crosses, such as 'Flavor Delight' and 'Flavor Supreme' and newer ones, which are marketed using his trademarked terms "Aprium" and "Pluot". Improved taste is a current goal of their breeding.

'Flavorich', with black skin and firm, flavorful, fine-textured orange flesh, is quite promising for fruit quality. Objectives in the southeastern United States (Georgia, Alabama, Texas, and Florida) include those of California plus additional disease resistance. Fruit firmness is somewhat less important because many local markets are available. Resistance breeding is going on for three primary diseases: bacterial leaf spot, fruit spot and twig canker caused by *Xanthomonas campestris* pv pruni (Smith) Dye; bacterial canker caused by *Pseudomonas syringae pv syringae* van Hall; and plum leaf scald caused by *Xylella fastidiosa* Wells. The first two diseases are problems in many other countries that are trying to grow Japanese plums, such as Australia, New Zealand, Italy, and South Africa (Topp *et al.*, 1989). Leaf scald is also a serious problem in Argentina and Brazil. In general, later bloom is more desirable but regions such as Florida (Topp and Sherman, 1990) and parts of Texas, Australia, and Brazil require even lower chilling requirements than those common in Japanese plums. 'Frontier' is the only California plum grown to any extent commercially in the southern United States. 'Fortune' is also surprisingly productive and moderately healthy in Georgia, in contrast to the shortlived 'Blackamber'. 'Angeleno' survives fairly well although the fruit rarely hangs on the tree long enough to ripen properly.

Weinberger in 1950s, able to release the wide spread varieties: 'Frontier', 'Friar' and 'Blackamber'. In California, there are also some private breeders also selected some important varieties such as the late-ripening 'Angeleno'. Eighteen new Japanese plum varieties and several varieties from inter specific crossings have been released since 2003 (Ramming 2006). A high internal fruit quality, high sugar content and a good balance of sugar and acid are becoming as important as the outer appearance of the fruits (Ramming 2006). The largest breeding programme outside California was located at Byron in Georgia. From 1964 to 1995, from more than 40,000 seedlings grown, eight varieties have been released (Okie and Ramming, 1999). A comparison of the plum varieties bred at Byron with 'Santa Rosa' was made by Okie (2006). At University of Florida, a low-chill breeding programme has been developed (Sherman *et al.*, 1992). and released cultivars 'Gulfbeauty' and 'Gulfblazze' in 1997.

In Japan, from five breeding programmes in traditional plum growing areas, they released some varieties with low chill requirements. Plum breeding in South Africa at the Infruitec Center for Fruit Technology from 1972 to 1993. was aimed at achieving large-fruited plums with resistance to diseases and good shipping quality. Eight varieties were developed. One of them, 'Celebration', is a black-skinned variety. In Australia, plum breeding was established by Topp in 1967 at the Horticultural

Research Station in Queensland. Also in Japan, plum-breeding activities are mostly made by private breeders. In China, the main objectives in variety improvement are late ripening time, large fruit size, good cropping, eating quality, post-harvest behaviour and resistance to spot and other diseases (Weisheng 2007). In Europe, breeding of Japanese plum is relatively new but is important in the future because of the increasing demand for large-fruited plums.

(c) Rootstock Breeding and Objectives

The most important breeding aims for rootstocks for European plum are

- ☆ Easy vegetative propagation (hardwood or softwood cuttings, micropropagation in vitro),
- ☆ Good compatibility with a broad range of varieties,
- ☆ Cold hardiness,
- ☆ Tolerance to calcareous soils,
- ☆ Tolerance to wet and dry soil conditions,
- ☆ Stability and durability of the grafting combination,
- ☆ Resistance against important biotic environmental factors like nematodes viruses, fungi, bacteria and phytoplasma,

Different *Prunus* species such as *P. cerasifera*, *P. domestica*, *P. spinosa*, *P. salicina*, *P. besseyi*, *P. tomentosa*, *P. pumila*,*P. americana*, *P. armeniaca*, *P. persica* and *P. dulcis* and its hybrids can be used as rootstocks for plums. 'Weito 6' and 'Weito 226' are selections of *P. tomentosa* made in Weihenstephan, Germany. These rootstocks are drought resistant and induce small, heavy cropping trees. However, they suffer in wet soils. Hybridisations of *P. tomentosa* and *P. domestica* are not successful. However, hybridisations of *P. tomentosa* with *P. besseyi* and other species are possible (Kask

Important Clonal Rootstocks for Plums and Prunes

Rootstok Name/No	*Species or Hybrid*	*Origin*
Myrobalan B	*P. cerasifera*	UK
Myrobalan 29C	*P. cerasifera*	USA
Adara	*P. cerasifera*	Spain
Myrabi	*P. cerasifera*	France
Mariana GF 8/1	*P. cerasifera* x *P. monsaniana*	France
Mariana GF 2624	*P. cerasifera* x *P. monsaniana*	USA
Mariana GF 8-5	*P. cerasifera* x *P. monsaniana*	South Africa
Pixy	*P. insititia*	UK
St. Julian A	*P. insititia*	UK
Ishtara (Ferciana)	*P. cerasifera* x *P.Salicina*	France
Julior (Ferdor)	*P. Insititia* x *P. domestica*	France
GF 677	*P. Persica* x *P. amygdalis*	France
Citation	*P. Persica* x *P.salicina*	USA

1989). For instance, the hybrid 'VVA-1' (*P. tomentosa* X *P. cerasifera*) was introduced by Eremin as a dwarfing rootstock for plum and apricot as well as VVA-2, a seedling originating from open pollination of VVA-1, and VSV-1, a hybrid between *P. incisa* and *P. tomentosa* (Kask 1989). Recently, a plum rootstock breeding programme was started at Technical University of Munich-Weihenstephan. The aim is to develop semi-dwarfing and dwarfing rootstocks, which are hypersensitive to PPV. Inter- and intra-specific hybridizations are carried out. If budsticks which are latently infected with PPV are grafted upon rootstocks showing a strong hypersensitive response against PPV, the budstick will either not grow or die after a short period of growth. In this way it is guaranteed that only trees free from PPV will leave the nursery. Hypersensitive rootstocks could also be used for scions hypersensitive against PPV.

8. CHERRY

(b) Breeding Objectives

The common objectives across the world are;

- ☆ Enhanced yield via self-fertile cultivars, increased fertility
- ☆ Reduced susceptibility to environmental damage and diseases at bloom and post-bloom; extending seasonality, especially for early ripeners, and
- ☆ Resistance to cracking.

(b) Achievements

Breeding efforts are going on in France, Germany, Hungary, Italy, Denmark, the UK, the Czech Republic, Romania, Estonia and Ukraine, Turkey, Japan, and China. In the last few decades there has been release of more than 140 novel cultivars that have provided the major thrust in extending the crop's seasonal calendar and improving fruit quality traits such as size, flavour, flesh firmness and colour. Fruit crack resistance has been enhanced and there are now a dozen cultivars with only 5-10 per cent of fruit affected by cracking due to rain-damages. Molecular studies have brought new light to bear upon the genetic control mechanisms of key tree and fruit traits and this should enhance future breeding efforts. The introduction of new self-fertile cultivars with mutated *S*-locus alleles has been a key factor in upgrading crop yields, especially in Italy. Notable releases include very early-season 'Primulat' and 'Early Big' (France); the self-fertile 'Sweet Early' and 'Grace Star' (Italy); the mid-season ripeners 'Georgia' (Italy) and 'Vera' (Hungary); the mid-to-late 'Kordia', 'Vanda', 'Techlovan' (Czech Republic) and 'Black Star' (Italy); and the late ripeners 'Regina' (Denmark) and 'Alex' (Hungary).

(c) Rootstock Breeding

Cherry rootstock breeding programs have produced a wide range of rootstocks selected for vigour, precocity and productivity. Based on the literature over the last 50 years, cherry rootstock breeding carried out in 17 countries resulting in more than 100 rootstocks. The most investigated species or groups are the *Prunus cerasus* and the group of inter specific hybrids (41 and 42 rootstock genotypes, respectively). Seed tree selection projects on *Prunus avium* L. and *Prunus mahaleb* L. occurred from 1960 through

Cherry Breeding Programmes and Updated Selection Process at different Breeding Institutes of Europe and Asia

Country	*Institution*	*Objectives*	*Latest Released Varieties*
Bulgaria	FGI Plodvid	Flavour, compact habit, self-Fertility, yield, early ripening	Vassilena, Daneliya, Stefaniya
Czechoslovakia	Research and Breeding Institute of Pomology, Holovousy	Fruit size, firmness, colour, early ripening, late ripening, high chill requirement	Aranca, Halka, Horka, Korvic, Marta, Silvana, Techlovan, Vanda, Vilma
Denmark	BAZ, Institute for Fruit Breeding, Dresden	Fruit size, firmness, colour, Early ripening, self-fertility, yield, early ripening, late bloom, cracking, brown rot	Nadino, Namare, Namati, Napruni
	FRI York	Fruit size, firmness early cropping, yield, cracking, , bacterial canker	Napruni
Estonia	Polli Research Centre of Horticulture, Karksi	Fruit size, firmness, yield, Early rípening, late ripening, cracking	Johanna, Karina, Regina
France	INRA, Station de Recherches Fruitières, Bordeaux UK – John Innes Institute, Norwich	Fruit size, firmness, colour, yield, early ripening, late ripening, cracking	Ferdiva, Ferdelices, Ferdouces, Ferlizac, Fermina, Ferpin, Fertard, Folfer
Hungary	Research Institute for Fruit growing and Ornamentals, Budapest	Fruit size, firmness, spur habit, brown rot, leaf spot	Alex, Aida, Anita, Carmen, Pal, Peter, Paulus, Petrus, Rita, Sandor, Vera
Italy	CRA – Instituto Sperimentale Frutticoltura, Ministry of Agriculture, Roma	Compact habit	Ferrovia Spur
	Dipartimento Colture Arboree, University of Bologna	Fruit size, firmness, colour, compact habit, spur habit, early cropping	Black Star, Blaze Star, Early Star, Grace Star, LaLa Star, Sweet Early
	Istituto Sperimentale Frutticoltura, Verona Province	Self-Fertility; 8 Early cropping; 9 yield, late ripening, late bloom, bacterial canker, easy fruit abscission	Bargioni 137, Enrica, Giulietta, Isabella

Contd...

Contd...

Country	Institution	Objectives	Latest Released Varieties
Lithuania	Lithuanian Institute of Horticulture, Babtai	Fruit size, flavour, compact habit, early ripening, uniform ripening, cracking, brown rot	Auste, Meda, Jurgita, Vasare, Vytenu rozine, Vytenu juodoji
Romania	Research Institute for Fruitgrowing, Pitesti, Iasi, Bistrita	Fruit flavour, self-fertility early ripening, cracking	Ana, Cetatuia, Catalina, Golia, Maria, Marina, Superb, Splendid, Tentant
England	East Malling Research Station, East Malling	Fruit size, firmness, flavor, colour, self-fertility; 8 early cropping; 9 yield, cracking; 17 brown rot,	Penny
Japan	YARC Yamagata	Fruit size, firmness, self-fertility, early ripening, uniform ripening, low chilling requirement	Benisayaka, Benihuhou, Benitemari
	KFTRC Aomori	Fruit size, firmness, flavour, colour, compact habit, spur habit self fertility, early cropping, cracking	J-no-shizukuYP Yamanasaha

the 1980s. Very few (Claverie, 1996; Hrotkó and Magyar, 1998) running seed tree selection projects have been reported recently. Clonal rootstocks from *P. avium* and *P. mahaleb* were not successful; however some *mahaleb* clones are less vigorous (Hrotkó and Magyar, 2004).

The most promising rootstock groups are the vegetatively propagated *P. cerasus* clones and the new inter specific hybrids. *P. cerasus* breeding projects in Germany (Weiroot series), Italy (CAB series) and Denmark (DAN series) showed that it is possible to select non-suckering *P. cerasus* as rootstocks in a wide range of vigour and compatibility (Schimmelpfeng, 1996; Sansavini and Lugli, 1996; Callesen, 1998). Other widely tested *P. cerasus* rootstocks are Edabriz (Edin *et al.*, 1996), Victor (Battistini and Berini, 2004), Masto de Montagna (Jiménez *et al.*, 2004), and some from Romania, Lithuania (Parnia *et al.*, 1997; Lanauskas *et al.*, 2004).

As *Prunus avium* hybrid 'Colt' was the first. The hexaploid version of 'Colt' (James *et al.*, 1987) proved to impart slightly more growth reduction but was not more precocious than the original one. The P-HL- series from the Czech Republic is supposed to be *P. avium* × *P. cerasus*. The clone P-HL-A is reported to be semi dwarf and very promising in the Czech Republic and Poland, but less successful or disappointing in other countries (Grzyb *et al.*, 2005). From among the Pi-Ku series (*P. avium* crosses), Pi-Ku-1 seems to be a promising moderate vigorous or semi dwarf rootstock (Wolfram 1996). The only successful hybridisation of *P. mahaleb* with *P. avium* was carried out in Oregon (Westwood, 1978) resulting in the MxM series. Two of these are considered as possible rootstocks for semi-intensive or intensive orchards (Edin *et al.*, 1996; Hrotkó *et al.*, 1999).

The most productive and extensive hybridisation project was carried out in Giessen, Germany, where thousands of hybrids *of P. avium, P. cerasus, P. fruticosa* and numerous Asian Prunus species were involved in the project (Gruppe, 1985; Walther and Franken-Bambenek, 1998; Franken-Bambenek, 2005).

Cherry Rootstock Breeding Programmes in World

Country	Institution	Series	Species/Genus
USA	University of California, Davis	UCMH series	*Prunus mahaleb*
	OregonForest Grove	M xM series	*Prunus mahaleb* x *Prunus avium*
	Michigan State University East Lancing	MSU series	*Prnus cerasus* hybrids
Canada	PARC Summer Land		*Prunus avium*
Germany	TUM, Weihenstephan	Weiroot series	*Prnus cerasus*
	JKI, Dresden	PI-KU series	*Prunus* hybrids
	Justus Liebeg University, Giessen	GIESELA series	*Prunus* hybrids
Belgium	CRA, Gambloux	GM series	*Prunus* hybrids
Czechoslovakia	Research and Breeding Institute of Pomology, Holovousy	PH series	*Prunus avium* x *Prunus cerasus*

Contd...

Contd...

Country	Institution	Series	Species/Genus
Poland	RIPF, Skierniewice	Propiel, Piast	*P. mahaleb*
Denmark		DAN series	
France	INRA, Station de Recherches Fruitières, Bordeaux UK – John Innes Institute, Norwich	Tabel Edabariz	*Prunus cerasus*
Hungary	Research Institute for Fruit Growing and Ornamentals, Budapest	Kornopay, Bogdany, Magyar	*Prunus mahaleb*, Inbreeding *P. mahaleb*, *P. fruticosa hybrids*
Spain	EEAD, Zaragosa	AdaraMasto de M series	*Prunus cerasifera*, *Prunus cerasus*
Italy	Dipartimento Colture Arboree, University of Bologna	CAB series	*Prunus cerasus*
	ICA Univerty of Bari	REAL series	
Lithuania	Lithuanian Institute of Horticulture, Babtai	P.3, P.7	*Prunus* hybrids
Romania	Research Institute for Fruitgrowing, Pitesti, Iasi, Bistrita	IP-C series	*Prnus cerasus* hybrids
England	HRI Station, East Malling	F 12/1, Charger Colt, Colt hexaploid	*Pruns avium, P. avium* x *P. Psuedocerosus*
Iran	KANRC Tehran University, Khorasan	K series	*P. mahaleb*
Russian Fed.	KBRS, Krymsk	KRYMSK series	*Prunus* hybrids

Important Rootstocks of Cherry

Country	Rootstocks Series		
	Giesela	*Weiroot*	*PI-KU*
Germany	Giesla 3 – Gi 209/1	Weiroot 10	PI-KU 1— piku 4.20
	Giesla 5 – Gi 148/2	Weiroot 11	PI-KU 3 – Piku 4.83
	Giesla 6– Gi 148/1	Weiroot 13	PI-KU 4 – Piuku 1.10
	Giesla 12– Gi 195/1	Weiroot 53	
		Weiroot 158	

9. PEACH

(a) Breeding Objectives

- ☆ Increased fruit size, fuller and more extensive blush, better skin ground colour, increased flesh-to-pit ratio.
- ☆ Low chilling requirement.
- ☆ Shelf-life, control of tree size and vigour.
- ☆ Reduction of postharvest disorders related to long-distance shipping

- Diseases and pest resistance (sharka PPV), powdery mildew, brown rot, leaf curl, *Xanthomonas* spp., and green aphid (the vector of PPV).
- *Phytoplasma* resistance.

(b) Achievements

Over the last 20 years, 60-70 peach and nectarine varieties have been released (Della Strada *et al.*, 1996; Fideghelli *et al.*, 1998). Of these, about 2/3 are peaches and 1/3 are nectarine varieties. Almost 50 per cent of the varieties released in the world are developed in the USA. Europe, mainly France and Italy, produces about 30 per cent of the varieties. Lesser amounts are developed in South Africa, Australia, Asia (mainly China and Japan), and Latin America (mainly Mexico and Brazil). Flat peaches are beginning to enter into the markets of Europe and North America. Although these have been grown in China for centuries, until recently their wide spread commercial use has been limited due to a lack of firmness and cracking problems. At present, breeding efforts are being made to develop a series of commercial cultivars in Nanjing (Ma *et al.*, 1999), Zhengzhou (Wang, 2001), and Beijing (Jiang *et al.*, 1999). An excellent series of flat peaches is the Ruipan series developed in Beijing by Jiang Quan and his collaborators. The variety that is commercially used in the USA is 'Saturn' developed by Dr. Hough at Rutgers University (New Brunswick, NJ). Morepantao (Peen too) varieties are being developed in California (USDA, Fresno), Florida (low chill), France and Italy (UFO series based on crosses with 'Saturn').

One of the alternate strategies to reduce the pesticide application is the use of resistant varieties to control diseases and pests. Several active breeding programmes are going on for resistance to brown rot (California, Mexico, Brazil, Italy), powdery mildew (Mexico, France, Italy), aphids (France, Italy), bacterial leaf spot (southeastern USA, humid South America, Asia), and plum pox (Europe). One of the most important objectives is to insert genes for resistance into a plant through genetic engineering. Once this is possible, this will lead to reduced pesticide use but the public acceptance of such varieties is still uncertain.

(c) Low Chilling Breeding

The active breeding programme for low chill peach and nectarine varieties was initiated in the transitional chill zone of California at the beginning of the last century. These programs released Babcock which was used along with other varieties of the Armstrong program to develop a range of nectarine and peaches. The Armstrong peach (Flamingo, June Gold, Robin and springtime) and the nectarines (Armking, Panamint, and Palomar) varieties contributed substantially to the programs in the lower chill zones as a source of germplsam.

In the late 1930s to 1960s, peach breeding programs begin in the USA (Florida, California, Georgia Louisiana and Texas), Brazil (Pelotas and Campinas), India (Saharanpur, Uttar Pradesh) and South Africa and few medium chill cultivars were released by the breeding programs in the transitional chill zone in California (Zaigar's Genetics and USDA, Fresno) and the Georgia (USDA). In the 1980s, new programs were initiated in California (Sun World, Bakersfield), South Georgia, (Attapulgus, cooperative program with the University of Georgia, University of Florida and USDA,

Byron, Georgia), Texas (low chill program in south Texas), and Mexico (Chapino and Queretaro). The US programs have been building on the germplsam developed by Florida, Georgia (USDA), Byron, GAO, and California. More recently, new efforts have been initiated in Taiwan (Taiwan Agricultural Research Institute, Taichung), China (Lirong Wang, Zhengzhou Fruit Research Institute), Thailand (Unaroj) Boonpraakob, Kasetsart University, Bangkok) and in Australia (Bruce, Top, Department of primary Industries, Namvbour, Quensland). Until recently, vast majority of low chill varieties have been developed and released by the public programs in the USA, Mexico and Brazil. This is in contrast with only 33 per cent of the nectarine and peach varieties released from 1990 until 1996 in the world were developed and released by the publicly funded programs (Fideghelli *et al.*, 1998).

10. APRICOT

(a) Breeding and Objectives

- ✰ Tree characteristics (rootstocks, tree vigor, growth habit, and productivity)
- ✰ Floral biology (flowering date, intensity, and fertility)
- ✰ Fruit characteristics (maturity, size, firmness, color, taste),
- ✰ Disease resistance and climatic adaptation (cold hardiness, chilling requirement, and spring frost)

Source of different Traits for Apricot Improvement

Trait	*Gene Source*
Low chilling (200-400 hrs)	Adedi, Amal, Arengi, Bedri, Carrot, Early Gold, Hamidi, Jazil, Louzi Precoce, Zaragosa
Large fruit size	Harcot, Goldrich, Hargrand, King Nugget, Perfection, Sundrop, Venus Large Early
Firm flesh	Goldrich, Hargrand, Harogem, NJA 32, Sundrop, Abate, Roksana
Drying types	Khurmai, Supkhani, Isfarak, Yerevani, Sateni, Hasanbey, Blenril, Moorpark, Hungarian Best, Babai, Royal, Musa, Blenheim
Canning types	Boccuciza Spinosa, Palummella, San Casstrese, Viragold, Budapest, Hargrand, Harlayne, Goldest, Harglow
Freezing types	Cafona, Don Gactano, San Francesco
Sweet kernel	Moorpark, Harcot
Early Maturing	Abid, Bedri, NJA 43, 53, 54, 87, Zaragozano, Jerseycot, Jubilee, Large Early, Ranna Samarkand
Late maturing	Begeron, Budapest, Comander, Litorol, Olimp, Sulmona

(b) Achievements

Hybridization was started in 1969 at Chaubattia (Seth *et al.*, 1976) crossing 'Charmagz' reciprocally with 'Shipley's Early', 'Kaisha', 'New Large Early' and 'St. Ambroise.' Out of hybridization programme three desirable hybrids were selected and released after evaluation. These apricot hybrids are 'Chaubatia Alankar', 'Chaubatia Madhu' and 'Chaubatia Kesri'.

Apricot Breeding Program in Major Countries

Country	Objectives (Besides the extension of the ripening season and fruit)
Austria	Resistance to PPV
Australia	Processing (drying)
Bulgaria	Resistance to PPV and diseases; adaptability; tree size
Czech Rep	Resistance to PPV and cold
France	Resistance to phytoplasms (ECA), PPV and bacterial diseases
Greece	Resistance to PPV; processing (canning)
Hungary	Resistance to cold, self-compatibility
India	Large fruit-size and processing drying
Italy	Resistance to PPV, diseases and cold; tree size
Japan	Resistance to bacterial diseases
New Zealand	Keeping quality (post-harvest)
Rumania	Resistance to cold and diseases
Russia	Resistance to cold and diseases
South Africa	Keeping quality (post-harvest); processing
Spain	Resistance to PPV; self-compatibility
Turkey	Processing (drying); resistance to brown rot and cold
Ukraine	Resistance to cold and to brown rot
U.S.A	High sugar content; processing (drying)

(c) Rootstocks, Breeding and Achievements

New apricot rootstocks should be tolerant or resistant to nematodes, diseases, insects, edaphic factors, with good propagation efficiency and rooting ability. The evaluation for the interaction between the root system and the canopy of the scion cultivar must be optimized to increase tree efficiency. The selection of inter specific hybrids for rootstocks is one solution to meet complex objectives in a rootstock-breeding program (Dosba 2003).One of the most intensely active areas of stone fruit rootstock breeding is the breeding for nematode resistance.

Many apricots (*P. Armeniaca* L) produced throughout the world are grown on seedling rootstocks of the same species (Crossa- Raynaud and Andergon 1987). Often such seed may be a mixture collected from several cultivars such as Franco used in Italy which is derived from Canino, Reale d Imola, Baracca, Prete and San Castrese apricot cultivars. In contrast the seedling rootstock Manicot (GF 1236) which was produced by INRA in France, is the single cultivar and is much more homogenous. Others rootstocks of such type are A470 and A474 from France, Haggith from Canada and Millioner 1 and Millioner 11 from Germany. Peach seedling is still used occasionally as rootstocks for apricots and seedlings such as GF 305, Missour and Rancho resistant have been tested widely. Seedling of Myrobalan (*P.Cerasifera*) is also utilized and almost all apricot producing countries have their own selections.

While selection of Myrobalan (*e.g.*, Myrobalan B and Myrobalan 29 C) are also used in some countries.

Adaptation to the environment; varieties for fresh and processing; fruit quality (size, firmness, aroma, flavor, skin color, sugar, and shape); disease resistance; MAS in combination with conventional breeding, mapping of the apricot genome; resistant or tolerant rootstocks and grafting compatibility with scion cultivars, and satisfying consumer preferences are some of the researchable issues that needs attention.

11. CONCLUSION

Rootstocks and interstocks are the integral part of modern fruit production. They have profound effect on fruit size and colour, plant height, tolerance to edaphic factors and biotic stresses. World over, identified rootstocks are used routinely and also new rootstocks are being identified and released for different purposes. However, in India development of rootstock is not adequate. New rootstocks and interstocks existing are mostly obtained from exotic sources and adopted after testing. The effort should be towards collection of local or indigenous wild germplasm in different tree and vine fruit crops. A beginning has been made to collect local rootstock for different temperate and fruit and nut crops from indigenous and exotic sources, which further requires standardization of their multiplication techniques. Thereafter, these need to be popularized under mission mode to take advantages in production and also to mitigate ill effect of biotic and abiotic stresses.

12. REFERENCES

Battistini, A. and Berini, E.S. 2004. Agronomic results of victor, a semi-dwarf cherry rootstock. *Acta Hort.* **658 :** 111–113.

Bell, R.L. 1991. Pears (*Pyrus*). *Acta Hort*, 290 : 657-700.

Bell, R.L., Quamme, H.A., Layne, R.E.C. and Skirvin, R.M. 1996. Pears *In*: Janick, J.; Moore, J.N. (eds) Fruit breeding vol 1: Tree and Tropical Fruits, John Wiley and Sons, New York, pp. 441-514.

Bellini, E. and Nin, S. 2002. Breeding for new traits in pear. *Acta Hort.* **596 :** 217-224.

Bobev, S. and Deckers, T. 1999. Field susceptibility to fire blight of pome fruits in Bulgaria. *Acta Hort.* **489 :** 221-224.

Callesen, O. 1998. Recent developments in cherry rootstock research. *Acta Hort.* 468 : 219–228.

Claverie, J. 1996. New selections and approaches for the development of cherry rootstocks in France. *Acta Hort.* **410 :** 373–375.

Crossa-Raynaud P and JM Audergon 1987. Apricot rootstocks. **In:** Rootstocks for fruit crops (Eds.) Rom, R.C. and Carlson, R.F.) *Wiley Interscience*. 295-320.

Della Strada, G., Fideghelli, C. and Grassi, F. 1996. Peach and nectarine cultivarsintroduced in the world from 1980 to 1992. *Acta Hort.* **374 :** 43-51.

Della Strada, G., Fideghelli, C. and Grassi, F. 1996. Peach and nectarine cultivarsintroduced in the world from 1980 to 1992. *Acta Hort.* **374 :** 43-51.

Dosba, F. 2003 Progress and prospects in stone fruit breeding. *Acta Hort.* **622**: 35–43.

Edin, M., Garcin, A., Lichou, J. and Jourdain, J.M. 1996. Influence of dwarfing cherry rootstocks on fruit production. *Acta Hort.* **410** : 239–243.

Fideghelli, C. and Loreti, F. 2009. Monografia dei portinnesti dei fruittiferi. *Centro di Ricerca per la Frutticoltura*, Rome.

Fideghelli, C., Della Strada, G., Grassi, F. and Morico, G. 1998. The peach industry in the world: Present situation and trend. *Acta Hort.* **465** : 29-39.

Fischer, M. and Mildenberger, C. 2000. New Naumburg/Pillnitz pear breeding results. *Acta Hort.* **538** : 735-739.

Franken-Bembenek, S. 2005. 'Gisela®5' rootstock in Germany. *Acta Hort.* **667** : 167–172.

Gruppe, W. 1985. An overwiew of the cherry rootstock breeding program at Giessen. *Acta Hort.* **169** : 189–198.

Grzyb, Z.S., Sitarek, M. and Guzowska-Batko, B. 2005. Results of a sweet cherry rootstock trial in Northern-Poland. *Acta Hort.* **667** : 207–210.

Hatton, R. 1917. Paradise apple stocks. *J. Royal Hort. Soc.* **42** : 361-399.

Hrotkó, K. 1999. Performance of Hungarian Mahaleb Selections. Annual Report of the. State Hort. Soc. Michigan for 1998. MSHS, Morrice, MI, pp.124–129.

Hrotkó, K. and Magyar, L. 1998. Inbreeding of Prunus mahaleb. *Acta Hort.* **468** : 393–400.

Hrotkó, K. and Magyar, L. 2004. Rootstocks for cherries from department of fruit science, Budapest. *Intl. J. Hort. Sci.* **10** : 63–66.

Hunter, D.M., Kappel, F., Quamme, H.A. and W.G., B. 2002a. 'AC Harrow Gold' pear. *HortScience* 37 : 224-226.

Hunter, D.M., Kappel, F., Quamme, H.A. and W.G., B. 2002b. 'AC Harrow Crisp' pear. *HortSci.* **37** : 227-229.

Jacob, H. 1998 Top, Topper and Tophit: Three new late ripening plum cultivars for a profitable market. *Acta Hort.* **478** : 165–168.

Jiménez, S., Garin, A., Albás, E.S., Bterán, J.A., Gogorcena, Y. and Moreno, M. 2004. Effect of several rootstocks on fruit quality of sunburst sweet cherry. *Acta Hort.* **658** : 353–358.

Jiang, Q., Guo, J. and Zheng, S. 1999. 'Ruipan No. 4 – a late ripening flat peach.(Chinese) *Acta Hort. Sinica*, **26** : 277.

Kask, K. 1989. The tomentosa cherry, *Prunus tomentosa* Thunb. *Fruit Var. J.* **53**: 50–51.

Kester, D.E. and T.M. Gradziel 1996. Almonds (Prunus). **In:** J.N. Moore and J. Janick (Eds.). Fruit Breeding. Wiley, New York, pp. 1–97.

Kviklys, D. and Kvikliene, N. 2008. Investigations on pear rootstocks at North European climatic conditions. *Acta Hort.* **800** : 671-674.

Lanauskas, J., Kviklys, D. and Uselis, N. 2004. Evaluation of rootstocks for sweet cherry cv. Vytenu Rozine. 8th Intl. Symposium on Integrating Canopy, Rootstock and Environmental Physiology in Orchard Systems, Programme and Book of Abstracts.

Okie, W.R. (2006) Breeding peaches and plums for the Southeast and beyond. Agricultural. Research Services, USDA-ARS, Byron, Georgia

Okie, W.R. and Ramming, D.W. 1999. Plum breeding worldwide. *Hort Tech.*, **9** : 162-176.

Okie, W.R. and Weinberger, J.H. 1996. Plums. In: Janick J, Moore JN (Eds.), Fruit Breeding, vol. 1. Tree and tropical fruits. John Wiley and Sons, Inc., New York

Preston, A.P. 1955 Apple rootstock studies: Malling Merton rootstocks. *J. Hort. Sci.*, **30**: 25–33.

Preston, A.P. 1967. Apple rootstock studies: Fifteen years results with some MIX crosses. *J. Hort. Sci.*, **42** : 41-50.

Ramming, D.W. 2006. Fruit Register list 43: Plum and Plum hybrids. USDA-ARS Crop. Diseases, Pests and Genetics, San Joaquin Valley Agricultural Sciences Center, Parlier, California.

Sansavini, S. and Lugli, S. 1996. Performance of the sweet cherry cultivar 'Van' on new clonal rootstocks. *Acta Hort.* **410** : 363–371.

Sherman, W.D., Topp, B.L. and Lyrene, P.M. (1992) Breeding low-chill Japanese-type plums for subtropical climates. *Acta Hort.* **317** : 149–153.

Sotiropoulos, T.E. 2006. Performance of the pear (Pyrus communis) cultivar William's Bon Chretien grafted on seven rootstocks. *Aust. J. Exp. Agr.* **46** : 701-705.

Topp, B.L. and D.M. Russell. 1989. Breeding early ripening Japanese plums. *Acta Hort.* **240** : 27–30.

Topp, B.L. and W.B. Sherman. 1990. Potential for low-chill Japanese plums in Florida. *Proc. Fla.* State *Hort. Soc.* **103** : 294–298.

Wertheim, S.J. 2002. Rootstocks for European pear: a review. *Acta Hort.* **596** : 299-309.

Wertheim, S.J., 1997.Useful differences in growth vigour within M.9 *Acta Hort.* **243** : 76-86.

Westwood MN, PB Lombard and HO Bjorstad 1976. Performance of Bartlett pear on standard and Old Home x Farmingdale pear rootstocks. *J. Amer. Soc. Hort. Sci.* **101** : 161-164.

Westwood, M.N. 1978. Mahaleb x mazzard hybrid cherry stocks. *Fruit Var. J.* pp. 32-39.

Wolfram, B. 1996. Advantages and problems of some selected cherry rootstocks in Dresden-Pillnitz. *Acta Hort.* **410** : 233–237.

Chapter 12

Temperate Fruits: All the Year Round

Vijay Singh Thakur

Vice Chancellor
Dr Y.S. Parmar University of Horticulture and Forestry,
Solan – 173 230, H.P.
E-mail: thakurvs1606@gmail.com

1. INTRODUCTION

The temperate fruits include mainly pome fruits (apple and pear), stone fruits (cherries, peach, plums, apricots and almond) nut crops (walnut, hazalnut, pecannut, chestnut and pistachio nut) and other temperate small fruits like strawberry, persimmon, kiwifruit, raspberries, blue berries, pomegranates and grapes. These temperate fruits are genetically adapted to temperate zone climatic conditions having pronounced seasonal cycles. The cultivation of temperate fruits encounter variable climatic conditions and changing seasons during annual growth cycle *viz.* spring, summer, monsoon, autumn and winter season. All temperate fruits initiate annual vegetative and reproductive growth cycles with the onset of spring and culminate with the fall of autumn and remain dormant in winter. The nature has bestowed the temperate fruits with the unique characteristics of shedding foliage in autumn, in order to impart protection against vegaries of the low winter temperature and spring frosts. These fruits are also referred as deciduous fruits because of this characteristic of autumn leaf shedding. In the process of evolution of these fruits, it is not only the deciduous nature of fruit plants acquired for protection against severity to winters but complementary requirement of winter chilling is also essential for initiating bud break during the forthcoming. This differential requirement of winter chilling for different temperate fruits and their cultivars extend the possibility of regulating flowering, fruit set, fruit maturing and ultimately the market availability of these fruits. The different temperate fruits and their varieties exhibit variation in the period of fruit maturation and ripening depending the agro-climatic conditions where they are being grown. Hence, these inherent properties of temperate fruits, make the availability of fresh fruits to the consumers during most part of the year. At present, there are numerous temperate fruits and a large number of varieties which are suitable for cultivation in the North West and North East Himalayan regions. Some of the temperate fruits have excellent keeping quality and storability especially apples and pears, which caters to the needs of consumers for almost all the year round as fresh fruits. However, the temperate fruits like apples, pears, strawberry, and peaches are also very suitable for post harvest value addition. The third group of temperate fruits is temperate nuts which are not only nutritious but high value low volume fruit, and have property of long storage life and ease in transportation to distant markets as dry fruits like walnut, pecannut and hazalnuts. The temperate fruit can be easily stored and marketed for another 2-3 years. Therefore, the wide varietal and genetic diversity of available temperate fruits being grown in the country, makes them easily available all year round as fresh fruits, stored fruits and dried or processed products. The north Indian hilly states, Jammu and Kashir, Himachal Pradesh and Uttarakhand with its temperate agroclimatic conditions have provided an appropriate base for production and availability of temperate fruits all year round.

As the apple industry has progressed in Himachal Pradesh during last five decades especially in high hills, the efforts have also been made in stone fruits cultivation simultaneously in mid and foot hills. In this course of time, a number of other fruits like pomegranate, strawberry, persimmon, nectarines, prunes, kiwifruit etc. were also added to fruit bowl of Himachal Pradesh. Keeping in view the agro ecological zones of Himachal Pradesh, there is a large opportunity and scope to diversify the existing fruit culture both by way of cultivation of site specific fruit crops as well as different varieties of a fruit crop with variable maturity period, so that the variety of temperate fruits are available all year round to fulfil the consumer requirements and market demands.

With the global warming, the potential apple growing areas especially in lower elevation with receding snow line have become marginal for apple cultivation. At the same time the opportunities for fruit cultivation have now emerged in very high elevations and dry temperate zone, like districts of Lahaul and Spiti and Pangi areas of District Chamba in Himachal Pradesh which have become potential areas for quality apple production. On the contrary, in lower elevations apple is being replaced by pears, stone fruits and other fruit crop of diversification like kiwifruit, pomegranate, nectarines.

The strategies are required to be worked out precisely for specific crops for a particular agroclimatic situation with respect to varietal selection, planting system and orchard layout, canopy management, soil moisture, nutrition and plant protection management and post harvest management of temperate fruits in order to accomplish dream of availability of temperate fruit around the year. Table 1 indicates in brief the strategy of availability of temperate fruit around the year.

Table 1: Production of available of Temperate Fruit around the Year

Fruits	*Jan*	*Feb*	*March*	*April*	*May*	*June*	*July*	*Aug*	*Sept*	*Oct*	*Nov*	*Dec*
Strawberry		x	x	x	x	x						
Peach				x	x	x	x					
Apricot					x	x	x	x				
Almond					x	x						
Plum and Prunes						x	x	x				
Cherry					x	x						
Pear					x	x	x	x	x			
Apple					x	x	x	x	x	x		
Walnut								x	x			
Pecannut								x	x			
Persimmon							x	x	x			
Kiwifruit									x	x	x	
Pomegranate							x	x	x	x		

Some of the important temperate fruits which can be suitably cultivated under Indian conditions for the year round availability as follows:

2. APPLE

Apple is the most important temperate fruit from the point of view of area under cultivation as well as production among the temperate fruits in the world and the whole temperate fruit industry revolves around it. Apple is being cultivated in 321900 ha area in the country with the production of 2203400 MT. (FAO, 2014). Apple accounts for a share of about 57 per cent of total area under temperate fruits and about 77 per cent of total temperate fruit production. The major apple producing states are Jammu and Kashmir, Himachal Pradesh and Uttrakhand falling in North West Himalayas. Apple is a temperate fruit crop which has been grown in variable agroclimatic and geographical conditions. At present, besides the flat areas of Kashmir Valley, it is grown very successfully from mid hills to high hills and dry temperate zone ranging from 4500 m amsl to 9000 m amsl in hilly conditions of Himachal Pradesh and Uttarakhand. Apple is a crop which has a short annual growth cycle from March to October. It is a crop of great promise with its high market value, a long shelf life and storability as well as very suitable to post harvest value addition. Due to diverse genetic diversity and availability of range of very early to late maturing cultivars, as well as adaptability to very diverse agroclimatic conditions from mid hills to very high hilly regions, apple fruits are available for marketing from June to October as fresh fruit. The apple fruits can be very well stored in cold storage or CA storage condition for 3-6 months *i.e.* October to March. Therefore, apple fruits are available for the consumers almost all around the year as fresh fruit or from cold storage. Traditionally, there were only a few varieties of apple which dominated the apple cultivation *viz.* Red Delicious Starking Delicious and Rich-a-Red and pollinizing varieties Golden Delicious and Red Gold since 1960s. The varietal situation has changed in recent times as there are a large number of varieties of apple available for cultivation with high yield potential and consumer preference.

The traditional standard Delicious varieties like Royal delicious and Red Delicious are being replaced by new high colouring strains and spur types mutants evolved from the Delicious cultivars, which excel in the appearance, colouration, early maturity as well as high productivity. The spur types have an added advantage of short stature, and high rate of spur developments as compared to standard types which and therefore, convenient for orchard management practices. Thus these spur type varieties are suitable for lower elevations and warm marginal areas, where the vegetative growth is excessive and poor red colour development in fruits and spur development. In recent times, new varieties Red Velox, Jeromine, Coe Red Fuzi, Scarlet spur II and Super Chief, Gale Gala and Redlumb Gala with high yield potential, better fruit quality and extended harvesting period and Shelf life are being imported to expand the germ plasm pool.

The introduction of some low chill, and very early maturing varieties like Anna and Dorsette Golden and very late maturing varieties like Red Fuzi, Coe Red Fuzi and Granny Smith has also led to the availability of fresh apples from June to October.

3. ALMOND

Almond is one of the oldest dry fruit crop and its kernel is a very high source of energy, nutritive value. Earlier almond cultivation was raised from trees of seedling origin especially in the Mediterranian countries, and production remained either static or declined with passage of time. American growers changed this scenario and cultivated almond on fertile irrigated sites with best of agronomic methods. In short span America became the world's leader in almond production, cultivating few outstanding cultivars like Non Pareil, Merced and Ne Plus Ultra. Almond cultivation has not reached desirable level in India so far. Almond cultivation in India is mainly confined to Jammu and Kashmir, Himachal Pradesh.

In Himachal Pradesh some indigenous selection like Spillo Badanjor Spillo, Jeori, JK selections have been in cultivation alongwith few exotic introductions cultivars, Non Pareil, Ne Plus Ultra were the only ones that were commercially successful. To have almond produce over a period of time cultivars that bloom late and mature early (prior to rains) needs to be cultivated on extensive scale. Cultivars/ selection like California Papershell, Pethick's Wonder, JKS-55, and Hybrid -15 have been successfully cultivated in north Indian plains especially Punjab.

Almond grows best on deep fertile soil, under wide range of climatic condtions with ample rainfall (1000-1200 mm a year) and around 100- 700 hours chilling. The fruit is very susceptible to spring frost. South facing slopes are best for almond culture and late blooming cultivars like Pariel, Ne Plus Ultra, Jordanolo, Etnea and Desmajo rojo are most suitable for places experiencing spring frosts

Cultivars with specific traits to be considered for cultivation are Early maturing (Tardy Non Pareil, Tuono, Fascionello, Fra Guilio Grande), Late blooming (Ferragnas, Genco, Filippo Ceo, Cristomorto, Tuono) and high shelling percentage (Vesta, Non Pareil, Ne Plus Ultra, Tardy Non Pareil).

For regular production selection of varieties that bear both on spurs and two year old wood, regular annual pruning for new vegetative growth production needs to be emphasized. The cultivars with no or slight tendency to biennial bearing, self fertile and compact cultivars like Genco, Tuono, Filippoceo Ferrante, Scorsa Verde, Sannicandro needs to introduced for evaluation.

Almonds are harvested either green or dry. At lower elevation there is advantage of harvesting the fruit prior to ripening as green almond and end use for confectioneries and health food supplements are available in the market from early May to Mid June. Almonds as dry fruits are available from August to October mainly from Jammu and Kashmir and the dry temperate zone of Himachal Pradesh.

4. APRICOT

Sweet kernel type apricot has great potential for commercial cultivation in high hills and dry temperate regions, whereas, bitter kernel type for mid hills areas. Apricot varieties EMA mature 7-10 days earlier than the commercially grown New Castle has promise of cultivation in mid hills. In dry temperate region CITH 1 and 2, Helman and Rakhail Karpo and exotic introduction Nari, Shakarpara have been found promising.

5. CHERRY

Cherry is a delicious temperate fruit crop which has attained an important position amongst the temperate fruits due to its high economic returns. Cherries are of two types sweet cherries (*Prunus avium*) and sour (*P. cerasus*), but sweet cherries are cultivated in Indian conditions due to its acceptability and demand in domestic conditions.

The cherry cultivation has not picked upto the extent as other temperate fruits due to specific agroclimatic requirements and availability of suitable cultivars, lack of quality planting material and poor shelf life. Cherry has high demand both in local and distant market, hence there is need to revive its cultivation by identifying major priority areeas like self fruitful cultivars, compact growth habit with regular and higher yield, cold hardy and cultivars resistant to cracking, precocious and suitable rootstocks.

Some useful gene sources with specific desirable traits like

Self fruitful cultivars - Stella, Compact Stella

Compact growth habit - Compact Stella, Compact Lambert

Firm fleshed - Brooks, Viscout

Rootstocks - Colt, F12/1, Gisela-5, Gisela-6

In India, cherries are grown primarily in Jammu and Kashmir and Himachal Pradesh but area and production has not expanded as other crops like apple due to its very specific agroclimatic requirements and suitability to specific locations only. Cherry is a small duration early season crop and is available in the market mainly from April to mid June. Some of the important old cherry cultivars in cultivation are Early Rivers (bears regularly, resistant to cracking), Bing (moderate to heavy crop, cracking problem), Early Purple Black Heart (ripens in April good yielder), Emperor Francis (flowers early good keeping quality), Royal Ann/Napolean (late flowering firm) still hold good from commercial point of view.

The new varieties with the desirable traits are also available now like Compact Lambert (Tree dwarf, less fruit cracking), Compact Stella (Universal pollinizer, less cracking, semi dwarf), Jubilee (Early, resistant to cracking), Salmo (Firm, heavy and regular bearer), Sam (High resistance to cracking, heavy yielder), Kristen (High quality and bearer, resist cracking), Viscount (Fruit large, 8.5 g, per fruit, free stone, good quality and resistant to cracking).

6. KIWIFRUIT

The kiwifruit has shown tremendous potential in the recent years and has been assessed as a nodal fruit for diversification of mid hill horticulture. It has gained much popularity, owing to its refreshing delicate flavour, pleasing aroma, high nutritive value, abundant antioxidants, health benefits and economic viability. The fruit is a rich source of vitamin C, phosphorus and potassium. Himachal Pradesh is the first state to demonstrate its commercial cultivation in the country. Recently, its commercial cultivation has been extended to Jammu and Kashmir, Arunachal Pradesh, Sikkim, Meghalaya and Manipur. As the apple has revolutionized the economy of farmers in high hills, kiwifruit is a boon to the farmers in mid and low hills. The demonstration on the improvement in quality production have enhanced the scope of kiwifruit in mid hills, where stone fruits namely; peach, plum and apricot have shown declining trend due to their perishable nature and poor marketability. The University of Horticulture and Forestry-Nauni, Solan has developed complete package of practices for this crop which has been adopted by other kiwi growing states of the country. There are tremendous scopes to expand the area under this fruit crop because of precocity and high yields; high nutritive and medicinal value; non-occurrence of serious insect, pests and diseases; wideradaptability; multiple uses and longer shelf life.

The kiwifruit is a deciduous vine which can be grown successfully in the sub-temperate, warm and humid climatic conditions at an elevation varying from 900 to 1800 m above mean sea level. However, some varieties like Hayward can perform well up to 2000 m above mean sea level. The commercially important cultivars are Abbott, Allison, Bruno, Hayward and Monty and the pollinizer of staminate cultivars are Allison, Matua and Tomuri. For pollination, one male plant is to be planted for every 9 female plants. All the varieties available in India are green fleshed but yellow fleshed and red fleshed varieties have been developed in other countries. Such commercial varieties of repute include Zesperi Gold, Greensill, Chiko, Saanichton-12, Hong Yang and Enza Red. The kiwifruit is novel fruit with good taste, long shelf life easily transplantable without much of loss and available in Indian markets from October to November from domestic production.

7. OTHER TEMPERATE NUTS

Besides almonds, walnut and pecan, the other temperate nut crops like pistachio, hazel nuts and chestnut, though grown only to a very limited scale also need to be explored to widen the scope of nut fruit availability to the consumers. The market demand of nut crops is on increase and therefore, due consideration needs to be given towards these crops not only for domestic consumption but was from the point of view of export potential. Cold temperate regions in Leh-Ladakh, Lahaul Spiti, Kinnaur and Pangi areas of Chamba district may be very suitable for pistachio cultivation as it withstands drought and adverse soil conditions much better than other nut crops.

Among the hazelnut, the larger European species *C. avellana* and *C. maxima* have commercial importance. Hazel nut (*Corylus columa*) also grows in Indian condition and is known as Winri, Wiri or Warrawi in Kashmir, Thangi, Thangoli in Pangi (Chamba, HP), Bhutia badam in Garhwal and Kapasi in Kumaon regions of Uttarakhand. The crops is also found in the north west Himalaya at 1700-3000 m above mean sea level in Pangi area of Himachal Pradesh. European hazelnut *C. avellana* has already been introduced in Pangi area (HP) and Kashmi and its performance is yet to be assessed. Some of the hazelnut cultivars of promise are Tombul, Palaz, Cakildak, Fosa (Turkey) Nacchione Mortarella, San Giovani (Italy, Negret, Panete, Gironell (Spain), Gaeam, Poongsil (Korea), C-80-15, A81-2, A84-206, A86-1 (China) which should be introduced to expand the germplasm pool for evaluation in Indian conditions.

In Himachal Pradesh, some chestnut (*Castanea sativa*) varieties were introduced at early times but failed to perform adequately, Cultivars being self-sterile susceptible to spring frost, lack of adequate pollination. Cultivars which have attracted the orchardists interest in recent time are Abundance Crane, Nanking, Orrin. Nut mature in the month of September to November but can be stored as dry fruits for long duration would not only fetch remunerative prices but also fulfill the need of market demands. Further the demand o nut crops is fast growing due to expansion of health food industry, confectionary and processing industry.

8. PEACH

The areas which have become marginal for apple and pear cultivation due to climate change and rise in temperature has identified potential areas for peach cultivation. July Elberta is the commercially cultivated variety, but it mature late in the season and its maturity sometimes coincides with monsoon rains thus results in production of poor qulity fruits due to infection of fruit flies. Some new varieties like Suncrest, Glo haven and early maturing varieties like Red haven and Sun Haven have shown promise in peach cultivation of Nectarine, Fuzzless peaches and its varieties like Snow Queen, Silver King and May Fire and Independence can be grown successfully in areas which are marginal for apple cultivation.

9. PEAR

Pear is a temperate fruit which follows apple in importance among the temperate fruits. It can be grown in more diverse agroclimatic conditions as compared to apple and is regarded as a hardy crop to grow which can tolerate vagaries of climate and stress conditions. Pear grows in wide range of climatic condtions, require 1200 chill hours for European cultivars and about 150 chill hours for Asian pears. Moderately hot climate are favourable with annual rainfall of about 100 cm. Land at higher elevations and on northern aspect with adequate moisture is best for pear production European as well as Oriental pears. In India both types of pears, European and Oriental pears are cultivated. The European pear have high chill requirement, better in quality and very remunerative suitable for cultivation in mid to high hill region. Presently to a number of subtropical regions, this has been possible through cultivation of low chill cultivars of Oriental pears (Asian native) like Kietter, Le conte, Baggugosha, Chinese sand pear and some of them are even cultivated in the high hills of south India.

A number of European cultivars (Bartlett, Max Red Bartlett, Anjou, Flemish Beauty, Beurre Hardy, Doyne du Comice, Conference) and Asian pears (Kietter, LeConte, Baggugosha and Gola) have performed well but new cultivars with staggered ripening, have better quality, adaptability to wide ecological conditions, regular bearing and compact growth suiting high density plantings needs to be introduced to enhance productivity.

Shelf life of pear is short as compared to apple, therefore fruit should be picked when green and firm to extend the storability of fruits. Delay in picking usually limits storage life and unsuitatable for transportation to distance markets. For extending the availability of pears in market there is a pool of varieties available with different maturity period and need to be adopted from commercial point of view. The important Early season varieties are- Early china, Laxton's Superb, Kashmiri Chojuro, Hosui, Kosui, Nijisseiki suitable to low hills and warmer areas, Mid season- Bartlett, Red Bartlett, Starkrimson, Clapp's Favourite, Flemish Beauty suitable for mid hills to high hills and Late season- Conference, Doyne du Comice, Eldorado suitable for high hills.

From diversification point of view the Asiatic pears are very suited for cultivation in low hill areas upto an elevation of 4000 msl. The fruits of such cultivars are available to the consumers from May to June end. The European pears are cultivated in typical temperature conditions from 4000-8000 msl and the buttery fruits in consistency are available from June to October. Pears have high adaptability to soil and climatic conditions and are not too sensitive to warm and drought conditions so can be easily cultivated in areas which are marginal for apple cultivation.

10. PECAN

After almond and walnut, pecan nut occupies a place of significance for its value among the temperate dry fruits. In India its cultivation is gaining popularity and attracted the attention of the growers. Among the present day cultivars more than 60 per cent are of seedling origin, and this cultivation scenario is fast changing by expansion in planting of new improved cultivars. For successful cultivation, choice of pecan cultivars is very essential and be based on prevailing climatic condition, market needs, incidence of diseases, cultural practices, chilling hours, and total days needed for a cultivar to mature are important consideration.

Pecan cultivation and production is largely influenced by minimum temperature, chilling hours met during winters, frost during spring and temperature during the growing season. Being a temperate fruit pecan also needs appropriate chilling hours at a specific level of temperature to foliate and produce fruit in a normal fashion. The chilling requirement of pecan varies from 500-1000 hours

The pecan expansion in India by and large has not progressed as it should have been, the prime reasons being no consideration given for appropriate selection of climatic and soil condition. The cultivars introduced failed to adapt to the sites where they were planted except Mahan and Mohawk. Hence the following priorities needs to be seriously followed to enhance pecan cultivation so that this prime nut crop shares its due importance in the production scenario in India. The earliest introduced pecan cultivars included Burkett, Mahan, Mohawk, though very good varieties but have failed in Indian conditions which has marginalized its cultivation. New cultivars more productive with wide adaptability and increased resistance to biotic and abiotic stresses needs to be introduced and planted at suitable sites to renew fresh interest among the orchardists. Some cultivars with improved pomological trials have been identified to sustain declining pecan production *i.e.* Stuart (quality high, scab resistant, precocious), Western (55-59 kernel per cent, precocious, prolific, self pollianted), Desirable (bears early, prolific, scab resistant), Wichita (Precocious, high resistant, quality excellent), Moreland (High quality, yield, disease resistance) and Ukulinga (Precocious, high yielding, 54 per cent kernel, less irregular bearing).

For better quality nuts be dried soon after harvest at room temperature with constant air condition to avoid mould development discolouration and rancidity. Proper storage for nut quality is important. Refrigerated storage preserves natural flavour, colour and texture, prevents moulding, insect damage, staleness, decreases rancidity. The Kernel moisture should be around 4 per cent for five texture and stability.

11. PERSIMMON

Persimmon is an attractive fruit of excellent quality and is being cultivated commercially in different parts of the world. In spite of the fact that the persimmon was introduced in India in 1920, this fabulous fruit remained underutilized though it can successfully be grown in warm temperate to sub tropical areas of Himachal Pradesh, J&K, Uttrakhand and north eastern states.Since persimmon is grown in the limited pockets of the country, majority of the people are not aware about this fruit. It is a deciduous fruit and is grown successfully in Himachal Pradesh at an elevation of 1000-1650 m above mean sea level. The Japanese persimmon has at least 1000 varieties world over, which shows wide variation in size, shape and colour. Very few cultivars like Hachiya, Hyakume, Fuyu and Jiro are grown in India. The growers are left with a limited choice for selection among astringent and non-astringent cultivars. In Himachal Pradesh, mostly the plantations are of astringent varieties. Fuyu is the most popular non–astringent variety. The fruits of astringent cultivars are inedible at harvest and needs treatments before consumption.

The availability of grafted plant material is a major limiting factor in the cultivation of persimmon. This is because of variable success obtained with different methods of vegetative propagation. In India, seedlings of *Diospyros lotus* (Amlook) are used as rootstocks for persimmon. Persimmon has many varieties, which bear both staminate and pistillate flowers, some bear only pistillate while, others produce staminate and pistillate flowers in different years.The trees are mostly planted in the scattered or neglected land, which generally do not receive any fertilizers, irrigation and proper attention. One of the main problems related to non-adoption of the fruit is astringency. The maturity indices are not well defined for the cultivars being grown. In view of popularizing and producing the quality fruits, harvest management warrants greater attention. The persimmon are available from September to November.

12. PLUM

In stone fruits, plum is main fruit crop and cultivar Santa Rosa is predominantly grown and accepted cultivar in mid hills. Efforts have been made for diversification of varieties in term of early and late in maturity with better fruit firmness and quality. The varieties like Red Beaut (early), Frontier and Mariposa (late), and Black Amber and Durate have been recommended. The European type plums, prunes has also shown promise in high hills and dry temperate areas.

13. POMEGRANATE

Pomegranate is a favourable table fruit in the subtropical regions and mainly grown in Maharashtra and Western India. But in recent time, the cultivation has been extended to North Indian states Punjab, Haryana and Himachal Pradesh and Uttarakhand. In India, it is considered as a crop of arid and semi-arid regions because it withstands different soil and climatic stresses. The wild sour pomegranate grows naturally in the vast tract of mid Himalayan slopes of Jammu and Kashmir, Himachal Pradesh and Uttrakhand. Locally,it is called as *Dalim,Daru,Daaru, and Dharan*.The mature fruits of Daru are harvested from its natural habitats and arils are dried to make *anardana*. The mid-hills and low shivalik hills provide congenial climate for its cultivation and are well adapted to the growing in wild and semi-wild conditions. It thrives best under hot dry summer and cold winter provided irrigation facilities are available. The pomegranate under cultivation in the country are either of seedling origin or the varieties introduced from neighbouring countries.Kandhari Kabuli with pinkish red fruits and blood red coloured grains and sweet taste has been found to be an excellent variety for mid-hills.Bhagwa,another variety has shown promise in lower hills. Though it is an evergreen cultivar but starts shedding leaves during winter after 3-4 years of planting in Himachal Pradesh conditions. The fruits mature in the first week of September. The size, colour and sugar acid blend are better in high temperature and warm weather conditions.

14. STRAWBERRY

Strawberry is a low volume and high cash crop and can be grown in the kitchen garden for home consumption. Among the soft fruits, strawberry is a most commonly and conveniently grown fruit in the plains as well as irrigated mid and high hill conditions. The strawberry cultivation has now gained momentum with large scale production on scientific lines. Many varieties are being grown in the hills of north India, Maharashtra (Mahableshwar) and south India (Conoor).Near big cities; it is being grown on large scale in Uttar Pradesh, Punjab and Haryana. In Himachal

Pradesh, it is cultivated in Kullu, Mandi, Kangra, Solan, Shimla and Sirmour districts. The Paonta valley in Sirmour district is known for high quality strawberry fruits and the most popular variety is Chandler. In Himachal Pradesh, high quality, disease free runner production is being done on large scale for supply to the subtropical areas. Mostly, the fruit is consumed as fresh fruit but now gained the importance in processing industry as well. The important strawberry processed products are jams, preserves and drinks. The fruits have distinct pleasant flavour and aroma. The flavour is used in confectionary, ice creams, soft drinks and in the preparation of chewing gums. This characteristic flavour of strawberry due to volatile esters which vary greatly among varieties. A fresh fruit contains vitamins and is also rich in minerals like potassium, calcium and phosphorus.

Strawberry grows well under temperate climate. Some cultivars are also grown in subtropical climate. Day light period of 12 hours or less and moderate temperatures are important for flower bud formation. Each cultivar may have a different day length and temperature requirement. Some cultivars are adapted to short days in autumn and hard winter (septentrional) or long days in autumn and moderate winter (meriodional). Senga Sengana, Redgaunlet and Gorella are grown under septentrional contions, while Tioga is an important cultivar of meriodional region. It can make flower buds under long day conditions. Besides, there are several day neutral strawberries like Selva, Fern, Muir, Hecker, Tristar and Tribute that have made their cultivation possible in different times of the year. Day neutral strawberries are higher yielders. Chandler is the most commercial variety in northern India and the fruits of this variety are of exceptionally high desert quality with outstanding colour and flavour. Temperature plays a critical role in the development of strawberry at a particular place. The availability of strawberry fruits depends uon the climatic conditions and the variety. In lower elevations, the fruits start maturing in last week of January to March in early maturity cultivars like Ofran and Winter Dawn. In mid hills, it comes in the market during March to May and in high hills during April to last week of June.

15. WALNUT

Among the nuts, walnut cultivation is the main stay of nut cultivation in India and the entire present day cultivation is of seedling origin and that to of Persian walnut. This seedling population over a long period of time has accumulated marked genetic variability in number of characters like shape, size, colour, hardness of shell, kernel flavour and colour. This diversity has provided a strong platform of walnut cultivation in India. Kashmir, followed by Himachal Pradesh and Uttarakahnd has exhibited enormous variability, which from time to time has been surveyed and superior genotypes based on tree, nut and kernel parameters were selected, isolated, multiplied and were recommended for cultivation. Selections like Karan Walnut and Bulbul (Kashmir) are known for their thin shell and high shelling percentage. In recent times other selections which have become popular are Hamdard, Suleiman and CITH series (CITH 1-10)

Despite suitable climatic conditions and genotypes of seedling origin the walnut cultivation has not attained the status which it deserves. Major constraints are lack of suitable cultivars based on microclimate, bearing habit, water requirement in relation to nut quality. Lack of adequate pollination (non synchronization), cultivars grown are terminal bearers and is a major hindrance in consistent and higher production of quality nuts. Walnut has wide adaptability to soil and climatic conditions and needs 700-1500 chill hours for successful growth, flowering and fruiting. Chilling, spring frost, rainfall and summer heat are important factors to be given due consideration.

To enhance walnut production certain priority areas needs special attention *i.e.* selection of precocious and lateral bearing cultivar, thin shelled high quality kernel (flavour, colour), high shelling percentage (55 per cent), selection of superior types from the indigenous plantation. Presently there are large number of cultivars that are compact, lateral bearing, precocious available for cultivation like Hartley, Payne, Franquette.

Some important walnut varieties which needs to be considered for future plantations are Hartley (high productivity, precocious), Paynee (High fruitfulness of lateral buds), Franquette (good pollinizer, medium of late blooming), Ashley (90 per cent lateral buds fruitful, early bearing, 54 per cent kernel), Serr (Mid blooming, 59 per cent), Sunland (58 per cent kernel mid season, very productive), Chico (excellent pollinizing cultivar with high quality Kernels), Vina (90 per cent lateral buds fruitful, high quality), Pedro (extended pollen shedding time), Chandler (49 per cent kernel, good seal), Tehama (mid season, good pollinizer).

The walnut harvesting is accomplished from mid September to October end begins earlier at warmer sites (low altitudes and later at higher altitudes (cold places). Walnuts are dry fruits with long self life and can be easily marketable for two years without rancidity.

16. CONCLUSION

It can be easily concluded for the foregoing information about the temperate fruit industry of India that despite the limited area with suitable temperate agroclimatic conditions in Northern Himalayan region extending from north west to north east of India there has been a significant growth in temperate fruit cultivation in past five decades especially in Himachal Pradesh, Jammu and Kashmir and Uttarakhand. The temperate fruits are the most remunerative crops for the hilly regions of the country which have poor soil depth, sleep slops, scarcity of irrigation facilities and marginal conditions unsuitable for field crop like cereals, pulses and oil seed crops. Keeping in view the huge market demands of temperate fruits, the country as well as the export potential, the expansion of temperate fruit industry has been boosted in these hilly states. Almost all the temperate fruits can be successfully grown in these states as temperate climatic conditions for the cultivation are available due to the vicinity of Himalayas fulfilling the chilling requirements of these fruits.

The variable agroclimatic conditions and available varietal diversity in the temperate fruits being grown in the country has made it possible to market these fruits almost the year round. Apple cultivation and production has been revolutionized by adoption of modern scientific orchard management practices, introduction of new varieties and rootstock and expansion of controlled atmospheric/ cold storages and temperature controlled transport system in the recent years. But still, the productivity of apple is far less as compared to developed western countries. Therefore, with more efforts and new technologies need to be generated and adopted to boos the productivity levels. Similarly, the pace of development in other temperate stone fruits and nut crops has also to follow the path of apple industry, to make the temperate fruit orcharding more remunerative and sustainable.

Chapter 13

Temperate Fruit and Nut Genetic Resources: Status and Priorities

K. Kumar[1], Amit Kumar[2] and Dinesh Singh[2]

[1]Department of Fruit Science,
Dr Y.S. Parmar University of Horticulture and Forestry,
Solan – 173 230, H.P.
[2]Division of Fruit Science
SKUAST (Kashmir), Srinagar – 191 121, J&K
E-mail: drkrishankumar@gmail.com

1. INTRODUCTION

Genetic resources comprising cultivated forms, land races and wild relatives are pivotal to any crop improvement, and temperate fruits and nuts are no exception. Almost all the commercially grown deciduous fruits in the hilly regions have relied upon the exotic introductions made over past more than a century period. The process is continuing. These have either been used directly as cultivars based upon their performance under a particular set of climatic conditions or they have remained a part of the germplasm holdings. In nut crops, however, there have been some efforts to harness genetic variability from seedling reared natural stands growing in semi-wild state. Of late, developmental agencies with funding from the Government have been instrumental in large scale import of planting material of temperate fruit species and directly distributing the growers for adoption. On one hand, this step aimed at replacing the age-old cultivars which carry poor commercial and market acceptance, on the other hand, the aim was to expand area under these crops with improved cultivars and rootstocks for improving overall productivity. As far as the status of temperate fruit and nut genetic resources is concerned, various SAUs and ICAR Institutes located in the north-west hilly regions are maintaining field germplasm collections augmented through direct import, field explorations and mutual exchange. In the past, attempts have been made by many to compile the total accessions available who also laid emphasis on the documentation, characterization, protection and conservation (Yadav, 1988; Sharma, 1993; Sharma and Kumar, 1994a; Sharma *et al.*, 2005; Kumar and Sharma, 2005a; Sharma and Kumar, 2006; Ahmad *et al.*, 2010; Gupta, 2011). Since these repositories include duplications and there is also many of the accessions carry synonyms and lack of updating, it may not be possible to draw exact figures of fruit-wise germplasm holdings in the country as a whole. Yet, based upon available information from published resources and personnel communications (Table 1) gives an update on the present status of temperate fruit and nut genetic resources in India. The current scenario of global hi-tech temperate fruit industry witnessing high productivity, leaves us many issues to address to make temperate fruit growing in India more remunerative and competitive. These issues are wide ranging and encompass low productivity, persistence with seedling rootstocks, dependence upon pesticides, non-suitability for processing, intellectual property regimes, *etc.* Now the question arises how to go about it especially in the light of climatic disturbances adversely affecting flowering and fruiting, coupled with inflow of fresh fruits from foreign markets due to liberalization. In general, the thrust has been to supplement the on-going efforts since the past several decades. As a result, we are at cross-roads ? For instance, over dominance of apple (monoculture), traditional farming, conventional mind-set, etc. are being felt as a major hindrance to species and varietal diversification and shifting to modern viable agro-technologies and farm practices. Well, given the perennial nature of temperate fruits and nuts, this appears to be a daunting task otherwise. Nevertheless, there exists opportunities in the form of relevant technologies and options available which needs to be managed effectively for harnessing maximum benefits.

On the front of genetic resources, based on past experiences, the future efforts should be targeted keeping in view the regional peculiarities, available know-how and resources. The major areas where the genetic resources can make impact are :

- ✰ Widening the genetic base of cultivated forms.
- ✰ Bringing untapped fruit crops in cultivation fold.
- ✰ Harnessing the existing genetic variability.
- ✰ Growing for specific use/specific environment.

2. WIDENING THE GENETIC BASE OF CULTIVATED FORMS

There is an immediate and absolute need to overcome monoculture. At present, commercial plantations of temperate fruits and nuts are over dominated by 'Delicious' apples, 'Bartlett' pears, 'Elberta' peaches, 'Newcastle' apricots, 'Santa Rosa' plums and others. Notwithstanding the attempts made culminating into recommendation of selected cultivars, excepting 'Fuji' and 'Gala' strains, most apple introductions made in last one or two decades belong to 'Delicious' group, and so is the case with pears and peaches. Evaluation studies are continuously being carried out by various workers at different centres to screen the available germplasm and select the individual genotypes exhibiting one or more desirable horticultural characters.

Table 1: Status of Temperate Fruit and Nut Genetic Resources in India

Fruit/Nut	*YSPUHF Solan*	*NBPGR Shimla*	*SKUAST-K Srinagar*	*SKUAST-J Bhadarwah*	*NBPGR Bhowali*	*IARI Shimla*	*CITH Srinagar*	*UUHF Bharsar*	*PAU Ludhiana*	*TNAU Kodaikanal*
Almond	83	15	48	08	–	25	26	07	–	–
Apple	203	144	219	26	29	73	210	40	08	11
Apricot	47	28	18	07	09	23	53	12	–	–
Cherry	28	07	22	03	–	13	28	03	–	–
Hazelnut	04	09	02	–	–	–	09	01	–	–
Kiwi	07	08	06	03	07	06	06	05	–	–
Peach	46	41	35	08	12	–	23	13	22	04
Pear	126	60	44	08	10	–	39	08	08	04
Pecan	15	11	08	–	–	–	02	–	–	–
Persimmon	04	05	05	–	03	–	03	–	–	–
Pomegranate	48	155	08	–	–	–	20	–	–	–
Plum	45	32	31	04	06	–	23	09	02	07
Ribes spp.	03	06	–	–	30	–	05	–	–	–
Rubus spp.	12	24	–	–	21	–	06	–	–	–
Strawberry	32	24	30	10	82	39	38	08	–	–
Walnut	69	83	42	08	27	–	248	16	–	–
Wild and minor fruits	21	39	09	–	20	12	35	04	–	–
Total	**793**	**691**	**527**	**85**	**256**	**191**	**774**	**126**	**40**	**26**

Source: Sharma and Kumar, 2006 ; Ahmad *et al.*, 2010; personal communications.

Table 2: Selected Reports on Evaluation of Exotic Introductions

Almond	Kumar *et al.*, 1993; 2012; Dalal *et al.*, 2004
Apple	Kumar and Verma, 2001; Majid *et al.*, 2006, 2008; Sharma and Chauhan, 2008; Kumar and Mir, 2012a, 2012b; Srivastava *et al.*, 2013; Hassan *et al.*, 2015
Apricot	Srivastava *et al.*, 2006; 2012; Sofi *et al.*, 2009
Cherry	Bhat *et al.*, 2002
Peach	Joshi *et al.*, 2005; Singh and Kumar, 2012; Singh *et al.*, 2014 Pear Sharma *et al.*, 1997; Kumar and Chandel, 2010
Plum	Singh *et al.*, 2011a; Kumar *et al.*, 2013
Pomegranate	Singh, 2012; Verma *et al.*, 2012
Strawberry	Joshi *et al.*, 2006; Das *et al.*, 2012; Singh *et al.*, 2012
Walnut	Sharma *et al.*, 2014 b, c

On-going systematic evaluation work has yielded very good information on the performance of these introductions and those which have performed better in one respect or the other *vis-a-vis* existing commercial counterparts are enlisted below :

Table 3: Performance Based Promising Temperate Fruit and Nut Exotic Introductions

Scions	
Almond	Nikitskyi, Prianyi, Waris, White Brandis
Apple	Super Chief, Scarlet Spur II, Gale Gala, Coe Red Fuji, Early Red One, Scarlet Gala, Anna (Low chilling)
Apricot	Ema, Harcot
Cherry	Bing, Napolean, Compact Stella, Lapins
Nectarine	Mayfire, Snow Queen, Red Gold, Independence, Silver King
Peach	Glohaven, Early Elberta, Suncrest
Pear	Sensation, Conference, Flemish Beauty, Ya-Li, Nijjeski (Low chilling)
Plum	Frontier, Red Beaut, Cacanska Zlato
Strawberry	Chandler, Selva, Camarosa, Sweet Charlie
Walnut	Franquette, Pieral, Lara, Chico
Rootstocks	
Apple	MM 106, MM111, Merton 793
Pear	BA 29, Quince C
Prunus	Myrocal. Montclar, Ishtara, Jaspi, Gisela 5

Still many of the introductions besides the above, have not found acceptance amongst the orchardists. Thus warranting more suitable cultivars with different blood to broaden the narrow genetic base. With the availability of genetically diverse and better performing individual genotypes, this vulnerable situation can be checked. Since genetic improvement of any crop species is a never ending process, as such there is and there will always be a priority to augment need-based germplasm holdings comprising improved genetic stocks for their utilization in future breeding programmes. Some of the fruit crop wise priorities which require immediate attention, subject to strictest possible quarantine measures, are as under :

Table 4: Objectives of Breeding Important Fruit and Nut

Fruit/Nut	*Desirable Characters*
Almond	Self-fertile, resistance to gummosis, late flowering, early maturing
Apple	Varieties meeting export standards, early colouring, sweet, scab resistant, suitability for processing, super-spur type, columnar growth habit
Apricot	Early maturing, improved fruit quality, drying-type, sweet kernels
Cherry	Wider adaptability, resistance to cracking, self-compatible, dwarfing
Peach	long shelf-life, resistance to *Phytoplasma,* leaf curl, nematodes, dwarfing
Pear	Coloured strains, low perishability, devoid of grit cells
Plum	All through sweet, firm, extended ripening season
Rootstocks	Dwarfing, tolerance to biotic/abiotic stresses, amenability to propagation
Strawberry	Large fruit size, ever - bearing, long shelf-life
Walnut	Precocious, lateral bearing, soft shelled, high oil content

Presented below is a selected list of genetically improved scion and rootstock cultivars given the importance of specific trait (s) which may be introduced, evaluated and utilized for enhancing productivity of temperate fruit and nut crops :

Table 5: Genetically Important Scions and Rootstocks

Scions	
Almond	Wood Colony, Winters, Durango, Ayles, Guara, Mancayo, Carmel, Sonora
Apple	Ambrosia, Ariane, MacExcel, Liaofu, Adams Apple, Supermac, W A 2, Co-op 31, Ariane, Crimson Gala, Spur York
Apricot	Robada, Orangered, Katy, Goldbar, Golden Sweet, Tomcot, Goldstrike
Cherry	Chelan, Attika, Tieton, Index, Attika
Nectarine	Arctic Glo, May Grand, Honey Kist
Olive	Arbequina Clone i-18, Arbosana Clone i-43, Koroneiki Clone i-38
Pecan	Hopi, Nacono, Lakota, Apache, Elliott
Peaches	Garnet Beauty, TexKing, Regina, Vulcan, P F Lucky 12, Reliance, Vinegold, Virgil, Red Globe
Pear	Orcas, Rescue, Cascade
Plum	Aphrodite, Black Splendor
Strawberry	Valley Sunset, V 151, Sapphire, Rubygem
Walnut	Sexton (UC90-31-10), Gillet (UC95-22-26), Forde (UC95-26-37), Vina, Howard
Rootstocks	
Apple	Pajam-1, Pajam-2, NAKB T-337, Geneva Series
Pear	Pyro, Pyrodwarf
Prunus	Hiwatha, Nickels, Viking

3. HARNESSING EXISTING GENETIC VARIABILITY

Considerable scope lies in naturalised gene pool of temperate fruit and nut species with respect to trait specific varied forms. Several surveys and explorations have been conducted in the past particularly to tap superior or elite trees from wild populations of pome and stone fruits in India (Kishore and Randhawa, 1993). In crops like walnut, almond, pecan, raspberry, hazelnut and wild pomegranate, the genotypes of seedling origin excelling in one or more desirable characters have been selected and conserved (Ahmad *et al.*, 2008; Dwivedi *et al.*, 2002; Kumar and Sharma, 2005b; Kumar *et al.*, 2005; Kumar, 2008; Rana *et al.*, 2007; Sharma and Kumar, 1994 b, 2001, 2005; Singh and Kumar, 2005; Singh *et al.*, 2009, 2011b, 2012; Kumar and Bhan, 2010; Bhat *et al.*, 2006; Negi *et al.*, 2007; Nagi *et al.*, 2014; Sharma *et al.*, 2014a; Zargar *et al.*, 2012). Even some of them are about to enter into commercialization particularly in nut crops. Notable examples are almond and walnut seedling selections with good nut and kernel quality, apricots with sweet kernels, self-fruitful almonds, lateral bearing walnuts, etc. Apricot is another candidate species in which the existing genetic variability needs to be exploited further particularly from wild apricot populations in the dry temperate regions (Kinnaur, Lahaul and Spiti, Ladakh) *e.g.*, drying type with sweet kernels, high oil content, etc. Wild peaches too, warrant attention as some forms may carry tolerance to biotic and abiotic stresses such as resistance to water logging, leaf curl, gummosis and other diseases/pests. In apple (Kumar and Sharma, 2005c), natural bud sports (mutants) keep evolving spontaneously and must be selected and commercialized as they carry additional advantage of being a variant from an already adapted genotype besides the trait for which it has been selected. Again the emphasis has to be lead on resistance to diseases (*e.g.* scab, powdery mildew), improved coloration, fruit quality, etc. In order to utilize the genetic resources (Sharma and Pramanick, 2012), diversity analysis is important to initiate breeding programmes. Several workers have conducted such studies to determine relatedness between a given set of genotypes using conventional (Kumar, 2008; Verma *et al.*, 2012) and molecular markers (Kaur *et al.*, 2010; Kashyap *et al.*, 2005; Sharma and Sharma, 2010; Sharma *et al.*, 2005; Ahmad *et al.*, 2012; Mir *et al.*, 2012; Bhat *et al.*, 2013).

4. BRINGING UNTAPPED FRUIT CROPS IN CULTIVATION FOLD

Only apples, pear, peach, plum, apricot and cherry are the candidate temperate fruits in which systematic commercial orchards exist. Nuts like walnut, almond and pecan are restricted to scattered plantations comprising one or few trees of seedling origin growing in semi-cultivated conditions. Available improved genotypes in nut crops coupled with standardization of vegetative propagation methods paves the way for systematic nut orchards with a view to diversify fruit cultivation in the hills. Kiwifruit stands out as a classical example of crop introduction in the recent times. Strawberry, pomegranate and olives are other examples worth quoting. The latter showing mixed response due to region-specific problems. Plumcots, a novel fruit crop (developed through hybridization between plums and apricots) introduced recently is being tested under mid-hill conditions of Himachal Pradesh and pistachio is being evaluated under dry temperate conditions. Still there are several temperate fruit species grown on commercial scale elsewhere, in which no sustained efforts have been made in India to introduce there and grow *e.g.* blueberries, blackberries, currants, saskatoon berries, hazelnuts, carob, *etc*. There are potential areas at higher altitudes in north-western Himalayas and unexplored north-eastern hills where these crops can make an impact to expand and strengthen the temperate fruit industry through crop diversification, besides varietal diversification as already stated above.

5. GROWING FOR SPECIFIC USE/SPECIFIC ENVIRONMENT

Till now, there has been a tendency amongst the horticulturists to grow any temperate fruit variety for multiple use (s). For an instance, in case of apple, the pre-dominant fruit crop of the region, no exclusive plantations exist of varieties suitable for processing whereas several varieties are available for the purpose. In sand pears, the American Pepsi could not yield fruitful results because of grit cells. On the other hand, in the context of global warming, the temperate fruits which were doing well 2-3 decades back are not doing so well due to insufficient chilling. As such, the high chill requiring temperate fruit cultivation is shifting to higher altitudes. This warrants the need to have pertinent genotypes fulfilling the environmental and marketing requirements. In the latter case, we must introduce genotypes with longer shelf-life (fit for storage and shipment) and suitable for processing (to counter perishability). For the lower areas, low to moderate chill genotypes need to be promoted already introduced and tested. For examples, cvs. like Anna, Michal, Shlomit in apple have shown promise under warmer areas. Several pear, peach and plum varieties which require low chilling have been found promising under sub-tropical conditions at low elevations (*e.g.*, 'Ya-Li', 'Nijjeski' pears, 'Satluj Purple' plum, 'Partap', 'Shan-i-Punjab' peaches).

6. CONCLUSION

The foregoing discussion conclusively indicates that voluminous information is now available which may be considered judiciously and need to be stablized the promising genetics resources under a given situation. The available gene pool should not be confined to mere collections, rather be brought under cultivation depending upon its suitability and having desirable character(s) or else be used as trait donors to develop genetically superior cultivars. Simultaneously, a mechanism be evolved to share the germplasm and related information between various stakeholders and free exchange of plant material and free access to information generated for the benefit of public at large. The board features of the mechanism could be :

- ☆ Inventorization and characterization of available gene sources.
- ☆ Development of web enabled database for public access.
- ☆ Establishment of National temperate fruit field gene banks.
- ☆ Modalities of material transfer and benefit sharing.
- ☆ Protection of indigenous genetic resources (registration).

7. REFERENCES

Ahmad, M.F., Wani, N., Hassan, G.A., Zafar, G. and Khan, I.A. 2008. Variability in quince (*Cydonia oblonga* Mill.) population from Baramulla district of Kashmir valley. *Indian J. Genetics*, **68** : 215-218.

Ahmad, N., Mir, J. I., Rouf, R., Rather, N. A., Rashid, R., Wani, S. H., Shafi, W., Mir, H. and Sheikh, M. A. 2012. SSR and RAPD analysis of genetic diversity in walnut (*Juglans regia* L.) genotypes from Jammu and Kashmir, India. *Physiol. Mol. Biol. Plants*, **18** : 149-160.

Ahmad, N., Lal, S., Singh, S. R. and Mir, J.I. 2010. Genetic resources of temperate fruits and nuts their conservation. In: *Souvenir-National Symposium on Conservation Horticulture*, March 21-23, 2010, Dehradun (Eds.), S.S.Singh, V.Singhal, K.Pant, S.K.Dwivedi, S.K.Kamal and P.Singh), pp. 56-68.

Bhat, A.R., Sharma, A.K., Ahmad, M.F. and Singh, S.R. 2006. A note on genetic variability in seedling origin of walnut (*Juglans regia* L.)in Kashmir valley. *The Orissa J. Hort.*, **34**: pp. 110-112.

Bhat, M.Y., Khajawall, M.H., Dalal, M.A., Dar, M.A.and Bhat, A.A. 2002. Physio-chemical variations of sweet cherry cultivars at harvest stage. *SKUAST J. Res.*, **4**: 77- 82.

Bhat. Z. A., Dhillon,W. S. and Singh, K. 2013. Genetic diversity studies on some pear genotypes using simple sequence repeats (SSRs) derived from apple and pear. *Indian J. Hort.*, **70** : 1- 6.

Dalal, M.A., Farooqui, K. D. and Das, B. 2004. Studies on varietal diversity in blooming, productivity and quality characteristics of almond germplasm in Kashmir valley. *Acta Hort.*, **662** : 151-155.

Das, B., Ahmed, N. and Attri, B.L. 2012. Performance of strawberry genotypes at high altitude temperate climate under different growing environments. *Prog. Hort.*, **44** : 242-247.

Dwivedi, D. H., Kareem, A., Dwivedi, S. K. and Mir, A. A. 2002. Physico-chemical characteristics of important apricot cultivars of the cold arid region of Ladakh. *Indian J. Hort.*, **59** : 112–118.

Gupta, S. 2011. Management of temperate fruit genetic resources in India. *Acta Hort.*, 918 : 71-80.

Hassan, A., Banday, F.A., Shiekh, M., Bisati, I.A. and Hussian, G. 2015. Evaluation of morphological and physio-chemical characteristics of some exotic cultivars of apple. *Ecology, Environment and Conservation*, **21**: 229-235.

Joshi, V.K., Shah, P.K. and Kumar, K. 2005. Evaluation of peach cultivars for wine preparation. *J. Food Sci. Tech.*, **42** : 83-89.

Joshi, V.K., Sharma, S. and Kumar, K. 2006. Technology for production and evaluation of strawberry wine. *Beverage and Food World*, **33** : 77-78.

Kashyap, S., Kaur, R., Kumar, K., Sharma, D. R. and Sharma, S. K. 2005. Molecular characterization and genetic diversity in *Fragaria* genotypes as revealed by randomly amplified DNA polymorphisms (RAPDs). *Acta Hort.*, **696** : 135-142.

Kaur, R., Sood, P., Vikal, Y., Kumar, K., Saxena, B. and Sharma, D.R. 2010. Genetic characterization of walnut (*Juglans regia*) by randomly amplified polymorphic DNA. Gene, *Genomes and Genomics*, **4** (Spl.Iss.1) : 32-36.

Kishore D.K. and Randhawa, S.S. 1993. Wild germplasm of temperate fruits. **In:** K.L. Chadha and O.P. Pareek (Eds.), *Advances in Horticulture - Fruit Crops*, Vol.1. Malhotra Publishing House, New Delhi, pp. 227-241.

Kumar, A. 2008. Genetic variability studies in wild apricot (*Prunus armeniaca* L.) collections from different districts of Kashmir. *Environ. Ecol.*, **27** : 1125-1128.

Kumar, A. and Bhan, S. 2010. Physico-chemical fruit characteristics of wild apricot (*Prunus armeniaca* L.) seedling accessions from higher altitudes of district Kinnaur. *Prog. Agric.*, 10(special issue) : 45-49.

Kumar, A. and Mir, M.Y. 2012a. Varietal assessment, heritability estimates and correlation studies in apple cultivars of South Kashmir. *J. Hort. Sci.*, **7** : 81-84.

Kumar, A. and Mir, M.Y. 2012b. Assessment of chilling requirement of different varieties of apple in South Kashmir. *Prog, Hort.*, **44** : 84-88.

Kumar, J. and Chandel, J. S. 2010. Performance of various pear in Kullu valley of Himachal Pradesh. *Indian J. Hill Agric.*, **1** : 164-167.

Kumar, J. and Verma, H. S. 2001. Performance of apple cultivars under low altitude conditions of Kullu valley. *Haryana J. Hort. Sci.*, 30(3 and 4) : 139-141.

Kumar, K., Sharma, R.L. and Sharma, G. 1993. Some late flowering and early maturing almond (*Prunus amygdalus* Batsch) germplasm. *Indian J. Pl. Gen. Res.*, **6** : 61-66

Kumar, K and Sharma, S. D. 2005a. Conservation and protection of fruit crop genetic resources: options and concerns **In:** *Intellectual Property Rights in Horticultural Crops* (Eds.), K. K. Jindal and Rajan Bawa), UHF, Nauni-Solan, pp. 273-280.

Kumar, K. and Sharma, S.D. 2005b. Self-compatible indigenous almond selections: characterization and assessment. *Acta Hort.*, **696** : 65-68.

Kumar, K. and Sharma, S.D. 2005c. Genetic Resources. In: *The Apple-Improvement, Production and Post-harvest Management* (Eds.), K.L.Chadha and R.P.Awasthi). Malhotra Publishing House, New Delhi, pp. 35-60.

Kumar, D., Ahmad, N. and Verma, M. K. 2012. Studies on high density planting in almond in Kashmir valley. *Indian J. Hort.*, **69** : 328-332.

Kumar, K., Sharma, R. and Sharma, S. D. 2005. Homogamy in persian walnut (*Juglans regia* L.) selections of indigenous origin from Himachal Pradesh, India. *Adv. Hort. Sci.*, **19** : 29-33.

Kumar, K., Tandon, S., Singh, D. and Sharma, V.K. 2013. Promising Japanese plums (*Prunus salicina* Lindl.) for mid-hills of Himachal Pradesh. *Indian J. Hort.*, **70** : 283-286.

Majid, S., Banday, F. A. and Rather, G. H. 2006. Characteristics of various exotic cultivars of apple under high density plantation. *SKUAST J. Res.*, **8** : 152-157.

Majid, S., Banday, F. A., Majid, B.A. and Nazir, N. 2008. Physio-chemical characteristics of various improved cultivars of apple under high density plantation on M_9 rootstock. *Environ. Ecol.*, 26(4A) : 1739-1741.

Mir, J. I., Ahmad, N., Rashid, R., Wani, S. H., Sheikh, M. A., Mir, H. I., Parveen and Shah, S. 2012. Genetic diversity analysis in apricot (*Prunus armeniaca* L.) germplasm using RAPD markers. *Indian J. Biotech.*, **11** : 187-190.

Nagi, M., Kumar, K. and Singh, D. 2014. Promising pecan {*Carya illinoensis* (Wang) K. Koch} seedling selections. *J. Res.* PAU, **51** : 116-118.

Negi, R.K., Sharma, S.D. and Kumar, K. 2007. Evaluating grapes (*Vitis* spp.) growing wild in Kalpa Valley of Himachal Pradesh. *Indian J. Crop Sci.*, **2** : 435-437.

Rana, J.C., Singh, D., Yadav, S.K., Verma, M. K., Kumar, K. and Pradeep, K. 2007. Genetic diversity collected and observed in persian walnut (*Juglans regia* L.) in Western Himalayan regions of India. *PGR Newsletter, Italy* No.**151** : 64-68.

Sharma, D., Kaur, R., Kumar, K. and Bhardwaj, S.V. 2012. Genetic diversity among selected genotypes of almond *Prunus dulcis* Miller D. A. Webb assessed by random amplified polymorphic DNA (RAPD) markers. *African J. Biotech.* **11:** 14877-14883.

Sharma, D. P. and Chauhan, J. S. 2008. Performance of apple cultivars under cold dessert condition of North-Western Himalayas. *Acta Hort.* **772** : 199-201.

Sharma, G., Kumar, K. and Sharma, R.L.1997. Promising pear germplasm for mid-hills of Himachal Pradesh. *The Hort. J.*, **10** : 7-14.

Sharma, G. and Sharma, N. 2010. Molecular characterization of diversity and relationship among almond [*Prunus dulcis* Miller (D.A. Webb)] cultivars and indigenous selections. *Indian J. Biotech.*, **9** : 378-383.

Sharma, R. and Kumar, K. 2005. Selecting walnuts from seedling tree populations in some villages of district Sirmour (HP). *Acta Hort.*, **696** : 85-90.

Sharma, R.L. 1993. Genetic resources of temperate fruits. In : KL Chadha and OP Pareek (Eds.), *Advances in Horticulture - Fruit Crops,* Vol.1. Malhotra Publishing House, New Delhi, pp. 243-266.

Sharma, R.L. and Kumar, K. 1994a. Temperate fruit crop improvement in India. **In** : *Progress in Temperate Fruit Breeding* (Eds.) H. Schmidt and M. Kellerhals). Kluwer Academic Publishers, Dordrecht, The Netherlands, pp. 149-156.

Sharma, R.L. and Kumar, K. 1994b. Genetic diversity and scope of walnut improvement in India. **In** : *Progress in Temperate Fruit Breeding* (Eds.), H. Schmidt and M. Kellerhals). Kluwer Academic Publishers, Dordrecht, The Netherlands, pp. 447-449.

Sharma, R. M., Kour, K., Singh, B., Yadav, S., Kotwal, N., Rana, J.C. and Anand, R. 2014a. Selection and characterization of elite walnut (*Juglans regia* L.) clone from seedling origin trees in North Western Himalayan region of India. *Australian J. Crop Sci.*, 8(2) : 257-262.

Sharma, S., Kumar, K. and Kumar, A. 2014b. Characterization of walnut (*Juglans regia* L.) for nut and kernel quality attributes. *Green Farming,* **5** : 866-868.

Sharma S, Kumar, K. and Kumar, A. 2014c. Vegetative growth and phonological characteristics of diverse walnut (*Juglans regia* L.) germplasm. *Environ. and Ecol.,* 32(4a) : 1475-1482.

Sharma, S.D. and Kumar, K. 2001. Preliminary evaluation of hazelnut seedling trees native to India. *Acta Hort.,* 556 : 29-35.

Sharma, S.D. and Kumar, K. 2006. Temperate Fruit Diversity : Status and Prospects. In: *Temperate Horticulture: Current Scenario* (Eds.) D.K.Kishore, Satish K.Sharma and K.K.Pramanick). New India Publishing agency, New Delhi, pp. 27-34.

Sharma, S.D., Kumar, K., Gupta, S., Rana, J.C, Sharma, B.D. and Rathore, D.S. 2005. Temperate Fruits. **In:** *Plant Genetic Resources: Horticultural Crops* (Eds.) B.S.Dhillon, R.K.Tyagi, S.Saxena and G.J.Randhawa). Narosa Publishing House, New Delhi, pp.146-167.

Sharma, Y.P. and Pramanick, K.K.2012. Utilization of plant genetic resources for the improvement of temperate fruit crops. *Indian J. Gen.,* **72** : 130-135.

Singh, D.B., 2012. Characterization and evaluation of pomegranate (*Prunica granatus* L.) germplasm under Indian arid ecosystem. *Indian J. Pl. Gen. Res.,* **25** : 139-145.

Singh D, Yepthomi, G. and Kumar, K. 2014. Performance of some low chill peach, *Prunus persica* (L.) Batsch germplasm accessions for fruit quality traits in Himachal Pradesh. *International J. Farm Sci.,* 4(3) : 72-80.

Singh, D. and Kumar, K. 2005. Walnut selections from Chamba district of Himachal Pradesh. *Indian J. Pl. Gen. Res.* **18** : 275-278.

Singh, D. and Kumar, K. 2012. Assessment of low chill nectarines (*Prunus persica* var. *nucipersica*) under mid-hill conditions of Himachal Pradesh. *Crop Improvement,* 39 (Spl.Issue) : 581-582.

Singh, D., Kumar, K. and Sharma, V. K. 2009. Evaluation of raspberry (*Rubus ellipticus* Smith.) genotypes growing wild in the North-western Himalayas for berry quality traits. *Indian J.Agric. Sci.* **79** : 913-916.

Singh, D., Sharma, V. K. and Kumar, K. 2011a. Evaluation and characterization of some exotic plum germplasm accessions for fruit quality traits. *J. Hill Agri.,* **2** : 59-62.

Singh, D., Sharma, V. K. and Kumar, K. 2011b. Selection of indigenous pecan [*Carya illinoensis* (Wang) K. Koch.] trees for better nut and kernel quality characters. *J. Hill Agri.,* **2** : 85-89.

Singh, S.R., Srivastava, K.K., Sharma, M.K., Singh, Lal and Sharma, V. K. 2012. Screening of strawberry (*Fragaria x ananassa*) vatieties under organic production system of Kashmir valley. *Indian J. Agric. Sci.,* **82**: 538-42.

Sofi, A.A., Verma, M. K., Ahmad, N., Verma, R. K., Kumar, D., Mir, J. I., Singh, D. B., Chaudhary, H., Lal, S., Singh, S. R., Srivastava, K. K., Shiekh, M. A., Pushkar, B. K. and Dar, G.A. 2009. New apricot varieties released. *Indian J. Hort.,* **66** : i-iii.

Srivastava, K.K., Das, B. and Banday, F. A. 2006. Evaluation of apricot genotypes for yield and quality attributes under Kashmir conditions. *Indian J. Pl. Gen. Res.* **19** : 134-136.

Srivastava, K.K., Sharma, M. K. and Bhat, S. K. 2012. Quality attributes of apricot (*Prunus armeniaca*) introductions for north-western Himalayan region. *Indian J. Agril. Sci.*, **82** : 78-80.

Srivastava, K.K., Ahmad, N., Das, B., Sharma, O. C., Singh, S.R. and Rather, J.A. 2013. Genetic divergence in respect to qualitative traits and their possible use in precision breeding programme of apple (*Malus x domestica*). *Indian J. Agril. Sci.*, **83** : 1217–1220.

Verma, M.K., Ahmad, N., Singh, A.K. and Awasthi, O.P. 2010. Temperate tree fruits and nuts in India. *Chronica Hort.*, **50**: 43-48.

Verma, M.K., Lal, S. and Ahmad, N. 2012. PCA and genetic divergence analysis in Indian pomegranate (*Punica granatum* L.) cultivars. *Prog. Hort.*, **44** : 265-270.

Yadav, I.S. 1988. Germplasm collection of pome, stone and nut fruits in India. *AICRP on Subtropical and Temperate Fruits*, CISH, Lucknow, pp. 72.

Zargar, K.A., Srivastava, K.K., Gulzaffar, Bhat, S.K. and Singh, S.R. 2012. Assessment of genetic diversity using quantitative and qualitative traits in hazelnut. *Indian J. Pl. Gen. Res.*, **25** : 298-303.

Chapter 14

Future Fruits for Temperate Regions: Production and their Nutraceutical Importance

D.B. Singh and J.I. Mir

ICAR-Central Institute of Temperate Horticulture,
Srinagar – 190 007, J&K
E-mail: deshbsingh@yahoo.co.in

1. INTRODUCTION

The identification of new species, their characterization, conservation and sustainable utilization is the key to improve agricultural productivity and sustainability, thereby contributing to national development, food security and poverty alleviation. Variation in climatic conditions and short term abrasions in weather parameters has raised levels of uncertainty, vulnerability and risk of investments in horticulture. In the face of warmer temperatures due to climate change, winter chill requirements will become harder to meet in many important temperate-fruit and nut-producing areas. Different agro ecological/phytogeographical regions of Western Himalayan Regions hold rich diversity in both the cultivated, underutilized and the wild temperate horticultural crops. The region is bestowed with climatic conditions suitable for growth and regeneration of all most kinds of temperate fruit crops. Due to this a wide range of natural population has been built up both at species and genotypic level in the region. Besides, natural physical mutagens have also helped in creation of variability in these fruits. In case of minor temperate fruits, very less attention has paid by the researchers to exploit their potential for commercial utilization. In the present context of climate change, the diversification of temperate fruit with various underutilized horticultural crops is one of the possible solutions to this major challenge as their cultivation can bridge the gap between increasing demands and supply of food.

The underutilized diversity of fruits and nuts has been categorized in to four groups as:

- ☆ Lesser utilized species confined to only local cultivation.
- ☆ Commonly cultivated species with several unidentified utilities.
- ☆ Species which have not been under cultivation practices but harvested from the wild.
- ☆ Species those which are regionally under cultivation but their value has been confirmed elsewhere under similar climatic conditions.

2. SCOPE OF UNDERUTILIZED HORTICULTURAL CROPS IN NORTH WESTERN HIMALAYAS (NWH)

In western Himalayas (WH), 80 per cent people depend upon agriculture and allied activities. Geographically, Jammu and Kashmir, Himachal Pradesh and Uttarakhand which comprises the western Himalayas are located at latitude 28°43-37°06 N and longitude 73°26 – 81°02 E, and altitude varies from 300 to 7,800 m amsl and is known for greater diversity of species of *Malus, Pyrus, Prunes, Vitis, Ribes, Rubus, Fragaria etc.* In North West Himalayan region, most of the area consists of difficult hilly terrain of different types (low, mid and high) which are unfit for cultivation of high input demanding crops. Such lands can easily be put to use for growing with low input in order to diversify the present day horticulture, and to meet the demands of increasing population, nutritional security and fast depletion of natural resources as well as the growing and changing human needs in the region. These untapped fruits are nature's wonderful gift to the mankind; indeed, they are life-enhancing medicines packed with vitamins, minerals, anti-oxidants and many phyto-nutrients (Plant derived micronutrients). They are an absolute feast to our sight, not just because of their colour and flavour but for their unique nutrition-profile that help human body free from diseases and keep it healthy. Therefore, it becomes possible to exploit the untapped potential of the region through location-specific horticulture and subsequently expanding the area under lesser known horticultural crops through adoption of modern scientific viable technologies. Underexploited horticultural crops have a vast potential for production of value-added products' with high therapeutic, medicinal values and antioxidant properties have high export potential.

3. CURRENT STATUS OF MINOR TEMPERATE FRUIT CROPS

(a) Indigenous Sources

Various ecological niches and variation in altitude in North Western Himalayas provided suitability for adaptation of minor temperate fruits.

Genus	Species
Pyrus	*P. pashia*, *P. Pashia* var. *kumaoni*, *P. pyrifolia*, *P. jacquemontiana*, *P. polycarpa*, *P. griffithii*
Prunus	*P.cerosoides*, *P. arminiaca*, *P. persica*, *P. cornuta*, *P. salicina*, *P. nepaulensis*, *P.wallichi*
Ficus	*F. carica*, *F. palmata*, *F. hispida*, *F. foreolata*, *F. nemorails*, *F. odorata*
Ribes	*R. alpestre*, *R. glaciate*, *R. griffithii*, *R. nigrum*, *R. rubrum*
Rubus	*R.ellipticus*, *R.foliosus*, *R. fruiticosus*, *R. lasiocarpus*, *R. nepaulensis*, *R. niveus*, *R. panniculatus*, *R. purpureus*, *R. ferox*, *R. rosaefolius*, *R. saxatilis*, *R. macilentus*
Sorbus	*S. cuspidata*, *S. foliolosa*, *S. micropyllaa*, *S. lanata*
Fragaria	*F. nubicola*, *F. indica*
Diospyros	*D. kaki*, *D. lotus*
Viburnum	*V. corilifolium*, *V. continofolium*, *V. grandiflorum*, *V. jacquemontii*, *V.mullaha*
Carissa	*C. congesta*, *C. grandflora*, *C. spinarum*
Hippophae	*H. rhamnoides*, *H. rhamnoides* var. *terkestania*, *H. salcifolia*, *H. tibetana*

(b) Exotic Sources

Some of the minor temperate fruit crop germplasm was introduced from various countries and have adapted well in the climate of Western Himalayas.

Genus	Species
Pineapple guava	*Feijoa slowing sellocoiniana*
Chinese *ber*	*Ziziphus jujuba*
Quince	*Cydonia oblanga*
Ceratonia	*Cerotonia siliquae*
Pistasia	*Pistisia vera*, *p. atlantica*
Rubus	*R.ellipticus*, *R. fruitcosus*, *R. lasiocarpus*,
Ribes	*Ribes* spp.
Cape gooseberry	*Physalis ixocarpa*, *P. peruvaina*, *P. pubescens*, *p. pruinosa*
Hazelnut	*Corylus malonocarpa*, *C. avellana*

4. IMPORTANT FEATURES OF IMPORTANT MINOR AND FUTURE TEMPERATE FRUIT CROPS

(a) Berries

These are considered soft fruit and include botanically different types of fruits such as blackberries, blueberries, strawberries, cranberries, gooseberries and currants and raspberries. These types are used as desserts as well as in processing. They are canned, frozen or made into jams, jellies or preserves. The juices are used in beverages and ice cream.

The most important vitamin in berries is vitamin C. Berries are also regarded as good source of β-carotene, thiamin, riboflavin and nicotinic acid. Anthocyanin is a major pigment in berries (Table 1).

(i) Blackberries (*Rubus fruticosus*)

Sweet, succulent blackberries are summer delicacies in the northern temperate regions. They too grow on shrubs known as "brambles". Binomially, brambles are a small perennial shrubs belonging to the vast *Rosaceae* family of bush berries, in the genus: *Rubus*. Depending upon cultivar type, blackberry bush can be classified into erect, semi-erect, and trailing types. Erect type bush generally features cane thorns, and spreads by root suckers (which send cane shoots) along the hedgerows. Whereas, trailing shrubs require trellis to support growth, and spread through fresh shoots known as *canes or primocanes*. During the second year, these primocanes become floricanes. White or pink flowers appear in the floricanes. Technically, blackberry is an aggregate fruit consisting of small drupelets arranged in circular fashion. Each drupelet composes of juicy pulp with single tiny seed. Each berry measures about 3-4 cm in length containing about 80-100 drupelets. *Loganberries* and *phenomenon-berries* are hybrids between blackberry and red raspberry. Several other bramble berries such as boysenberry, nessberry, youngberry, marionberry etc., are hybrids of dewberry, blackberry, and wild raspberry cultivars.

(ii) Blueberries (*Vaccinium* spp.)

Sweet, juicy blueberries are rich in natural pro-anthocyanin pigment anti-oxidants. These tiny, round blue-purple berries have long been attributed to the longevity and wellness of indigenous people living around subarctic regions in the Northern hemisphere.Botanically, it is a deciduous shrub belonging within the family of *Ericaceae*, in the genus, *Vaccinium*. Broadly, *Vaccinium* species are classified according to their growth habit as high-bush and low-bush berries. High-bush blueberry (*Vaccinium corymbosum*) is a highly branched, erect deciduous shrub with rich foliage. It grows upto 10-12 feet tall in cultivated orchards and bears clusters of small, cream-white flowers during spring, which subsequently develop into tiny berries after about two months. In the wild, high bush-blueberry is found on the edges of marshes, lakes, ponds, and streams. Low-bush blueberry (*Vaccinium angustifolium*) is a short, erect plant that grows about one-two feet in height and spread through underground rhizomes. Under the cultivated farms, it is grown as two-year cycle crop since the whole plant is either mowed down or burnt to allow new shoots that appear only during the next season.Both species require well-drained sandy, acidic soil to flourish. The shrub prefers open sunny conditions and intolerant of shade. In general, the berries can be ready for harvesting when they turn completely blue from green-pink, become soft, juicy, and sweeter. Traditionally, bluberries gathered by handpicking, and therefore, require intense labor. Soon after the berries separated from the shrub, they sorted out and transported to cold facility for storage.

(iii) Black Currants (*Ribes nigrum*)

Black currants are one of very popular, summer season berries. Indeed, they are incredibly rich in several valuable health benefiting phyto-nutrients, and anti-oxidants

Table 1: Chemical Composition of Berries

Constituent	*Blackberries* (*Rubus* sp.)	*Cranberries* (*Vaccinium* sp.)	*Gooseberries* (*Ribes* sp.)	*Raspberries* (*Rubus* sp.)	*Strawberries*) (*Fragaria* sp.
TSS (per cent)	15.2	13.0	11.1	13.9	10.2
Total sugars (per cent)	4.3	3.5	4.6	1.57-5.34	5.0
Acidity (per cent)	0.68-1.84	2.90-3.17	1.21-2.91	0.74-3.62	0.52-2.26
Ascorbic acid (mg/100 g)	20	11-33	20-50	19-38	89
β-carotene (mg/100 g)	0.1-0.59	0.02-0.58	0.18	0.05-0.08	0.15
Thiamine (mg/100 g)	0.03	0.03	–	0.02-0.03	0.03
Riboflavin (mg/100 g)	0.034	0.02	0.024-0.03	0.03-0.09	0.027
Nicotinic acid (mg/100 g)	0.40	0.10	0.30	0.40-0.9	0.6
Minerals (per cent)	0.5	0.20	0.40	0.5	0.5
Phosphorus (mg/100 g)	23.9	11.2	19.0	22	23
Potassium (mg/100 g)	208	119	170	168	161
Sodium (mg/100 g)	3.7	1.8	1.2	1.0	1.5
Calcium (mg/100 g)	63.3	14.7	18.5	22	22
Magnesium (mg/100 g)	29.5	8.4	8.6	–	11.7

that are vital for our health. Blackcurrant is a small shrub belonging to the family of *Grossulariaceae*, in the genus; *Ribes*. Currants are native to central and northern Europe and Siberia. The currant (*Ribes*) plant is a fast growing, deciduous, small shrub reaching about 5-6 ft in height. In general, currants grow best in the regions where summers are humid but winters severe and chilling. During the season, *Ribes nigrum* bears pendulous chain of small berries. Each currant berry has a size of about 1 cm in diameter, very dark purple, almost black with a glossy skin, and a persistent calyx at its apex. It can carry about 3-10 tiny, edible seeds. Currants can come in different colors. Berries of red, white and pink currants feature translucent pulp, sweeter in taste. They, however, are considered inferior to blackcurrants in terms of antioxidant power. Jostaberry are hybrids of *R. nigrum* and other wild gooseberries.

(iv) Cranberries (*Vaccinium macrocarpon*)

Unique, wild and natural by habitat, cranberries are rich in phyto-nutrients (naturally derived plant compounds), particularly proanthocyanidin antioxidants, which are essential for all-round wellness. The berries are indeed containing numerous chemical substances that may offer protection from tooth cavities, urinary tract infection, and inflammatory diseases. This berry-plant is described as an evergreen dwarf, creeping shrub or a low-lying trailing vine, belonging in the family of *Ericaceae*, in the genus: *Vaccinium*, and sub genus: *Oxycoccos*. In their natural habitat, the plant grows vigorously in acidic sandy bogs throughout the cooler parts of the Europe, Northern states in the United States and Canada. The plant is actually a dwarf, creeping shrub, or vine, which runs upto 10 to 20 cm in height. It features slender, wiry, not so thick, woody stems bearing small, evergreen leaves. Cranberry season generally lasts from October until December. The fruit is small, round, red color berry. Each berry features four centrally situated tiny seeds enclosed inside capsules. The berry is very acidic in taste, having pH in the range of 2.3 to 2.5.

(v) Raspberry (*Rubus idaeus*)

Wonderfully delicious, bright-red raspberry is among the most popular berries to relish! They are rich sources of health promoting plant-derived chemicals, minerals, and vitamins that are essential for optimum health. Botanically, raspberry is a small shrub belonging to the family *Rosaceae*, in the genus *Rubus*. It grows very well under temperate climates. The berry is native to the Europe but today widely cultivated in many temperate regions all over the world under supervised farms. Chief producers of raspberries are Poland, United States, Germany, and Chile. Several sub-species of raspberries exist. However, the most popular commercial cultivar in practice is red-raspberry. This type actually is the hybridization of *R. idaeus* (*European raspberry*) and *R. strigosus* (*American raspberry*) types. Technically, the whole berry is an aggregate of small "drupe" fruits, which are arranged in circular fashion around a hollow central cavity. Each tiny drupelet features small juicy pulp with a single, tiny whitish-yellow seed. Raspberries have a taste that varies by cultivar, and ranges from sweet to acidic, a feature quite similar to strawberries. Raspberry has conical shape, weighs about 2-4 g and contains 80-100 drupelets arranged in concentric whorls. While the most common type of raspberry (*Rubus idaeus*) is red-pink in color, hybrids actually come in a range of colors including black, purple, orange, yellow and white.

(vi) Gooseberries (*Ribes uva-crispa* L.)

Gooseberries are small, round to oval berries of European origin. They grow in the wild all over the temperate climates of Europe, North America, and Siberia. Botanically, they related very closely to currants, and belong to the same family of Grossulariaceae, in the genus, *Ribes*. Packed with pigment antioxidant polyphenolics and vitamins, the berries come in different color, flavor, and shapes. Gooseberries (*R. uva-crispa* L.) are one of the four wild Ribes species (*R. alpinum* L., *R. rubrum* L. and *R. petraeum Wulf.*) growing in the Northern Hemisphere. As in currants, gooseberry grows best in regions where summers are humid but winter is severe and chilling. Gooseberry plant is a fast growing, small deciduous shrub growing about 4-6 ft in height, featuring sharp thorns all along its woody branches. The bush begins fruiting 2-3 years after plantation. Berries come in many shapes, colors, and taste. They can be round, oval, pear-shaped or elongated, green, white, yellow, purple, red-brown or black color, sweet and tart. Their outer surface can be smooth or fuzzy (hairy) with conspicuous veins. Inside, a berry may hold 15-30 tiny edible seeds. In general, the berries measure 1-2 cm in width and weigh about 4 g to 10 g.

(vii) Strawberry (*Fragaria* x *ananassa*)

Delicious, rich-red, sweet, yet gently tangy strawberries are among the most popular berries. These berries are native to Europe, however, nowadays cultivated in many temperate regions all over the world as an important commercial crop. Botanically, the plant is a low-growing runner (creeper) belonging to the family of *Rosaceae*, in the genus: *Fragaria*. Strawberry is a small, low-lying, spreading shrub. It bears small white flowers which eventually develop into small conical, light green, immature fruits. They turn red up on maturity with each berry featuring red pulp with tiny, yellow color seeds piercing through its surface from inside. Its top end carries a green leafy cap and stem that is adorning its crown. Each berry features conical shape, weighs about 25 grams and measures about 3 cm in diameter. The berries have the taste that varies by cultivar type, and ranges from quite sweet to acidic. Although wild varieties are sometimes available in the market, large-scale production uses the modern "plasticulture" system. In this method, raised beds are formed each year, fumigated, and covered with plastic which prevents weed growth and crop spoiling.

(b) Kiwi (*Actinidia chinensis*)

Kiwi fruit, also known as *Chinese gooseberry*, is one of the delicious fruits with full of promising health promoting phyto-chemicals, vitamins and minerals. This widely recognized, wonderfully unique fruit is native to eastern Chinese "Shaanxi" province. And for the same reason, this exotic fruit is recognized as the national fruit of China. Kiwifruit plant is a semitropical, deciduous, large woody vine belonging within the family of Actinidiaceae in the Genus, *Actinidia*. The journey of this humble gooseberry from mainland China to Oceania, and thence to across the world is quite interesting. During early years of twentieth-century kiwifruit seeds were carried to New Zealand by missionaries where it has become naturalized. From New Zealand, the fruit spread all across the world, and now grown at a commercial scale in many

European and as far as in California in North America. During each season which lasts from September until November, the kiwi vine bears numerous oval shaped, fuzzy, brown colored fruits. Each kiwi berry measures approximately a large size hen's egg, and weighs up to 125 g. internally; its flesh is soft, juicy, emerald green with rows of tiny, black, edible seeds. Fruit texture is similar to strawberry or sapodilla, and the flavor resembles a blend of strawberry and pineapple fruits. Apart from the common green kiwifruit variety, several other cultivars have been in practice. "Hardy kiwi fruit" (*A. arguta*) also known as *baby kiwi,* is much smaller than "Fuzzy kiwi fruit" (*A. deliciosa*). Its size is similar to that of a large grape, with smooth, edible skin. Inside, "hardy kiwi fruit" resembles "fuzzy kiwi-fruit" in color, texture, but possess further intense flavor and sweetness. "Gold Kiwifruit," developed by hybrid technique by agricultural research department in New Zealand, has a smooth, sparse hairs, bronze skin, a pointed cap at one end and distinctive golden-yellow flesh with less tart and more tropical flavor than green-kiwifruit. Commercially, it commands a higher market value than common green kiwifruit.

(c) Quince (*Cydonia oblonga*)

Quince fruit is a member in the *Rosaceae* family of pome-fruits. Quince is rarely eaten raw but used in cooking where just a small section of this fruit would impart the whole recipe with a pleasant fruity aroma. This flavorful fruit is native to the Asia Minor regions, and once popular in the households as delicacy but these days has taken backseat in the modern day kitchens. Binomially, quince is the only fruiting tree in the genus: Cydonia. Quinces are medium sized semi-tropical deciduous trees, growing to about 10 to 15 feet in height. Pink-white flowers appear during the late spring and early summer, which develop into golden color pear-shaped fruits. The fruit is larger than average apple and bumpy; appear somewhat like large guava, avocado, or as short-necked pear fruit. Its fuzzy surface is smooth as in peaches. The quince fruit weighs about 250-750 g or more in some varieties. Inside, its flesh is light yellow, gritty and has several seeds at its center as in apples. Raw quince has intense fruity smell, and together with its bright yellow color instantly attracts the fruit lover's attention. However, raw fruits, even after ripen, generally astringent and tart in taste.

(d) Persimmon (*Diospyros virginiana*)

Persimmon fruit is a golden yellow, round or oval, flavorful, smooth textured delicacy from far East Asian origin. Its sweet, delicious flesh is packed with several health promoting nutrients such as vitamins, minerals, and anti-oxidants vital for optimum health. Botanically, persimmons belong to the family of *Ebenaceae*, in the genus: *Diospyros*. This delicate fruit is native to China. From China, tt spread to Korean peninsula and Japan very long time ago, and later was introduced to California during the middle of the 19th century. Persimmons are either multi-trunked or single-stemmed deciduous trees, which may grow up to 25 ft in height. They grow best in areas where winters are moderate and summers relatively mild. Persimmon trees classified broadly into two general categories: those that bear "astringent fruit" (whilst unripe) and those that bear "non-astringent" fruits. An astringent cultivar, which is commonly cultivated in Japan known as *"Hachiya,"* is high in *tannins* and must be allowed to ripen fully until it attains jelly-soft consistency before being fit to eat. A non-astringent persimmon, on the other hand, contains less tannin and can be eaten while it is crispy, as in apples. Astringency can be removed by treating the fruit with carbon dioxide or alcohol. During each season, the tree bears numerous fruits that vary by cultivar from spherical to heart to flatten or squash-like in shape. They also greatly vary in size from as little as a few ounces to more than a pound. The color of the fruit varies from light yellow-orange to dark orange-red. The entire fruit is edible except for the seed and calyx.

Astringent varieties: Eureka, Hachiya, Saijo, Tamopan, Tanenashi, Triumph, *etc.*

Non-astringent varieties: Fuyu (Fuyugaki), Gosho/Giant Fuyu/O'Gosho, Imoto, Izu, Jiro, Maekawajiro, Okugosho, Suruga, *etc.*

(e) Chinese Ber (*Ziziphus jujuba* Mill.)

Chinese ber *was* originated in central and southern China. It requires a small amount of chilling during winter to set this fruit, however, it can withstand in wide range of temperature (below freezing to 34°C). Average fruit weight varies 4 - 14 g with 16-21 percent total soluble solids.The ripened fruit can be stored at room temperature for 2 to 3 weeks or may be dried on candied jujube can be eaten like dry apricot.

(f) Pecan Nut (*Carya illinoensis* Koch)

Are classified in family Jugalandaceae is called as 'Queen of nuts'. It has a native of southern east United States of America. Fruit mature in September –October months. Alternate fruit bearing is more pronounced in pecan nuts. Hundred gram of Kernel is containing 71g fat, 92g protien, 14.6g carbohydrates and 2.3g fiber. Fruit also contain calcium, phosphorus, potash, iron, manganese and vitamin A.

(g) Husk Tomato (*Physalis* spp.)

They belong to the family Solanaceae and genus *Physalis.* It is herbaceous annual or perennials, mostly native of tropical north and South *America.* it is called 'Husk tomato', 'strawberry tomato', 'Tomatallo', and winter cherry' in Europe and north America, while cape –Goose berry' or Raspberry in northern part of the country. The species *P. ixocarpa* and *P. peruvaina* produces large fruit as compare to other species. Average fruit weight varies from 30 -60 g.fruit can be stored easily at ambient temperate up to three weeks during winter. Total soluble solid varies from 6.0 to 9.5 percent.

(h) Tree Tomato (*Cyphomandra betacea*)

Fruit belongs to family Solonaceae. It is a native of Peru and extensively cultivated in the whole Andern region. It is a self pollinated plant. Its occurrences in India is noticed in mid hilly area of Uttaranchal and Himachal Pradesh. Fruits are of medium size (80-100g) green before maturity, and fully ripned fruit is red with purplish strips.

(i) Hazelnuts or Filbert (*Corylus* spp.)

Nut belongs to the family Betulaceae. The common hazelnuts are classified as *Corylus americana*, the beaked hazelnut as *Corylus cornuta*, the European filbert as

Corylus avellana pontica and the cobnut as *Corylus avellana granis*. The giant filbert is classified as *Corylus maxima*. In India, Hazel nut (*C. Colurna*) is known as "Winri", "Wiri", "Warrawi", "Wuriya" in Kashmir, Thangi, Thangoli in Pangi (Chamba, HP), "Bhutia badam" in Garhwal and Kapasi in Kumaon region of Uttarnchal Pradesh.The plants are usually 2-3 m tall, each plant has separate male flowers borne in catkins and female flower borne in clusters.

(j) Sea Buckthorn (*Hippophae* spp.)

Locally called "Ames", "Chharma", "Dhurchuk", "Gartsak" and "Sarlais" is one of the wild resources in the higher Himalayas. It belongs to the family Elaegnaceae and genus *Hippophae,* grows abundantly in higher Himalayas and other countries like China, Russia and Mongolia. It is deciduous thorny shrub or small tree of 9- 12 m height above the ground with profuse branching and has a life span of 100-150 years. It grow naturally on the river side's and sun facing slopes in part of Ladakh, Chamba, Kinnuar, Lahul- Spiti, Kullu in Himachal Pradesh at an elevation of 7000 to 12,000 feet. It can resist winter temperature as low as -43°C. The fruit of sea buckthorns starts ripening from first week of October and remaining on the plant up to December. The fruits are orangish –yellow in colour and contain 90-95 percent yellow coloured sour juice. The fruits of sea buckthorn contain 12.1 percent protein, 12.3 percent fat, 9.4 percent carbohydrates 150 mg/100 g calcium, 50 mg/100 g phosphorus and 116 mg/100 g iron. The seed oil is also rich in vitamins K and E, corotenoides, flavonoides, steroids, linoleic and linolenic acid. Oil extracted from seeds, pulp, tender branches and leaves is used for making various life saving medicines and drugs to combat various diseases.

(k) Chilgoza (*Pinus gerardiana* Wall.)

It grows in abundance in Kinnaur district of Himachal Pradesh where it is called as "RI", "Rhi" or "Shangti". The seed are cylindrical, 2-3 cm long and 10-12 cm in diameter at base. The seed are eaten raw or roasted. The cones are plucked before they open and are heated to make scale expand and release the seed. The kernel contains 52.3 per cent oleic acid and 42.8 per cent linoleic acid.

5. PROCESSING AND PRODUCT DIVERSIFICATION OF BERRIES AND OTHER UNTAPPED TEMPERATE FRUITS

Berries and other untapped and underutilized temperate fruit crops are used as desserts as well as in processing. They are canned, frozen or made into jams, jellies or preserves. The juices are used in beverages and ice-cream (Table 2).

6. FUTURE FRUITS AND NUTRACEUTICALS

The term nutraceuticals, is derived from the words "nutrition" and "pharmaceutical", is a food or food product that provides health and medical benefits, including the prevention and treatment of disease. A nutraceutical is demonstrated to have a physiological benefit or provide protection against chronic disease. Their bioactive ingredients, the phytochemicals, sustain or promote health and occur at the intersection of food and pharmaceutical industries. They play a crucial role in maintaining optimal immune response, such that deficient or excessive intakes can

Table 2: Processing and Medicinal Value of Untapped Fruits of Western Himalayan Regions

Sl.No.	Name of Fruit	Products Prepared	Medicinal Value
1.	Blackberries	Juice, wine, jam and syrup	Diarrhoea, gout, arthritis, blackberry shoots boiled into a tea is said to have been used for mouth irritations such as canker sores
2.	Blueberries	Jams, juices, syrups and baked products. Also dairy, confectionery and beverages	Antiseptics, diuretics, anti-inflammatory, anti-hyperglycemic and anti-carcinogenic
3.	Chinese ber	Eaten as a snack, or with tea, canned, jujube tea, juice, vinegar, pickles and wine.	Antifungal, antibacterial, antinuclear, anti-inflammatory, sedative, antispastic, antifertility/contraception, hypotensive and antinephritic, cardiotonic, antioxidant, immune stimulant, and wound healing properties
4.	Cranberries	Concentrate, sweetened dried cranberries, frozen cranberries, sliced cranberries, cranberry puree, cranberry puree concentrate, powdered cranberry, sauce, jams, jellies *etc.*	Antiseptics, diuretics, anti-inflammatory, anti-hyperglycemic and anti-carcinogenic
5.	Currants	Jam, jelly, eaten raw	Kidney diseases and menstrual and menopausal problems, enhancer to assist women in becoming pregnant, effective treatment for pre-menstrual syndrome
6.	Gooseberries	Jam, jelly, juice, sauce and cupcakes	Appetizing, tonic, diuretic, laxative, useful in inflammations, enlargement of the spleen and abdominal troubles. The fruit is considered to be a tonic, diuretic and purgative in the Punjab.
7.	Hazelnut	To make pralines, chocolates, paste, hazelnut butter, flour	Anti cancer medication
8.	Husk tomato	Eaten raw, salads, jams, jellies, dried and pie.	Used as a remedy for abscesses, coughs, fevers and sore throats
9.	Japanese persimmon	Eaten raw, dried, fresh and cooked	Stomach ailments and diarrhea, fever, blood pressure, cough, prevents heart attack
10.	Medlar	Eaten fresh, salad, pie, jam, jelly and chutney	Soothing the throat and is a popular ingredient for cough drops, soothe the digestive and respiratory systems
11.	Pecan nut	Eaten fresh, used in sweet deserts, savory dishes and pecan pie	Lower the risk of gall stones, reduce high cholesterol

Contd...

Table 2–*Contd...*

Sl.No.	*Name of Fruit*	*Products Prepared*	*Medicinal Value*
12.	Pineapple guava	Juice, salad, sauce, eaten raw, used as dessert fruit, slices, candies, preserves, jellies and marmalade	Therapeutic mechanisms against cancer, bacterial infections, inflammation and pain.
13.	Quince	Jam, jelly, quince pudding, murabah, wine, brandy and Cider	Treat sore throat and to relieve cough, relieves intestinal discomfort. remedy for pneumonia and lung disease
14.	Raspberries	Candy jellies, jam, canned, used as ingredient in the preparation of sauces, candies, yogurts, ice creams and infusions.	Antiseptics, diuretics, anti-inflammatory, anti-hyperglycemic and anti-carcinogenic.
15.	Seabuckthorn	Juice, syrup, skin cream, pie, jam, lotions, liquors, juice as baby food and beer	Treatment for inflammatory disorders, cancer prevention, effective in bone marrow
16.	Strawberries	Juice, nectar, wine, puree and wine	Anti-oxidant, anticancer, for healthy Immune system

have negative impact on health. Phytochemicals, are non-nutritive plant chemicals that have either defensive or disease protective properties. They are nonessential nutrients and mainly produced by plants to provide them protection. Dietary intake of phytochemicals may promote health benefits, protecting against chronic degenerative disorders, such as cancer, cardiovascular and neurodegenerative diseases. Majority of foods, such as whole grains, beans, fruits, vegetables and herbs contain phytonutrients/phytochemicals. These phytochemicals, either alone and/ or in combination, have tremendous therapeutic potential in curing various ailments. Phytochemicals with nutraceutical properties present in food are of enormous significance due to their beneficial effects on human health since they offer protection against numerous diseases or disorders such as cancers, coronary heart disease, diabetes, high blood pressure, inflammation, microbial, viral and parasitic infections, psychotic diseases, spasmodic conditions, ulcers, osteoporosis and associated disorders. Majority of untapped fruits of future importance, such as strawberry, blackberry, blue berry, currants, crane berry, gooseberry, rosehips, kiwi, lingon berry, red currant, rasp berry, sour cherry, apricot, peach etc contain phytochemicals of nutraceutical importance. These phytochemicals, either alone and/or in combination, have tremendous therapeutic potential.

Carotenoids, tocopherols, ascorbates, lipoic acids and polyphenols are strong natural antioxidants with free radical scavenging activity. Endogenous antioxidants enzymes like super oxide dismutase (SOD), catalase, glutathione peroxidase, glutathione reductase, minerals like Se, Mn, Cu, Zn, vitamins A, C and E, carotenoids, limonoids and polyphenols exert synergistic actions in scavenging free radicals. The natural antioxidant system is mainly classified into two categories namely *in-vitro* and *in-vivo* antioxidants. Free radical scavengers act as hydrogen donors, electron donor, peroxide decomposer, singlet oxygen quencher, enzyme inhibitor, synergist, and metal-chelating agents. Both enzymatic and non-enzymatic antioxidants exist in the intracellular and extra cellular environment to detoxify ROS. To provide maximum intracellular protection, these scavengers are strategically compartmentalized in the cell and offer protection that can be classified as follows:

(i) First line of defense

It comprises preventing antioxidants that act by quenching of O_2-, decomposition of H_2O_2 and chelation of metal-ions and enzymes such as superoxide dismutase (SOD), catalase, glutathione peroxidase and non enzymatic molecules.

(ii) Second line of defense

The antioxidants belonging to this category of defense include glutathione (GSH), mainly ascorbic acid, alpha-tocopherol (vitamins C and E), carotenoids and flavonoids.

(iii) Third line of defense

Third line antioxidants are complex group of enzymes for repairing damaged DNA, damaged proteins, oxidized lipids and also stop chain propagation of peroxyl lipid radicals.

(iv) Fourth line of defense

A fourth line is related immune system and comes in operation when all other protections fail.

Among the phytochemicals/phytonutrients mentioned as potentially providing health benefits are polyphenols, flavonoids, isoflavonoids, anthocyanidins, phytoestrogens, terpenoids, carotenoids, limonoids, phytosterols, glucosinolates and fibres.

(a) Phytochemicals and their Health Benefits

(i) Polyphenols

Epidemiological studies provide convincing evidence that diet rich in antioxidants is associated with a lower incidence of degenerative diseases. Fruits such as apple, grape, pear, cherry and various berries are rich sources of polyphenols (Tables 3 & 4). Red wine or a cup of coffee or tea contains about 100 mg polyphenols. Their total dietary intake may be about 1g per day, which is about 10 times higher than that of vitamin C and 100 times higher than those of vitamin E and carotenoids. Ferulic acid is associated with dietary fiber linked with hemi cellulose of the cell wall by means of ester bonds. Caffeic acid in the form of caffeoyl esters and coumaric acids are common in apples, pears, and grapes. Additionally, apples and pears are rich in chlorogenic acid and grapes in gallic acid. Apples contain high levels of quercetin among fruits. Grain-derived products are especially significant in human diet as they have higher concentration of phenolic acids in the outer layers of kernel that constitute the bran. Most of the phenolic acid derivatives are hydrolysable tannins and are usually esterified with glucose. Citrus fruits are major sources of flavonones and hesperidin is found in abundance (120-250 mg/l) in orange juice.

Anthocyanins are pigments of fruits such as cherries, plums, strawberries, raspberries, black berries and red currant and their content varies from 0.15 to 4.5 mg/g in fresh berries.

People who consume traditional diets rich in soy and tea rarely experience breast, uterus and prostate cancer. Although there are a range of potentially antimutagenic fruits, vegetables and cereals but their intake is generally below the level essential to protect from various mutagens.Polyphenols historically have been considered as anti-nutrients by nutritionists, because some, *e.g.* tannins, have adverse effects such as decreasing the activities of digestive enzymes, energy, protein and amino acid availabilities, mineral uptake and having other toxic effects. Detection of the antioxidant activities of many polyphenols has reunited opinion toward the health benefits provided by many of these compounds. The most important dietary phenolics are the phenolic acids (including hydroxybenzoic and hydroxycinnamic acids), polyphenols (hydrolysable and condensed tannins) and flavonoids, the latter being the most studied group. Phenols protect plants from oxidative damage. They have also been studied extensively as antioxidant protectants for human beings and play beneficial role in reducing the risk of coronary heart disease, diabetes, hypertension and some types of cancer.

Table 3: Fruits a Source of Antioxidant Polyphenols

Polyphenols	Sources
Flavonoids (Anthocyanidins): Cyanidin 3-glycosides, Malvidin, Delphinidin, Pelargonidin	Blue berries, black berries, cranberry, raspberry, black currant, black grape, straw-berries, cherries, plums, pomegranate, juice, red wine
Flavonoid glycosides: Rutin, Hesperidin, Naringin	Orange, orange juice, lemon, grapefruit, tangerine juice
Flavanones: Naringenin, Eriodictyol, Hesperetin	Grapes, citrus fruits and their juices, tangerine juice, peppermint
Flavanols: Morin, Procyanidins Prodelphinidins, Catechin, Epicatechin and their gallates	Apricots, apples, grapes, peaches, pears, plums, raisins, berries, cherries, red wine
Anthoxanthins (Flavonols): Myricetin, Fisetin, Quercetin, Kaempferol, Isorhamnetin	Cherry, tomatoes, spinach, celery, onions, peppers, sweet potato, lettuce, broccoli, kale, buckwheat, beans, apples, apricots, grapes, plums, berries, currants, cherries, juices, ginkgo biloba, red wine, tea, cocoa
Phenolic acids: Caffeic acid, Chlorogenic acid, Ferulic acid,p-coumaric acid, Sinapic acid, Ellagic acid, Gallic acid	Lemon, peach, cider, Strawberry, raspberry grape juice, pomegranate juice bluberry, cranberry, pear, cherry, cherry juice, apple, apple juice, orange, grapefruit
Tannins: Catechin, Epicatechin polymers, Ellagitannins, Proanthocyanidins, Tannic acids	Pomegranate, walnuts, peach, olive, plum, chick pea, peas, grape seeds and skin, apple juice, strawberries, raspberries, blackberry, lentils, haricot bean, red wine, immature fruits

Prakash *et al.* (2012).

Table 4: Quantities of Antioxidant Polyphenols (mg/100g) and their Class in some Commonly Used Fruits

Source/Class of Phenols	*mg/100g*	*Source/Class of Phenols*	*mg/100g*
Black grape/Flavonol	1.5-4.6	Blueberry/Flavonol	3.2-17
Orange/Flavanones	21-70	Grape juice/Flavanones	10-66
Lemon juice/Flavanones	5-31	Cherry/Catechins	5.2-23
Chocolate/Catechins	46-61	Grape/Catechins	3-18
Peach/Catechins	5-15	Apple/Catechins	10-43
Red raspberry/Catechins	2-48	Strawberry/Catechins	2-50
Apricot/Catechins	10-25	Blackberry/Catechins	9-11
Black currant/Anthocyanins	130-400	Strawberry/Anthocyanins	15-75
Red wine/Anthocyanins	20-38	Black grape/Anthocyanins	30-750
Cherry/Anthocyanins	35-450	Plum/Anthocyanins	2-27

Prakash *et al.* (2012).

There is inverse association between intake of flavonoids and coronary heart disease mortality. Flavonoids in regularly consumed foods appeared to reduce the risk of death from coronary heart disease.

(ii) Anthocyanidins

These are water-soluble flavonoids that are aglycones of anthocyanins. These compounds are among the principal pigments in fruits and flowers. The color of these pigments is influenced by pH and metal ion complexes. The flavonoids, anthocyanidins are antioxidants *in vitro*, and might be expected to have antioxidative and anti-mutagenic properties *in vivo*. They have been found potent antioxidant activity for isolated anthocyanidins (aglycons and glycosides) and complex plant samples in the ferric reducing ability assay.

(iii) Phytoestrogens

These are non-steroidal phytochemicals (Table III) quite similar in structure and function to gonadal estrogen hormone. They offer an alternative therapy for hormone replacement therapy (HRT) with beneficial effects on cardiovascular system and may even alleviate menopausal symptoms. They are potential alternatives to the synthetic selective estrogen receptor modulators (SERMs), which are currently applied in HRT. They have antioxidant effects due to their polyphenolic nature, anti-carcinogenic, modulation of steroid metabolism or of detoxification enzymes, interference with calcium-transport and favorable effects on lipid and lipoprotein profiles On the basis of chemical structure phytoestrogens can be classified as flavonoids, isoflavonoids, coumestans, stilbenes, and lignans. They occur in either plants or their seeds.

Flavonoids have similar structure to estrogen and have the capacity to exert both estrogenic and anti-estrogenic effects and provide possible protection against bone loss and heart diseases. They share structural features with estrogen, in the sense that the presence of particular hydroxyl groups that can be positioned in a stereo chemical alignment virtually identical to one of the estrogen. Populations in China, Japan, Taiwan and Korea are estimated to consume high quantities of isoflavones and women of these countries complain fewer incidences of osteoporosis and related health problems, especially hot flushes, cardiovascular diseases, lower incidence of hormone dependent breast and uterine cancer. Although there are two isomers of resveratrol, *cis* and *trans*, but only the *trans* form has been reported to be estrogenic. It is found only in the skin of red grapes, in green grapes and white wine very low levels of *trans*-resveratrol are found.

(iv) Terpenoids

The terpenes, also known as isoprenoids, are the largest class of phytonutrients in green foods, and grains. Their importance to plants relates to their necessity to fix carbon through photosynthetic reactions using photosensitizing pigments. Animals have evolved to utilize these compounds for hormonal and growth regulatory functions (vitamin A) and, as it is now being understood, the presence of these molecules in animal tissues also provides a measure of protection from certain diseases, especially those related to chronic damage and growth dysregulation. Terpenes have a unique antioxidant activity in their interaction with free radicals. They react with free radicals by partitioning themselves into fatty membranes by virtue of their long carbon side chain.

(v) Carotenoids

They are highly pigmented, yellow, orange and red, are present in fruits and vegetables, and when consumed by birds are incorporated into the yolk of eggs. Carotenoids comprise two types of molecules, carotenes and xanthophylls. Carotenes are tissue specific in their biological activity and beta-carotene has vitamin A activity. Beta-carotene, lycopene and lutein, protect against uterine, prostate, breast, colorectal and lung cancers. They may also protect against risk of digestive tract cancer. The xanthophyll types of carotenoids offer protection to other antioxidants, and they may exhibit tissue specific protection. Zeaxanthin, cryoptoxanthin and astaxanthin are members of the xanthophylls.

(vi) Limonoids

These are terpenes present in citrus fruit. They provide chemotherapeutic activity by inhibiting Phase I enzymes and inducing Phase II detoxification enzymes in the liver. D-Limonene, the commonest monocyclic monoterpene, found within orange peel oil, inhibits pancreatic carcinogenesis induced in experimental models and also provides protection to lung tissue.

(vii) Phytosterols

These are another important terpene subclass. Two sterol molecules that are synthesized by plants are beta-sitosterol and its glycoside. In animals, these two molecules exhibit anti-inflammatory, anti-neoplastic, anti-pyretic and immuno-modulating activity. Phytosterols were reported to block inflammatory enzymes, for example by modifying the prostaglandin pathways in a way that protected platelets. Phytosterols compete with cholesterol in the intestine for uptake, and aid in the elimination of cholesterol from the body. Saturated phytosterols appear to be more effective than unsaturated compounds in decreasing cholesterol concentrations in the body. Their actions reduce serum or plasma total cholesterol and low-density lipoprotein (LDL) cholesterol. Competition with cholesterol for absorption from the intestine is not unexpected as the structure of plant sterols is similar to that of cholesterol.

(viii) Glucosinolates

They are present in cruciferous vegetables and are activators of liver detoxification enzymes. Consumption of cruciferous vegetables offers a phytochemical strategy for providing protection against carcinogenesis, mutagenesis and other forms of toxicity of electrophiles and reactive forms of oxygen. The sprouts of certain crucifers, including broccoli and cauliflower, contain higher amounts of glucoraphanin (glucosinolate of sulforaphane) than do the corresponding mature plants. Crucifer sprouts may protect against the risk of cancer more effectively than the same quantity of mature vegetables of the same variety. The mechanism of the protective effects is thought to involve the modulation of carcinogen metabolism by the induction of Phase 2 detoxification enzymes and inhibition of Phase 1 carcinogen-activating enzymes, thereby possibly influencing several processes related to chemical carcinogenesis. A reducing effect on tumor formation has been shown in rats and mice, and studies carried out in humans using high but realistic human consumption amounts of indoles and brassica vegetables have shown putative positive effects on health. Indole-3-carbinol is a glucosinolate metabolite that inhibits organ-site carcinogenesis in rodent models. Its preventive effect on human mammary carcinogenesis may be due in part to its ability to regulate cell cycle progression, increase the formation of anti-proliferative estradiol metabolite and induce cellular apoptosis.

(ix) Polysaccharides

Traditionally, polysaccharides are used as thickening, emulsifying and stabilizing agents. Nowadays, a huge hyll group. Their nutraceutical properties are able to be used in food as fat replacers, dietary fibres or prebiotics. The physiological activities include prevention of cancers, obesity, cardiovascular diseases, inhibition of tumor growth, neutralize the side effects of chemotherapy and radiation, normalize blood pressure, help to balance blood sugar, combat autoimmune disease, act as an anti-inflammatory, lower cholesterol and blood lipids, improves liver function and increase calcium absorption. Many mushrooms, fruits, vegetables, legumes, certain cereals and grains, such as oats and barley, have beneficial polysaccharides.

(b) Role of Phytochemicals in Health and Diseases

The nutraceuticals industry is now expending billions of dollars each year for the purchase of health-related foods and supplements. The respective health benefits are based on science and ethics, for health claims for functional foods, and presence of certain phytochemicals. They are constituents of plants and have certain pharmacological and/or physiological effects in the ethnomedical treatment of various disorders. Phytochemicals play an important role in human health as antioxidants, antibacterial, antifungal, anti-inflammatory, anti-allergic, antispasmodic, chemopreventive, hepatoprotective, hypolipidemic, neuroprotective, hypotensive, prevent aging, diabetes, osteoporosis, cancer and heart diseases, induce apoptosis, diuretic, CNS stimulant, analgesic, protects from UVB-induced carcinogenesis, immuno-modulator and carminative. Capsaicin, the pungent ingredient present in red pepper and ginger, has anti-carcinogenic and anti-mutagenic effects. In humans, curcuminoids, the polyphenolic phytochemicals, acts as an anti-inflammatory and cancer preventive drug. Osteoporosis is related to multiple factors including aging, hormone deficiency and diet. Most of the studies suggest that phytoestrogens are somewhat effective in maintaining bone mineral density (BMD) in postmenopausal women and to alleviate the osteoporosis and associated disorders. Evidence from several human studies demonstrates that certain dietary phytoestrogens can produce estrogenic effects in the postmenopausal women, including estrogen-like effects on vaginal cytology and reductions in hot flushes. In postmenopausal women, cardiovascular diseases (CVDs) are one of protein and flaxseed have the ability to lower total and LDL cholesterols and raise HDL cholesterol resulting in reduced risk of CVDs. There is evidence to support the hypothesis that phytoestrogen consumption contributes to the lower incidence of CVDs in Asian countries.

A large number of epidemiological studies had shown that people who consume high amounts of isoflavonoids (phytoestrogens) in their diets have lower incidences of various types of cancers including breast, prostate and colon cancer. A high plasma concentration of the mammalian lignan, enterolactone, is correlated with a reduced risk of breast cancer. Similar correlations have also been found between dietary intakes of isoflavonoids and lignans and thyroid, ovarian, and breast cancers in pre- and postmenopausal women. The incidence of hormone dependent tumors is lower in Asia and Eastern Europe where consumption of phytoestrogens is higher as compared to others. Breast, ovarian, prostate, and colon cancer show a negative correlation with phytoestrogen intake when compared with mortality rates due to cancer. The epidemiological, animal, and cell-line data suggest that phytoestrogens may play a protective role against the development of prostate and breast cancer. It has been reported that increased consumption of beans, lentils and peas, tomatoes, and dried fruits was associated with significantly decreased prostate cancer risk.

Diets rich in phytonutrients may supply a variety of phytoestrogens such as isoflavones, resveratrol, lignans etc, capable of producing a range of pharmacological effects. In females, life is affected by a variety of estrogen-related conditions such as osteoporosis, cognitive and cardiovascular decline, increased risk of breast cancer and other symptoms that decrease the overall quality of life. Phytoestrogens appear to have physiological effects in humans, with the most supportive data being related to the effects of soy protein supplements on lipids and lipoproteins and on vascular function. Therefore, postmenopausal women who have the greatest breast cancer risk should be encouraged to increase their phytoestrogen intake.

Structure-Activity Relationship (SAR) is an approach designed to find relationships between chemical structure and biological activity. As such it is the concept of linking chemical structure to a chemical property or biological activity including toxicity. The analysis of SAR enables the determination of the chemical groups responsible for evoking a target biological effect in the organism. The SAR approach therefore assumes that the structure of a molecule contains the features responsible for its physical, chemical, and biological properties of the phytochemicals responsible for their therapeutic importance in nutraceuticals.

7. FUTURE PERSPECTIVES

To strengthen the work on genetic improvement following strategies would be requiring for effective implementation and to achieve goals.

- ☆ Systematic planning for explorations for minor temperate fruit including wild texa particularly for those crops, whose economic importance is known for commercial exploitation.
- ☆ Need based introduction of new crops, new varieties with specific traits.
- ☆ Mostly minor temperate fruit crops are hardly and resilient, once established in field gene bank then agronomic practices should be developed considering the climatic condition suited to them.
- ☆ Priority should be to conserve the material in field gene bank and protocol should also be developed for *in vitro* conservation and cryopreservation as well.

- ☆ *In situ* conversation by the various government agencies and policy planners to protect the ecological niches, where large variability is existing *e.g.* hazelnut in pang, dry type apricot in Leh and Ladakh and Kinnuar chilgoza in Kinnuar, and walnut in Jammu and Kashmir.
- ☆ Molecular characterization of underutilized temperate fruits will provide the information on identification of use full genes and other molecular markers, which will help in genetic improvement work.
- ☆ There is need to investigate value addition useful for products and processing, neutraceutical, ornamental and other potential applications.

8. CONCLUSION

The North West Himalayan region is bestowed with the most favourable climatic conditions for the production of under-exploited horticultural crops. Besides this, quality seeds, varieties and hybrids of these horticultural crops could be produced and exported. The increase in area and production of these underexploited horticultural crops will not only provide nutritional security and save money on import but also export of fresh horticultural crops and seed in further expected to boost region economy. The use medicinal plants by indigenous people to treat different ailments has a long history. Very recently the scientific data supported nutritional and medical importance of phytonutrients/phytochemicals for the prevention and treatment of several diseases. Careful studies are necessary on the various phytochemicals for their roles in the prevention of chronic degenerative diseases. The ever-widening choice of food ingredients makes it possible for food designers to provide food choices that meet the public's expressed desire for healthy food. Other aspects of determining the role of phytochemicals in functional foods include consumer attitudes, any competitive advantage for manufacturers producing functional foods and identification of those areas of research needed to produce foods with the desired health effects. The future of nutraceuticals of both plant and animal origin holds exciting opportunities. Different kinds of promising indigenous/ local fruits should be exploited or improved exotic fruit cultivars should be cultivated which shall provide a greater scope for extended supply of quality fruits. Safeguard losses due to disease or insect pests, fruits for processing and to generate higher per capita income. These horticultural crops also provide many fold employment opportunities in agro-based industries, packaging and storage preservation, canning and transportation.

9. REFERENCES

Anonymous, 1996. Report on International Technical Conference on Plant Genetic Resources. Leipzig, Germany, 17-23 June.

Costantino, L., Albasini, A., Rastelli, G. and Benvenuti, S. 1992. Activity of polyphenolic crude extracts as scavengers of superoxide radicals and inhibitors of xanthine oxidase, *Planta Med*. **58**: 342.

Dillard, C.J. and German, J.B. 2000. Review Phytochemicals: nutraceuticals and human health. *J. Sci. Food Agric.*, **80**: 1744-1756.

Dorgey, T. 1998. Seabuckthorn-A wonder plant in Ladakh. Paper presented in Symposium on "Role of women in Eco-development in Ladakh" held at Divisional Forest Office, Ladakh.

Gurmeet, P. 1997. Report on the use of Ladakh Flora in Amchi System of Medicine. Amchi Research System, Ladakh.

Joshi, B.D. and Rana, J.C. 1994. Collection of temperate fruit and other wild relatives in North West Himlayas. *IBPGR, News Letter*, 15.

Kumar, V. 2003. Seabuckthorn –a potential bio resources in Himalaya. *Invention Intelligence*, **38**: 150-167.

Maikuri, R.K. and Singh, A. 1994. Ames (*Hippophae rhamnoides*) from Garhwal Himlaya-a success story. *Himlaya Payavaran News Lett.* **6**: 9-12

Namiki, M., 1990. Antioxidants/antimutagens in food, Crit. Rev. *Food Sci. Nutr.* **29**: 273.

Prakash D., Upadhyay, G., Pushpangadan, P. and Gupta, C. 2011. Antioxidant and free radical scavenging activities of some fruits. *J. Complement Integrt. Med.*, **8**: 1-19.

Prakash, D and Gupta, K.R. 2009. The antioxidant phytochemicals of nutraceutical importance. *The Open Nutraceuticals J.*, **2**: 20-35.

Prakash, Dhan., Gupta, Charu and Sharma, Girish 2012. Importance of phytochemicals in nutraceuticals. *J. Chinese Med. Res. Dev. (JCMRD)*, **1**: 70-78.

Sharma, B.D. and Rana, J.C. 1998. Genetic resources of temperate fruits in Himachal Pradesh. *Presented in the International Tree Science Conference*, New Delhi, 10-13, April, 1998.

Sharma, B.D., Rana, J.C. and Joshi, B.D. 1997. Agri-diversity in North – Western Himlayas. *Indian J. Plant Genet. Resour.* **10**: 63-67.

Sharma, B.D., Rana, J.C. and Yadav, S.K. 2001. *A Glance at Temperate Fruits Gene Bank.* NBPGR, New Delhi. 1-19.

Sharma, S.D., Kumar, K., Gupta S., Rana, J.C., Sharma, B.D. and Rathopre, D.S. 2005. Temperate Fruits. pp.146-167. **In:** *Plant Genetic Resources: Horticultural Crops.* (Dhillon, B.S. *et al.*, Eds.), Narosa Publishing House.

Chapter 15

Potential Rootstocks for Temperate Fruits in India

M.S. Mankotia, B.S. Thakur and Neena Chauhan

Dr Y.S. Parmar University of Horticulture and Forestry,
Regional Horticultural Research and Training Station,
Mashobra, Shimla – 171 007, H.P.
E-mail: msmankotia2006@yahoo.co.in

1. INTRODUCTION

Rootstock is one of the most important components of any orchard system in temperate fruit production. The choice of a suitable rootstock for a high performing tree is based on the compatibility with the scion; growth characteristics precocity influence on fruit quality; pest and disease resistance and ability to tolerate adverse climatic as well as soil conditions and most important is its availability of rootstock. Therefore it is essential to consider the points while selecting the rootstock for temperate fruit crops.

There are two types of rootstocks are available under temperate fruit orchard systems – seedling and clonal. The seedling rootstocks are derived from the seeds. Seed can be collected from wild or cultivated trees. However, it is better to collect seeds from isolated virus tested trees. Clonal rootstocks are multiplied by layering or cuttings; however, they can be multiplied by micro propagation. To reduce chances of root suckering and burrknotting of micro propagated plants it is recommended that cutting hedges or layer beds should first be made from such plants and these should be used for propagation through conventional methods.

2. ROOTSTOCKS FOR DIFFERENT CROPS

(a) Apple

Seedling rootstocks are most commonly used rootstocks in India. In general the nurserymen usually use a mixture of seeds from different sources. However, the seedlings of diploid cultivars (Golden Delicious, Yellow Newton, Wealthy, McIntosh. Granny Smith, Maharaji and Red Gold) give satisfactory results in comparison to triploid cultivars (Chadha,1962; Kanwar,1987). The resultant grafted plants are vigorous in nature with good anchorage. However, lack of uniformity in the orchards and poor precocity are the major drawbacks of such type of plantations.

Over the last century there has been a worldwide trend away from seedling rootstocks to clonal rootstocks; since clonal rootstocks import desirable characters such as dwarfing and precocity to the scion cultivar. In 1920s, Sir Ronald Hatton of the East Malling Research Station in the United Kingdom collected rootstocks from all over Europe, categorized them and then released the Malling series of apple rootstocks. This lead to commercial development of orchard systems based on smaller trees (Walker, 1980). A planned breeding program between the John Innes Institute at Merton, England, and the East Malling research station was initiated to produce woolly apple aphid resistance rootstocks resulted in Merton Immune (MI) series and later the Malling Merton (MM) series. Rootstocks from these programs are still used as the primary rootstocks throughout the world.

In the 1930s as research results from clonal dwarfing rootstock trials were obtained, planting systems based on smaller trees and clonal rootstock were developed in Holland and England (Wertheim, 1981). The tree canopy was lowered to only 50 cm above the ground. The tree height was reduced from 6 to 8 m to 4 to 5 m and the tree density was increased from 70 to 100 tree trees/ha to 250-350 trees/ha. In the 1950s and 60s systems utilizing semi-dwarf rootstocks at densities from 300-500 trees/ha were developed. This system had earlier production from the use of semi dwarfing rootstocks; easier management due to the smaller trees and higher orchard productivity. Since 1970s, higher and higher tree densities have been used in most apple growing regions of the world which has been made possible by the wide spread adoption of dwarfing clonal rootstocks.

Over the last 5 to 6 decades, as growers worldwide have used the M, MI and MM series of rootstocks, their limitations have become apparent. These have included lack of winter hardiness, lack of resistance to collar and root rot, susceptibility to fire blight, burrknots, poor anchorage, root suckers, sensitive to apple replant disease and brittle graft unions. In the years to come, it is likely that a new group of improved rootstocks from the breeding programs will replace the existing series rootstocks and become the primary commercial rootstocks around the world.

Efforts to develop improved apple rootstocks through breeding have been done in the United States (CG or G, MAC, OAR, Arkansas and Minnesota series), Canada (KSC, Ottawa and Vineland series), the United Kingdom (M, M1, MM and AR series), Germany (J.9 and Pillnitzer series), Sweden (Alnarp, Bemali and BM series), Russia (Budagovsky series), Poland (P series), the Czech Republic (IT-E series), Israel (MH series), Romania (Voinesti series) and Japan (JM series).

In Russia, the Michurin College of Horticulture introduced the Budagovsky series with improved cold hardiness. The most widely known rootstock from Budagovsky series is B.9. The other rootstocks from B series tested outside Russia are B.491 (M.27 size), B.146 and B.469 (M.9 size), B.490 (MM.106 size) and B.118 (MM.111 size).

The Polish programme were initiated in 1954 at Research Institute of Pomology and Floriculture, Skierniewice focussed on development of rootstocks with improved cold hardiness (Zagaja, 1980); the P series rootstocks beside cold hardiness have dwarfing effect with considerable resistance to crown rot. However, trials in the United States have indicated that these rootstocks are susceptible to fire blight and woolly apple aphid.

In Canada, the program to develop rootstocks originated at Horticultural Research Institute of Ontario, Vineland. The Vineland (V series) rootstocks were obtained from the open pollinated crab apple "Kerr" by Dr A. Hutchinson. Among the V series, V.1 and V.3 were found to be promising. The dwarfing effect was comparable with M9 with similar cropping efficiency. Further, V.3 was found to be without any sign of suckering whereas others in V series produced root suckers to a greater or lesser extent (Ferree *et al.*,1995).

In Germany, the Pillnitz program has released four rootstocks: Supporter 1, Supporter 2, Supporter 3 and Supporter 4. The objectives of this program were to develop rootstocks with ease of propagation, improved anchorage and resistance to biotic and abiotic stresses. However, none of these has gained wide spread acceptance (Robinson, 2005). In the Czech Republic, the program was initiated to develop rootstocks with improved propagation and has released eight rootstocks (IT-E series) (Robison, 2005). Horticulture Research International-East Malling has developed a new series of apple rootstocks (AR series) with the objectives to develop rootstocks with a vigour range between M.27 and M.9; replacement of M.26 and replacement of MM.106 that has resistance to phytophthora rot. Among released rootstocks, AR.86-1-25 (M-116) which seems to be replacement for MM.106 has been found very promising (Webster *et al.*, 1986). Breeding program at Morioka (Japan) have developed 10 rootstocks in the Japan-Morioka series. The most promising clones reported are JM1, 2,5,7 and 8. In USA the CornelI-Geneva breeding program was focused on developing rootstocks with improved orchard characteristic for better adaptation to the biotic stresses such as fire blight and crown rot. Rootstocks which have been released are G.65 (M.27 size) G.16 (M.9 size), G.11 and G.202 (M.26 size) G.30 (M.7 size), CG-3041 (M.9 size) and CG-5935 (M.26 size). Based on its effect on tree size control in relation to that on seedling, the clonal rootstocks have been classified into 5 categories (Webster and Wertheim, 2003). The details of each category alongwith the rootstocks is discussed below:

(i) Super dwarfing rootstocks

Malling 27 (M.27) is the most dwarfing rootstock, which was bred in HRI – East Malling, UK with the parentage M.13 x M.9. Size of the trees grafted on M.27 is about 15-20 per cent of the seedling. It is best suited for ultra high density plantings, however, its performance is below average in poor, infertile, shallow or drought prone soils.

This rootstock induces heavy fruit set resulting in smaller fruit size of the produce. Moreover, the trees on this rootstock are poorly anchored. It is sensitive to winter cold and woolly apple aphid. Two rootstocks B.146 and P.22 also falls under this category and are extremely dwarfing and produce very poor fruit size when grown without supplementary irrigation (Webster and Hollands, 1999). Rootstock M.27 and other recently introduced rootstocks of similar vigour are best suited for vigorous scion varieties planted in high density planting systems on deep, highly fertile soils with adequate irrigation facilities. Other rootstocks in this category showing preliminary promise are G.65 (Robinson *et al.*, 1999), J-TE-G (Dvorak, 1988); AR.10-2-5, AR.628-2 and AR.69-7 (Webster and Tobutt, 2001).

(ii) Dwarfing rootstocks

Rootstock M.9 is the traditional rootstock of this category. It produces trees about 25-35 per cent the size of the seedling. It is highly precocious, productive and produces large fruit size. Although it has poor anchorage and needs support, the growers in the western countries have adopted it with great success. It has been successfully used in wide ranges of climatic and soil conditions (Barritt *et al.*, 1995; Ferree *et al.*,1995). There are about two dozen clones of M.9 with slightly different vigour and propagation characteristics but they all have similar yield efficiency (Engel, 1986).

Although M.9 is used around the world with great success in high density plantings, its greatest weakness is its susceptibility to fire blight. It is also susceptible to woolly apple aphid and to apple replant diseases and is not well adapted to climates with high soil temperatures. Over the last 50 years breeders of apple rootstocks have tried to produce rootstocks with similar vigour and attributes to M.9 with additional advantages with respect to winter cold damage (Russian, Polish and Canadian rootstock selections) and to fire blight (Geneva rootstock selections).

Rootstock B.9 (Russian rootstock selection) has the vigour slightly larger than M.9 with winter hardiness and good yield efficiency (Autio *et al.*, 2011a). The anchorage is slightly better than M.9 and resistant to collar rot, however, it is sensitive to woolly apple aphid and is difficult to propagate (Barritt *et al.*, 1995; Kruczynska and Czynczyk, 1998; Webster and Hollands, 1999). In 2002 NC-140 apple rootstock trial, the maximum yield efficiency was recorded in Gala cultivar grafted on B.9 rootstock in comparison to Gala on other rootstocks (Autio *et al.*, 2011b).

Rootstock Mark (previously MAC 9), a bred at Michigan State University in the USA showed great promise initially with excellent yield precocity and efficiency (Carlson and Perry, 1986; Perry, 1990). However, it performed very poor in some trials, especially those planted on hot dry soils. It is very sensitive to drought (Fernandez, 1997) and to crown gall and develops a swelling at or just below ground level, which reduces tree performance on some sites (Stover and Walsh, 1994). Other rootstocks in this vigour category, that have shown initial promise are AR.680-2, AR.295-6 and AR.486-1 from the HRI- East Malling programme and G.16 and CG-3041 from the Cornell Geneva programme (Webster and Wertheim, 2003).

Rockstock G.16 is a fully dwarfing rootstock with tree growth and vigour similar to vigorous clones of M.9. Precocity and cumulative yield was similar or slightly better than M.9. Wide soil adaptability and some tolerance to replant diseases are other advantages of G.16. The major drawback of this rootstock was found to be its sensitivity to some latent viruses in scion wood. CG-3041 is a fully dwarfing rootstock with vigour similar to M.9. It was reported to be highly resistant to fire blight and phytophthora. It is more precocious and productive as compared to M.9 with excellent fruit size. From the results obtained from trials, it appears that CG-3041 will be a possible replacement for M.9 in near future. (Robinson, 2005).

(iii) Semi dwarfing rootstocks

Malling 26 (M.26) is the traditional rootstock under this category. Where climatic and soil conditions are unsuited for M.9, apple growers requiring dwarfed trees often turn to slightly more invigorating clonal rootstocks such as M.26 or other rootstocks of similar vigour potential producing trees about 35-50 per cent the size of seedling. M.26 exhibits much better cold tolerance than M.9 (Ferree and Carlson, 1987) and for this reason has often been chosen for areas where winters are occasionally severe. Its performance is highly variable around the world. In wet climates with good soils, tree size can be quite large. In replant soils or in dry climate it can be almost as small as M.9. It is extremely susceptible to fire blight and is also susceptible to woolly apple aphids. M.26 is very susceptible to phytophthora root rot and does not tolerate wet soils. It is not drought tolerant and requires irrigation on light soils. It is also very susceptible to replant disease. M.26 also have a tendency to produce burr-knots on the above ground root stem shank. Fruit breeders have tried to develop new rootstock with similar vigour and cold tolerance to M.26 without burr-knotting and improved induction of yield precocity and efficiency (Webster and Wertheim, 2003).

Other rootstocks showing promise under this category are CG-5935 and Geneva 202. CG-5935 (cross of Ottawa 3 x Robusta 5) is a semi dwarfing rootstock that produces a tree slightly larger than M.26. It is the most precocious and productive semi dwarf rootstock. It has similar efficiency to M.9 alongwith excellent fruit size and wide crotch angles besides being cold tolerant (Robinson, 2005). It is highly resistance to fire blight and phytophthora, but its resistance to woolly apple aphid is unknown. From the various studies, it appears that CG-5935 will be possible replacement for M.26.

Geneva 202 (semi-dwarf) also falls under this category. In addition to fire blight resistance, it also has good resistance to woolly apple aphid which is an important rootstock pest in many warm climates (Robinson, 2005). Under New Zealand conditions, it has been much more productive than M.26 and seems to be one of the best rootstocks for alternative to M.9 and M.26 in climates that have problems of woolly apple aphid infestations. The early performance of apple trees on Geneva 202 under South African apple replants disease condition was better than other rootstocks under testing (Bezuidenhout *et al.*, 2014).

(iv) Semi-vigorous to vigorous rootstocks

The traditional rootstocks under this category are M.7, MM.106 and MM.111. The tree size on these rootstocks vary from 60-80 per cent of seedling. M.7 is a free standing and productive rootstock. It has wide soil adaptability. It is quite tolerant to fire blight, phytophthora rot and to replant disease. However, it is not very precocious, produces small fruit size, prone to suckering and is quite winter tender. MM.106 is a highly productive and precocious rootstock and has performed very well under New Zealand conditions (Robinson, 2005). However, it is very susceptible to phytophthora rot. It is also susceptible to tomato ring spot virus which results in the brown line disorder with Delicious. MM.111 is a moderately productive rootstock that produces a tree slightly larger than MM.106 or M.7. It has wide soil adaptability and is quite tolerant to drought. However, it is not precocious and trees grow excessively vigorously at full bearing. Recently, several rootstocks have been tested from the Cornell Geneva rootstock programme that has shown early promise. Among CG-4814, CG-7707 and G.30, the rootstock G.30 was found to be more yield efficient (Autio *et al.*, 2011a).

(v) Very vigorous rootstocks

The rootstocks under this category produce tree size 90-100 per cent of seedling. The traditional rootstock are MM.109 and Merton 793. MM.109 is a very vigorous rootstock, but have average yield precocity and efficiency. It performs well on drought soil (Parry, 1965) but due to poor anchorage and sensitivity to collar and crown rot, it is rarely used as a rootstock in apple orchards (Preston, 1966). Merton 793 produces tree slightly smaller than MM109, but induces poor yield precocity with average yield efficiency when the tree comes to full bearing. The advantages of Merton 793 are good anchorage with few suckers and resistant to woolly apple aphid and crown rot. This rootstock is very suitable for replant situations on poor soils (Webster and Wertheim, 2003).

In India, elaborate rootstocks trials were initiated on M and MM series of apple rootstocks soon after their introduction during 1960s at Mashobra and Kotkhai in Himachal Pradesh, at Shalimar in Jammu and Kashmir and Chaubattia in Uttrakhand (Chadha, 1978). Promising rootstocks have been identified and recommended for commercial use in different states depending upon varying agro-climatic conditions. These included M.9, M.26, M.7, MM.106, MM.111 in Himachal Pradesh (Chadha, 1978; Thakur *et al.*, 1986); M.2, M.4, M.7 and M.9 in Jammu and Kashmir (Chadha, 1978); and M.13 and MM.106 in Uttrakhand (Om *et al.*, 1976; Om and Pathak 1983).

The IARI Regional Horticultural Station, Shimla, collected a number of wild and indigenous species for their evaluation as rootstocks for commercial cultivars. Out of the 6 *M. baccata* types, *M. baccata* Rohru was observed to be more dwarfing to M.9, whereas, *M. baccata* Shillong, *M. baccata* Khrot, *M. baccata* Gaibong and *M. baccata* Dhak were as dwarfing as M.9. *M. baccata* Srinagar was observed to have similar vigour as compared to MM.106 (Randhwa and Kishore, 1986). *M. baccata* var Himalaica appears to hold promise as a suitable rootstock for apple because of compatibility, dwarfing characteristics and resistance to pests (Srivastava *et al.*, 1977). *M. sikkimensis* is a dwarfing rootstock found in Sikkim. It is resistant to collar rot (Randhawa and Ram ,1976) and powdery mildew (Ram and Randhawa, 1979), but extremely sensitive to latent viruses in scion (Randhawa and Kishore, 1986). *M. dirangensis* growing wild in Arunachal Pradesh is another potential rootstock for apples (Joshi and Singh, 1980).

(b) Pear

Generally, seedling rootstocks are the most widely used. In Eastern Europe, Taiwan and India, local seedlings are used; while, seedlings of commercial cultivars particularly Williams and Winter Nelis are used in Canada, USA, New Zealand, South Africa and Chile. In Japan, seedling stock of various cultivars of *P. pyrifolia* are used. In China, the seedling stocks of *P. calleryana*, *P. Betulifolia* and *P. ussuriensis* are used. The seedling stocks of *P. calleryana* are also used in the United States and Australia; whereas, *P. betulifolia* is also used in the United States and Israel. In India, *P. pashia* and *P.pashia* var. Kumaoni, *P. pyrifolia* and local cultivars are used as a seed source for pear rootstocks.

(i) Pear (Pyrus) clonal rootstocks

Some *Pyrus* rootstock selections that are clonally multiplied for commercial use are discussed below:

a. **OH x F Series**: The Old Home x Farmingdale (OH x F) series of rootstocks originated in the United States. OH x F 40, 51, 69, 87, 97, 217, 282 and 333 are the major rootstock selections. Most of these selections are vigorous. The Horner series of rootstock was developed from crosses of OH and F rootstocks. This series offers the potential for producing smaller and more productive pear trees with larger fruit. Trees on H-4 and H-10 selections of the series have shown promising results (Mielke and Smith, 2002). Under warm climate of Israel, the performance of OH x F series rootstocks (67, 97, 333, and 153) was much better than low vigour stocks such as OH x F 217, BA 29, Quince C (Stern and Doron, 2011).

b. **BPI**: BPI originated in South Africa and is reported to have vigour similar to Q-A and BA 29 and good yield efficiency. However, it is highly susceptible to pear decline and fire blight (Wertheim, 1998).

c. **Pyrodwarf**: Pyrodwarf originated from a cross between Old Home and Bonne Luise d' Aranches. It reportedly has 50 per cent low vigour as compared with *P. calleryana* D6 seedling. It is compatible with most of European and some East-Asian pears. It is tolerant to water logging and is winter hardy.

d. **Pyriam**: Pyriam is a clonal rootstock developed by INRA in France through open pollination. It is seen as a potential replacement for Quince BA 29 in South-East France. It has good graft compatibility with William pear and low susceptibility to fire blight. Pyrian induces slightly higher vigour than BA29 but has equal productivity and fruit size (Simard and Michelesi, 2002).

e. **BM 2000**: It originated in Australia as a result of open pollination of likely parents, Williams and Packham. It is described as having medium vigour as compared with D6.

(ii) Quince Rootstocks

All the available quince rootstocks are clonal and are being used for pear production for many years in Europe. BA 29, Quince A, Quince Sydo, Quince Adams and Quince C are the most commonly used rootstocks. Many evaluation have been carried out with quince rootstock. However, there is often some variation in results between sites and scion cultivars. Therefore generally BA 29 is considered the most vigorous followed by Quince A and Quince Sydo (both approximately 75 per cent of seedling) and than Quince Adams. Quince C is the least vigorous at approximately 60 per cent of seedling. Quince C and Quince Adams have the highest yield efficiency compared to other Quince rootstocks. Annual and cumulative yields were greater in pear cultivar Conference when grafted on quince rootstocks (Quince A, Quince Adams, Quince Sydo and Quince C) in comparison to Pyriam rootstock. Yield efficiency values were greater in all quinces than Pyriam. The fruit size was largest on trees with Sydo and Quince A as a rootstock (Iglesias and Asin, 2011).

More recently, three other promising quince clones have emerged – Quince EMH (developed at East Malling), C132 (Selection from the Causascus region of Russia) and Eline (a Romanian Selection sourced from Fleuren Nurseries in the Netherlands). These rootstocks are generally considered to perform similar to Quince C in terms of vigour control and yield efficiency. However, in some traits they have exhibited traits that may make them more attractive than Quince C, such as improved fruit size (EMH and C132) and reduced russetting (Eline).

(c) Sweet Cherry

(i) Mazzard (*Prunus avium*) rootstocks

Mazzard rootstocks are still the most popular rootstocks for sweet cherries throughout the world. They are easy to raise from seed and invariably show excellent compatibility with scion cultivars. Mazzards are very vigorous rootstocks which produce sweet cherry tree with poor yield precocity. Seedlings and also some clones of Mazzard frequently delay fruit ripening by several days when compared with Mahaleb and other species and hybrid rootstocks. The Mazzards are difficult to propagate through vegetative means and in the past Mazzard seed was collected from the fruiting orchards which were variable in growth and inconsistent in their tolerance to winter cold. This prompted researchers in several countries to select improved lines of seedling Mazzards for use as rootstock.

There are two vegetatively propagated selections of Mazzard released from East Malling – F.12/1 and Charger. Rootstock F.12/1 is more vigorous than Mazzard seedlings in many situations and shows very good compatibility with sweet cherry cultivars. It exhibits useful resistance to bacterial canker but is sensitive to crown gall. Trees on F.12/1 are reported to make growth late in the season and may therefore acclimatize poorly in some locations, leading to winter cold damage.

Charger is more resistant to bacterial canker and form trees slightly small than those on F.12/1.

(ii) Mahaleb (*Prunus mahaleb*) rootstocks

In comparison to Mazzards, the principal merits of Mahalebs are their better adaptability to calcareous or droughty soils. Trees on Mahaleb also ripen their wood earlier in the autumn than Mazzards and show better resistance to zinc deficiency and lime induced chlorosis. Many seedling selection have been made in the US, France, Germany, Hungary, Ukraine and Moldavia.

St Lucie 64 (SL.64) is the vegetatively propagated selection of Mahaleb. Trees on SL.64 thrive best on well drained soils but also recommended for use in sites suffering from cherry replant syndrome.

(iii) Sour cherry (*Prunus cerasus*) rootstocks

Sour cherry rootstocks have the advantage of being colder hardy than Mazzard or Mahaleb stocks and they also tolerate wet soils better. However, they perform poorly on droughty and high calcareous soils. Some types exhibits inconsistent compatibility with several sweet cherry scions, such as Sam and Van. Seedlings of sour cherry are usually very variable as rootstocks and are not commonly used for this purpose.

Tabel (Edabriz), Weiroot series, Vladmir, Stockton Morello and CAB selections are vegetatively propagated selections of sour cherry. Unlike many other sour cherry rootstocks, Edabriz exhibited good compatibility with most of the sweet cherry cultivars. Trees on Edabriz are dwarfed on some sites attaining only 15-20 per cent the size of similar trees on F.12/1 (Edin, 1989). Growth and cropping of Lapins grown in heavy soil on Edabriz were found to be least in vigour as compared to other rootstocks (Cmelik *et al.*, 2004).

(iv) Prunus hybrids as cherry rootstocks

Colt is a hybrid between *P. avium* and *P. pseudocerasus* compatibility is generally very good on colt, although occasionally problems are experienced with the cultivars Sam and Van. Trees on colt crop precociously and on favourable sites very productive. Fruit size is often larger from trees grown on colt compared with F.12/1. One problem with colt is its sensitivity to cold damage when grown in areas experiencing low winter temperatures. Further, Colt require fertile, well-drained and adequately irrigated soils if it is to perform well. Poor growth and chlorotic symptoms are observed on trees grown in shallow, dry and highly calcareous soils.

M x M series of rootstocks were selected in Oregon, USA from open-pollinated seedlings of *P. mahaleb*. On the basis of their growth habit and leaf size most are thought to be hybrids with *P. avium*. M x M 2, 39, 60 and 97 show some resistance to *Phytophthora* spp. Clones 14, 39, 60 and 97 have shown more resistance to winter cold injury and canker than Mazzard and Colt.

The Gisela series originated from breeding program at Giessen, Germany. The promising selections on the basis of compatibility, reduction of scion vigour and good cropping have been released. Gisela 1, Gisela 4, Gisela 5, Gisela 10 and Gisela 12 appears to be most promising.

GM series originated from Gembloux, Belgium have been evaluated for their ease of vegetative propagation, compatibility with cherry cultivars, influence on scion vigour and cropping, resistance to bacterial canker and other diseases, propensity to sucker and suitability to different soil and climatic conditions. Four clones (GM 8, GM 9, GM 61/1 and GM 79) have been found promising.

Pi-Ku 4.20 and Pi-Ku 4.17 originated from Dresden-Pillnitz, Germany have shown promising results in cherry cropping and inducing intermediate vigour.

(v) Factors considered for selecting a cherry rootstock

i. **Graft compatibility**: It has been known that Mazzard rootstocks are compatible almost with all scion cultivars and even today no evidence exists to contradict this statement (Long and Kaiser, 2010). However, Mahaleb was found to be incompatible with several scion cultivars, resulting in tree death.

   a. Some new rootstocks such as Weiroot 13 have shown incompatible symptoms with several scion cultivars, and Colt with Sam and Van. Gisela rootstocks (5, 6 and 12) have not shown scion incompatibility (Long and Kaiser, 2010).

ii. **Cold Hardiness**: It has been reported that many scion cultivars on Mahaleb rootstocks acclimatize earlier in winter than on other rootstocks. The parent material of the Gisela rootstocks, *P. cerasus* and *P. canescens* are both hardier than Mazzard and impart hardiness to the Gisela rootstocks. Colt which is a popular rootstock has been found to be less hardy than Mazzard.

iii. **Anchorage**: It has been reported that trees on Gisela 6 and sometimes Gisela 5 often tilt away from the prevailing winds. This is especially true for more top-heavy central leader trees. Support through stakes or a trellis system is desirable. In other commercial rootstocks, the anchorage of the trees is adequate and do not require support.

iv. **Damage by fungal pathogens**: Several species of *Phytophthora* attack the roots of cherries and may on some sites cause severe problems (Wicks *et al.*, 1984). Mazzard, sour cherry and colt rootstocks have shown some tolerance to these root pathogens whereas Malaleb being the most susceptible. Leaf spot disease is often very severe in some cherry production regions. Mahalebs are tolerant while sour cherries are mostly very susceptible.The hybrid rootstock VP-1 (*P. cerasus* x *P. maacki*) is reported to be resistant to leaf spot (Kolesnikova *et al.*, 1985).

v. **Damage by bacterial pathogens**: Crown gall is a serious problem in cherry production. It reduces growth of cherry trees planted on droughty soils, particularly where trees are grown in areas experiencing very hot summers. The Mazzard F.12/1 and colt are particularly sensitive to crown gall infections of the roots, while Mazzard seedlings are less susceptible (Breton, 1980). Hybrids between *P. avium* x *P. Canescens* were very susceptible to crown gall, whereas those between *P. cerasus* and *P. fruiticosa* were not (Schmidt, 1989).

vi. **Precocity, consistency and abundance of scion yields**: Sweet cherry trees on vigorous Mazzard or Mahaleb rootstocks exhibit poor cropping precocity, trees often taking upto 12 years to come into full production on F.12/1. Trees on colt and some of the size controlling rootstocks are in comparison much more precocious. This increased precocity is not, however, necessarily linked with dwarfing effect of the rootstock. The Belgian selections Inmil and Damil both dwarfing, often induce poor cropping precocity in the scions worked upon them.

vii. **Drought tolerance**: Mahaleb rootstocks are generally much more tolerant to drought conditions than Mazzards; sour cherry and colt rootstocks are particularly sensitive to such conditions. Several of the M x M hybrid rootstocks also show good tolerance to drought conditions. Many of the shallow – rooted dwarfing rootstocks also exhibit poor tolerance to drought conditions.

(d) Peach and Nectarines Rootstock

Peach seedlings have been and still are the principal rootstock source of peach on a worldwide. The seed sources can be conveniently divided into three groups: those from the wild types, those from commercial cultivars and those from special rootstock selections (Rom, 1983). The wild types are usually obtained from peach trees that are growing wild or in a nearly wild state. In India, the seed source for propagating rootstock is wild peach. Commercial cultivars, used for processing or drying are also seed sources of peach. Halford and Lovell are of great importance as peach seedling rootstocks in Canada, the United States and Mexico (Rom, 1983). The third group consists of selections that have little or no commercial value other than as rootstock seed sources for peach and nectarine.

Interspecific hybrids of peach x almond (GF556, GF 677) are of importance in France and other countries of Southern Europe, especially in the Mediterranean basin. They are clonally propagated and are especially useful on alkaline soils because they are resistant to chlorosis, have some tolerance to wet and dry soils, and are especially useful in replant situations (Rom, 1983; Rom, 1984). Nemaguard, a seed propagated interspecific hybrid (*P. persica* x *P. davidiana*) is used in California and Florida. Resistance to root knot nematodes is the primary reason for its wide acceptance in regions with mild winters. However, its adverse effect on peach scions with respect to cold hardiness and bacterial canker limit its use to cooler regions (Layne, 1987).

Plums and plum hybrids are the most commonly used rootstock for peach other than peach seedling and peach inter specific hybrids (Rom, 1983). They are better adapted to wet and water logged soils (Layne, 1987). Other *Prunus* spp. also used as rootstocks for peach includes *P. davidiana, P. amygdalus, P. armeniaca, P. insititia, P. salicina, P. tomentosa* and *P. besseyi* (Rom, 1983).

The following factors considered for selecting a peach rootstock

(i) Adaptability to biotic and abiotic soil factors

Ring nematode has been linked directly to the onset of peach tree short life (PTSL) syndrome. The rootstocks such as Lovell, Halford, Bailey and Nemaguard are moderately to highly susceptible to this nematode. Only the rootstock Guardian BY 520-9 developed in the United States have survived better than others under field tests in South Carolina and Georgia (Okie *et al.*, 1994; Reighard *et al.*, 1997). In California, Viking rootstock has shown good tolerance to ring nematode and Bacterial canker (Reighard, 2008).

Root-knot resistant rootstocks reported are Barrier 1 and Cadaman (*P. persica* x *P. davidiana*), Penta and Tetra (*P. domestica*), Adesoto 101 (*P. insititia*), plum hybrids Myran (*P. cerasifera* x *P. salicina*), Ishtara (*P. cerasifera* x (*P. persica* x *P. cerasifera*). Julior (*P. insitita* x *P. domestica*), Hiawatha (*P. besseyi* x *P. salicina*), peach – almond hybrids, Garnem and Felinem, and Pumiselect (*P. pumila*). (Esmenjaud *et al.*, 1997; Fernandez *et al.*, 1994; Pinochet *et al.*, 2002; Nicotra and Moser, 1997).

(ii) Soil texture, pH and fungal soil pathogens

Peach rootstocks are well adapted to sandy, gravelly or loamy soils. In contrast, peach seedling rootstocks are generally not adapted to poorly drained, heavy clay or to calcareous soils where pH is above 7.5. Peach x almond rootstocks such as Titan, Hansen 536, Altas and Viking (interspecific hybrids) are tolerant to saline and alkaline soils. (Reighard, 2008).

On heavy or poorly drained soils, peach seedlings are susceptible to phytophthora rot. Similarly, almost all stone fruit rootstocks are susceptible to the oak root rot fungus. A rootstock- Sharpe, from USDA-Byron, Georgia, USA is reported to have good tolerance to this fungus (Reighard, 2008).

(e) Almond

Almond seedling rootstocks are the most widely used rootstock in the world. In India, the almond plants are generally raised on the seedlings of bitter almond, wild peach and occasionally on the seedlings of *Behmi* (*P. mira*). Seeds from bitter almond trees have been characteristically used by nurserymen as sources of seeds. Trees from such sources are thought to be more vigorous and generally superior to trees from sweet-kernelled seeds.

Almond seedling rootstocks are adapted to resist drought and calcareous soil conditions. They appear to be able to tolerate stress better than peach root systems. On the other hand, almond seedlings require well drained soils and are highly susceptible to soil saturation. Almond seedling rootstock are susceptible to various soil diseases, including crown gall, oak root fungus and crown rot.

Peach seedlings are the dominant rootstock for almond where the soils are slightly acidic and irrigation facilities are sufficient. Cultivars growing on peach seedling rootstocks are vigorous and tend to grow more than comparable trees on almond seedlings. Thus, trees come into bearing early and are more productive at younger ages than trees on almond. Peach seedlings are sensitive to drought and calcareous soils and do best under irrigated but well drained, slightly acidic soils.

The peach-almond hybrid rootstocks are best adapted for trees growing in well drained, dry and calcareous conditions. In Europe, most interest has been in GF-677, a chance seedling discovered in France in mid 1940s. This rootstock has become the dominant rootstock for almond as well as peaches in France since 1975. Its use has spread to other parts of the Mediterranean basin (Kester and Grassely, 1987).

Similarly, GF-557 selected from the hybrid population is also root knot nematode resistant. Yield efficiency of the trees on hybrids is greater than trees of parental almond and peach under comparable conditions (Kester and Grasselly, 1987).

(f) Apricot

Apricots are propagated commercially by grafting or budding on seedling rootstocks. The rootstocks used vary greatly from one country to another. In India, wild apricot (Chuli) and wild peach seedlings are generally used as rootstocks. Wild apricot seedlings are the commercially used rootstocks as the graft union on this rootstock is good and the trees are more vigorous as compared to wild peach rootstocks. Wild peach is a satisfactory rootstock for light soils and dry conditions but sometimes the graft union may get enlarged or become rough (Tomar and Rehalia, 2002).

The use of rootstocks other than *Prunus armenica* is restricted by problems of graft incompatibility. Winter cold damage is not usually a problem in India with trees on wild apricot and even high soil temperature are not harmful. Apricot and peach seedling rootstocks are limited to light, well drained, neutral pH soils that are relative low in lime. In areas with heavy soils, a high water table, or late or long rains, root asphyxiation and oxygen stress limit the use of apricot and peach rootstocks (Southwick and Weis, 1998).

Size controlling rootstocks have been developed to reduce vegetative vigour of rootstocks and scion while maintaining high production. Citation, a plum x peach hybrid developed by Zaiger Genetics, Modesto, California and Ishtara (Bernhard and Duquesne, 1988) produce small apricot trees in certain situations. Torinel plum, myrobalans and Mycrocerasus rootstocks were also developed for their size controlling ability. Except Torinel, other size controlling rootstocks may not affect fruitfulness (Southwick and Weis, 1998).

(g) Plum

Majority of the plum cultivars grown in India belong to Japanese plum (*P. salicina*). These plums are more vigorous, productive, precocious in bearing and disease resistant than the *P. domestica* (European plums). Main rootstock used for plum in India is wild apricot seedling. In the western countries plums are usually produced on Myrobalan (*P. cerasifera*) seedlings. In UK, Holland and Scandinavia, the seedlings or clonal selections of St. Julien (*P. insititia*) are more popular. At present, increased interest has been shown in clonal rootstocks for plums and several new selections have been developed. Most of these selections are invigorating and are reported to be resistant to particular problems (anoxia and high pH), although few (Pixy, Ferlenain and Maridon) are dwarfing and more suited to use in high density plants (Webster and Wertheim, 1993).

(h) Walnut

English or Persian walnut (*Juglans regia* L) is grown through the temperate regions of the world. Rootstock usage varies by region and degree of development of the walnut industry. Propagation of improved selections or cultivars on seedling rootstocks have become increasingly common in the United States and parts of Europe. In many countries the majority of bearing English walnuts are of seedling origin (Popov, 1983). However, the current trends are clearly towards increased rootstock usage to take advantage of active cultivar selection programme.

Currently all rootstocks for *J. regia* are derived from open pollinated seed and the nursery practice for all rootstock species is similar. *J. hindsii* seedlings are the most common rootstock for *J. regia* in California (USA). However, this rootstock is intolerant of heavy soils, water logging, replant situations and soils with low available phosphorus and zinc coupled with high levels of lime. The positive attribute is its low susceptibility to *crown gall* and *Armillaria*. *J. regia* is probably the most common rootstock for walnut cultivars worldwide because of its general availability in walnut-growing regions. Other species which have been tried as rootstock for English walnut but with little success are *J. california*, *J. microcarpa* and *J. major*.

Paradox walnut is a hybrid seedling produced from a cross of *J. hindsii* (Northern California black walnut) and English walnut. It can be used as a seedling or clonal rootstock. VX211 is a clonal selection from Paradox walnut with an exceptional vigour. It is tolerant to nematodes, resistant to phytophthora and has excellent survival in replant orchards. RX1 clonal walnut rootstock is a cross between Texas black walnut and English walnut. It produces smaller trees and has better survival in replant orchards.

3. CONCLUSION

In India, the apple producing areas are distributed in varying soil, climatic and topographical situations. Under ideal conditions, the high density planting can be adopted by using traditional dwarfing to semi dwarfing rootstocks (M.9 and M.26). However, these rootstocks are lacking in some desirable attributes such as good anchorage and resistance to woolly apple aphid. G.16 and CG-3041 are most suitable replacement for M.9. Geneva 202 is the most promising rootstock to replace the traditional M.26 as it possesses additional advantages of being resistance to woolly apple aphid, fire blight and better adaption to replant situations. With the changing trends of growing standard Delicious cultivars to spur types, the scope of semi vigorous to vigorous rootstocks (M.7, MM.106, MM.111 and Merton 793) for use in semi intensive plantation has increased. However, these rootstocks have some limitations hence there is need of rootstocks with vigour greater than M.26, precocity, yield productivity and good fruit quality, besides strong resistance to aerial or soil borne pests and diseases. G.30 and M.116 are the most promising under this category.

Traditionally pears are grown on seedlings rootstocks (*P. pashia* and *P. pyrifolia*). The main drawback of this combination is high vigour resultant tree with lack of precocity. At present, the quince rootstocks seem to be alterative, despite the problems of graft incompatibility with some cultivars, poor tolerance to winter cold and alkaline soils. These problems can be overcome by substituting with various *Pyrus* species as rootstocks. Unfortunately, most of these are very difficult to propagate through vegetative means and generally produce trees of very large stature. Pyrodwarf has shown promising results in Germany and needs to be tested under Indian conditions.

The priorities in rootstock selection for sweet cherries are for dwarfed trees suitable for high density systems of management and for growing under protective shelters to alleviate bird and rain damage. Colt is the widely used as a rootstock; however, the resultant trees are vigorous in nature. Moreover, it is susceptible to crown gall and poorly drained soils; a suitable rootstock is needed to replace Colt. Dwarfing rootstocks developed in the western countries need to be tested in wide range of environmental conditions. Wild peach seedlings are most commonly used rootstocks for peaches and nectarines. In western countries, the *Prunus* species other than *P. persica* are also used as rootstocks for peaches and nectarines; particularly on poor heavy soils. In addition, the peach x almond hybrids (GF-667, GF-567) provide tolerance to highly alkaline soils, to anoxia in poorly drained soils and occasionally to some nematodes. Other *Prunus* hybrid between *P. perscia* and *P. davidiana* provide strong nematode resistance. Wild peach and almond seedlings are traditionally rootstocks for almond. However, the peach x almond hybrids such as GF-677 have additional advantages under biotic stress conditions. Plum and apricots are widely grown on wild apricot seedlings. However, there is a need in both fruit crops to have a dwarfing rootstocks suited for high density plantings. The other important need will be rootstocks which provide good graft compatibility with all of the principal scion cultivars. Resistance to nematodes will also be essential for some situations.

4. REFERENCES

Autio, W., Robinson, J.L., Cline, J., Crassweller, R.M., Embree, C.G., Hoover, E., Lang, G., Masahni, J., Parker, M.L., Perry, R., Reighard, G.L. and Warmund, M. 2011a. Performance of several semi dwarfing rootstock with Fuji and McIntosh as scion cultivars in the 1999 NC-140 semi dwarf apple rootstock trials. *Acta Hort.* **903**: 42.

Autio, W., Robinson, T.L., Cowgill, W., Hampson, C., Kushad, M., Lang, G., Masabni, J., Miller, D.D., Parra-Quezada, R.A., Perry, R. and Rom, C. 2011b. Performance of Gala apple trees on supporter 4 and different strains of B.9 and M.26 rootstocks as part of the 2002 NC-140 apple rootstock trait. *Acta Hort.* **903** : 40.

Barritt, B.H., Konishi, B.S. and Dilley, M.A. 1995. Performance of three apple cultivars with 23 dwarfing rootstocks during 8 seasons in Washington. *Fruit Var. J.* **49**: 158-170.

Benzuidenhout, C.M., Van Schnoor, L. and Cook, N.C. 2014. Rootstocks evaluated for apple replant tolerance. *Acta Hort.* **1058** : 70.

Breton, S. 1980. Le cerisier, Centre Technique Interprofersionel des Fruit er Lequmes, Paris, pp. 21-31.

Carlson, R.F. and Perry, R.L. 1986. Mark apple rootstock. *HortSci.* **21** : 165.

Chadha, K.L. 1978. Three decades of research on fruits. III. Temperate Fruits. *Indian Hort.* **23** : 9-13.

Chadha, T.R. 1962. Some studies on the germination of apple seeds of some commercial and local varieties grown in the Kullu Valley. *Himachal Hort.* **3** : 5-10.

Cmelik, Z., Druzic, J., Duralija, B. and Bencic, D. 2004. Influence of clonal rootstocks on growth and cropping of Lapins sweet cherry. *Acta Hort.* **658** : 16.

Dvorak, A. 1988. Breeding of rootstocks J-TE and their influence on growth and productivity of different cultivars *Acta Hort.* **224** : 325-329.

Edin, M., Masseron, A., Tronel, C. and Claverie, J.1989. Dwarfing and semi dwarfing rootstocks for cherry trees: preliminary results of experiments in France. *Fruit Belge* **57** : 235-241.

Engel, G. 1986. Virus-free and virus tested M.9 selections. *Acta Hort.* **160** : 79-82.

Esmenjand, D., Minot, J.C, Voisin, R., Pinochet, J., Simard, M.H. and Salesses, G. 1997. Differential response to root-knot nematodes in *Prunus* species and correlation genetic implications. *J. Nemat.* **29** : 370-380.

Ferree, D.C. and Carlson, R.F. 1987. Apple rootstock. In rootstocks for fruit crops : Rom, R.C. and Carlson, R.F. (Eds.). *Wiley Interscience*, New York, pp. 107-144.

Ferree, D.C., Hirst, P.M., Schmid, J.C. and Dotson, P.E. 1995. Performance of three apple cultivars with 22 dwarfing rootstocks during 8 seasons in Ohio, *Fruit Var. J.* **49** : 171-178.

Fernandez, C., Pinochet, J., Esmenjaud, D., Salesses, G. and Felipe, A. 1994. Resistance among new *Prunus* rootstocks and selections to root-knot nematodes in Spain and France. *HortSci.* **29** : 1064-1067.

Kanwar, S.M. 1987. Apples – Production Technology and Economics. *Tata McGraw* Publishing Company Limited, New Delhi, pp. 88-98.

Kester,D.R. and Grasselly, C. 1987. Almonds Rootstocks. In 'rootstocks for fruit crops' (eds Rom, R.C. and Carlsop, R.F.). *A Wiley- Interscience* Publication, John Wiley and Sons, pp. 265-293.

Kolesnikova, A.F., Ossipov, Yu.V., Kolesnikov, A.I. and Osipov, Yu. V. 1985. New hybrid rootstock for cherries. *Acta Hort.* **169** : 159-162.

Kruczynska, D. and Czynczyk, A. 1998. The effect of dwarf rootstocks and interstems on growth, yield and mineral leaf content of three apple cultivars : growth and yield. *Zeszyty Naukowe Iinstytutu Sadownictwa I Kwiaciarstwa* **5** : 25-26.

Iglesias, I. and Asin, L. 2011. Agronomical performance and fruit quality of Conference pear grafted on clonal Quince and pear rootstocks. *Acta Hort.* **903**: 59.

Joshi, M.C. and Singh, S.M. 1980. New crab apple *Malus dirangensis* from Arunachal Pradesh. *Plant Sci.* **12**: 21-22.

Layne, R.E.C. 1987. Peach rootstocks. **In:** '*Rootstocks for Fruit Crops*' (Eds.), Rom, R.C. and Carlson, R.F.). A Wiley – interscience publication, *John Wiley and Sons*, pp. 185-216.

Long, L.E. and Kaiser, C. 2010. Sweet cherry rootstocks. A Pacific Northwest Extension Publication. PNW 619. September 2010, pp. 1-8.

Mielke, E.A. and Smith, L. 2002. Evaluation of the Horner rootstocks. *Acta Hort:* **596** : 325-330.

Nicotra, A. and Moser, L. 1997. Two new rootstocks for peach and nectarines. Penta and Tetra. *Acta Hort.* **451** : 269-271.

Okie, W.R., Reighard, G.L., Beckman,T.G., Nyczepir, A.P., Reilly,C.L., Zehi, E.L. and Newall, Jr, W.L. 1994. Field screening *Prunus* for longevity in the south eastern United States. *HortSci.* **29**: 673-677.

Om, H. and Pathak, R.K. 1983. Influence of rootstocks on growth and productivity of some apple cultivars. *Indian J. Hort.* **40** : 168-172.

Om, H., Pathak, R.K. and Srivastava, R.P. 1976. Apple rootstocks studies – growth and cropping of apple cultivars on Malling and Merton clones. ICAR Fruit Research Workshop, Ranikhet, UP 17-20 November, 1976.

Parry, M.S. 1965. Field trials of Malling – Merton rootstocks. **In** : *Annual Report of East Malling Research Station for 1964.* Maidstone, UK, pp. 91-96.

Perry, R.L. 1990. Mark in the rootstock arsenal. *Compact Fruit Tree,* **23**: 1-3.

Pinochet, J., Fernandez, C., Cunill, M., Torrents, J., Felipe, A., Lopez, M.M., Lastra, B. and Penyalver, R. 2002. Response of new interspecific hhybrids for peach to root knot and lesion nematodes and crown gall. *Acta Hort.* **592** : 707-716.

Ram, R.D. and Randhawa, S.S. 1979.Resistance of different species of pome and stone fruits to powdery mildew incited by *Podosphaera leucotricha*. *Sci. Cult.* **45**: 256.

Randhawa, S.S. and Kishore, D.K. 1986. Preliminary evaluation of native *Malus baccata* types as rootstocks. **In**: *Advances of Research on Temperate Fruits* (Eds.), Chadha, T.R., Bhutani, V.P. and Kaul, J.L.). pp. 83-93.

Randhawa, S.S. and Ram, R.D. 1976. Standardization of propagation techniques. ICAR Fruit Research Workshop, Ranikhet, 17-20, November. 1976.

Reighard, G.L. 2008. New and emerging rootstocks. Presented at the Mid Atlantic Fruit and Vegetable Convention and Trade show, Hershey, Pa. Jan. 30 2008.

Reighard, G.L., Newall, W.C., Beckman, T.G., Okie, W.R., Zehr, E.I. and Nyczepir, A.P. 1997. Field performance of Prunus rootstock cultivars and selections on replant soils in South Carolina. *Acta Hort.* **451** : 243-250.

Robinson, T.L. 2005. Rootstock as a key component to High density orchard. *Compact Fruit Tree* **36** : 9-13.

Robinson, T.L., Cummins, J.N., Hoying, S.A., Johnson, W.C., Aldwinkle, H.S., Norelli, J.L., Momol, M.T. and Saygili, H. 1999. Orchard performance of fire blight-reistant Geneva apple rootstocks. *Acta Hort.* **489** : 287-294.

Rom, R. 1984. A new generation of peach rootstocks. Proc.43rd Nat. Peach Council *Annual Convention*, pp. 59-68.

Rom, R.C. 1983.The peach rootstock situation: An international perspective. *Fruit Var. J.* **37** : 3-14.

Popov, S. 1983. Similarity of the fruit of mother trees and their progeny when propagating *unglans regia* by seed. Gorskostopanska Novka. **20**: 12-19.

Preston, A.P. 1966. Apple rootstock studies: fifteen years results with Malling-Merton clones. *J. Hort. Sci.* **41** : 349-360.

Schmidt, H. 1989. Observations on the incidence of crown gall on cherry hybrid rootstocks *Erwerbsobstbau*, **31** : 42-43.

Simard, M.H. and Michelesi, J.C. 2002. Pyriam a new pear rootstock. *Acta Hort.* **596** : 351-355.

Southwick, S.M. and Weis, K.G. 1998. Selecting and propagating rootstocks to produce apricots. *HortTechnology* **8** : 164-170.

Srivastava, R.P., Pathak, Bana, D.S. and Pandey, V.S. 1977. Utilization of some important fruit trees growing wild in the Himalayan region. (In) Fruit Breeding in India, (Ed. Nijjar, G.S.) *Oxford and IBH Publicity* Co., New Delhi. pp. 1-8.

Stern, R.A. and Doron, I. 2011. Performance of 'Coscia' pear (*Pyrus communis*) on nine rootstocks in the north of Israel. *Acta Hort.* **903** : 60.

Stover, E.W. and Walsh, C.S. 1994. Tissue proliferation of Mark rootstock Shanks: effect on tree growth and correlation with site factors. *HortSci.* **29** : 475.

Thakur, G.C., Awasthi, R.P. and Chadha, T.R. 1986. Rootstock trials on apple at Mashobra. ICAR Fruit Research Workshop, Dapoli, 18-20 December 1986.

Tomar, C.S. and Rehalia, A.S. 2002. Apricot cultivation in India – Problems, prospects and future strategies to enhance productivity. In compendium of summer school on 'Enhancement of Temperate Fruit Production in Changing Climate (Eds.), Jindal, K.K. and Gautam,D.R.) Chapter 17 : 136-144.

Walker, C.D. 1980. The development of intensive orchards in England: A Historical note. *Acta Hort.* **114** : 309-317.

Webster, A.D. and Hollands, M.S. 1999. Apple rootstock studies: comparison of Polish, Russian, USA and UK selections as rootstocks for the apple cultivar Cox's Orange Pippin (*Malus domestica* Borkh.). *J. Hort. Sci. and Biotech.* **74** : 367-374.

Webster, A.D., Smith, R.A. and Watkins, R. 1986. Apple rootstock studies I. Preliminary evaluations of several MM106 x M.27 hybrids. *J. Hort. Sci.* **61** : 429-437.

Webster, A.D. and Tobutt, K. 2001. Breeding and selection of new apple rootstocks at Horticulture Research International-East Malling. *Acta Hort.* **557** : 189-192.

Webster, A.D. and Wertheim, S.J. 2003. Apple rootstocks. In 'Apples: Botany, Production and Uses' (Eds.), D.C. Ferree and I.J. Warrington), *CAB International*. pp. 91-124.

Webster, A.D. and Wertheim, S.J. 1993. Comparisons of species and hybrids for European plum cultivars. *J. HortSci.* **68** : 861-869.

Wicks, T.J., Bumbierrs, M., Warcup, J.H. and Wallace, H.R. 1984. Phytophthora in fruit orchards in South Australia. *Biennial Report of the Waite Agricultural Research Institute*, 1982-1983, 1984, 147.

Wertheim, S.J. 1981. High density planting. Development and current achievements in the Netherlands, Belgium and Germany. *Acta Hort.* **114** : 318-327.

Wertheim,S.J.1998. Pear rootstocks. **In**: Rootstock guide- Apple, pear, cherry, European plum. Publication No 25, December, 1998, pp. 61-84.

Zagaja, S.W. 1980. Performance of two apple cultivars on P. series dwarf rootstocks. *Acta Hort.***114** : 162-169.

Chapter 16

Recent Advances in Propagation and Nursery Management of Temperate Fruits and Nuts

J.S. Chandel and Pramod Verma

Department of Fruit Science,
Dr Y.S. Parmar University of Horticulture and Forestry,
Solan – 173 230, H.P.
E-mail chandeljs@yahoo.co.in

1. INTRODUCTION

The climatic conditions of North-Western Himalayas are congenial for commercial cultivation of temperate fruits. Temperate fruits are mainly grown in the states of Jammu and Kashmir, Himachal Pradesh and Uttarakhand and on a limited scale in North-Eastern states of Arunachal Pradesh, Nagaland, Meghalaya, Sikkim and Manipur. In India, temperate fruits occupy an area of 5,45,790 ha and production is 26,33,600 MT (Anon, 2013). The temperate fruit industry in India confronted with the problems of low productivity, year-to-year fluctuation in production and poor quality produce. Several factors are attributed to be associated with these problems, but the use of low grade and poor quality plantings materials for the establishment of an orchard are of major concern. The nursery plants are said to be the foundation of an orchards and ultimate success in fruit growing depends upon the quality of plant material used for establishing the orchards. The use of healthy, true-to-the type, disease, insect and virus free planting material is, therefore, the basic pre-requisite for the success of an orcharding and quality fruit production. Establishment of bud-wood bank of commercial and high yielding varieties is the prerequisite for the production of quality planting material. Propagation and nursery managements are also important aspects of quality nursery production. The fruit plants do not come true-to-the type when raised from seed and therefore, nursery plants are generally raised through vegetative methods of propagation either on their own roots by cuttings or layering techniques or on the roots of other plants (rootstocks) by budding or grafting techniques. Majority of temperate fruits are generally propagated on rootstock genotypes rather than on their own roots because to have the advantage of rootstock genotypes like on growth control, precocity, cropping and resistance to insect-pests and diseases. Among the temperate fruits, nut crops are very difficult to propagate and excellent success has been achieved with the use of new propagation techniques in nut crops.

2. PRE-REQUISITE FOR NURSERY ESTABLISHMENT AND PRODUCTION OF QUALITY PLANTING MATERIAL

(a) Establishment of Bud-Wood Bank and Mother Tree Blocks

The most important step for the production of quality nursery plants is establishment of mother tree orchards of most promising and improved cultivars preferably at the production areas. The bud-wood iscion bank can be established through introduction of promising and improved varieties from abroad and selection of meritorious trees of promising cultivars from the existing research institution/ plantations. Mother trees are planted at close spacing of 2 m x 2 m or 3 m x 3 m depending up on type of fruit crop. Proper record of variety is made by making maps in the nursery register. Trees are hard pruned every year to developed hedges for the production of scion wood in bulk. Trees are frequently irrigated and fertilized with higher dose of nitrogen to make have vegetative growth. Rigorous plant protection measures and indexing for viruses are followed for control of insect-pests and diseases. Separate tools like secateurs, pruning saw, knife and pick axe *etc.* are kept to use in mother tree scion Bank orchard, which are also sterilized before their use to keep the mother plants free from viruses.

(b) Seed Orchard

In temperate fruits generally seeds of wild species are preferred for raising the seedling rootstocks. However, seeds of some commercial varieties like Golden Delicious and Granny Smith in apple, Bruno in kiwifruit, Burkett in pecan nut are also used. Therefore, to meet the requirement of seeds, the rootstock orchard of crab apple, Golden Delicious and Granny Smith cultivars for apple, wild peach for peach, Kainth and Zarainth for pear, wild apricot for apricot and plum, Persian walnut (very hard shell walnut) for walnut and bitter almond for almond are to be established.

(c) Establishment of Clonal Rootstock Bank

Due to various advantages of clonal rootstocks over seedling rootstocks and problem of replanting, the orchardists are demanding nursery plants of apple, pear and stone fruits on clonal rootstocks. To meet the requirement of farmers, the virus-free mother stool beds of promising rootstocks of apple like EMLA-111, EMLA-106, EMLA-7, Merton 793, EMLA- 9, pear like Quince and BA29, stone fruits like GF 677.

Istara, Myrocal, Marianna, Myrobalan, Colt and Giesela are to be established and maintained in the nursery for large scale multiplication.

(d) Selection of Site and Soil

Nursery site should be selected in the vicinity of potential fruit production locations in an isolated area. Nursery area should be flat, fertile and rich in organic matter. The soil should be deep, loam to sandy loam with the pH of 6.5 to 7.5. Heavy and clay soils are not suitable. Site should be well connected with road, having assured irrigation facilities. Nursery area should be frost-free receiving adequate sun shine and preferably on north western aspects.

(e) Propagation

Temperate fruits are commercially propagated through asexual method of propagation; however, sexual method is generally employed for raising seedlings rootstocks. The propagation of fruit plants may be classified in two broad categories.

(*i*) Propagation of rootstocks

(*ii*) Propagation of scion cultivars

The rootstock propagation is accomplished through seed in case of seedlings rootstock and through layering, cutting and tissue culture in case of clonal rootstocks. The scion cultivars are commonly is propagated through grafting and budding methods.

(i) Propagation of rootstock

a. **Seedlings:** Seedling rootstocks are commercially used on which scion variety is grafted or budded. In apple, seeds of crab apple (*Malus baccata*) or self-pollinizing varieties like Golden Delicious and Granny Smith are used for raising seedling rootstocks. Bandana and Chandel (3), while evaluating the seedling rootstocks of different apple varieties found that seedling rootstocks of Crab apple, Red Gold and Granny Smith varieties are better in terms of growth of rootstock and scion variety grafted on them, which also resulted in production of more saleable plants under protective conditions in a year. Seedlings of Kainth (*Pyrus pashia*) and Zarainth (*Pyrus serotina*) are used as a rootstock for pear. Kainth seedling without interstock and Quince-C rootstock with interstock (Beurre Hardy) were observed best and were compatible with most of the commercial pear cultivars (Chauhan, 7).

   Stone fruits are propagated on their own seedlings, *e.g.* peach on wild peach, plum and apricot on wild apricot, almond on bitter almond or behmi and cherry on Paja (*Prunus cerasoides*) seedling or on clonal rootstocks. Seeds of wild walnut for walnut, Burkett variety of pecan nut and Bruno variety of kiwifruit are used for raising seedling rootstocks. Seeds should be collected from fully ripened fruits, extracted by hand or machine should not contain pulp on the skin and dried under shade. Seed are treated with bavistin (3 g/kg seed) to be stored for some time before sowing.

b. **Seed Stratification**: Seeds of deciduous fruit trees generally do not germinate until they get chilling treatment. Exposing the seeds to low temperature (0 to 5°C) for a considerable period helps in breaking the dormancy thus resulting in stimulation of germination. Seeds of temperate fruits are dormant, which require stratification (low temperature) treatment to break dormancy (Table 1) Seeds are placed in layers in moistened sand kept in wooden box. This box is kept at low temperature a little above the freezing point for one to six months depending upon the type of seed.

Table 1: Seed Stratification Period for different Temperate Fruits Seed

Fruit Plant	*Effective Temp. (°C)*	*Time Period (Days)*	*Germination(%)*
Apple	2-7	60-70	70-80
Apricot	3-7	45-50	75-80
Cherry	0.5-7	90-100	80-85
Peach	1-7	70-75	55-60
Pear	0.5-7	35-40	75-80
Pecan nut	0.5-7	70-75	60-65
Walnut	1-6	70-75	70-75
Kiwifruit	2-5	50-60	75-80

Pre-sowing treatments of seeds with certain chemicals and growth regulators reduced the stratification period, enhancing the seed germination and stimulated seedling growth. Treatment of seeds with GA_3 and thiourea alone or in combination with cold stratification is effective in improving germination in temperate fruits. Though GA_3 treatment is effective in breaking dormancy and stimulating germination of un-stratified seeds but is most effective when used in combination with stratification. In a study on seed germination of peach, the highest seed germination (95 per cent) occurred when the seeds were soaked in GA_3 at 100 ppm for 24 h and stratified at 5°C for 15 days, indicating that GA_3 helps in reducing the stratification period. The exogenous application of GA increases the level of endogenous GA-like substances and antagonizes the effect of inhibitors. GA also stimulates the synthesis of alpha-amylase and ribonuclease in the seeds, thus helps in quick and better germination. In peach, use of 5,000 ppm thiourea, 5,000 ppm GA_3 + thiourea + etherel increased seed germination. Negi and Ananda (18) observed that pre-sowing treatment with 200 ppm GA_3 and BA reduced the stratification period and improves seed germination in wild peach.

The germination of apricot seeds can be improved by removal of kernel from shell, stratification and GA_3 or kinetin treatment. The seeds are soaked in GA_3 5000 ppm or kinetin 5 ppm before planting increased seed germination and seedling growth. Recently, some PGPR have shown great promise in seed germination and growth of seedlings. The effect of seed treatment with *Bacillus megaterium* inoculants on vegetative growth of four month old apple seedlings from net house and six month old seedlings from nursery under field conditions was studied by Sharma (21). He found that *Bacillus megaterium* significantly increased the shoot, leaf and root growth than un-inoculated control. The effect of seed bacterization was more pronounced when the seed treatment was given after stratification as compared to the seed treatment given before stratification. Significant increase in germination percentage by the use of plant growth promoting rhizobacteria in apple seed (Mahajan, 17) Bandana (2) found significant increase in seed germination and growth of apple seedlings under protective conditions with seed soaking in 100 ppm GA_3 and PGPRs *Bacillus licheniformis* and *B. methylotropicus.*

Germination responses to temperature, medium and gibberellic acid (GA_3) treatments were studied by Celik *et al.* (4) in kiwifruit. The seeds treated with four GA_3 concentrations (0, 2,000, 4,000 and 6,000 ppm) were sown in trays with peat moss, perlite + humus and soil mixture and subjected to the temperatures of 20°, 25°, 30° and 35°C with bottom heating, under controlled conditions. Seeds sown in peat moss and subjected to the temperature of 35°C with bottom heating reached the maximum germination percentage of 99.17.

c **Seed Sowing** :Pre-stratified seeds of apple and pear are sown at a depth of 3-5 cm in rows 10 –15 cm apart in the seed beds, which are thoroughly ploughed and manured before sowing. The seeds of peach, apricot, almond, pecan nut and walnut are sown 3-5 cm deep at a distance of 15-20 cm seed to seed and 20-25 cm apart in rows. After sowing, the seed beds are mulched with 10 cm thick dry grass and light irrigation is given to avoid desiccation of stratified seeds. As the seedlings start germinating, the mulch material is removed. When seedling attain 5-6 cm height and having 3-4 leaves then they are transplanted from seed beds to nursery beds at a spacing of 10-12 cm seedling to seedling and 15-20 cm apart in rows. The nursery beds are prepared by thoroughly ploughing and 60-80 t/ha well rotten farm yard manure along with 50 kg P_2O_5 is mixed in the nursery beds before transplanting of seedlings.

In high hills areas experiencing sufficient snowfall in winter, seeds are sown directly in the field during December, however, sometimes poor germination occurs due to heavy rainfall.

(ii) Clonal rootstocks

Clonal rootstocks are generally multiplied through stooling or layering, however it can also be propagated through cuttings and tissue culture techniques.

a. **Stooling or Layering**: Layering or stooling is the operation of rooting stems and shoots while they are still attached to the parent plant. The clonal rootstocks of temperate fruits are usually propagated by stooling or trench layering depending upon their inherent vigour. The vigorous rootstocks are best raised from layers, while the weaker ones are raised from stools. Layering requires more space and care than stooling but gives four times more rooted shoots per plant.

For the establishment of stool bed, healthy stock plant of clonal rootstocks are planted 40 to 45 cm apart in a row and 2 m spacing is kept between the rows. Plants are cut back to 40-60 cm and allow growing for one year. The new growth starts in the spring, plants are cut to 2.5 cm above the ground level. The shoots develops from the crown and later these shoots have grown 10-15 cm, loose soil, sand, cocopeat, FYM mixture is used to cover each shoot to one half of its height. When shoots have grown 20-25 cm, a second hilling operation take place. The third and final hilling operation is made in mid-summer when the shoots have developed 45 to 50 cm. Girdling the base of the shoots by wiring or application of 2,000 ppm IBA stimulate rooting. Khatik (15) reported that IBA @ 2,500 ppm was found best treatment for inducing better root system in apple clonal rootstock Merton 783 in sand + cocopeat + vermicompost (1:1:1) media (Fig. 1), whereas the media with IBA 1,500 PPM +NAA 3,000 PPM resulted maximum average root diameter. By the end of growing season, the stool shoots are separated from the parent plant and rooted layers are transplanted in the nursery beds.

2500 ppm IBA | Sand: Cocopeat : vermicompost media | Soil media (Contol)

Fig. 1: Effect of IBA and Growing Medium on Rooting in Mound Layers.

In case of trench layering, the stock plants are planted in a trench 60 to 70 cm apart in a straight line at an angle of 45°. The distance between the rows is kept 2 meters. The plants are allowed to grow during the first year to establish a good root system. During the winter season, the plants along with 3-4 strong laterals are brought down to a horizontal level and pegged carefully. Buds are covered with 2.5 cm of soil before they emerge. Subsequent application of rooting medium such as sawdust, sand, FYM and soil are added periodically as in case of stooling and rooting layers are separated from the mother plant at the end of growing season. The stool beds or trench layers can be used for 15-20 year with proper handling.

b. **Cuttings**: Both hardwood and semi-hardwood stem cuttings can be employed for the multiplication of clonal rootstocks, but the use of hardwood cutting is more common since these are easy to prepare, not readily perishable and do not require special equipments or attention as in semi hardwood cuttings. The hard-wood stem cuttings are made between late fall to early spring depending on the planting time, whereas,the semi-hard wood cuttings are usually taken during mid or late summer along with leaves. The length of stem cuttings is usually 15 to 30 cm or more with basal cut about 1 cm below node and the top cut 1 cm above node. The semi-hard wood stem cuttings are planted in mist chamber or glass house having high humidity. The clonal rootstocks of different stone fruits showed marked differences in their rooting propensity. Planting time and use of growth regulators especially IBA influence root induction in rootstocks to various levels. Percentage of rooting in rootstocks can be enhanced by planting the cutting in December and treating them with 1000 to 2000 ppm IBA for 30 seconds. Ersoy *et al.* (12) studied the effect of IBA applications on rooting of softwood top cuttings of M9 apple clonal rootstock. The highest rooting of cutting was found with 2,500 ppm IBA with 85-90 per cent humidity level. Verma (24), while studying the effect of size of cuttings and concentration of growth regulators on rooting in hardwood cuttings of Merton793 of apple found that 35 cm long and with 1.25-1.50 cm diameter cuttings treated with 2,500 ppm IBA gave highest rooting of 65 per cent under shade net conditions.

(iii) Micro-propagation

Micro-propagation or tissue culture refers to *in vitro* and aseptic culture of various plant part such as meristem tips, isolates cell or protoplast in artificial grown medium. Although, the clonal rootstocks of stone fruits are propagated commercially by mound and trench layering, yet the rate of multiplication is low. For rapid mass multiplication of clonal rootstocks of temperate fruits, micro propagation or tissue culture technique can be utilized to increase the rate of multiplication in a limited space. During the recent years, successful accomplishment of micro propagation through tissue culture technique has been made in clonal rootstock of apple and cherry. Sharma *et al.* (22) has been made an attempt not only to develop a protocol for rapid multiplication of colt rootstock of cherry, but also to make micro propagation a cost effective proposition. They obtained multiple shoots on MS medium containing 0.5 mg/l each of BAP and kinetin along with 3 per cent table sugar. These shoots readily rooted on MS medium containing 1 mg/l IBA. Well rooted plants were transferred to a mixture of sand and soil after 1-2 weeks to the nursery. They recorded nearly 100 per cent survival of rooted plants in the nursery beds mulched with dry grass and adequately watered. In the regeneration of peach rootstock.

(c) Propagation of Scion Cultivars

The majority of temperate fruits are commercially propagated by grafting and budding, however in some fruits the method of cutting and micro-propagation can be employed for raising nursery plants.

(i) Collection and treatment of scion wood

For grafting, the scion wood is collected during dormant season from the identified and labeled mother trees. Generally one year old shoots are collected and each separate bundle must be carefully labeled for variety. The scion wood is to be stored for one or two months, therefore, labels and marking must be durable. Scion woods are stored in moist chamber or in cool chamber at 0-2°C. Sticks can also be packed in moist sphagnum moss buried in soil or in the trenches at cool and shady place. Before grafting, the scion wood are dipped in a solution of tree spray oil, bavistin, chloropyriphos and then dried in the shady place, which helps to disinfect the scion wood.

For summer budding, shoots are collected from healthy, disease and virus free mother plants at the time of budding. Remove all the leaves immediately to reduce water loss. For longer distance transportation, the scion wood are wrapped in wet sphagnum moss and then wrapped in thin polythene.

(ii) Grafting

Though several methods *viz.*, tongue, whip, cleft, and root grafting are employed for the propagation of temperate fruits, but tongue and cleft grafting during February-March have given the best results. In tongue grafting, there is considerable cambial contact between stock and scion and it heals quickly thus makes a strong union. Negi and Ananda (18) obtained highest bud take success (81 per cent) with tongue grafting in almond, when performed in second fortnight of February than with shield and chip methods of budding and they recommended tongue method of grafting for commercial propagation of almond under mid hill conditions of Himachal Pradesh. Under cold desert of Ladakh, Dwivedi *et al*. (10) used whip, tongue and cleft method of grafting during February and March and reported highest bud-take success with tongue and cleft method of grafting. They recommended tongue and cleft method of grafting performed in first week of February for commercial propagation of apricot in Ladakh region.

Walnut is very difficult to propagate and both grafting and budding technique are used with varying level of success. Gautam (14) obtained 50-70 per cent bud-take success with tongue grafting in walnut. Rezaee *et al*. (20), while working on top working of walnut trees with nine different methods of grafting or budding showed that, the modified bark grafting performed at mid-April showed the highest grafting success (100 per cent) and 96.3 per cent survival.

During recent years *hypocotyl* grafting technique was introduced in practice (Gandev and Dzhuvinov, 13). In this method, grafting takes place during vegetation with growing tip of soft wood cutting on cleft. Grafted plants should be kept in about 80-90 per cent relative humidity and temperature of 26 - 27°C. In Bulgaria, Gandev and Dzhuvinov (13) obtained 83 per cent graft survival, the grafted plants being kept in a greenhouse under controlled temperature.

(iii) Budding

Shield and chip budding is most widely used for propagating temperate fruits because of its simplicity and efficiency Nearly 100 per cent bud-take success has been reported in various stone fruits by T-budding and 85 to 92 per cent bud-take success in walnut with chip and 95 per cent in pecan nut with patch and annular method. Among the temperate fruits, walnut is very difficult to propagate and showed only 40-50 per cent bud-take success in grafting. The studies on the propagation of walnut through summer budding methods revealed that chip budding performed during

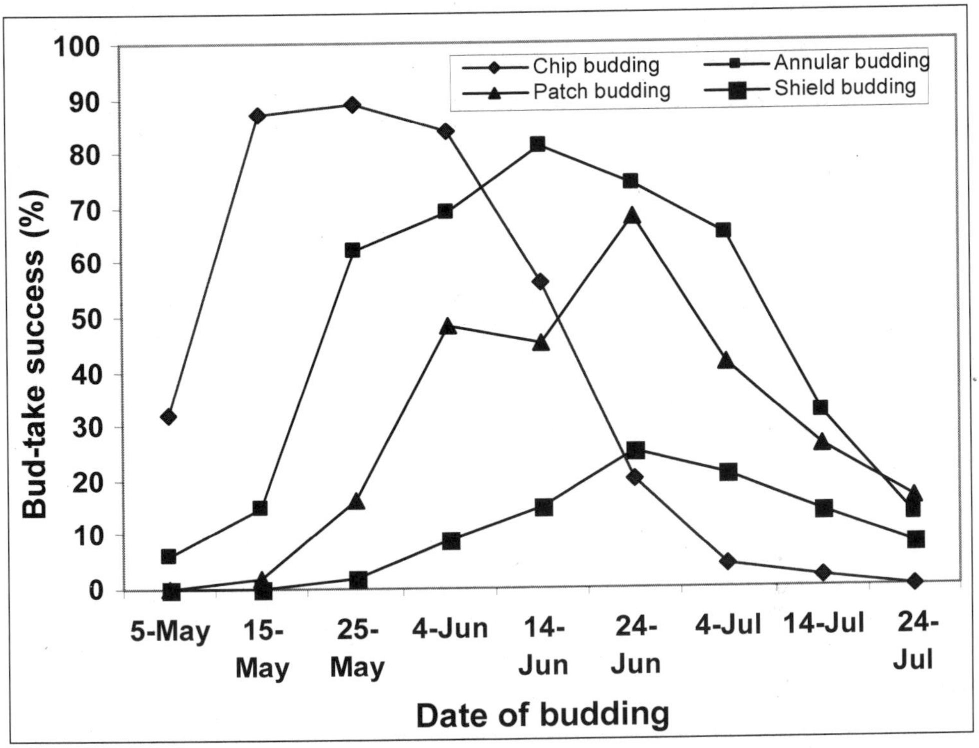

Figure 1: Effect of different Methods and Time of Budding on Bud-take Success in Walnut.

Plate 2: Chip Budded Plants in the Nursery (92 per cent success).

Mid May to first week of June gave as high as 85 to 92 per cent bud-take success (Chandel *et al.*, 6) under open field conditions (Figure 1 and Plate 2.).

Ebrahimi *et al.* (11) studied different budding methods (patch, shield and chip) and compared under controlled and field conditions using different native varieties of Persian walnut (*Juglans regia* L.) The highest success rate was obtained with patch (91.0 per cent) followed by shield (31.1 per cent) and chip (19.1 per cent) budding under greenhouse conditions versus 25.0, 15.0 and 10.0 per cent under field conditions, respectively.

The traditional production of grafted walnut trees in the nursery needs 2-3 years, which is long-term and labour-consuming, with low land efficiency in propagation. In order to improve the speed of walnut propagation thick-shell, fresh walnut was harvested, and was seeded directly in mid-September. The development of root system fully formed before the soil was frozen in the winter, the seedlings started to grow early in spring of next year and attain buddable size in May-June. The optimum period for budding was mid May to Early June for Chip and in July for patch method and bud take success varies from 80 to 92 per cent. This technique could shorten the walnut propagation period to 1 year, and reduce costs and improve the economic benefits of propagation of selected cultivars (Chandel, 5). Similarly, Xin *et al.* (25) also suggested a rapid grafting technique using direct sowing of fresh seed and budding the rootstock with patch method in June–July with 90 per cent bud take success.

Table 2: Time and Method for Commercial Propagation of different Temperate Fruits

Fruit Crop	*Recommended Propagation Method*	*Time of Operation*
Apple	1. Tongue grafting 2. Chip budding	February–March February and July
Pear	1. Tongue grafting 2. Chip budding	February July
Almond	1. Tongue Grafting	February
	2. Chip budding	Mid-March
Apricot	1. Tongue Grafting 2. T-budding 3. Chip budding	February May–June First week of September
Cherry	1. Tongue Grafting	February–March
Peach	1. Tongue Grafting	January–February
	2. T-budding	May–June
Plum	1. Tongue Grafting 2. Chip budding	January–February February
Walnut	1. Tongue grafting 2. Chip budding 3. Patch budding 4. Hypocotyls grafting	February–March Mid May to First week of June June–July April
Pecan nut	Patch and annular budding	June–July
Kiwifruit	Tongue grafting and Chip budding Hard-wood cuttings	February February

Scions of Fuji, Galaxy Gala, Golden Reinders, Pink Lady and Summer Red apple varieties were grafted by the whip grafting and chip budding techniques on M_9 apple rootstocks. The grafting success in greenhouse was 88 per cent and in the external environment was 65 per cent (Koyuncu and Ersoy, 16). Chip budding done on 15 March and 31 March were most successfully resulting in highest graft success, scion growth and maximum yield of saleable plants in Ladakh, therefore, it is recommended that second fortnight of March is the best time for vegetative propagation of apple through chip budding in cold arid conditions of Ladakh (Dwivedi and Stobdan, 9). Bandana (2) reported 98 to 100 per cent bud-take success in mid July with chip budding in apple under polyhouse conditions.

(iv) Cutting

Although the temperate fruits are commercially raised by budding or grafting on their own seedling rootstocks but some fruits are propagated by cuttings. Owing to variability in rootstock, the problem of incompatibility or of poor graft union is experienced and when the plants are grown on their own roots such problems are eliminated. Kiwifruit is commercially propagated by Hard-wood and semi-hard wood cuttings. Propagation through cutting is most rapid and easiest method. The ideal cutting is 0.5 to 1.0 cm thick with relatively short internodes about10-15cm in length having at least 3-4 buds. Cutting should be taken from middle portion of current season's growth shoot in June-July in case of semi-hard cuttings and one year old shoot during January-February in case of hard wood cuttings. After preparation, the cuttings are dipped in 4000-5000 ppm IBA solution for 10-15 seconds and then planted in mist propagation chamber or in open field condition for rooting. A study was conducted to elucidate the effect of pre-conditioning treatments, IBA and collection time on the rooting of semi-hardwood cuttings of kiwifruit under intermittent misting was studied by Pratima and Rana (19). The application of pre-conditioning treatments *viz.* blanching and girdling in combination with IBA increased rooting in hardwood and semi-hard wood cuttings. The maximum rooting percentage (70.0 per cent) was recorded in semi-hardwood cuttings treated with blanching + girdling + IBA @ 2000 ppm.

Chhukit (8) observed highest rooting, number of adventitious roots, length of longest root, total root length and total biomass of the roots in cuttings planted in rooting media sand + forest leaf compost + cocopeat (1:1:1), which was found the best medium for raising hardwood cuttings for kiwifruit (Plate 3).

Treatment with IBA 5000 ppm and planting in growing medium of sand : forest leaf compost : cocopeat (1:1:1)

Plate 3: Rooting in Hardwood Cuttings of Kiwifruit.

4. RAPID METHOD FOR PRODUCTION OF NURSERY PLANTS UNDER POLYHOUSE CONDITIONS

Nursery plants of temperate fruits are produced in open field conditions, which generally takes 2-3 year to produce grafted or budded plants. However an agro-technique has been developed to produce grafted or budded plants of apple and walnut in one year under protected condition. The stratified seeds are sown in beds during mid February in polyhouse. The seedling attain bud able size in July and budded with chip method from first week to second week of July, which gave as high as 94-99 per cent bud take success in apple and 82-85 per cent success in walnut. The budded plants become saleable during same year. Similarly, the disease and virus free planting material of strawberry can also be produced under polyhouse conditions.

Sowing of stratified seeds (in mid February). Chip budding (in July) Saleable plants (in December)

Plate 4: Production of Apple Nursery in a Year-under Polyhouse Conditions.

(a) Nursery Management of Temperate Fruits

Nursery management is one of the most important aspects and plays an important role in the production of healthy and quality nursery plants of temperate fruits. Most of the future problems in horticulture may be eliminated if the plants are raised on scientific lines. Although the nursery production has increased tremendously during the recent years but scientific emphasis on the management aspects of temperate fruits nurseries especially the stone fruits is rather low. Thus the due attention should be paid on every aspects of nursery management. The important operations followed in nursery after grafting and budding are:

(i) Removal of polythene

This is an important operation, which must be done timely. After 4 weeks of sprouting as the grafted plants attain 30-40 cm height and the graft union start swelling, polythene strips is loosened by the sharp cut on the back side of the graft. In case of chip budding, it must be removed after 3 weeks.

(ii) Deshooting

After grafting and budding, the shoots generally arise from the rootstocks below the graft union. These shoots should be removed immediately after sprouting at regular intervals; otherwise it results in failure of grafting and budding.

(iii) Staking

After removal of polythene, staking is must since the union are tender in summer and the plants blown off with higher wind velocity. Wooden sticks are generally used for staking of grafted plants.

(iv) Single stem

The Indian orchardists need single straight stem nursery plants for plantation, whereas in European countries feathered grafted plants of 3-4 year old are preferred for plantation. In grafted or budded plants no scion shoots are allowed to grow and they are pinched off from the growing scions

(b) Mulching and Nursery Raising Techniques

Mulching of nursery plant is very essential especially in nursery established at water scarce areas to promote growth of plants. Immediately after grafting, nursery plants are mulched with 10 cm thick dry grass or with black polythene, which helps in soil moisture conservation and control weed population in the nursery. Singh and Gautam (23) studied, the effect of five nursery raising techniques *i.e.* seedlings raised in polythene bags and transplanted after germination by chopping the tap root, transplanting of young germinating seedlings from nursery beds; raising of seedlings by restriction of root system by placing polythene sheet at 20 cm depth; under cutting of one year old and newly germinating seedlings was studied on lateral root development vegetative growth of walnut seedlings. The number of lateral roots, vegetative growth and survival in the field after transplanting was maximum in seedlings in which undercutting was done at germinating time.

(c) Irrigation

Irrigation is very essential for good growth of nursery plants of apple. Irrigation should be given at frequent intervals or alternate days. Watering should be done in evening time to avoid evaporation. Irrigation should be given through sprinkler systems.

(d) Manuring and Fertilizations

The requirement of manure and fertilizers depends largely on the type and fertility of soil. Well rotten FYM at the rate 60-80 tons per hectare along with 30 kg of P_2O_5, 60 kg K_2O and 60 kg of N per hectare is applied in the nursery beds before sowing of seeds or transplanting of seedlings. Remaining half dose of nitrogen (60 kg/ha) is applied with the onset of rains (July).

(e) Hoeing and Weeding

The nursery beds should be kept always clean or free from weeds by manual weeding. Nursery beds should be dug 5-10 cm deep at least 3-4 times with the help of forks to loosen the soil.

(f) Plant Protection

Nurserymen have to be very careful in regular plant protection measures against powdery mildew, root rot, premature leaf fall, woolly apple aphid, crown gall, San

Jose scale and Defoliating beetles with the application of recommended fungicides and insecticides. Besides this, the nursery site should be changed every year to reduce the infestation of insects and diseases.

5. REFERENCES

Anonymous. 2014. *Indian Horticulture Databse.* National Horticulture Board, MoA, GoI, Gurgaon.

Bandana. 2014. Studies on the nursery production of apple (*Malus domestica* Borkh.) under protected conditions. M.Sc. thesis, Dr. Y. S. Parmar University of Horticulture and Forestry, Nauni, Solan, Himachal Pradesh.

Bandana and Chandel, J.S. 2015. Effects of varieties on seed germination, seedling growth and growth of budded plants of apple under protected conditions. *Int. J. Farm Sci.* **5**: 74-82.

Celik, H., Zenginbal, H. and Ozcan, M. 2006. Enhancing germination of kiwifruit seeds with temperature, medium and gibberellic acid. *HortSci.* **33**: 39–45.

Chandel, J.S. 2015. Annual report. Department of Fruit Science, Dr. Y.S. Parmar University of Horticulture and Forestry, Nauni, Solan, Himachal Pradesh.

Chandel, J.S., Gautam, D.R. and Sharma, N.C. 2006. Chip budding: an excellent method of propagation of walnut (*Juglans regia* L.). *Acta Hort.* **705**: 335-339.

Chauhan, J.K. 2006. Studies on the effect of rootstock and interstock on morphological, physiological and anatomical characteristics of pears. Ph.D. thesis, Dr. Y.S. Parmar University of Horticulture and Forestry, Nauni, Solan, Himachal Pradesh.

Chhukit, K. 2009. Studies on vegetative propagation of kiwifruit (*Actinidia deliciosa* Chev.). M.Sc. thesis, Dr. Y.S. Parmar University of Horticulture and Forestry, Nauni, Solan, Himachal Pradesh.

Dwivedi, S.K. and Stobdan, T. 2011. Standardization of optimum time of chip budding in apple under cold arid condition of Ladakh. *Prog. Hort.* **41**: 210-212.

Dwivedi, S.K., Singh, B. and Paljor, E. 2000. Studies on vegetative propagation of apricot through grafting in Ladakh. *Indian J. Hort.* **57**: 39-41.

Ebrahimi, A., Vahdati, K. and Fallahi, E. 2007. Improved success of persian walnut grafting under environmentally controlled conditions. *Int. J. Fruit Sci.* **6**: 3-12.

Ersoy, N., Kalyoncu, I.H., Aydin, M. and Yilmaz, M. 2010. Effects of some humidity and IBA hormone dose applications on rooting of M9 apple clonal rootstock Softwood top cuttings. *African J. Biotech.* **9**: 2510-2514.

Gandev, S. and Dzhuvinov, V. 2006. Performance of hypocotyl grafting of walnut under uncontrolled temperature conditions. *Acta Hort.* **705**: 351-353.

Gautam, D.R. 1990. Studies on the winter and summer vegetative propagation techniques of walnut (*Juglans regia*). *Acta Hort.* **284**: 27-31.

Khatik, P.C. 2010. Effect of auxin and rooting media on multiplication of apple clonal roostock (Merton 793). M.Sc. thesis, Dr Y. S. Parmar University of Horticulture and Forestry, Nauni, Solan, Himachal Pradesh.

Koyuncu, F. and Ersoy, N. 2011. Nursery growing in controlled greenhouse and orchard by using various grafting methods in some apple (*Malus* x *domestica*) varieties. *J. Food Agric. Environ.* **9**: 243-246.

Mahajan, R. 2007. Growth promotion of apple seedlings by application of plant growth promoting rhizobacteria. M.Sc. thesis, Dr. Y. S. Parmar University of Horticulture and Forestry, Nauni, Solan, Himachal Pradesh.

Negi, K.S. and Ananda, S.A. 1996. Effect of different grafting methods on the development of almond nursery trees. *J. Hill Res.* **9**: 367-370.

Pratima, P. and Rana, V.S. 2011. Effect of pre-conditioning treatments, IBA and collection time on the rooting of semi- hardwood cuttings of kiwifruit, *Actinidia deliciosa* Chev. *Int. J. Farm Sci.* **1**: 30-36.

Rezaee, R., Hassani, D. and Vahdati, K. 2014. Long term trials on top working of walnut trees in Iran. *Acta Hort.* **1050**: 197-200.

Sharma, A. 2007. Growth promotion of apple seedlings with plant growth promoting rhizobacteria. M.Sc. thesis, Dr. Y. S. Parmar University of Horticulture and Forestry, Nauni, Solan, Himachal Pradesh.

Sharma, D.R., Chauhan, P.S., Kaur, R. and Srivastava, D.K. 1992. Micropropagation of colt- a semi dwarf rootstock of cherry. *Indian J. Hort.* **49**: 209-212.

Singh, G. and Gautam, D.R. 2005. Effect of nursery raising techniques on lateral root development of walnut seedlings. *Acta Hort.* **696**: 185-189.

Verma, P. 2013. Effect of preconditioning treatments, size of cuttings, plant growth promoting rhizobacteria and IBA on rooting in cuttings of apple (*Malus* x *domestica* Borkh.) clonal rootstock Merton 793. M. Sc. thesis, Dr. Y. S. Parmar University of Horticulture and Forestry, Nauni, Solan, Himachal Pradesh.

Xin,G., Zhang, J.D., Wang, H.M., Wang, H., Hu, Y.Q. and Zhang, H.Y. 2014. A rapid grafting technique using direct sowing of fresh walnut seeds. *Acta Hort.* **1050**: 177-180.

Chapter 17

Status and Strategies for Meeting Planting Material Requirement in Temperate Fruit and Nuts

F.A. Banday, M.K. Sharma and M.S. Mir

Division of Fruit Science,
Sher-e-Kashmir University of Agricultural Sciences and Technology of Kashmir,
Srinagar – 190 025, J&K
E-mail: headfruitscience@gmail.com

1. INTRODUCTION

In India, horticulture has emerged as one of the potential agricultural enterprise in accelerating the growth of economy over the years. Its role in the country's nutritional security, poverty alleviation and employment generation programmes is becoming increasingly important. On account of significant production increases in horticultural crops across the country, a 'Golden Revolution' is in the offing and India has emerged as a leading player in the global scenario. As a result of a number of thoughtful research, technological and policy initiatives and inputs, horticulture in India, today, has become a sustainable and viable venture for the small and marginal farmers. Therefore, there is a great scope for the horticulture industry to grow and flourish.

In India, total production of fruits is estimated to be 81.285 million tonnes from 6.98 million hectares (2012-13). During 2014-15, India exported fruits worth Rs. 2771.32 crores. However the overall productivity of the fruit orchards is very low as is compared to horticulturally advance countries. Majority of the orchards are old and to increase the productivity replacement of old orchards using improved cultivars and planting of additional area would be required. A large number of high yielding varieties are now available but have not reached the orchardists for want of planting material. Although the farmers are less sensitive to price when they are sure of quality planting material, its exclusiveness and that monetary compensations will accrue from higher production.

Among the fruit crops, a temperate fruit crops namely apple, peach, pear, plum, apricot, almond and walnut occupy major share in North Western Himalayan states of the country. Jammu and Kashmir, Himachal Pradesh and Uttarakhand are major producers of these fruit crops with cumulative annual production of approximately 33.00 lakh MT over an area of over 5.25 lakh hectares for all the major temperate fruit crops with a meager productivity of 6.29 MT/ha. The areas under temperate fruit crops has also been expanding in hilly tracts of North Eastern Himalayan is states like Sikkim, Arunachal Pradesh, and Nagaland, and some hilly parts of western Ghats such as Ooty and Kodakanal. Total production of apple in the country is about 24.98 lakh MT from an area of about 3.05 lakh ha. In case of stone fruits, total production is approximately 170 thousand MT over an area of about 58.0 thousand hectares with 2.93 MT/ha productivity. The area under nut fruits is about 145 thousand hectares with a production of about 2.54 MT with a productivity of only 1.75 MT/ha. Among the fruit crops grown in India, apple is the fourth major fruit crop with a productivity of around 7.00 MT/ha, whereas, in leading apple growing countries productivity as high as 60-70 MT/ha is common. In the past, apple industry experienced a slow growth rate in production and productivity, but in the last 5 years, good growth has been recorded, respectively. There is a wide gap of production and productivity levels among different apple growing states in India, while, Kashmir is leading with a productivity of 11 MT/ha, North Eastern states are far behind with only about 3 MT/ha. Even within the state, a marked variation has been observed in production trend as evident in Kashmir division of Jammu and Kashmir with a productivity range of 5.84 MT/ha in district Budgam to 20.71 MT/ha in district Baramullah. However, the major emphasis is to enhance the overall production and productivity of temperate fruits with an area expansion on large scale under superior cultivars following improved management systems as lack of quality planting material of fruit crops has been identified as one of the major factors for such a poor performance of majority of temperate orchards in India. Approximate estimation for total annual requirement for quality planting material of different temperate fruit crops is about 25-30 lakhs in Jammu and Kashmir, about 20-25 lakhs in Himachal Pradesh, about 10-12 lakhs in Uttarakhand and about 7-10 lakhs in north eastern and other states under conventional planting system. Production of high quality fruits shall not only facilitate major domestic market share, but also increase the export share, keeping in view various norms of good agricultural practices codex standards and trade barriers.

The present status of temperate fruit orchards in India can be summarized as below:

- ☆ Cultivation of only few traditional cultivars.
- ☆ Monoculture within cultivars of same crop.
- ☆ Lack of diversification with available improved varietal spectrum.
- ☆ Traditional density orchards.
- ☆ Poor rate of area expansion under superior cultivars.

- ☆ Low proportion of pollinizers.
- ☆ Lack of sufficient pollinators.
- ☆ Poor orchard management.
- ☆ Poor canopy management.
- ☆ Lack of diversification with other fruit types like kiwifruit, olive, nectarine, prunes, berries etc.
- ☆ Marginalization of temperate fruits in many areas due to climate change and lack of availability of low chilling or other superior cultivars of stone and nut fruits as an alternative.
- ☆ A drastic shift from fruit cultivation to vegetable cultivation in the low to mid hill region in some states.
- ☆ Lack of systematic plantation of nut crops because of non-availability of planting material.
- ☆ Prevalence of old, exhaustive, seedling as well as scattered plantation, especially of nut fruit crops like walnut, almond, pecan nut, chestnut etc. which show variability and lack of uniformity in fruit quality.
- ☆ Large scale use of seedling or inferior rootstocks due to lack of sufficient quantity of clonal rootstocks.
- ☆ Majority of the orchards have outlived their life and now need to be replaced by new plantations.
- ☆ Senile orchards need rejuvenation by improved planting material.
- ☆ Majority of the cultivars are inferior and are not suitable for high density planting system.
- ☆ Import of quality temperate fruits from horticulturally advanced countries is giving tough competition to our domestic fruit growers.

Different temperate zones alongwith altitude and climatic features of different states, where temperate fruits are being grown are given in Tables 1-5.

Table 1: Temperature Fruit Production Zones in Jammu and Kashmir

Zone	*Altitude (amsl)*	*Climatic Feature*	*Crop*
Zone II Doda, Rajouri, Poonch parts of Udhampur	Mid to high altitude 750-1200 m	Sub-tropical temperate transition zone	Plum, peach, apricot, olive, pistachio, and apple, almond, walnut and strawberry
Zone III Srinagar, Ganderbal, Budgam, Anantnag, Pulwama, Kulgam, Baramullah and Kupwara	Mid to high altitude temperate zone 1200-2500 m	Wet and cold winter	Apple, pear, almond, cherry, walnut, plum, peach, apricot, olive, pecan nut, pistachio nut, kiwi, grape, chestnut and strawberry.
Zone IV Leh and Kargil	High altitude zone: 2500 to 4000 m	Cold arid zone, extremes of temp. (-40°C in winter), dry climate with low rain-fall and heavy snowfall	Apricot, sea buckthorn, apple and walnut

Das *et al.* (2012)

Table 2: Temperate Fruit Production Zones in Himachal Pradesh

Zone	*Altitude (amsl)*	*Climatic Feature*	*Crop*
Sub tropical, sub-mountain and low hills (Una, Bilaspur, Kangra and Hamirpur)	Upto 914 m	Very hot summer and mild winter	Sub-tropical fruit crops and low chilling pear, peach and plum etc.
Sub temperate, sub humid mid hills (Solan, Sirmour, Kangra, Chamba, Mandi and Kullu)	915-1523 m	Summer is not very hot winter intense to moderate and light snowfall	Very much suitable for stone fruits, kiwifruit, olive and pear, moderately suitable for apple and walnut
Wet temperate high hills (Shimla, Mandi, Chamba and Kullu)	1524 -2472 m	True temperate climate with mild summer and intense winter, heavy snow and rainfall	Pome fruits, cherry, peach, plum, apricot, nut fruits, olive and kiwi fruit
Dry temperate high hills and cold arid zones (Lahul and Spiti, Kinnaur and some parts of Chamba)	Above 2472 m	Dry climate, less rain-fall, very intense winter with heavy snow fall and absence of summer	Apple, almond, apricot, walnut, hop, sea-buckthorn and *chilgoza*

Das *et al.* (2012)

Table 3: Temperate Fruit Production Zones in Uttarakhand

Zone	*Altitude (amsl)*	*Climatic Feature*	*Crop*
Tarai and lower Doon	Upto 600 m	Humid sub-tropical climate	Low chilling peach, pear and plum
Upper Doon, Bhawar and lower Shivalik	600-1200 m	Sub-tropical to sub-temperate climate	Peach, pear, plum and apricot
Middle Garhwal-Kumaon	1200-1800 m	Sub temperate to temperate climate	Peach, plum, apricot, apple, almond and walnut
Upper Garhwal and Kumaon	1800-2400 m	Wet temperate climate with moderate snowfall	Apple, peach, plum, apricot, almond and walnut
Cold zone (Munsiari, Badrinath areas *etc.*)	2400-3600 m	Wet temperate to dry temperate climate with heavy snowfall and severe cold	Apple, walnut and seabuck-thorn

Das *et al.* (2012)

Table 4: Temperate Fruit Production Zones in Western Ghat

Zone	*Altitude (amsl)*	*Climatic Feature*	*Crop*
Nilgiri hills	2000 to 2600 m	Sub-temperate	Low to medium chilling cultivars of apple, peach, plum, pear, strawberry and kiwi fruit.
Kodaikanal	>2000 m	Sub-temperate	Low to medium chilling cultivars of apple, peach, plum, pear, strawberry and kiwi fruit

Das *et al.* (2012)

Table 5: Temperate Fruit Production Zones in North Eastern States

Zone	*Altitude (amsl)*	*Climatic Feature*	*Crop*
Arunachal Pradesh: Tawang, Dirang, Bomdila, Shergoan, Diban valley, Upper Subansiri, Anini, North Eastern parts of Lohit and Northern part of East Siang **Meghalaya:** Upper Shillong, Mawphlong and Mairang **Mizoram:**Blue Mountain, Halkhan,Turpang, Nauzuarzo and Tiang **Nagaland** :Tuesongand, Zonebota,Vangkung areas of Mokachung. **Sikkim:**Gnathing, Chhangu, Serrathong, Lima Zema, Karponag, Bordong, Resi, Kangdem, Melli Param, Lachem Hilley and Yaksum	There are three altitude ranges in North Eastern region for growing temperate fruit crops: 1. 5000 (at Sela pass) to 3000 m (at Tawang) in Arunachal Pradesh 2. 1300 to 2500 m (in areas like Naga hills, Miker hills and hills and plateau of Shillong) 3. Around 750 m (plains and sub tropical climate in lower hills and plains)	Climatic feature comprise colder areas in the higher hills and mild cold to hot in the lower hills and plains: ☆ Typical temperate type climate in the higher hills ☆ Mild temperate to sub tropical climate in the mid hills. ☆ Higher hills of Arunachal Pradesh and Sikkim are typical temperate type where winters are very cold, experiences snowfall and summer is mild.	☆ Apple, pear, peach, plum, cherry, pistachio nut, almond, apricot, walnut, chestnut and kiwifruit in the higher hills ☆ Low chilling peach, pear, plums and passion fruit in sub temperate to sub tropical areas. ☆ High chilling superior apple cultivars alongwith other high chilling temperate fruit crops are very much suitable for cultivation in higher hills.

Das *et al.* (2012)

2. NEED FOR QUALITY PLANTING MATERIAL

With the shift from subsistence farming to commercial farming practices, the temperate fruit production in India has shown an upward trend, yet the inherent potential has not been fully realized. The main reason for this is the non-availability of quality planting material of superior cultivars, which is required for planting in new areas to expand the area under different fruit crops under different densities, gap filling and replacement planting to change old and senile orchards to new ones. At present the total number of nurseries available in India is 6724 (Table 6). The sector-wise status of fruit nurseries in India is as under.

Table 6: Status of Fruit Crop Nurseries in Public and Private Sectors (2006-07)

Sl.No.	*Sector*	*No. of Nurseries*
1.	Public sector	1,594
2.	Private sector	4,607
3.	SAUs/ICAR	138
4.	Model nurseries established under NHM	385
	Total	**6,724**

Singh *et al.* (2008)

Table 7: State-wise Planting Material Requirement of Temperate Fruits and Nuts per Year (Under conventional system)

Sl.No.	*Sector*	*Plant Material Requirement (Lakh)*
1.	Jammu and Kashmir	25-30
2.	Himachal Pradesh	20-25
3.	Uttarakhand	10-12
4.	North eastern and other states	7-10
	Total	**62-77**

Singh *et al.* (2010)

Planting of poor quality low yielding planting material in orchards is often responsible for low quality fruits. The trees on seedling rootstocks give heterogeneous performance and the average yields are much lower.

Kashmir has initiated thinking about shifting towards high density planting system. With the concept of high density planting (HDP), *i.e.* planting of more number of plants than optimum through manipulation of tree size for the enhancement of productivity by maintaining a balance between vegetative and reproductive growth without impairing the plant health, the productivity of the orchards can be increased manifolds. If we want to transform the fruit industry by using high density rootstocks, the requirement will be as follows:

Table 8: Estimated Number of Plants Required for Conversion of 33 per cent of Orchard Area into High Density Orchard

Crop	*Area (ha) 2014-15 under Kashmir Division*	*Area (ha) to be Converted to HDP (33 per cent of current area)*	*Change in Density with Use of HDP (trees/ha)*	*Total Plants Required in Five Years (in lakhs)*
Apple	1,44,733	48,000	1,000	480.000
Cherry	3461	1,200	1,111	13.332
Pear	7,900	2,600	625	16.250
Total				**509.582**

Division of Fruit Science, SKUAST-Kashmir.

3. CONSTRAINTS IN NURSERY PRODUCTION PROGRAMME

Although much has been achieved in standardizing the techniques for the propagation and raising of plant nurseries, but the programme is faced with following constraints:

(*a*) **Lack of soil and irrigation facility** : Availability of sufficient area with good soil and irrigation facilities are the basic requirements for establishing a fruit nursery. Majority of the nurseries are raised at the same location for many years due to non-availability of sufficient land, which results in perpetuation and infection of soil borne diseases.

(*b*) **No budwood mother orchard/bank** : Most of the fruit nurseries do not maintain separate budwood mother orchard as a source of scion/budwood for multiplication. As a result the nurseries collect scion wood/bud wood from different sources and some of them are not authentic. The farmers also take scion shoots from juvenile or old and diseased orchards.

(*c*) **No pedigree records** : Fruit nurseries take the budwood from the trees where no pedigree record is available. As a result the quality of the plant material as well as its produce is not good.

(*d*) **No rootstock banks** : Rootstock is the basic unit and the most important input for the production of quality planting material. This is determines the success of the orcharding. However most of the fruit nurseries do not maintain separate rootstock banks for the production of uniform plant material. As a result the nurseries collect seed/rootstock from different sources resulting in poor quality of nursery plants produced. Orchardists also procure rootstocks from unknown sources which also many times are infected with diseases and pests.

(*e*) **Plant protection measures**: Most of the fruit nurseries do not follow proper plant protection measures. As a result of this, the nursery plants are affected by various pests and diseases, which impair the health of the plant and yield poor quality fruits.

(*f*) **Skilled manpower**: The nursery production work in temperate fruits is suffering because of lack of trained manpower for propagation and other nursery operations.

(*g*) **Lack of infrastructure**: Infrastructural facilities such as controlled mist-chambers, Hi-tech polyhouse, soil health presenting techniques like soil fumigation equipment, irrigation facilities, bottom heat equipment, budding/grafting machines and isolation houses, virus-indexing, imuno diagnostic kits *etc.* are not available in our fruit nurseries.

(*h*) **Improper implementation of nursery registration act** : The production and sale of fruit nursery plants is governed by nursery registration act. This means that good quality disease-free nursery plants are produced and supplied to the orchardists but in practice, fruit nurseries adopt unscrupulous practices in the propagation and production of nursery plants and supply poor quality nursery plants to the orchardists because of lack of proper and strict enforcement and provisions of this act.

4. PROPAGATION METHODS AND ROOTSTOCKS RECOMMENDED FOR TEMPERATE FRUITS AND NUTS

Propagation methods have been standardized for larger multiplication of temperate fruit crops and nuts. The rootstocks have also been standardized for the production of quality planting material. The commercial methods followed under each fruit and nut crops and the rootstock standardized are showing in Table 9:

Points to be Considered for the Production of Quality Planting Material

Following points should be kept in mind to produce quality nursery plants.

- ☆ Soil should be fertile.
- ☆ Nursery should be fenced.
- ☆ Nursery should be preferably on the road side or accessible through vehicles.
- ☆ Nursery should be well labeled.
- ☆ Periodical soil testing of the nursery should be conducted.
- ☆ Adequate amount of manures and fertilizers should be added to the nursery.
- ☆ Fertilizers should uniformly be mixed with the soil, because direct contact with the delicate roots of stocks may damage them.
- ☆ Nursery should have assured irrigation facility.
- ☆ Fresh seeds extracted from healthy fruits should be used for raising of seedling rootstocks.
- ☆ Nursery floor should be well managed.
- ☆ Dried/diseased seedlings should be discarded.
- ☆ Extra or damaged roots of rootstocks are trimmed to ensure the straight penetration in soil.

Table 9: Commercial Method and Time of Propagation and Recommended Rootstocks for Planting Material Production of Temperate Fruits

Crop	*Commercial Propagation Method (s)*	*Time of Propagation*	*Recommended Rootstock(s)*
Apple	Tongue/wedge grafting T-budding	Grafting : February-March Budding : July-August	Seedling : Crab Clonal : Dwarf : M-9 Semi-dwarf : MM-106 Semi-vigorous : MM-111
Pear	Tongue/wedge grafting T-budding	Grafting : February-March Budding : July-August	Wild pear Quince-C
Peach	Tongue grafting T-budding	Grafting : February-March Budding : July-August	Wild peach, wild plum, wild apricot
Plum	Tongue grafting T-budding	Grafting : February-March Budding : July-August	Wild plum, wild apricot
Apricot	Tongue grafting T-budding	Grafting : February-March Budding : July-August	Wild apricot, wild plum
Cherry	Tongue grafting T-budding	Grafting : February-March Budding : July-August	Sour cherry, colt, Mazzard, Mahaleb
Kiwifruit	Hard wood cuttings Soft wood cuttings	Hard-wood cuttings: March Soft-wood cuttings: June	–
Grape	Hardwood cuttings	February-March	–
Strawberry	Runners	October-November upto February-March	–
Almond	Tongue grafting T-budding	Grafting : February-March Budding : July-August	Bitter almond and wild peach
Walnut	Tongue grafting Patch budding	Grafting : Tongue grafting in last week of February under polyhouse and in second week of March under field conditions	Seedling
Chestnut	Tongue grafting T/patch budding	Grafting: March Budding : August	Seedling

- ☆ Budwood and graftwood should only be taken from bearing, disease-free mother plants having known pedigree.
- ☆ All the rootstocks, scion material and grafted plants are kept covered with moistened gunny sheets under shade to avoid desiccation.
- ☆ Scion sticks with 3-4 buds of 8-10 cm length and 8-10 mm diameter should be grafted on the rootstocks of matching diameter.
- ☆ Budding/grafting should be practiced at a height of 15 cm above the ground level.
- ☆ Grafted plants are planted at 8-10 cm spacing in the trenches (10-15 cm depth) and soils are generally pressed from both the sides carefully to place the plants upright and straight.

- ☆ Care should be taken that graft union is always above the soil surface and upper roots are not kept exposed above the soil.
- ☆ Care should be taken that graft union is not disturbed at the time of planting.
- ☆ Good feathered plants should be developed.
- ☆ Knife should be sterilized with alcohol or sodium hypochlorite and should regularly be washed with surface disinfectants.
- ☆ Seedlings should be treated with Carbendazine @ 1 g per litre of water at the time of transplanting.
- ☆ Recommended plant protection measures should be followed for the control of insect pests and diseases.
- ☆ Layout and nursery map should be prepared indicating the blocks of each cultivar.
- ☆ After completion of the planting, plants should be watered regularly.
- ☆ Mulching with leaf mould or grass straw or black polyethylene conserve soil moisture and restrict weed growth.

5. STRATEGIES FOR PRODUCTION OF QUALITY PLANTING MATERIAL

Establishment of fruit plant nurseries without any proper objectives shall not solve the basic purpose of production of quality planting material. Nurserymen should know all the technical aspects which include guidelines and procedures to be followed before establishing any nursery, technological knowhow regarding the type of planting material to be produced and how to produce, maintenance of the quality standards and finally sale or distribution to the end users. Production of quality planting material of fruit crops should follow the basic points mentioned below:

(a) Creation of Budwood Bank

Bud wood bank of improved and commercial cultivars of different fruit plants should be established in different districts of the state. These should be maintained scientifically under the supervision of farm university, KVKs and state department of horticulture for the supply of quality budwood for raising of fruit plant nurseries to government and private sectors for further multiplication. Cultivars selected must be of superior types, should have good market demand and recommended by recognized agencies. Selection of mother plant for collection of bud wood is most critical for production of authentic, healthy and disease free planting material. As the plant health, fruit quality and productivity depends on health and genuineness of the mother plant, therefore it should have known pedigree in respect to health, bearing and quality. The selected plants should be virus-indexed and only disease-free plants should be used as source of bud wood. Budwood should be taken from matured wood of current season growth or next to last growth. Budwood taken should be kept in moist sphagnum moss and gunny bags to avoid desiccation.

Criteria for selection of cultivar as mother plant should be as follows:

- ☆ It should be genetically true-to-the type.

- ☆ It should have consistency in better performance in terms of vegetative growth.
- ☆ It should have regular flowering behavior.
- ☆ It should be a regular bearer, best quality with all good pedigree records.
- ☆ It should be precocious and prolific bearer.
- ☆ It should have freedom from most of the diseases/pests.
- ☆ It should have novelty in fruit quality and storage life.
- ☆ It should be healthy and free from diseases, pests and physiological disorders.
- ☆ It should have consumer preference and commercial viability in the local and distant markets.

Establishing a bud wood or scion wood block of temperate fruit crops is similar as in case of any commercial orchard plantation. The basic approach is comprised of a base map of the complete plantation plan, layout of marked area, demarcation of fruit specific blocks, fruit specific spacing depending upon the planting and training system, pit digging and planting.

(b) Raising of Seed Blocks or Seed Orchards

Seed blocks or seed orchards of crab apples like Siberian crab, Snow Drift, Manchurian, Golden Hornet, *Malus floribunda* etc. or bitter/wild almond, wild apricot or peach, small and hard shelled walnut, *behmi, kainth, paja* and wild cherries etc should be raised for seed collection.

(c) Establishment of Rootstock Banks/Rootstock Multiplication

Several rootstocks with immense potential have become the integral part of the temperate fruit nursery. The temperate fruit industry of apple, pear, peach, plum, apricot, walnut, almond *etc.* is based on the use of good rootstocks. There is the need to use only good quality rootstocks in temperate fruits. Most of the temperate fruits and nuts are propagated by vegetative means as these fruits come into bearing earlier when propagated by vegetative means as compared to seeds. Rootstocks exert profound influence on the vigour, productivity, quality of fruits, longevity of scion, adaptability to soil and climatic conditions and to different pests and diseases. Selection of proper rootstocks for a particular area greatly determines the success of the orcharding. High density rootstocks are generally multiplied by stooling and layering for obtaining true to type plants.

Characteristics of Clonal Rootstocks

Following are the desired trails:

- ☆ Uniformity.
- ☆ Size control.
- ☆ Precocity.

- ☆ Production efficiency.
- ☆ Resistance to pests, diseases and adverse climatic conditions.

However the clonal rootstocks are also confronted with following problems:

1. Specific requirements.
2. Specific to soil and climatic conditions.
3. Needs more care and better management.

The ideal rootstock should have the following characteristics :

- ☆ Freestanding with well developed root system.
- ☆ Induce precoiety.
- ☆ Good anchorage.
- ☆ Easy to propagate.
- ☆ Dwarfing in nature.
- ☆ Graft compatible with the scion cultivar.
- ☆ Resistant to known diseases and insect-pests.
- ☆ Widely adaptable.
- ☆ Free from suckering.
- ☆ High yielding.
- ☆ Free from soil problems existing in the area.
- ☆ Free from replanting problem.

Several components of future commercial temperate fruit growing will depend on successful use of rootstocks for better scion compatibility, canopy architecture, fruit quality, nutrient absorption, water use efficiency and adaptation under the influence of climate change. Furthers, there is urgent need to develop new rootstocks, which are well adapted to the local conditions. Rapid and large scale multiplication of available apple clonal rootstocks, introduction or import and collection of pome, stone and nut fruit rootstock, which are not available in India through proper agencies and multiplication of these clonal rootstocks and their propagation with proper indexing prior to large scale multiplication is of utmost importance. For the production of seedling rootstocks, the seeds should be collected from the cultivar, which produces large number of plump and viable seeds which produce vigorous and sturdy seedlings.

(d) Certification of Planting Material

Proper certification of the planting material of all the cultivars which are to be planted in the bud-wood bank regarding authenticity of cultivars, virus-indexing or testing for any other diseases and insect pests should be carried out.

(e) Planting Material Production under Protected Cultivation

Technology have been standardized for large scale multiplication of temperate fruits with assured success. A success of about 60-70 per cent have been observed when grafting of walnut was performed under zero-energy high humidity polyhouse condition. This method has been found as superior as well as convenient as compared to *hot cable callusing technique* in the Division of Fruit Science, SKUAST-K. Techniques have also been standardized for propagation of kiwifruit and grape through hardwood cuttings in zero-energy high humidity polyhouse conditions in the Division of Fruit Science, SKUAST-K, Shalimar.

(f) Production of Zone Specific Planting Material

Certain zones produces extraordinary quality fruits due to their peculiar climatic factors and these have been known only for the production of particular variety of a particular fruit crop. There is great demand of quality planting material of a particular variety in these zone, hence the efforts should be made for the production of quality planting material of the particular variety/varieties in these zones.

(g) Micro-propagation Facilities

This is an established method of producing large number of identical disease free plants. The protocols have been standardized for large scale multiplication of apple clonal rootstocks, *viz.* M-9, MM-106 and MM-111, kiwifruit, cherry rootstocks (Colt, Mazzard, Mahaleb), almond and pear by the Division of Fruit Science, SKUAST-K, Shalimar. Use of ELISA has been effectively demonstrated in detecting the viral diseases which requires to be translated in a practical mechanism.

(h) Growing Media

Growing media such as soil, sand, perlite, vermiculite, cocopeat *etc.* are used by nursery men in plant propagation. The media should hold the cuttings firm during propagation and should retain and supply sufficient moisture to the cuttings. The medium should also be free from weed seeds, diseases and nematodes. The propagating media can be easily pasteurized by heat, chemicals or solar radiations.

(i) Soil Testing of Nursery

Soil of the nursery should be tested at regular interval so that any deficiency of the nutrient in the soil can be added at an appropriate time to improve the health of nursery plants.

(j) Good Care and Handling of Nursery Plants

Nursery should be thoroughly cultivated and kept free from weeds. Nursery should be well watered and it should neither be so dry nor waterlogged. It should be well manured and fertilized. The nursery should be regularly inspected for different intercultural operations.

(k) Pest and Disease Management

Adoption of pest and disease management in planned way is essential for the production of quality planting material. The nursery should be regularly inspected for different plant protection measures.

(l) Information about New Cultivars

The nurserymen should be well informed about the newly released or recommended cultivars or any promising cultivars, which are to be released in near future. The nurserymen should be informed about the characteristics of these cultivars as well as their future scope so that the nurserymen can prepare well in advance for raising the quality planting material of these cultivars.

(m) Information about New Technology

Nurserymen should always be early adopter of any new technology. He should maintain a close liaison with farm university, KVKs as well as States Department of Horticulture to get the information about the latest technologies developed in the field of nursery production. He should have an eye on market and demand.

(n) Accreditation of Nurseries

Availability of true to type, quality planting material is crucial for success of fruit orcharding. This necessitates having a net-work of fruit nurseries which conform to Model Nursery Standards in terms of infrastructure, quality of planting materials and adoption of good nursery management practices. State governments should ensure supply of quality for fruit crops by enactment of Fruit Plant Nursery (Regulation) Act and enforcement of its provisions through licensing of fruit plant nurseries. However, all the States have not enacted Fruit Nursery Acts. As per the information available in report of the working group on Horticulture for the XI Five Year Plan, at present only 8 states have adopted Nursery Act and in 9 states some system of registration/monitoring exists for nurseries whereas in other states there is no nursery act at present. Therefore, National Horticulture Mission has taken initiative to specify the Nursery Standards by specifying Infrastructure required for setting up of Model Horticulture Nursery etc. and thus accrediting the existing nurseries on the basis of nursery infrastructure, production system and quality parameters of planting material and good nursery management practices in a comprehensive manner. A nursery should function as a reliable source of supply of quality planting material for fruit crops. To ensure availability of good quality planting material, NHB has started a system of registration of horticulture nurseries on voluntary basis.

6. POLICY CHANGES AND PROGRAMME IMPLEMENTATION FOR ENSURING PRODUCTION AND SALE OF QUALITY PLANTING MATERIAL

- ☆ There is need for developing crop specific progeny blocks at State Agricultural Universities, at district level by States Department of Horticulture, ICAR institutes and KVKs.
- ☆ There should be a quality control enforcement by evolving a uniform system by legislating a Central Nursery Act.
- ☆ Possibility of formulating uniform nursery rules for all the states/UTs under the proposed Seed Act, 2004 should be explored.

- A mechanism should be evolved to procure the scion material only from the nurseries accredited by National Horticulture Board with atleast two stars.
- Model nursery act should be prepared by Ministry of Agriculture and should be circulated to all state governments/UTs for enforcement.
- Mechanism should be evolved for continuous capacity building programme/skill development of the nursery growers in propagation and nursery management through Agricultural/Horticulture universities and KVKs.
- Model nurseries need to be established by State Agricultural Universities/ ICAR institutes/State Department of Horticulture/KVKs in order to make them the source of identified and genuine plant material on large scale.
- There is need to involve the nursery associations in the certification process and also in the procurement of planting material of new varieties and rootstocks.
- Accreditation of fruit nurseries through NHB should be speed up.
- Quarantine mechanism available in the country should be followed in true letters and spirit.
- Roadside sale of plant material of unknown pedigree should be banned.
- Supply of appropriate ratio of pollinizers alongwith plant material should be made compulsory.

7. INSTITUTIONS AND DEPARTMENTS FOR TECHNICAL GUIDANCE

The institutions involved to provide technical guidance to nurserymen regarding production of quality planting material of temperate fruit crops are as followed:

- Sher-e-Kashmir University of Agricultural Sciences and Technology of Kashmir, Shalimar-190025, Srinagar (J&K), Research Stations and KVKs.
- Central Institute of Temperate Horticulture, Srinagar
- Sher-e-Kashmir University of Agricultural Sciences and Technology of Jammu, Chatha, Jammu-180009 (J&K), Research Stations and KVKs.
- Dr Y.S. Parmar University of Horticulture and Forestry, Nauni, Solan-173230 (H.P.), Research Stations and KVKs.
- C.S.K. Himachal Pradesh Krishi Vishvavidyalaya, Palampur-176062, Research Stations and KVKs.
- IARI Regional Station, Shimla, Amartara-171004, Himachal Pradesh.
- Department of Horticulture, J&K Govt., Rajbagh, Srinagar-190008 (Jammu and Kashmir), all district horticulture offices.
- Department of Horticulture, Government of Himachal Pradesh, Naubahar, Shimla-171002, H.P., all district horticulture offices.

- G.B. Pant University of Agriculture and Technology, Pantnagar-263145, Uttarakhand, Research Stations and KVKs.
- Uttarakhand Horticulture and Forestry University, Bharsar-246123, via. Chaipal Ghat, Pauri Garhwal, Uttarakhand, Research Stations and KVKs.
- Central Institute of Temperate Horticulture, Regional Station, Mukteshwar, Nainital-263138, Uttarakhand.
- Vivekananda Parvatiya Krishi Anusandhan Sansthan, Almora-263601, Uttarakhand.
- Department of Horticulture and Food Processing, Govt. of Uttarakhand, Chaubatia, Ranikhet-263651, Uttarakhand.
- Central Agricultural University, Imphal, Manipur, College of Horticulture and Forestry, Passighat, Arunachal Pradesh and college of PG studies, Umiam, Meghalaya.
- ICAR Research Complex for NEH Region, Umroi Road, Ri-Bhoi-793103, Meghalaya
- Horticulture Departments in all the north eastern states.
- National Horticulture Board, Ministry of Agriculture and Farmers Welfare Government of India 85,Institutional area, Sector-18, Gurgaon - 122015 (Haryana).

8. REFERENCES

Das, B., Ahmed, N., Attri, B.L. and Singh, P. 2012. Production of quality and elite planting material of temperate fruit crops. Central Institute of Temperate Horticulture, Regional Station, Mukteshwar, Nainital, Uttarakhand.

Singh, A.K., Patel, V.B. and Singh, S.K. 2008. Propagation of quality planting material in fruit crops for higher profitability. **In:** *Income and livelihood security through horticultural development*, J.N.L. Srivastava and Lallan Singh (Eds.). IFFCO Foundation, New Delhi, pp. 73-81.

Singh, S.R., Shagoo, P.A. and Jan, Nusarat. 2010. Innovative techniques for quality planting material production of temperate fruits. **In:** *Advance Cultivation Techniques of Temperate Horticultural Crops*, 17-28 May. pp. 25-28.

Chapter 18

Micropropagation in Temperate Fruits and Nuts

Rekha Chaudhury and S.K. Malik

Tissue Culture and Cryopreservation Unit,
ICAR- National Bureau of Plant Genetic Resources,
Pusa Campus, New Delhi – 110 012
E-mail: rscientist58@gmail.com; rekha@nbpgr.ernet.in

1. INTRODUCTION

In vitro micropropagation, encompassing the ability to establish shoot tip cultures, proliferate shoots, induce rooting, and acclimatize the resulting plantlets, has been successfully applied on commercial scale to different fruit crops with huge potential in the future. These techniques had numerous applications to fruit crops beginning nearly 80 years ago with embryo rescue techniques for stone fruits (see Zimmerman and Swartz, 1994). From this beginning, the method has been applied successfully to produce commercially acceptable early-ripening peach and nectarine cultivars. As most species and commercial cultivars are genetically heterozygous, their genetic integrity needs to be maintained through vegetative propagation. *In vitro* cultured plants can be mass propagated within limited space and under disease-free conditions. Application of *in vitro* culture methods to temperate fruit crops has also been to eliminate disease-causing viruses from strawberries (see Zimmerman and Swartz, 1994). Meristem-tip culture has since become an integral part of virus-indexing programs for a number of fruit crops including apple (see Brown and Maloney, 2005) in conjunction with thermotherapy. In few cases micrografting of meristem tip onto an *in vitro*-grown seedling is being done similar to that done in citrus.

2. APPLICATIONS OF *IN VITRO* PROPAGATION

The utility of *in vitro* propagation is mainly for:

- ✰ Production of pathogen-free plants, for direct planting purpose or also for further micropropagation.
- ✰ For production of clonal rootstocks
- ✰ Rapid multiplication of selected genotypes for breeding purpose.
- ✰ Large-scale production of new cultivars for more rapid introduction into commercial trade.
- ✰ Propagation of plants for which conventional vegetative methods are ineffective or unprofitable.
- ✰ Production of selected plants for F_1 hybrid seed production
- ✰ As prerequisites to exploiting somaclonal variation and induced mutations, and for the development of transgenic plants

3. METHODOLOGY AND EXPLANTS

Micropropagation may be achieved by either organogenesis or somatic embryogenesis. Former is largely influenced by the genotype, physiological state of the explant, age of the explant and in vitro environment. Embryogenesis is the development of embryogenically-competent somatic calls, under the influence of specific phytohormones, similar in pattern to development of a zygotic embryo.

Micropropagation has been used to produce both rootstocks and self- rooted cultivars of most major temperate tree fruits (Hutchinson and Zimmerman, 1987, see Litz, 2005). Commercial production has been mainly of rootstocks for peach, alongwith that for apple, pear, cherry, apricot and plum (Zimmerman, 1991).

Micropropagation in *Prunus* has been done usually using shoot tips and nodal cuttings and large scale commercial micropropagation of selected virus-free genotypes of peach rootstocks has been reported (Battistini and De Paoli, 2002).Zygotic embryos are also in use in vitro for embryo rescue and micropropagation for raising peach and nectarine.

Lately *in vitro* culture has been applied for genetic improvement of fruit crops. The applications include production of hybrid plants from fused protoplasts, somaclonal variation and mutagen-driven changes in ploidy in regenerated plants, haploid plants from anther culture, and transfer of specific genes *via Agrobacterium*-mediated transformation.

Nut crops have been generally recalcitrant to *in vitro* conditions whereas apple, pear, cherry, peach, plum, stone fruits and olive and good candidates for *in vitro* culturing.

4. DIVERSE CULTURE MEDIA: OPTIMIZATION FOR BEST RESPONSES IN MICROPROPAGATION

The Murashige and Skoog (MS) medium has proved to be widely adaptable and has been used for micropropagating apple, pear, peach, cherry, plum and other stone fruits, raspberry, blackberry, currant, gooseberry and strawberry. Woody Plant Medium (WPM) has been used for blueberry and also for a number of the non-cricaceous fruit crops as well, but is not as commonly used as MS. Anderson's medium has been used for blueberry and also for apple and raspberry. For strawberry, knop's is a commonly used medium. Lepoivre medium has been in usage for many tree fruits particularly in Europe. Damiano *et al* (2000) applied an innovative temporary immersion technique, a method based on the complete immersion of the proliferating explants in a liquid medium, alternating with a dried period of *in vitro* culture of wild pear (*Pyrus communis* var pyraster L.). This led to improved proliferation ability of shoots (by 13 times) obtained from leaves through adventitious regeneration and the plants revealed an excellent stem elongation and a higher rooting ability. In another study, attempts were made to define the appropriate types and concentrations of mineral nutrients for shoot proliferation of pear (Reed *et al.*, 2013). Five mineral nutrient factors were created from MS salts – NH_4NO_3, KNO_3, mesos (Ca-Mg-Cl-Mn-SO4-PO4), metals (Zn-Mn-Cu-Co-Mo-B-I), and Fe-EDTA. Each factor was varied over a range of concentrations and mineral nutrient formulations were identified for each genotype that resulted in significantly better growth.

Micropropagation techniques for over 20 pear (*Pyrus*) cultivars belonging to seven species have been reported by Bell and Reed (2002). While most published methods use Murashige and Skoog (MS) basal nutrient medium, or slight modifications thereof, Lepoivre (LP) and Driver-Kuniyuki Walnut (DKW) media, have improved shoot proliferation rates. Solid media gelled with agar, sometimes in combination with gellan gum, have traditionally been used, but two-phase liquid overlay or intermittent liquid immersion techniques reportedly greatly increased the shoot proliferation.

Lima-Nishimura *et al.* (2003) reported higher growth and multiplication and lower occurrence of hyperhydric shoots in apple rootstocks 'Jonagold' and 'Marubakaido' when cultured in media enriched with 0.2 per cent xyloglucon, isolated from seeds of *Hymenaea coubari*. Use of bioreactor for apple micropagation has been reported by Chakrabarty *et al.* (2003) as a cost-effective commercial method.

5. ROLE OF *IN VITRO* CULTURE IN GERMPLASM CONSERVATION

Ex situ conservation of genetic variability of cultivated plants and their wild relatives is the responsibility of the National Bureau of Plant Genetic Resources (NBPGR). In this regard biotechnological approaches are presently in use for *in vitro* mass clonal propagation, virus indexing, *in vitro* conservation and cryopreservation of germplasm. Many national and international organizations have used *in vitro* techniques and applied it to more than 1100 species, as a complementary conservation method to field genebanks for conservation of vegetatively propagated germplasm.*In vitro* repository at NBPGR, the unique multi-crop repository of the country, maintains large collection (more than 1900 accessions) of various tropical and temperate crops of 158 species. At National Cryogenebank more than 11,000 accessions (totaling 797 species) of diverse non-orthodox and orthodox seed species, dormant buds and pollen have been successfully cryostored for the last 26 years. In vitro recovery of explants (embryos and embryonic axes), shoot tips, meristems, somatic embryos and cells after cryostorage is standardized/ optimized for raising the plantlets for transfer to the field. Dormant buds cryotorage has been initiated for the first time in the country at NBPGR. Core collection of mulberry comprising diverse *Morus* species has been successfully crystored and recovery done *in vitro*. *In vitro* raised cryostored plants have been transferred to field conditions. Genetic stability of mulberry plants was assessed using ISSR markers after cryostorage for various periods and no changes were observed when compared to fresh controls.

6. CLONAL CRYOBANK USING DORMANT BUDS FOR FUTURE PROPAGATION

In view of necessity for creating duplicate collection of temperate species that are presently maintained in the field, cryobanking of dormant buds of mulberry, apricot and walnut to create a base collection for long-term preservation has been ongoing at NBPGR. Plant genetic resources of these species including quality planting material of elite varieties and obsolete varieties are being tested for degree of cold hardiness during peak winters. Different accessions have been found to have different kinetics of acclimatization. The extent of acclimation that the explant possessed when collected had a direct relation with the success in cryopreservation. The survival and recovery percentage of diverse accessions has been tested under *in vitro* conditions or grafted in case nodal sections/grafts are cryostored. It was observed that most of the accessions, which survived LN exposure, had inherent dormant or semi-dormant nature, whereas, accessions that failed to recover after LN exposure were non-dormant or semi-dormant in nature. On an average the survival under *in vitro* culture remained high above 50 per cent in most of the accessions and about 40 per cent in field grafting. A total of 275 accessions belonging to *Morus indica, M. alba, M. latifolia, M cathayana, M laevigata, M.nigra, M. australis, M bombycis, M multicaulis* and *M rotundiloba* have been successfully cryostored buds (Choudhary *et al.*, 2015b). Regeneration of cryopreserved mulberry germplasm accessions after 1-3 years of storage indicated no survival loss. Following cryostorage, use of recovery conditions like dark incubation followed by under reduced light for several days was studied. For apricot and walnut, the two-step freezing and encapsulation-dehydration was attempted with explant recovery under *in vitro* conditions. Work on cryobanking of almond buds with subsequent regeneration *in vitro* has been successful at NBPGR (Choudhary *et al.*, 2014a and b). Micropropagation techniques have been applied with encouraging initial results to serve as a viability criteria for control and frozen explants. The effect of phenolic compounds on *in vitro* regeneration response in *Juglans regia* for different genotypes were analyzed and use of antioxidants like citric acid, PVP and activated charcoal improved the survival and growth.

In general, juvenile explants e.g embryos, cotyledons, hypocotyles or bud explants from seedlings and dormant buds (Choudhary *et al.*, 2015a) are more responsive to *in vitro* regeneration, than mature tissues.

(a) *In vitro* Conservation

In vitro conservation involves the maintenance of explants in tissue culture conditions under controlled, pathogen-free environment. The relative advantages are the storage and distribution of disease-free propagating material. *In vitro* methods can be employed for short- and medium- term conservation; the key strategy being reduction of growth and increase of subculture interval. NBPGR maintains about 300 accs of temperate fruits (Anonymous, 2015) in *in vitro* repository at NBPGR.

(b) Cryopreservation

Cryopreservation, is the storage of biological materials at ultra-low temperatures (-180°C to –196°C. Cryopreservation reduces cell and tissue deterioration by virtually halting metabolism and, hence, ensures theoretically 'infinite' period of storage without any loss in viability or vigour, or any changes in genetic makeup. More than 700 accessions of temperate fruits and nuts are being cryobanked at NBPGR and efforts are on for refinement of procedures for recovery *in vitro* of conserved embryos, embryonic axes (Malik and Chaudhury 2010) and dormant buds (Rao *et al.*, 2007; 2009) of temperate species (Fig. 1 and 2).

Fig. 1: *In vitro* Regeneration of Plantlets of *Juglans regia* and *Prunus persica* from Cryobanked Embryonic Axes.

Fig. 2: *In vitro* Regeneration of *P. dulcis* Plantlets in Culture Medium Raised from Cryopreserved Embryonic Axes.

A: Shoot growth initiation after 15 days of culturing; B: Well grown rooted plantlet after 35 days of culture in medium and C: Plantlet after one week of transfer to soilrite.

(c) Success after Micropropagation

There is more to micropropagation than just *in vitro* regulation of plantlets. Before and after micropropagation are just as important as in vitro regeneration. In apple, Diaz-Perez *et al.* (1995) examined acclimatization and subsequent gas exchange, water relations, survival and growth of micropropagated plantlets after transplanting to soil. High relative water content was important for plant survival and growth. Rooting and acclimatization of micropropagated apple shoots was done using IBA with high survival (70-100 per cent) by Bolar *et al.* (1998). In another study (Lane *et al.*, 2003) micrografted shoot tips from in vitro cultures to transplanted root stock plants. This can thus be used for propagating transgenic clones for field testing.

Irrespective of the micropropagation method, morphologic and genetic stability of the produced plants needs to be carefully examined. Heritability of the somaclonal variation, if any is required to be investigated. Genetic stability in mulberry after cryostorage and regeneration in vitro has been tested (Rao *et al.*, 2007; Choudhary *et al.*, 2013).After plantlet development *in vitro*, acclimatization *ex vitro* is done in greenhouse and field condition. Testing and deployment of clonal material are also important steps. In view of clonal variation in micropropagated trees of apple on orchard planting, reported by Zimmermann (1997), suitable precautions need to be taken.

7. REFERENCES

Brown, S.K. and Maloney, K.E. 2005. *Malus* X *domestica* apple In : Biotechnology of Fruit and Nut Crops (ed R.E. Litz) (Biotechnology in Agriculture series 29),CABI, Wallingford, Oxfordshire, UK, 475-511 pp.

Battistini, A. and De Paoli, G. 2002.Large scale Micropropagation of several peach rootstocks. *Acta Hort.* **592**: 29-34.

Bell, R.L. and Reed B.M. 2002. *In vitro* tissue culture of pear: Advances in techniques for micropropagation and germplasm preservation. *Proc. 8th IS on pear* (Eds. Corelli-Grappadelli *et al.*) *Acta Hort.* **596**: ISHS, 2002.

Chakrabarty, D., Hahn E.J., Yoon, Y.J. and Paek, K.Y. 2003. Micropropagation of apple rootstock M.9 EMLA using bioreactor. *J.Hort. Sci. Biotech.* **78**: 605-609.

Chaudhury, R. *et al.* 2001. Germplasm conservation: *In vitro* repository and cryobank. **In:** *National Bureau of Plant Genetic Resources: A Compendium of Achievements.* (Dhillon, B.S., Varaprasad, K.S., Srinivasan, K. Singh, M., Archak, S., Srivastava, U. and Sharma, G.D. Eds.). National Bureau of Plant Genetic Resources, New Delhi.

Choudhary, R., Chaudhury, R., Malik, S.K. and Sharma, K.C. 2015a. An efficient regeneration and rapid micropropagation protocol for almond using dormant axillary buds as explants. *Indian J. Exp. Biol.* **53**: 462-467.

Choudhary, R., Chaudhury, R. and Malik, S.K. 2015b Development of an efficient regeneration and rapid clonal multiplication protocol for three different *Morus* species using dormant buds as explants. *J. Hort. Sci. Biotech.* **90**: 245–253.

Choudhary, R., Malik, S.K., Chaudhury, R., Sharma, K.C. 2014a. Long-term conservation of dormant buds of *Prunus dulcis* (Miller) D.A. Webb. using three different new cryotechniques. *Romanian Biotechnological Letters* **19**: 9583-9592.

Choudhary, R., Malik S.K., Chaudhury, R., Sharma, K.C. and Saran, P.L. 2014b. Microscopic study on freezing behaviours in dormant buds of almond (*Prunus dulcis* Miller). *Prog. Hort.* **46** : pp. 1255-1258

Choudhary, R., Chaudhury, R., Malik, S.K., Kumar, S. and Pal, Digvender 2013. Genetic stability of mulberry germplasm after cryopreservation by two-step freezing technique. *African J. Biotech.* **12**: 5983-5993.

Damiano, C., Caboni, E., Frattarelli, A., Giorgioni, M., Liberali, M., Lauri, P. and D'Angeli, S. 2000. Innovative micropropagation of temperate fruit trees: the case of pear. *Acta Hortic*. 530, 181-186, DOI: 10.17660/*ActaHort*. 2000, 530, 20.

Lima-Nishimura, N., Quoirin, M., Naddaf, Y., Wilhelm, H.M., Ribas, L.L.F. and Sierakowski, M.R. 2003 A xyloglucan from the seeds of a native Brazilian species *Hymenaea coubari* for micropropagation of Marubakaido and Junagold apples. *Plant Cell Rep.* **21**: 402-407.

Litz, R.E. 2005. *Biotechnology of Fruit and Nut Crops*, Biotechnology in Agriculture Series, No 29 CABI Publishing, UK.

Malik, S.K. and Chaudhury, R. 2010. Cryopreservation of seeds and embryonic axes of wild apricot (*Prunus armeniaca* L.)*Seed Sci. and Tech.* **38** : 231-235.

Rao, A., Chaudhury, R., Kumar, S., Velu, D., Saraswat, R.P. and Kamble, C.K. 2007. Cryopreservation of mulberry germplasm core collection and assessment of genetic stability through ISSR markers. *Int. J. Indust. Entomol.* **15** : 23-33.

Rao, A., Chaudhury, R., Malik, S.K., Kumar, S., Radhakrishanan, R. and Qadri, S.M.H. 2009. Mulberry biodiversity conservation through cryopreservation. *In Vitro Cellular Dev. Biol. –Plant*, **45** : 639.

Reed, B.M., Wada, S., DeNoma, J., Evans, T.J. and Niedz, R.P. 2013. Improving *in vitro* mineral nutrition for diverse pear germplasm. *In Vitro Cellular Dev. Biol. - Plant* 06/2013; **46**: S64. DOI: 10.1007/ s 11627 - 013-9504-1.

Zimmermann, R.H. 1997. Orchard variation in micropropagated trees of ' Redspur Delicious' apple. *HortSci.* **32**: 935-936.

Zimmerman, R.H. and Swartz, H.J. 1994. *In vitro* culture of temperate fruits. In Plant Cell and Tissue Culture. I.K. Vasil and T.A. Thorpe (Eds.), Kluwer Academic Publishers, Dordrecht.

Chapter 19

Diagnostics of Major Virus Diseases of Apple and Production of Virus Tested Planting Material

Anil Handa, Santosh Watpade, Ajay Brakta, Neeraj Sharma, Usha Sharma, P.D. Thakur and M.N. Adarsh

Department of Plant Pathology
Dr Y.S. Parmar University of Horticulture and Forestry,
Solan – 173 230, H.P.
E-mail: anilhanda@msn.com

1. INTRODUCTION

Among different temperate horticulture crops, Apple (*Malus* x *domestica* Borkh.) is the most important fruit crop of temperate regions of the world. In India, apple is mainly cultivated in North Western Himalayan region which include states of Himachal Pradesh, Jammu and Kashmir and Uttarakhand in addition to the North Eastern hilly states of the country. Any losses occurring would affect the economy and livelihood of growers for whom apple is the only cash crop. Interestingly, all the major apple producing states of India display the huge variation in terms of productivity ranging from 3.7 MT/Ha in Uttarakhand to 8.6 MT/ha in Jammu and Kashmir with Himachal falling somewhere in between 3.9 MT/ha (Anonymous, 2013). Apple diseases particularly, virus and virus like diseases have a major impact on apple production worldwide and often become dominant yield limiting factor in orchard management. Viral diseases are of vital importance and a number of viruses have been reported by different workers from different apple growing areas of the world, *viz., Apple chlorotic leaf spot virus* (ACLSV), *Apple mosaic virus* (ApMV), *Apple stem pitting virus* (ASPV), *Apple stem grooving virus* (ASGV) (Desvignes *et al.*, 1990; Campbell, 1963; Posnette *et al.*, 1963; Zahn, 1996), *Carnation ring spot virus* (CRSV), *Cherry rasp leaf virus* (CRLV), *Sowbane mosaic virus* (SoMV), *Tomato ring spot virus* (ToRSV), *Tobacco ring spot virus* (TRSV), *Tobacco necrosis virus* (TNV), *Tomato bushy stunt virus* (TBSV) (Nemeth, 1986), *Prunus necrotic leaf spot virus* (PNRSV) (Sanchez Navarro and Pallas, 1997; Chandel *et al.*, 2008) and *Little cherry virus 2* (LChV-2) (Eppler *et al.*, 2001). Among these viruses only ApMV, ACLSV, ASPV and ASGV are of major importance and rest are minor viruses and do not cause major economic loss to the yield. In additions to these viruses, apple has also been reported to be infected by some viroids *viz., Apple scar skin viroid* (ASSVd) and *Apple fruit crinkle viroid* (Koganezawa *et al.*, 1983; Di Serio *et al.*, 2003; Koganezawa *et al.*, 2003).

Effective selection and propagation of disease free planting is the only way to tackle viral diseases in apple and this is highly dependent on the technology for quick and reliable detection of viruses and related pathogens. Therefore, reliable detection of viral infections is of paramount importance not only in the diagnostic and epidemiological studies of a disease, but also in indexing plants for producing virus free planting stocks in the management of viral diseases. Biological, serological and molecular methods of detection by employing indicator plants, enzyme-linked immunosorbent assay (ELISA) and reverse transcription polymerase chain reaction (RT-PCR) respectively, are being mainly used all over the world in the various virus indexing and certification programmes of apple (Mink,1989). However, for virus indexing in mass scale, ELISA is widely employed because of its rapidity, accuracy and sensitivity (Nemeth, 1986). In this chapter we compiled available literature for important viruses of apple.

2. IMPORTANT VIRUS DISEASES

(a) Apple Chlorotic Leaf Spot Virus

The economic importance of ACLSV is largely due to its worldwide distribution and its capacity to induce severe graft incompatibilities in some *Prunus* combinations, causing major problems in nurseries. Although most strains are latent in fruit trees, others can be responsible for russetting, top working disease and lethal decline of apple on some rootstock varieties (Desvignes and Boye, 1989). The virus causes serious disease in stone fruits (Nemeth, 1986; Desvignes and Boye, 1989; Sutic *et al.*, 1999). Although a natural spread in the field has been detected, the natural mode of spread is unknown (Brunt *et al.*, 1996).

Apple chlorotic leaf spot virus was first reported in *Malus* from USA (Mink and Shay, 1959; Lister *et al.*, 1965). ACLSV was described in detail by Luckwill and Campbell (1963) from USA; Cropley (1968); Lister (1970) from Holland; Nemeth (1986) from Hungary; Mink (1989) from Japan; Brunt *et al.* (1996) from UK; Semenas and Koukharchik (2000) from Belarus. It is distributed worldwide in pome and stone fruits, including apple, apricot, cherry, plum, peach, pear and quince (Lister, 1970; Nemeth, 1986; Dunez and Delbos, 1988; Mink, 1989; Adams *et al.*, 2004). The occurrence of chlorotic leaf spot disease on Golden Delicious cultivar of apple was recorded at Regional Fruit Research Station, Mashobra, Himachal Pradesh on the basis of graft transmission of symptoms but nothing was mentioned about the association of any particular virus (Nagaich and Vashisth, 1965).

Symptoms can be highly variable and depend on the virus strain and the host species and cultivar infected. The virus could produce line patterns, chlorotic rings, chlorosis, stunting and dark green mottle symptoms (Bar Joseph *et al.*, 1979; Yanase *et al.*, 1975). In sensitive *Malus* cultivars, symptoms can include chlorotic leaf spots and/or ring or line patterns on foliage, asymmetric leaf distortion, premature leaf drop, stunting, terminal dieback, inner bark necrosis and xylem pitting and local bark necrosis surrounding the inoculated buds (Nemeth, 1986; Mink, 1989). Gilmer (1962) from New York State Agricultural Research Station, Geneva, has reported that ACLSV caused severe phloem necrosis in stem tissue of Spy 227 scion after one year of inoculation. Terminal dieback and mild stem pitting symptoms were also observed on some of the inoculated plants.

Malformation and reduction in leaf size and chlorotic rings or line patterns were accentuated by ACLSV infection (Nemeth, 1986). The virus caused top working disease of apple trees grown on *Malus prunifolia* var. *ringo* root stocks in Japan (Yanase, 1974; Yanase *et al.*, 1979). Some ACLSV strains caused russetting and lethal decline of apple on certain rootstock varieties (Desvignes and Boye, 1989).

ACLSV is the most important virus among all apple viruses which causes huge losses to apple crop (Nemchinov *et al.*, 1995). In China, Liu *et al.* (2002) revealed that in top worked apple orchards, 41.7- 88.55 per cent of trees were infected with top work disease resulting in a tree mortality of 15.2-70.0 per cent. Infected trees grew normally in the first 1-2 years, but fibrous and main roots started to die gradually, shoot growth was limited and fruits were small. The existence of three latent viruses ASPV, ACLSV and ASGV in the infected trees was confirmed. These viruses infect many commercial apple cultivars with an infection rate of up to 80- 100 per cent and cause yield loss of up to 40 per cent in China (Wu *et al.*, 1998). Cembali *et al.* (2003) also reported losses of the order of 30 per cent in fruit yield of Golden Delicious cultivar of apple due to mix infection of ASGV, ASPV and ACLSV.

No natural vector is reported for ACLSV so far, but there are a few reports for natural transmission of the virus via root grafting (Gilmer *and Brase*, 1961). The virus is easily graft-transmissible to woody hosts and also sap-transmissible to certain herbaceous hosts. Susceptible herbaceous hosts of the virus include *Chenopodium quinoa*, *Chenopodium amaranticolor*, *Nicotiana occidentalis*, *Phaseolus vulgaris* etc. *Malus platycarpa*, *Malus sylvestris* cv. R12740-7A, *Malus sylvestris* cv. Spy 227, *Malus sylvestris* cv.Virginia Crab, *Cydonia oblonga* (Quince) C7/1, *Pyronia veitchii*, *Pyrus communis*, *Crataegus* spp., *Prunus armeniaca*, *Prunus spinosa*, *Prunus persica*, *Prunus domestica etc.* are some woody hosts of the virus (Bar Joseph *et al.*, 1979; Cropley, 1963; Cropley, 1968; Dunez *et al.*, 1972; Lister and Bar Joseph, 1981; Pena Iglesias and Gonsalves, 1975; Posnette and Ellenberger, 1975; Yanase, 1974; Yanase *et al.*, 1975).

Mc Crum (1965) concluded that virus could spread from infected trees of indicator (R 12740-7A) to adjacent healthy indicator plants through direct mechanical contacts. Shoots which were in direct contact of the infected plants expressed initial symptoms. Smith (1972) reported transmission of ACLSV from peach to peach through the contaminated budding knife. Unconfirmed reports of role of *Eudorylamoid* nematode in transmission requires further confirmation (Brunt *et al.*, 1996). Garcia Ibarra *et al.* (2010) reported that ACLSV infection affected the germination of the apricot seeds however, it was confirmed that ACLSV was not seed transmitted as reported earlier by Sutic *et al.* (1999). The virus is thus transmitted by grafting of infected planting material, mechanical inoculations and unclean horticultural practices.

(b) Apple Mosaic Virus

It is a member of subgroup III of the *Ilarvirus* genus in the *Bromoviridae* family (Alrefai *et al.*, 1994), which includes Prunus necrotic ringspot virus (PNRSV) and Rose mosaic virus (Rybicki, 1995). Apple mosaic is one of the oldest known and most widespread diseases caused by apple viruses. Apple mosaic virus is economically important and common pathogen in commercial cultivars (Mink, 1989, Stouffer, 1989).

Apple mosaic virus (ApMV) present worldwide and is an economically important and common pathogen in commercial apple cultivars (Mink, 1989). First reported in *Rosa* spp. and *Malus domestica* from the USA by White (1928) and Bradford and Joley (1933). It is common pathogen in commercial cultivars of apple and is economically important because considerable losses have been reported in apple due to its infection. Singh *et al.* (1979) from India studied the effect of apple mosaic virus on the growth, yield and quality of apple in 30 year old ApMV infected and healthy apple trees and found that shoot growth, fruit set, fruit weight, yield/tree and fruit ascorbic acid content were reduced by ApMV infection.

Most commercial cultivars are known to be affected with variable severity of symptoms. Golden Delicious and Jonathan are severely affected, whereas Winesap and Mclntosh are only mildly affected. Except in severe cases, infected trees can still produce a crop and yield reductions may vary from 0 to 50 per cent. In some cultivars, bud set is severely affected. Apart from that the presence of ApMV was found to reduce the growth of apple trees (Chamberlain *et al.*, 1971), increase the height of the climacteric and decrease the content of malic acid (Makarski and Agrios, 1973). It was also reported to decrease girth of trees (Thomsen, 1975), decreased bud take by 3-20 per cent along with stunted growth (Rebandel *et al.*, 1979), reduced the quality and quantity of pollen (Lemoine, 1982) in infected apple trees. Cambeli *et al.* (2003) reported 46 per cent reduction in yield in Golden Delicious and 9 per cent reduction in Mclntosh cultivar due to infection with ApMV.

In trials comparing the yields of healthy Golden Delicious apple trees with trees infected with several latent viruses, rubbery wood, flat limb and a mild strain of apple mosaic virus, either separately or together, the average yields of 4 year old healthy trees were almost double to those of diseased trees (Wood, 1974). Zawadzka (1983) found decreased growth and total yield (9 to 42 per cent) in 5 of 6 cultivars due to infection of ApMV. Rubbery wood was also reported to decrease the yield without affecting the fruit size.

Regardless of the host, infection by ApMV is characterized by chlorotic or, more often, bright yellow discolourations of the leaves in the form of blotching, mottling, vein banding or yellowing, ringspots, line and oak leaf patterns, seldom accompanied by evident deformation of the leaf blades (Fulton, 1972, 1980). Pale to bright cream coloured yellow areas develop on apple leaves as they expand during early spring. The mosaic areas may be irregular in outline or may occur in bands along major veins. These areas often become chrome yellow or white as the season progresses. Affected leaves may be interspersed with normal leaves on individual shoot. Leaves which exhibit strong mosaic symptoms in the early season may develop large necrotic areas during period of high temperature or increase sunlight. Such leaves may drop prematurely (Fridlund, 1989). In infected plants symptoms are usually outstanding in spring but tend to fade away as the season progresses, and are little or not evident on summer vegetation (Halk *et al.*, 1984; Imed *et al.*, 1997).

The virus is not transmissible through insect vector but is transmitted by pollen, vegetative propagation from infected trees or by mechanical sap inoculation to herbaceous hosts (Nemeth, 1986). Apple mosaic virus was transmitted by both natural and artificial root grafting between apple seedlings in glasshouse experiments and also by natural root grafting in the nursery. Simple intertwining of the roots of adjacent plants was not sufficient to transmit the disease (Dhingra, 1972). ApMV in nature is transmitted most frequently through infected planting material (Sutic *et al.*, 1999). The virus is not transmitted by *Cuscuta campestris, C. gronovii, C. subinclusa* (Fulton, 1952) or *C. reflexa* (Nagaich and Vashisth, 1965).

The virus has a wide host range. Experimentally or naturally, it has infected over 65 species in 19 families (Fulton 1952, 1965). It preferentially infects woody hosts such as blackberry, raspberry (*Rubus* sp.), apricot, cherry, almond, (all *Prunus* sp.), roses (*Rosa*), mountain ash (*Sorbus aucuparia*), horse chestnut (*Aesculus hippocastanum*), red horse chestnut (*A.* x *carnea*) and hop (*Humulus lupulus*) (Brunt *et al.*, 1996). ApMV has also been reported from hazel nut (*Corylus avellana*) (Aramburu and Rovira, 2000) and strawberry (Tzanetakis and Martin, 2005). Under experimental conditions susceptibility to infection is reported in host species of Apocynaceae, Corylaceae, Leguminosae, Papilionoideae, Rosaceae and Solanaceae families. Species susceptible to experimental virus infection include *Catharanthus roseus, Corylus avellana, Malus sylvestris, Nicotiana tabacum, Phaseolus vulgaris, Cucumis sativus* (cucumber), *Torenia fournier, Vinca rosea* (periwinkle), *Vigna sinensis* (cowpea). *Malus sylvestris* (apple) cv. Lord Lambourne and Jonathan develop a prominent mosaic and are recommended as woody indicators.

(c) Apple Stem Grooving Virus

Apple stem grooving virus (ASGV) is wide spread in its occurrence, and has been described as a latent virus of apple because it produces no symptoms on most commonly used commercial rootstocks and varieties (Nemeth, 1986). It was identified as a causal agent of apple top working disease when infected apple trees were grafted onto rootstocks *of Malus sieboldii* and *Malus sieboldii* var. *arborescens* (Yanase, 1983).

Apple stem grooving virus is a flexuous filamentous virus estimated to be 600-700 nm x 12 nm in size, encapsidating a positive-sense single stranded RNA, 6496 nucleotides in length excluding the poly (A) tail at the 3′end of the genome (Lister, 1970; Yoshikawa *et al.*, 1992; Yoshikawa and Takahashi, 1992 and Hirata *et al.*, 2003). ASGV is a member of the plant virus family *betaflexiviridae*, and the type member of the genus *capillovirus.*

Apple stem grooving virus (ASGV) is one of the economically important latent viruses that are distributed in apple production areas worldwide (Liu *et al.*, 2013). Initially ASGV was reported to infect *Malus sylvestris* cv. Virginia Crab (Lister *et al.*, 1965; De Sequeira, 1965 and 1967; Waterworth and Gilmer, 1969). Several surveys conducted by different scientists across the globe have shown that these viruses are widely distributed in most of the pome fruit growing areas of Europe (Kundu, 2003; Massart *et al.*, 2009; Mathioudakis *et al.*, 2006; Paunovic and Jevremovic, 2008), America, Asia, (Birisik *et al.*, 2008; Dhir *et al.*, 2009; Salem *et al.*, 2005), Australia and New Zealand (Rodoni and Constable, 2008). ASGV has been also reported from Bulgaria, China, Egypt, India, Italy, Japan, Korea, Netherland, Portugal, Slovenia, South Africa and probably occurs wherever apple trees are cultivated (Liebenberg, 2013).

ASGV infection could cause 12 to 30 per cent reduction in yield of 'Golden Delicious' (Cambeli *et al.*, 2003). Maxime *et al.*, 2004 studied the influence of *Apple stem grooving virus* (ASGV) on tree growth of various apple cultivars in the nursery in Romania. Results revealed that the average tree height of the 14 cultivars infected was 23.4 per cent lower, and the average diameter 13.7 per cent smaller than in case of healthy trees.

Apple stem grooving virus (ASGV) is generally known to cause latent infection in most of the commercial cultivars of apple (Nemeth, 1986). However, *Apple stem grooving virus* (ASGV) produces stem grooving, swelling of graft unions, graft union incompatibility, top working disease and graft union necrosis symptom when infected cultivar is grafted upon *Apple stem grooving virus* (ASGV) sensitive cultivars *M. sieboldii or M. pumila* cv. Virginia Crab (Lister, 1970; Yanase, 1974; 1983). Symptoms such as chlorotic leaf spot and brown line may develop immediately above graft union (Welsh and Van der Meer, 1989). In apple and pear ASGV has been associated with tree decline and graft union necrosis in sensitive combinations of scion and rootstock (Kundu, 2002b). Susceptible *Malus* species include severe xylem pitting and grooving with pegs protruding on inner bark face, phloem necrosis, reduced vigour of the canopy and an overall decline of the plant. ASGV produces chlorotic leaf spots, stem grooves and pits, union necrosis and swelling of the stem above the graft union symptoms on 'Virginia Crab'. Scions and interstocks did not usually show wood symptoms. Plants grafted with infected material displayed poor budwood welding, developed poorly and either died at the nursery or declined in the orchard (Nickel *et al.*, 2001).

(d) Apple Stem Pitting Virus

The virus particles of ASPV are flexuous filamentous of 1250 nm long with ssRNA genome of 9306 nt (Jelkmann, 1994). The virus particle contained a single-stranded positive sense RNA genome of 9.3 kb consisting of a single polypeptide coat. Its genome consists of a poly (A) tail. Virions do not contain lipids or carbohydrates (Martelli and Jelkmann, 1998). ASPV consists of a single species of RNA of 3.1×10^6 and a major coat protein of 8,000 (Koganezawa and Yanase, 1990). The genomes of two ASPV isolates (PA66 and IF38) have been completely sequenced and found to consist of 9306 and 9237 nt, respectively, excluding the 3′ poly (A) tail (Jelkmann, 1994; Yoshikawa *et al.*, 2001).

Apple stem pitting virus (ASPV) is a widespread pathogen of pome and stone fruit trees in all over the world (Nemeth, 1986). First reported in *Malus sylvestris* from midwest of U.S.A. by Smith (1954). Since then the virus has been detected all over the world where apple trees are grown which, include Australia, Western and Eastern Europe, Asia and South Africa.

The incidence of the virus infection is usually symptoms less (Nemeth, 1986) and occurs together with other pome fruit viruses in mixed infection (Kundu, 2003; Leone *et al.*, 1998). ASPV is known to cause a disease associated with the tree decline, stem pitting, graft union necrosis and vein yellows in apples and pears (Welsh and Uyemoto, 1980; Desvignes *et al.*, 1990; Kundu, 2002). Like ASGV, ASPV also causes virus symptoms on 'Virginia Crab' such as stem pitting of xylem that generally stops at the graft union. ASPV also causes fluted (grooved) fruit in 'Virginia Crab' and epinasty. Depressions and grooving develop on the stem and the limbs below the union, and longitudinal sutures appear on the bark. In its mild form, symptoms develop only after the second year following inoculation. Symptoms consisting of pitting of the woody cylinder, or epinasty and decline, develop in some ornamental *Malus* species and in Virigina Crab and Spy 222 indicators. It causes apple Spy 227 epinasty and decline disease in Spy 222 indicators, while apple dwarf disease in *M. platycarpa*. Moreover, a close relationship was reported between ASPV and the causal agent of necrotic spot and vein yellows diseases of pear (Lemoine, 1979). Apple spy 227 epinasty and decline virus (Gilmer, 1962; Mink *et al.*, 1971), hawthorn ring pattern mosaic virus, pear necrotic spot virus (Yanase *et al.*, 1989; Koganezawa and Yanase, 1990), pear stony pit virus (Koganezawa and Yanase, 1990; Nemeth, 1986), pear vein yellows virus (Yanase *et al.*, 1989) are few synonyms for ASPV.

Although ASPV is considered to be latent on most commercial apple varieties several cases have been reported of symptoms on apple associated with ASPV infections like green crinkle and star crack symptoms (Desvignes *et al.*, 1999), stem pitting symptoms of Virginia crab, epinasty and decline of Spy 277 (Stouffer, 1989) and top working disease when an infected rootstock is grafted on the susceptible rootstocks *M sieboldii* and *M. prunifolia* (Jelkmann, 1997). Symptoms on pear indicator cultivars include vein yellows, decline and stem pitting (Leone *et al.*, 1998).

ASPV is transmitted by mechanical inoculation, by grafting but not transmitted by contact between plants, seed, pollen and vectors. *Apple stem pitting virus* is transmitted by mechanical inoculation and grafting, no insect vector has been identified to transmit the virus (Yanase *et al.*, 1989). The virus is transmitted mechanically to herbaceous hosts such as *Nicotiana occidentalis* 37B, causing systemic vein yellowing and leaf necrosis and to *N. occidentalis*, causing necrotic local lesions, *M. sylvestris* cv. Virginia Crab causing stem pitting, *M. sieboldii* MO65 causing dieback, inner bark necrosis and necrotic leaf spot and *Pyrus communis* cv. Nouveau Poiteau causing necrotic leaf spot. (Stoufer, 1989; Koganezawa and Yanase, 1990). The virus was transmitted by inoculation of sap from stem pitting-diseased apples to *Nicotiana occidentalis* and partially characterized by Martelli and Jelkmann (1998).

Natural host range of ASPV is restricted to pome fruits including apple, pear, quince, hawthorn and rowan (Martelli and Jelkmann, 1998; Sutic *et al*, 1999; Kundu and Yoshikawa, 2008).The following plant species from 7 plant families are susceptible to ASPV under experimental conditions: *Celosia cristata, Chenopodium murale, Chenopodium quinoa, Crataegus, Cucumis sativus, Gomphrena globosa, Malus platycarpa, M. sieboldii, M. sieboldii* var. *arborescens, M. sylvestris, Nicotiana occidentalis* (cv. 37B), *Nicotiana occidentalis* ssp. *obliqua, Nicotiana glutinosa, Nicotiana rustica, Physalis floridana, Pyronia veitchii, Pyrus communis, Sesamum indicum, Sorbus mitchelli, Tetragonia tetragonioides Celosia argentea, Ocimum basilicum* and *Lactuca sativa* (Yanase, 1974; Stouffer, 1989; Koganezawa and Yanase, 1990).

The virus is only transmitted by grafting to woody host plants, causing apple top working disease when grafted on the susceptible rootstocks (*Malus prunifolia* var. ringo, *M. siebolddii* and *M. siebolddii* var. *arborescens*) and vein yellow and necrotic spots in pear (Jelkmann, 1997; Desvignes, 1999). In indicator hosts for the disease, the virus can cause symptoms including stem pitting in Virginia crab, epinasty decline and stem pitting in Spy227, dwarfing in *Malus platycarpa* and stem pitting at graft union in R-127407A (Leone *et al.*, 1998). Natural host range includes *Malus sylvestris, Malus sieboldii, Malus sieboldii* var. *arborescens, Pyrus communis, Crataegus* spp., and *Sorbus mitchelli.*

3. DETECTION OF APPLE VIRUSES

In case of diseases of plants exact detection of causal organism is very important to device management strategy. Viral diseases are difficult to diagnose since they cant not be cultured on artificial medium. Generally three thechniques are used for diagnosis of viruses *viz.*, biological indexing, serological and molecular.

(a) Biological Indexing

It is simple thechnique and on one host one can detect many viruses also no need of exact knowledge of pathogen requires. The major limitations of biological indexing are its application length (3 years), its cost in terms of glasshouse space and labour intensity and sometimes symptoms are difficult to interpret. Neverthless, this method is still used for detecting pathogens because it is sentistive and viruses of unkown etiology can be detected for which no precise information is available. The virus could be detected by mechanical sap transmission to the herbaceous indicators like *Chenopodium quinoa, Nicotiana glutinosa* and *Phaseolus vulgaris* cvs Pinto and Alsa (James, 2008). Widly used woody indicators for detection of viruses in apple are Russian var. R12740-7A *Malus platycarpa,* Spy 227, *Malus pumila* cultivars Virginia Crab, Radiant and R 12740 7A.

Mink and Shay (1959) reported that Russian var. R12740-7A and its progeny proved to be valuable indicators of apple chlorotic leaf spot (ACLSV), apple mosaic and stem pitting viruse and while indexing virus infection in an apple orchard. Similarly, *Malus platycarpa* was reported to induce line pattern symptoms on leaves following bud inoculation from symptomless commercial cultivars of apple and they named it as Platycarpa line pattern virus (Luckwill and Campbell, 1959). Cation and Carlson (1960) indexed 215 apple trees for the detection of virus infection on a biological indicator Russian variety R12740-7A and reported that 56 per cent of the trees were carrying infection of apple chlorotic leaf spot. Gilmer and Brase (1961) detected the association of ACLSV of apple with decline of Spy 227 rootstock. Gilmer (1962) has reported that ACLSV caused severe phloem necrosis in stem tissues of Spy 227 scion after one year of inoculation. Terminal dieback and mild stem pitting symptoms were observed on some of the inoculated plants. Chairez and Lister (1973) reported the symptoms of ACLSV on woody indicator R12740-7A and also on herbaceous indicator *C. quinoa* along with wide spread occurrence of ACLSV in Golden Delicious, Granny Smith and Starking cultivar of apple. Paunovic (1988) used woody indicators R12740-7A and *Malus platycarpa* to detect the ACLSV presence by observing symptoms evoked by the virus on these indicators. Transmission to woody indicator plants was carried by using double budding method. R12740-7A indicator reacted to most of the isolates by producing characteristic symptoms of chlorotic blotches, asymmetric leaf distortion, puckering and stunting of shoot growth. However, *Malus platycarpa* was found to exhibit symptom to few isolates. Lister *et al.* (1964) reported for the first time the back transmission of ACLSV from systemically infected *C. quinoa* to woody indicator Russian var. R12740-7A by approach grafting. Myrta *et al.* (2004) in Albania conducted studies on the detection of ApMV and other viruses in apple by using *Malus pumila* ('Virginia Crab' and 'Radiant'), *Pyronia veitchii*, *Malus pumila* 'V. Crab', 'R 12740 7A', 'Spy 227', 'Lord Lambourne', 'Golden Delicious', *Pyronia veitchii*, *Pyrus communis* woody indicators. In Turkey biological indexing was conducted with indicators, *viz.*, *Malus pumila* cultivars Virginia Crab, Radiant and R 12740 7A against apple viruses (Birisik *et al.*, 2008).

Biological detection of ApMV has been carried out by different workers by employing indicator plant species. Apple mosaic virus was transmitted by both natural and artificial root grafting between apple seedlings in glasshouse experiments and also by natural root grafting in the nursery. Simple intertwining of the roots of adjacent plants was not sufficient to transmit the disease (Dhingra, 1972). Dhingra and Niazi (1973) detected the infection of ApMV in 11 cultivars of apple by using budding and grafting method.

A number of diagnostic woody indicator host species have been identified as indicator species for ASGV infection including; Cotoneaster, *Malus pumila* cv. Virginia Crab, *M. sieboldii* MO-63, *M. sylvestris*, *Citrus excelsa*, *Poncirus trifoliata* (Nemeth, 1986; Welsh and Van der Meer, 1989). Virus particles occur singly or in aggregates in the cytoplasm of mesophyll and phloem parenchyma cells (Ohki *et al.*, 1989) and are transmitted through grafting and mechanical inoculation. Gilmer and Uyemoto (1972) reported the sap transmission of chlorotic leaf spot and stem grooving viruses from infectious Zucchini squash sap to 8 day old McIntosh apple seedlings. An *Apple chlorotic leaf spot virus* isolate was similarly transmitted from infectious *Chenopodium* sap. Plese *et al.* (1975) studied pathological anatomy of trees affected with apple stem grooving virus. Under the influence of apple stem grooving virus elongated grooves appeared on the stem surface of *Malus sylvestris* cv. Virginia Crab and of *Cotoneaster bullata*. Sawamura and Osada (1980) observed that ASGV induced mottling in *Nicotiana glutinosa* and specific symptoms consisting of necrotic lesions in inoculated leaves and systemic symptoms with tip distortion and epinasty on *Chenopodium quinoa*. Minoiu *et al.* (1983) detected *Apple stem grooving virus* in 4 per cent of 180 apple trees by using Virginia Crab K-6 as an indicator. Howell and Mink (1995) reported a new woody plant indicator *Malus micromalus* clone GMAL 273 for rapid detection of apple stem grooving virus after screening of 201 *Malus* clones. It displayed particularly distinctive foliage symptoms within 14-21 days of inoculation. This is in contrast to the 6-8 months frequently required for reliable ASGV reactions to appear on Virginia Crab indicators grown under the same conditions. Posnette *et al.* (1976) reported biological detection of ASPV by using Virginia Crab as woody indicator and concluded that intermediare grafting gave better result than budding.Minoiu (1977) studied presence of apple stem pitting virus in different cultivars through biological indexing and found that Virginia Crab K-6 was the most suitable indicator plant.Yanase and Yamaguchi (1982) in their finding reported *Malus sieboldii* MO-65 as a new indicator plant for apple stem pitting virus. Yang *et al.* (1997) used Russia Clone 12740-7A, SPY227 and Virginia Crab for the detection of the latent viruses.Leone *at al.* (1998) confirmed the identity of apple stem pitting virus with pear vein yellows virus from symptoms on apple and pear indicators after back transmission from *Nicotiana occidentalis*. Isolates of apple stem pitting virus (ASPV) from diseased apple trees were maintained in *N. occidentalis* then back transmitted mechanically to apple seedlings and indexed by double budding on apple and pear indicators.

Silva *et al.* (2008) evaluated the suitability of biological multiple indexing and RT-PCR detection of latent viruses in apple plants. The method consisted of grafting five indicator species and two buds of the plant to be indexed on to one rootstock. Results showed that the sensitivity and efficacy of the MBI is adequate for reliable detection of apple stem grooving virus, (ASGV), *Apple stem pitting virus* (ASPV) and apple chlorotic leaf spot virus (ACLSV) of mother plant stock in certification programs. Sample analysis by RT-PCR confirmed the results of the diagnosis of ASPV and ASGV by multiple biological indexing.

(b) Serological Detection

Serological techniques are based upon the specific recognition of viral antigens by antibodies, and are frequently preferred because of their speed, specificity and simplicity. The methods have been routinely incorporated into indexing programs where virus free plant materials are required. Clark and Adams (1977) detected ACLSV in apple leaves by using double antibody sandwich form of ELISA (Voller *et al.*, 1976 a,b). Flegg and Clark (1979) reported the detection of ACLSV by modified procedure of alkaline phosphatase (ALP) based direct double antibody sandwich (DAS) ELISA in its different natural hosts including apple. Fuchs *et al.* (1979) reported the use of ELISA for the detection of apple chlorotic leaf spot virus (ACLSV), apple stem grooving virus (ASGV) and tomato bushy stunt virus (ToBSV). Detienne *et al.* (1980) in France used DAS ELISA to detect 9 different strains of ACLSV isolated from 13 fruit tree species with special extraction buffers.

Fuchs (1983) compared the different serological methods for detecting ACLSV in apple. He recommended the latex test for detection of ACLSV in petals of apple trees. However, ELISA had been reported to be reliable in the detection of this virus in forced leaf buds and petals. Rankovic and Vuksanovic (1983) investigated the detection of ACLSV by ELISA technique in different plant parts of 40 apple cultivars grafted on different root stocks and seedlings. ACLSV was detected in buds, leaves, petals and fruits of all cultivar grown in Yugoslavia as standard grafting material. However, it was absent only in some indigenous varieties and one variety of foreign origin grafted on seedling. Fuchs *et al.* (1985) while comparing 51 isolates of ACLSV in precipitation drop test and latex test using three antisera reported that most of the isolates did not differ significantly in their antigenic specificity. Results of investigation indicated the feasibility of using monovalent antisera for serological detection of ACLSV in apple, pear, plum and cherry in Germany. Legrand (1986) utilized the indirect ELISA for detection of ACLSV in extract of both herbaceous hosts and fruit trees. He reported that indirect ELISA was more appropriate than DAS ELISA in the qualitative detection of ACLSV and the same technique could be used for the detection of fruit tree viruses in infected herbaceous plants.

Direct DAS ELISA and indirect ELISA were also used to detect different strains of ACLSV and apple stem grooving virus (ASGV) from different *Malus* species in Japan. Of the six ACLSV strains tests, five were detected by direct DAS ELISA from *Malus* species. While less than half of the test samples from apple and pear infected with ASGV gave positive results. Direct ELISA was considered by the investigators unsuitable for routine detection of ASGV as specificity for the homologous strains of ASGV was too great. Modified indirect ELISA by using F $(ab')_2$ fragment was reported as sensitive as direct ELISA for the detection of homologous virus strains and was also the most sensitive for detection of heterologous strains (Yanase *et al.*, 1986). Poul and Dunez (1988) reported the production of monoclonal antibodies against ACLSV isolate from *Prunus domestica* and further use of monoclonal antibodies in the detection of purified virus. While comparing the detection of ACLSV by the use of monoclonal antibodies and polyclonal antibodies under DAS ELISA, it was found that monoclonal antibodies were more than 10 fold sensitive over polyclonal antibodies. Fuchs *et al.* (1988) studied the serological detection of viruses of pome or stone fruits including ACLSV. In their investigations based on long term virus detection experiments, drew up a calendar of details of months in which ELISA test could be used to get 90 per cent detection reliability in specified plant tissue.

Similarly during the last decades numerous reports on the serological detection of ACLSV by using DAS-ELISA or other variants of ELISA are figured here in the literature (Fidan,1994; Choueiri *et al.*, 1993; Varveri and Bem, 1995; Kryczynski *et al.*, 1995; Tozzi *et al.*, 1995; Turk, 1996; Polak *et al.*, 1997; Varveri *et al.*, 1997; Lemoine and Michelesi, 1997; Chuangui *et al.*, 1997; Chunmin *et al.*, 1997; Lessa *et al.*, 1998; Zhang *et al.*, 1998; Myrta *et al.*, 2001; Jarrar *et al.*, 2001; Cieslinska and Kaminska, 2002; Freitas *et al.*, 2004; Sertkaya *et al.*, 2004; Ulubas and Ertunc, 2005). Fuchs (1979) detected ASGV reliably by ELISA from flowers and leaves of apple shoots in addition to other viruses of temperate fruits. In detection from different plant parts fruits were less suitable as a test material for reliable detection of the virus. Rankovic and Vuksanovic

(1983) in Yugoslavia investigated ELISA technique in detecting apple stem grooving virus (ASGV) from different plant parts by using bark, leaf buds, leaves, petals and fruits. Newly developed leaf buds and petal tissues were observed to be suitable in successful detection of the virus. The studies of Fuchs *et al.* (1985) on serological detection of ASGV and ACLSV isolates by using homologous and heterologous antisera titres for analysis indicated the feasibility of using monovalent antisera for serological detection. Machita *et al.* (1986) reported the detection of apple stem grooving virus by enzyme-linked immunosorbent assay (ELISA). The investigator detected ASGV reliably from leaves, leaf buds and petals by ELISA method from apple tree. Yanase *et al.* (1986) detected apple stem grooving virus in apple by direct double antibody sandwich method and modified indirect ELISA (F(ab')2 procedure. Fidan (1994) used ELISA to detect apple mosaic virus (ApMV), apple chlorotic leaf spot virus (ACLSV) and apple stem grooving virus (ASGV) in apple leaves in Turkey. Out of 220 samples tested, 63 (28.6 per cent) were infected with ACLSV and 52 (23.6 per cent) with ASGV.

Wu *et al.* (1998) compared three ELISA methods for the detection of apple chlorotic leaf spot virus and apple stem grooving virus. Protein A sandwich enzyme-linked immunosorbent assay (PAS-ELISA) has previously been used as a routine method for the detection of ACLSV and ASGV and for screening of virus-free materials generated from virus elimination programmes. The standard and modified version of the double antibody sandwich ELISA (DAS-ELISA) were compared with PAS-ELISA for their effectiveness in detecting ACLSV and ASGV from tissue culture of commercially important apple cultivars. The standard DAS-ELISA successfully detected ASGV but failed to detect ACLSV. Moreover, the modified DAS-ELISA and PAS-ELISA were equally good in detecting both viruses. Polak and Zieglerova (2001) from Czech Republic reported detection of ASGV by ELISA. The virus was found to be commonly distributed in all the apple growing areas of the country. Some cultivars were infected up to 100 per cent, while several others were found to be virus-free. A high incidence of infected trees appeared both in older (20 years) and younger (7 to 8 years) orchards. Corvo and Barros (2001) studied the different techniques of DAS-ELISA for reliable detection of *Apple stem grooving virus* (ASGV) in the growing season. In these attempts they have conducted a series of experiments using different apple cv. Bravo de Esmolfe tissues including winter budsticks, flowers, leaves and growing shoots. Procedures adapted from the modified version of DAS-ELISA were compared with standard DAS-ELISA procedures. The former were clearly advantageous due to their simplicity, increased rapidity and adequacy to detect ACLSV and ASGV throughout extended sampling periods, including dormancy. Postman and Cameron (1987) from National Clonal Germplasm Repository, Corvallis, USA, reported seven ApMV infected *Corylus avellana* cultivars which were detected by ELISA. ApMV infected trees may be symptomless or may exhibit chlorotic ring spots and line patterns on the leaves of the plant. Fuchs *et al.* (1988) gave details of the months in which ELISA tests on specified plant tissues give at least 90 per cent detection reliability in the host-virus systems of apple chlorotic leaf spot virus, apple mosaic virus and apple stem grooving virus.

The latex test and ELISA permitted 100 per cent detection of relatively high concentration of these viruses in forced or dormant buds of leaves and petals. The reliability of serological detection decreased with the use of older leaves, particularly with ApMV (Gruntzig and Fuchs, 1988). The distribution of apple mosaic virus (ApMV) was studied in the main growing areas of the Czech Republic. ELISA was used to detect ApMV in symptomatic apple plants in almost all the gardens tested (Polak, 1994). ELISA test were used to detect apple mosaic ilarvirus (ApMV), apple chlorotic leaf spot trichovirus (ACLSV) and apple stem grooving capillovirus (ASGV) in apple leaf extracts. Out of 220 samples collected, 34 (15.45 per cent) were infected with ApMV, 63 (28.6 per cent) were infected with ACLSV and 52 (23.6 per cent) with ASGV. Apple trees which gave positive reactions to ELISA were also tested on herbaceous test plants (Fidan, 1994). Incidence of apple mosaic ilarvirus (ApMV) was surveyed in 140 apple orchards in Tarragona and Girona provinces. Mosaic symptoms on leaves were observed during spring. ELISA analysis revealed that most trees of cv. Golden Delicious were infected with the mosaic disease (Rovira and Aramburu, 1998). Myrta *et al.* (2004) in Albania conducted studies on the detection of ApMV and other viruses in apple by using ELISA. Author found that all the 140 samples of apple were carrying infection of ApMV along with ACLSV and ASGV.

Apple mosaic virus (ApMV) infection on apple (*Malus pumila*) trees were observed in orchards in the central Anatolia Region of Turkey with prominent symptoms of pale to bright cream spots, mosaic, and yellowing on leaves. Leaf samples from apple cultivars Anna, Gala, Gold Star, Golden Delicious, Granny Smith, Jonathan, Starking and Rajka were collected and tested for the presence of ApMV with specific antisera using DAS-ELISA. ApMV was found in 68.3 per cent of 461 samples tested (Akbas and Ilhan, 2005). A survey was conducted to determine the distribution of ApMV in Tokat province in Turkey by using DAS-ELISA. One hundred sixteen plant samples were collected from 24 orchards in Tokat and were tested by DAS- ELISA which revealed the presence of ApMV in 6.9 per cent of the leaf samples. The virus was detected in 15.09 per cent of the apple plants at two orchards (Ylmaz *et al.*, 2005). Birisik *et al.* (2008) studied the incidence of apple (*Malus domestica*) viruses in southern Anatolia, Turkey, between 2002 and 2005. A total of 108 orchards and 10 varietal collections were visited in the districts of Adana, Antalya, K. Maras and Osmaniye. Some 413 samples of leaves and/or dormant cuttings were obtained from apple trees. Sanitary testing was conducted by ELISA, biological indexing, and RT-PCR. All samples were tested by ELISA for the presence of apple chlorotic leaf spot virus (ACLSV), apple stem grooving virus (ASGV) and apple mosaic virus (ApMV). The overall virus infection rate as revealed by ELISA was 18.8 per cent. Svoboda and Polak (2010) used leaves, flower petals, dormant buds and phloem tissues for detection of ApMV and reported the highest relative virus concentration in young leaves in April before flowering.

(c) Molecular Detection

The polymerase chain reaction (PCR) is an extremely sensitive and specific technique (Saiki *et al.*, 1988) for detection and identification of the plant pathogens. It is widely used for the diagnosis of plant diseases, allowing detection of very small amount of the disease agent in the infected plant and also for the cloning of genomic fragments of the pathogens (Henson and French, 1993). RNA containing viruses require RT step to generate cDNA prior to PCR amplification (RT-PCR). Specificity of PCR is based on the use of oligonucleotide primers that are complimentary to the regions flanking the target nucleic acid sequence to be amplified and the PCR usually requires purification of the target nucleic acid. Another modification of RT-PCR used sometimes for detection of plant viruses is immunocapture reverse transcriptase polymerase chain reaction (IC-RT-PCR). Candresse *et al.* (1995) used polyvalent PCR based assay for the detection of apple chlorotic leaf spot virus and to achieve maximum sensitivity an immune capture step was carried out directly in PCR tube before the reverse transcription and PCR reaction. This assay proved very useful for the detection of this virus in a variety of plant material and from variety of woody hosts. Similarly Nemchinov *et al.* (1995) used specific primers for ACLSV for detection of the viral coat protein gene in the total nucleic acids extracted from ACLSV infected apple and peach. These primers were utilized for amplification in RT-PCR and IC-RT-PCR. Krizbai *et al.* (2001) characterized the Hungarian isolates of apple chlorotic leaf spot virus by using IC-RT-PCR. Menzel *et al.* (2002) used multiplex RT-PCR assay for detection of four apple viruses with coamplification of plant mRNA as internal control by using four virus specific primer pairs and one primer pair specific for mitochondrial nad5 gene. This was the first report on the use of a specific internal RNA control from total nucleic acids. Menzel *et al.* (2003) also developed a sensitive and reliable multiplex RT-PCR-ELISA technique for the detection of apple chlorotic leaf spot virus, apple stem pitting virus, apple mosaic virus and apple stem grooving virus. Rana *et al.* (2007) characterized the Indian isolate of ACLSV based on ELISA and RT-PCR studies.

During the recent past, RT-PCR, IC-RT-PCR, multiplex RT-PCR or pentaplex RT-PCR techniques was reported to be successful in the detection of ACLSV alone or in combination with other apple viruses in case of mixed infection (Pasquini *et al.*, 1998; Yoshikawa *et al.*, 2000; Jan *et al.*, 2003; Lebas *et al.*, 2004; Al-Rwahnih *et al.*, 2004; Ulubas and Ertunc, 2005; Foissac *et al.*, 2005; Serce and Rosner, 2006; Caglayan *et al.*, 2006; Polak *et al.*, 2006; Polak *et al.*, 2008; Fan *et al.*, 2009; Watpade *et al.*, 2012; Watpade *et al.*, 2013).

The complete nucleotide sequence of ApMV has been characterized from several parts of the world (Alrefai *et al.*, 1994; Shiel *et al.*, 1995; Shiel and Berger, 2000; Petrzik and Lenz, 2002). Roussel *et al.* (2004) developed RT-PCR protocols suitable for a routine diagnosis of latent and ILAR viruses in fruit tree certification. This technique was simplified by using crude plant extracts instead of total RNA preparations, and by the analyses of pooled samples. Various other workers (Seigner, 2000; Petrzik and Lenz, 2002; Hou *et al.*, 2004; Petrzik, 2005; Caglayan *et al.*, 2006; Polak *et al.*, 2006; Yardmc *et al.*, 2008; Thokchom *et al.*, 2009) also reported the use of RT-PCR for successful detection of different temperate fruit viruses. Several RT-PCR techniques have been described for ASGV detection either in single target assays (Kirby *et al.*, 2001; Clover *et al.*, 2003), or in multiplex assays (Foissac *et al.*, 2005; Roy *et al.*, 2005). Fruits, leaves, flowers, bark and roots tissue have been used for RT-PCR detection of ASGV. RT-PCR assays were found more reliable than bioassays or ELISA for ASGV detection (James, 1999; Kirby *et al.*, 2001; Menzel *et al.*, 2003; Kundu *et al.*, 2003 and Hassan *et al.*, 2006). Daniel *et al.* (1998) developed a colorimetric RT-PCR (or RT-PCR ELISA) test that combined the sensitivity of PCR and the convenience of ELISA. Marinho *et al.* (1998) reported the detection of apple stem grooving virus in dormant apple trees with crude extracts as templates for one-step RT-PCR. Kirby *et al.* (2001) compared the bioassays and laboratory assays for Apple stem grooving virus. The standard double-budding assay (a field assay using Virginia Crab indicator buds) and the glasshouse test with the indicator *Malus micromalus* were compared with ELISA and immunocapture PCR for the detection of apple stem grooving virus (ASGV) in 102 apple trees and 3 oriental pear trees. Twenty-two trees were positive for ASGV by both ELISA and IC-PCR but three of these trees were negative by Virginia Crab, three were negative by *M. micromalus* and one was negative by both these bioassays. The infected trees were re-tested by IC-PCR and ELISA in a second year; the IC-PCR results were confirmed but two of the 22 infected trees were negative by ELISA. On this evidence, IC-PCR was found to be more reliable assay for ASGV than the slow and expensive bioassays. Kundu (2002a) successfully used the reverse transcription polymerase chain reaction (RT-PCR) assay for the detection of *Apple stem pitting virus* (ASPV) and *Apple stem grooving virus* (ASGV) in four apple cultivars (Idared, Spartan, Vista Bella, and Stark Earliest). Kundu *et al.* (2003) also detected apple stem grooving virus in different tissues of apple trees throughout the year in different plant parts by RT-PCR and ELISA. Detectable amounts of ASGV were generally found in all tissues (bark, dormant buds, petals and leaves) tested by RT-PCR from January to mid-June. Leaves during flowering (in May) were the most suitable tissues for the virus detection by both methods (RT-PCR and ELISA). The leaves collected in summer (June, July and August) or other tissues such as bark, dormant buds and petals were not reliable for ASGV detection by ELISA.

A pentaplex reverse-transcription polymerase chain reaction (Pentaplex RT-PCR) in a single tube for the simultaneous detection of the pome fruit viruses: *Apple stem pitting virus* (ASPV), apple stem grooving virus (ASGV), apple chlorotic leaf spot virus (ACLSV) and apple mosaic virus (ApMV) was developed by Polak *et al.* (2006). This is the first report of the simultaneous detection of all four viruses and host mRNA as an internal specific control. Pentaplex RT-PCR was applied successfully throughout the year, using different plant organs (leaves or dormant buds). The sensitivity of detection by monoplex and pentaplex RT-PCR assays was comparable. Negi *et al.* (2010) detected ASGV in different temperate fruit crops of Himachal Pradesh in India through RT-PCR as well as ELISA. Similarly, Rana *et al.* (2011) also detected the ASGV and other viruses of apple in her studies on the occurrence of different viruses under natural conditions through the use of ELISA and at molecular level. Rana *et al.* (2011) reported the presence of apple mosaic virus (ApMV), apple stem grooving virus (ASGV), *Apple stem pitting virus* (ASPV), Apple chlorotic leaf spot virus (ACLSV), the major apple viruses and Prunus necrotic ringspot virus (PNRSV), a minor apple virus, at the molecular level in India. Similarly, several other RT-PCR techniques have been described by different workers either in single target assays or in multiple assays (Kirby *et al.*, 2001; Clover *et al.*, 2003; Nickel *et al.*, 2004; Foissac *et al.*, 2005; Roy *et al.*, 2005; Zhang *et al.*, 2005; Guo *et al.*, 2005; Zheng *et al.*, 2006).

A sensitive and reliable method based on the reverse transcription polymerase chain reaction (RT-PCR) has been developed for the detection of ASPV in infected plants by several workers (Hadidi *et al.*, 1998; Kundu, 2002; Kundu 2003; Yuxuan *et al.*, 2005; Komorowska *et al.*, 2010; Li *et al.*, 2010 and Dhir *et al.*, 2010). Hadidi *et al.* (1998) reported PCR-detection of *Apple stem pitting virus* (ASPV) from pome fruit hosts and sequence variability among viral isolates. A relatively simple and rapid PCR assay was developed for the detection of *Apple stem pitting virus* (ASPV) from apples and pears. The assay allowed accurate diagnosis of ASPV from leaf, bark and bud tissues of diseased apple and pear trees. Two multiplex reverse transcription-polymerase chain reaction (RT-PCR) assays with specific co amplification of plant mRNA as an internal control from total nucleic acids for the parallel detection of apple chlorotic leaf spot virus, *Apple stem pitting virus* (ASPV), apple mosaic virus and apple stem grooving virus were developed by Menzel *et al.* (2002), Menzel *et al.* (2003). Kundu (2002) described the application of RT-PCR assay for the detection of ASPV and ASGV in apple orchard (Czech Republic). ASGV and ASPV occurred in as many trees tested. Mixed infection by ASGV and ASPV was found in all tested cultivars.

Polak *et al.* (2008) from Research Institute of Crop Production, Czech Republic detected four pome fruit viruses in germplasm collection in the Czech Republic by using ELISA as well as pentaplex RT-PCR. A total of sixty-eight accessions covering native and foreign cultivars were tested for the presence of the main pome fruit viruses: ASPV, ASGV, ACLSV and ApMV. Regardless of the cultivars or the origin, different combinations of mixed infections of viruses were found in infected samples. All positive samples detected by ELISA were confirmed by pentaplex RT-PCR; however, RT-PCR revealed more infected trees than by ELISA. ACLSV and ASPV were the most prevailing viruses in apple and pear, the two viruses were detected in almost all mixed infections. Twenty-eight out of 29 apple accessions representing fifteen cultivars were carrying mixed infection. The most prevailing viruses of apple were ACLSV (96.5 per cent), ASPV (89.7 per cent) and ASGV (34.5 per cent) whereas infection of ApMV was detected only in one apple tree. Birisik *et al.* (2008) studied the incidence of apple (*Malus domestica*) viruses in southern Anatolia, Turkey,between 2002 and 2005. A total of 108 orchards and 10 varietal collections were visited in the districts of Adana, Antalya, K. Maras and Osmaniye. Some 413 samples of leaves and/or dormant cuttings were obtained from apple trees. Sanitary testing was conducted by ELISA, biological indexing, and RT-PCR. All samples were tested by ELISA for the presence of apple chlorotic leaf spot virus (ACLSV), apple stem grooving virus (ASGV) and apple mosaic virus (ApMV). The overall virus infection rate as revealed by ELISA was 18.8 per cent. The prevailing viruses were ACLSV (10.6 per cent), ASGV (5.0 per cent) and ApMV (3.1 per cent). Biological indexing was conducted with following indicators: *Malus pumila* cultivars Virginia Crab, Radiant and R 12740 7A against apple viruses in Turkey. Biological indexing revealed higher rates of ACLSV (46.8 per cent), ASGV (60.8 per cent) and *Apple stem pitting virus* (ASPV; 54.5 per cent) infection. RT-PCR tests also confirmed the presence of ASPV, ASGV and ACLSV detected previously by ELISA and biological indexing. This preliminary survey reveals high rates of virus infection in apple in eastern Turkey.

4. VIRUS FREE QUALITY PLANTING MATERIAL

Apple viruses are threat for commercial cultivation of apple which is evident form literature. Fortunately no natural vector is reported to transmit these viruses till date. Hence dessimination of virus free planting material is certainly only important management practice for apple viruses. Since these viruses can transmit from either infected scion or rootstock used for grafting both should be free from viruses. In India mostly seedling rootstocks are used and these viruses are not seedborne therefore identification and manitainance of virus free elite mother block of scion varieties will be suffiecient to produce virus free planting material. In case of clonal rootstock indexing is nessesary as requires for scion variety. Virus indexing of mother plants should made mandatory for all apple nurseries across country. Also random indexing of grafted plants will further increase accuracy of the process. This all will lead to increase in production, average productivity of apple in country and ultimately improve economical status of our farmer.

5. CONCLUSION

Necessary laboratory facilities and trained manpower is required for the selection of healthy stocks with the use of suitable diagnostics. Healthy stock is then further multiplied in different phases for distribution to the growers. Most of the developed countries (Europe, North America and other developed countries) have well defined certification programmes to tackle the viral diseases. Therefore, there is a need to develop a similar type of certification in other countries involved in the cultivation of fruit trees. It could be possible with the collaborative and co operative efforts from the different government and private agencies including fruit growers. Even certified planting material of virus free status may be imported at the initial stage that can be maintained and later multiplied with proper care in isolated pockets for use in the propagation to raise disease free plantation.

6. REFERENCES

Adams, M. J., Antoniw, J. F., Bar Joseph, M., Brunt, A. A., Candresse, T., Foster, G. D., Martelli, G. P., Milne, R. G. and Fauquet, C. M. 2004. The plant virus family flexiviridae and assessment of molecular criteria for species demarcation. *Arch Virol.* **149**: 1045-1060.

Akbas B. and Ilhan, D. 2005. Widespread distribution of apple mosaic virus on apple in Turkey. *Plant Dis.* **89**: 1010.

Al Rwahnih, M., Torturo, C., Minafla, A., Saldarelle, P., Mysta, A., Pallas, V. and Savino, V. 2004. Molecular variability of apple chlorotic leaf spot virus in different hosts and geographical regions. *J. Pl. Pathol.* **86:** 117-122.

Alrefai, R. H., Shiel, P. J., Domier, L. L., D'Arcy, C. J., Berger, P. H. and Korban, S. S. 1994. The nucleotide sequence of apple mosaic virus coat protein has no similarity with the other Bromoviridae coat protein genes. *J. Gen. Virol.* **75**: 2847-2850.

Anonymous, 2013. http: / /www.nhb.gov.in

Aramburu, J. and Rovira, M. 2000. Incidence and natural spread of Apple mosaic virus in hazel in north-east Spain. *Pl. Pathol.* **49**: 423-427.

Bar Joseph, M., Garnsey, S. M. and Gonsalves, D. 1979. The closterovirus: A distinct group of elongated plant viruses. *Adv. Virus Res.* **25**: 93.

Birisik, N., Myrta, A., Hassan, M. and Baloglu, S. 2008. A preliminary account on apple viruses in Mediterranean Region of Turkey. *Acta Hort.* **781**: 125-130.

Bradford, F. C. and Joley, L. 1933. Infectious variegation in the apple. *J. Agric. Res.* **46**: 901-908.

Brunt, A. A., Crabtree, K., Dallwitz, M. J., Gibbs, A. J., Watson, L. and Zurcher, E. J. 1996. Apple mosaic Ilarvirus. Plant Viruses Online: Descriptions and Lists from the VIDE Database.

Caglayan, K., Serce, C. U., Gazel, M. and Jelkmann, W. 2006. Detection of four apple viruses by ELISA and RT-PCR assays in Turkey. *Tur. J. Agric. For.* **30**: 241-246.

Cambeli, T., Folwell, R. J., Wandschneider, P., Eastwell, K. C. and Howell, W. 2003. Economic implications of a virus prevention program in deciduous tree fruits in the US. *Crop Prot.* **22**: 1149-1156.

Campbell, A. I. 1963. The effect of some latent virus infections on the crop and cropping of apples. *J. Hort. Sci.* **38**: 15-19.

Candresse, T., Lanneau, M., Revers, F., Macquarie, G., German, S., Dunez, J., Grasseau, N. and Malinovsky, T. 1995. An immuno capture PCR assay adapted to the detection and the analysis of molecular variability of apple chlorotic leaf spot virus. *Acta Hort.* **386:** 136-177.

Cation, D. and Carlson, R. F. 1960. Determination of virus entities in an apple scion/ rootstock test orchard. *Quart. Bull. Michigen Agric. Exp. Stat.* **43**: 435-443.

Cembali, T., Folwell, R. J., Wandschneider, P., Eastwell, K. C. and Howell, W. E. 2003. Economic implications of a virus prevention programme in deciduous tree fruits in the US. *Crop Prot.* **22**: 1149-1156.

Chairez, R. and Lister, R. M. 1973. A comparison of two strains of apple chlorotic leaf spot virus. *Phytopathol.* **63**: 1458-1464.

Chamberlain, E. E., Atkinson, J. D., Hunter, J. A. and Wood, G. A. 1971. Effect of apple mosaic virus on growth and cropping of "Freyberg" apples trees. *NZ J. Agric. Res.* **14**: 936-943.

Chandel, V., Rana, T., Handa, A., Thakur, P. D., Hallan, V. and Zaidi, A. 2008. Incidence of *Prunus necrotic ring spot virus* on *Malus domestica. Ind. J. Phytopath.* **156**: 382-384.

Choueiri, E., Digiaro, M., Minifra, A. and Savino, V. 1993. A survey of peach virus in Apulia. *Adv. Hort. Sci.* **7**: 61-64.

Chuangui, Y., Tian, Y. and Luo, X. 1997. A study on the fast detection of latent apple viruses. *J. Beijing. Univ.* **19**: 87-91.

Chunmin, L., Yaqin, W., Zhang, D. and Wang, X. 1997. Study on the occurrence of apple chlorotic leaf spot virus in stone fruits. *Chinese Fruits.* **4**: 12-13.

Cieslinska, M. and Kaminska, M. 2002. Evaluation of enzyme-linked immunosorbent assay (ELISA) modifications for detection of apple chlorotic leaf spot virus. *Phytopathol. Polonica,* **25**: 59-68.

Clark, M. F. and Adams, A. N. 1977. Characteristics of the microplate method of enzyme-linked immunosorbent assay for the detection of plant viruses. *J. Gen. Virol.* **34:** 475-483.

Clover, G. R. G., Pearson, M. N., Elliot, D. R., Tang, Z., Smales, T. E. and Alexander, B. J. R. 2003. Characterization of a strain of apple stem grooving virus in *Actinidia chinensis* from China. *Pl. Pathol.* **52**: 371-378.

Corvo, L. M. and Barros, M. T. F. 2001. The use of a simplified DAS-ELISA procedure for large-scale detection of apple chlorotic leaf spot and apple stem grooving viruses in apple trees over extended sampling periods. *Acta Hort.* **550**: 269-274.

Cropley, R. 1963. The association of sap transmissible virus with apple chlorotic leaf spot. *Pl. Dis. Rep.* **48:** 195-197.

Cropley, R. 1968. Comparison of some apple latent viruses. *Ann. App. Biol.* **61:** 361-422.

Daniels, J., Marinho, V. L. A., Kummert, J. and Lepoivre, P. 1998. Development of colorimetric RT-PCR tests for apple stem grooving virus detection in apple trees. *Acta Hort.* **472**: 105-111.

De Sequeira, O. A. 1965. Recent work on some sap transmissible virus from apple-II. Apple cucumber virus and E-36 virus. *Zastita Bilja,* **16:** 247-250.

De Sequeira, O. A. 1967. Studies *on a* virus causing stem grooving and graft-union abnormalities in Virginia Crab apple. *Ann. App. Biol.* ***60****: 59-*66.

Desvignes J C and Boye R. 1989. Different diseases caused by chlorotic leaf spot virus on the fruit trees. *Acta Hort.* **235**: 31-38

Desvignes, J. C., Boye, R., Cornaggia, D. and Grasseau, N. 1999. Virus Diseases of Fruit Trees. Paris, Editions centre technique interprofessionnel des fruits et legumes, 202p.

Desvignes, J. C., Boye, R., Cornaggia, D. and Grasseau, N., 1990. Maladies virus des arbres fruiters. Ctifl Paris, France.kunda

Detienne, G., Delbos, R. and Dunez, J. 1980. Use and versatility of the immunoenzymatic ELISA procedure in the detection of different strains of apple chlorotic leaf spot virus. *Acta Phytopathol.Hungarian Acad. Sci.* **15**: 39-45.

Dhingra, K. L. 1972. Transmission of apple mosaic by natural root grafting. *Indian J. Hort.* **29**(3/4): 348-350.

Dhingra, K. L. and Niazi, F. R. 1973. Indexing of apple varieties for the presence of apple mosaic virus. *Indian Phytopathol.* **26**: 315-318.

Dhir, S., Tomar, M., Thakur, P. D., Ram, R., Hallan, V. and Zaidi, A. A. 2010. Molecular evidence for Apple stem pitting virus infection in India. *Pl. Pathol.* **59**: 393.

Dhir, S., Tomar, M., Thakur, P. D., Ram, R., Hallan, V. and Zaidi, A. A. 2009. *Apple stem pitting virus* diagnosis on apple in India. *Ind. J. Virol.* **20**: 45.

Di Serio, F., Malfitano, M., Alioto, D., Ragozzino, A. and Flores, R. 2003. Apple dimple fruit viroid. **In:** Viroids. (Eds.), Hadidi A, Flores R, Randles J W and Semancik J S. CSIRO Publishing, Collingwood, Australia, pp. 146-149.

Dunez, J. and Delbos, R. 1988. Closteroviruses. **In:** European Handbook of plant diseases. (Eds.), Smith I M, Dunez J, Lelliott R A, Phillips D H and Archer S A. pp. 5-7.

Dunez, J., Marenaud, G., Delbos, R. P. and Lansac, M. 1972. Variability of symptoms induced by the apple chlorotic leaf spot virus - A type of CLSV probably responsible for bark split disease of prune trees. *Pl. Dis. Report*, **56:** 293-295.

Eppler, A., Ryback, M. and Kountrias, A. 2001. The distribution of little cherry disease in Germany: LChV1 AND LChV2 in cultivars of sweet cherry and in some other species. *Mededelingen Faculteit Landbouwkundige en Toegepaste Biologische Wetenschappen, Universiteit Gent*. **66**: 147-149.

Fan, X. D., Dong, Y., Zhang, Z. P., Li, L. and Pei, G. Q. 2009. Multiplex RT-PCR assay for simultaneous detection of three latent apple viruses. *Acta Hort. Sinica*, **36**: 1821-1826.

Fidan, U. 1994. Indexing of apple trees for apple mosaic virus, apple chlorotic leaf spot virus and apple stem grooving virus by ELISA. *J. Turk. Phytopathol.* **23**: 127-132.

Flegg, C. L. and Clark, M. F. 1979. Detection of apple chlorotic leaf spot virus by a modified procedure of enzyme linked immunosorbent assay. *Ann. Appl. Biol.* **91**: 61-65.

Foissac, X., Dumas, L. S., Gentit, P., Dulucq, M. J., Marais, A. and Candresse, T. 2005. Polyvalent degenerate oligonucleotides reverse transcription polymerase chain reaction. A polyvalent detection and characterization tool for trichoviruses, capilloviruses and foveaviruses. *Phytopathol.* **95:** 6.

Freitas, A. J., Novaes, Q. S., Betti, J. A., Rezende, J. A. M., Kuniyuki, H. and Yuki, V. A. 2004. DAS-ELISA detection of apple chlorotic leaf spot and apple stem grooving virus in apple petals. *Fitopatologia Brasileira*, **29**: 572.

Fridlund, P. R.1989. *Virus and viruslike diseases of pome fruits and non infectious disorders.* WSU Press, USA. 330p.

Fuchs, E. 1983. Comparison of different serological methods for detecting apple stem grooving virus and apple chlorotic leaf spot virus. *Arch. Gartenbau*, **31**: 237-245.

Fuchs, E., Gruntzig, M. and Alkai, B. 1988. Serological detection of mechanically transmissible viruses of pome and stone fruits. *Nachrichtenblatt fur de Pflanzenschutz in der DDR* **42**: 208-211.

Fuchs, E., Gruntzig, M. and Merker, D. 1985. Serological comparison of different isolates of apple stem grooving virus and apple chlorotic leaf spot virus. *Arch. Phytopathologie Pflanzenschutz* **21**: 427-435.

Fuchs, E., Merker, D. and Kegler, G. 1979. Detection of apple chlorotic leaf spot virus, apple stem grooving virus and tomato bushy stunt virus with ELISA. *Archiv fur Phytopathologie und Pflazenschutz* **15**(6): 421-424.

Fulton, R. W. 1972. Apple mosaic virus. CMI/AAB Descriptions of Plant Viruses, No. 83. Wellesbourne, UK: *Assoc. Appl. Biol.* pp. 4-7.

Fulton, R. W. 1980. Ilarviruses **In**: Handbook of Plant Virus Infections and Comparative Diagnosis. (Eds.): Kurstak E. North Holland, Amsterdam. Elsevier, pp. 377-421.

Fulton, R. W. 1965. A comparison of two viruses associated with plum line pattern and apple mosaic. *Zastita Bilja* **16**: 427–430.

Fulton, R. 1952. Mechanical transmission and properties of Rose mosaic virus. *Phytopathol.* **42**: 413-416.

Garcia-Ibarra, A., Rubio, M., Dicenta, F. and Martinez-Gomez, P. 2010. Evaluation of resistance to *Apple Chlorotic Leaf Spot Virus* (ACLSV) in controlled greenhouse conditions in the apricot breeding programme of CEBAS-CSIC in Murcia (Spain). *Acta Hort.* **862**: 487-490.

Gilmer, R. M. 1962. Virus induced anatomical changes in apple wood. *Phytopathol.* **52**: 928-953.

Gilmer, R. M. and Brase, K. D. 1961. The association of chlorotic leaf spot virus of apple with decline of Spy 227 rootstocks. *Phytopathol.* **51**: 457-461.

Gilmer, R. M. and Uyemoto, J. K. 1972. Sap transmission of chlorotic leaf spot and stem grooving viruses to apple. *Phytopathol.* **62**: 585-586.

Guo, L. X., Xiang, B. C., Chen, H. Y., Duan, W. J. and Zhu, S. F. 2005. Sequence analysis of the capsid protein gene of KRL-1 isolate of Apple stem grooving virus. *J. Fruit Sci.* **22**: 576-578.

Hadidi, A., Nemchinov, L. and Faggioli, F. 1998. PCR-detection of apple stem pitting virus from pome fruit hosts and sequence variability among viral isolates. *Acta Hort.* **472**: 67-73.

Halk, E. L., Hsu, H. T., Aebig, J. and Franke, J. 1984. Production of monoclonal antibodies against three ilarviruses and alfalfa mosaic virus and their use in serotyping. *Phytopathol.* **74**: 367-372.

Hassan, M., Myrta, A. and Polak, J. 2006. Simultaneous detection and identification of four pome fruit viruses by one tube pentaplex RT-PCR. *J. Virol. Meth.* **133**: 124-129.

Henson, J. and French, R. 1993. The polymerase chain reaction and plant disease diagnosis. *Ann. Rev. Phytopathol.* **143**: 369-373.

Hirata, H., Lu, X., Yamaji, Y., Kagiwada, S., Ugaki, M. and Namba, S. 2003. A single silent substitution in the genome of Apple stem grooving virus causes symptom attenuation. *J. Gen. Virol.* **84**: 2579–2583.

Hou, Y. L., Yang, J. L., Dong, Y. F., Zhang, Z. P., Wang, S. H. and Wu, P. 2004. Detection techniques for apple mosaic virus disease by RT-PCR. *China Fruits*, **6**: 5-6.

Howell, W. E. and Mink, G. I. 1995. *Malus micromalus* clone GMAL 273: a new woody plant indicator for rapid detection of apple stem grooving virus. *Acta Hort.* **386**: 439-443.

Imed, A., Boscia, D., Boari, A., Saldarelli, P., Digiaro, M. and Savino, V. 1997. A comparison of apple mosaic virus isolates from Prunus trees and production of specific monoclonal antibodies. *EPPO Bulletin*, **27**: 563-564.

James, D. 1998. Factors affecting the detection of apple stem grooving virus by PCR analysis. *Acta Hort.* **472**: 119-124.

Jan, F. J., Wu, Z. B., Kuo, A. J., Zheng, Y. X., Chang, H. H., Su, C.C. and Yang, Y. S. 2003. Detection of Apple chlorotic leaf spot and Apple stem grooving viruses by the method of reverse transcription-polymerase chain reaction. *Pl. Pathol. Bull.* **12**: 10-18.

Jarrar, S., Myrta, A., Terlizzi, B. and Savino, V. 2001. Viruses of stone fruits in Palestine. *Acta Hort.* **550**: 245-248.

Jelkmann, W. 1997. Apple stem pitting virus. **In:** Filamentous viruses of woody plants. (Eds.): Vesely D and Monette P. Research Signpost, Trivandrum, India, pp.133-141.

Jelkmann, W. 1994. Nucleotide sequences of apple stem pitting virus and of the coat protein gene of a similar virus from pear associated with vein yellows disease and their relationship with potex and carlaviruses. *J. Gen. Virol.* 75: 1535-1542.

Kirby, M. J., Guise, C. M. and Adams, A. N. 2001. Comparisons of bioassays and laboratorys assay for apple stem grooving virus. *J. Virol. Meth.* **93**: 167-173.

Koganezawa, H. and Yanase, H. 1990. A new type of elongated virus isolated from apple trees containing the stem pitting agent. *Pl. Dis.* **74**: 610-614.

Koganezawa, H., Yang, X., Zhu, S. F., Hashimoto, J. and Hadidi, A. 2003. Apple scar skin viroid in apple, **In**: Viroids. (Eds.): Hadidi, A. Flores, R., Randles, J. W. and Semancik, J.S. CSIRO Publishing, Melbourne. pp. 137-141.

Komorowska, B., Malinowski, T. and Michalczuk, L. 2010. Evaluation of several RT-PCR primer pairs for the detection of Apple stem pitting virus. *J. Virol. Meth.* **168**(1/2): 242-247.

Krizbai, L., Ember, I., Nemeth, M., Kolber, M., Pasquini, G. and Faggioli, F. 2001. Characterization of Hungarian isolates of apple chlorotic leaf spot virus. *Acta Hort.* **550** : 291-295.

Kryczynski, S., Szyndel, M. S., Paduch, C. E. and Glowacki, M. 1995. Virus infection status of some apple propagative material plantations in Poland. *Phytopathologiae Polonica* **22**: 77-84.

Kundu, J. K. 2002a. The application of RT-PCR assay for the detection of Apple stem pitting virus and Apple stem grooving virus in four apple cultivars. *Pl. Prot. Sci.* **38**: 13-17.

Kundu, J. K. 2002b. Simplification of RNA preparation procedure for RT-PCR in detection of pome fruit tree viruses. *Pl. Prot. Sci.* **38**: 252-254.

Kundu, J. K. and Yoshikawa, N. 2008. Apple stem pitting virus. **In:** *Characterization, Diagnosis and Management of Plant Viruses: Horticultural Crops.* (Eds.): Rao, G. P., Myrta, A. and Ling, K. L. Studium Press, Houston. pp. 1-15.

Kundu, J. K., Svoboda, J. and Polak, J. 2003. Detection of Apple stem grooving virus in different tissues of apple trees throughout the year. *Pl. Prot. Sci.* **39**: 93-96.

Lebas, B. S. M., Elliott, D. R., Vanderbrink, R., Ochoacorona, F. M., Tang, J. Z. and Alexander, B. J. R. 2004. Apple chlorotic leaf spot virus infection induces plum pox virus like symptoms on plum in New Zealand. *Acta Hort.* **657:** 1221-1225.

Legrand, G. 1986. Utilization of indirect enzyme linked immunosorbent assay for the detection of plant viruses. *Mededelingen van de Facutteit Landbouwwetenschappen Rijksuniversiteit Gent.* **51**: 783-789.

Lemoine, J. 1982. The effect of virus diseases on the behaviour of the ornamental apples. *Perpetue Evereste and Golden Gem. Revue Horticole.,* **43**: 45-47.

Lemoine, J. and Michelesi, J C. 1997. Incidence of contamination in an apple orchard. *Arboriculture Fruitiere,* **509**: 43-48.

Leone, G., Lindner, J. L., van der Meer, F. A., Schoen, C. D. and Jongedijk, G. 1998. Symptoms on apple and pear indicators after back-transmission from green house conditions. *European. J. Pl. Pathol.,* **133**: 857-863.

Lessa, A. O., Castro, L. A. S. and Daniel, J. 1998. Occurrence of apple chlorotic leaf spot virus in apple orchards of the states of Santa Catarina and Rio Grande do Sul, Brazil. *Fitopatologia Brasileira,* **23**: 505.

Li, L. L., Dong, Y. F., Zhang, Z. P., Zhang, Z. H., Fan, X. D. and Pei, G. Q. 2010. RT-PCR detection and molecular variability of Apple stem pitting virus. *Acta Hort. Sinica,* **37**: 9-16.

Liebenberg, A. 2013. Influence of latent apple viruses on *Malus sieboldii* derived apple proliferation resistant rootstocks. Ph.D. Thesis. The Faculty of Biosciences. Universität Heidelberg.

Lister, R. M. 1970. Apple chlorotic leaf spot virus. CMI/AAB Description of plant viruses. No. 30. Commonwealth Agriculture Bureaux, Kew.

Lister, R. M., Bancroft, J. B. and Nadakavukaren, M. J. 1965. Some sap transmissible viruses from apple. *Phytopathol.* **55**: 859.

Lister, R. M., Bancroft, J. B. and Shay, J. R. 1964. Chlorotic leaf spot from a mechanically transmissible virus from apple. *Phytopathol.* **54**: 1300-1301.

Liu, P., Lei, Zhang., Huawei, Zhang., Honghong, Jiao. and Yunfeng, Wu. 2013. Detection and molecular variability of Apple stem grooving virus in Shaanxi, China. *J. Pytopathol.* **161** : 445-449.

Liu, Y., Schiff, M. and Kumar, S. P. 2002. Virusinduced gene silencing in tomato. *Pl. J.,* **31:** 777-786.

Luckwill, L. and Campbell, A. 1959. *Malus platycarpa* as an apple virus indicator. *J. Hort. Sci.* **34**: 248 - 252.

Luckwill, L. C. and Campbell, A. 1963. *Commonw. Bur. Hort. Pl. Crops. Tech. Commun.* No. **30**: 789.

Machita, I., Saito, A., Fukushima, C. and Tanaka, Y. 1986. Detection of apple stem grooving virus by enzyme-linked immunosorbent assay (ELISA). *Ann.Rep. Soc. Plant Protec. North Japan,* **37**: 99-102.

Makarski, J. S. J. and Agrios, G. N. 1973. Respiration, organic acid, and sugar composition of apple fruits collected from apple mosaic virus or russet ring virus-infected trees. *Phytopathol.* **63**: 1483-1488.

Marinho, V. L. A., Kummert, J., Ruffard, G., Colinet, D. and Lepoivre, P. 1998. Detection of apple stem grooving virus in dormant apple trees with crude extracts as templates for one-step RT PCR. *Pl. Dis.* **82**: 785-790.

Martelli, G. P. and Jelkmann, W. 1998. Foveavirus a new plant genus. *Arch.Virol.* **143:** 1245-1249.

Massart, S., Brostaux, Y., Barbarossa, L., Batlle, A., Cesar, V., Dutrecq, O., Fonseca, F., Guillem, R., Komorowska, B., Olmos, A., Steyer, S., Wetzel, T., Kummert, J. and Jijakli, M. H. 2009. Interlaboratory evaluation of two Reverse-transcriptase Polymeric Chain Reaction based methods for detection of four fruit tree viruses. *Ann. Appl. Biol.***154**: 133-141.

Mathioudakis, M. M., Maliogka, V. I., Dovas, C. I., Vasilakakis, M. and Katis, N. I. 2006. First record of the *Apple stem pitting virus* (ASPV) in Quince in Greece. *J. Pl. Pathol.* **88**: 225-225.

Mc Crum, R. C. 1965. Spread of apple chlorotic leaf spot virus from tree to tree. *Pl. Dis. Repor.* **49**: 958–959.

Menzel, W., Jelkmann, W. and Maiss, E. 2002. Detection of four apple viruses by multiplex RT-PCR assays with co-amplification of plant mRNA as internal control. *J. Virol. Meth.* **99**: 81-92.

Menzel, W., Zahn, V. and Maiss, E. 2003. Multiplex RT-PCR-ELISA compared with bioassay for the detection of four apple viruses. *J. Virol. Meth.* **110**: 153-157.

Mink, G. I. 1989. Apple chlorotic leaf spot virus. **In**: *Virus and virus-like diseases of stone fruits and simulating noninfectious disorders.* (Ed.): Pullman W A. College of Agriculture and Home Economics, Washington State University, pp. 8-11.

Mink, G. I. and Shay, J. R. 1959. Preliminary investigations on apple virus diseases. *Proc. Indian Nat. Sci. Acad,* **68**: 88.

Mink, G. I., Shay, J. R., Gilmer, R. M. and Stouffer, R. F. 1971. Latent virus of apple. II. Symptoms in woody indicators and strain variation. *New York Agric. Exp.Stat.***1**: 9-15.

Minoiu, N. 1977. Research on the apple stem pitting virus. *Lucrarile Stiintifice ale Institutului de Cercetari pentru Pomicultura Pitesti,* **6**: 205-210.

Minoiu, N., Isac, M., Parnia, P., Casavela, S., Kegler, H., Schmidt, H. B., Richter, J. and Fuchs, E. 1983. Etiology and control of apple stem grooving virus. *Lucrarile Stiintifice ale Institutului de Cercetare si Productie Pentru Pomicultura Pitesti,* **10**: 215-222.

Myrta, A., Abou, G. N., Terlizzi, B., Savino, V. and Al-Rwahnih, M. 2001. Viruses and viroids of stone fruits in Jordan. *Bulletin EPPO,* **31**: 95-98.

Myrta, A., Terlizzi B., Stamo, B., Al-Rwahnih, M., Savino, V. and Carraro, L. 2004. A preliminary account of the presence of pome fruit viruses in Albania. *Acta Hort.* **657**: 55-58.

Nagaich, B. B. and Vashishth, K.S. 1965. Apple mosaic in Shimla Hills. *Indian Phytopathol.* **16**: 354-361.

Negi, A., Rana, T., Kumar, Y., Hallan, V., Ram, R., Hallan, V. and Zaidi, A. A. 2010. Analysis of the coat protein gene of Indian strain of apple stem grooving virus. *J. Biochem. Biotech.* **19**: 91-94.

Nemchinov, L., Hadidi, A., Candresse, T., Foster, J. A. and Verderevskaya, T. 1995. Sensitive detection of apple chlorotic leaf spot virus from infected apple or peach tissue using RT-PCR, IC-RT-PCR or multiplex IC-RT-PCR. *Acta Hort.* **386:** 50-57.

Nemeth, M. 1986. Virus, mycoplasma and rickettsia diseases of fruit trees. Akademiai Kiado. *Budapest,* 841p.

Nickel, O., Fajardo, T. V., Jelkmann, W. and Kuhn, G. B. 2001. Sequence analysis of the capsid protein gene of an isolate of Apple stem grooving virus and its survey in Southern Brazil. *The Fitopatologia Brasileira,* **26**: 655–659.

Nickel, O., Fajardo, T. V., Trivilin, A. P., Targon, M. L. N. P. and Machado, M. A. 2004. Production and use of polyclonal antibodies to the coat protein of Apple stem grooving virus expressed in *Escherichia coli. Acta Hort.,* **657**: 35-40.

Ohki, S. T., Yoshikawa, N., Inouye, N. and Inouye, T. 1989. Comparative electron microscopy of *Nicotiana occidentalis* confirm the identity of *Apple stem pitting virus* with *pear vein yellows virus. Acta Hort.,* **472**: 61-65.

Pasquini, G., Faggioli, F., Pilotti, M., Lumia, V. and Barba, M. 1998. Characterization of apple chlorotic leaf spot virus isolate from Italy. *Acta Hort.,* **472:** 195-202.

Paunovic, S. 1988. Properties of two apple chlorotic leaf spot virus isolates. *Acta Hort.* **235:** 39-47.

Paunovic, S. and Jevremovic, D. 2008. Comparative results of detection of pome fruit viruses by different methods. *Acta Hort.,* **781**: 147-153.

Pena-Iglesias, A. and Ayuso-Gonzales, P. 1975. Preliminary identification of the viruses producing Spanish apricot pseudo pox (viruela) and apricot mosaic diseases. *Acta Hort.,* **44**: 255-265.

Petrzik, K. 2005. Capsid protein sequence gene analysis of Apple mosaic virus infecting pears. *Eur. J. Pl. Pathol.* **111**: 355-360.

Petrzik, K and Lenz O. 2002. Remarkable variability of apple mosiac virus capsid protein gene after the nucleotide position 141. *Arch. Virol.* **147**: 1275-1285.

Plese, N., Hoxha, E. and Milicic, D. 1975. Pathological anatomy of trees affected with apple stem grooving virus. *Phytopathoogy Zeitschrift*, **82**: 315-325.

Polak, J. 1994. Distribution of apple mosaic virus in apple gardens of the Czech Republic. *Ochrana Rostlin*, **30**: 85-89.

Polak, J., Hassan, M. and Myrta, A. 2006. Simultaneous detection and identification of four pome fruit viruses by one-tube pentaplex RT-PCR. *J. Virol. Meth.* **133**: 124-129.

Polak, J., Hassan, M. and Paprstein, F. 2008. Detection and distribution of four pome fruit viruses in germplasm collection in the Czech Republic. *Acta Hort.* **781**: 113-118.

Polak, J., Zeiglerova, J. and Bouma, J. 1997. Diagnosis and distribution apple chlorotic leaf spot virus of pome fruits in Czech Republic. *Zahradnictvi*, **24**: 89-94.

Posnette, A. F. and Ellenberger, C. E. 1975. Bark-split- A virus disease of Plums. *Ann. Appl. Biol.*, **45**: 74-80.

Posnette, A. F., Campbell, A. I. and Cropley, R. 1976. Indicators and indexing. *Acta Hort.*, 275-323.

Posnette, A. F., Cropley, R. and Ellenberger, C. E. 1963. The effect of virus infection on the growth and crop of apple, pear and plum tree. *Phytopathologia Mediterranea* **2**: 158-160.

Postman, J. D. and Cameron, H. R. 1987. Apple mosaic virus in US Filbert germplasm. *Plant Dis.*, **71**: 944-945.

Poul, F. and Dunez, J. 1988. Use of monoclonal antibodies for the detection of apple chlorotic leaf spot virus. *Acta Hort.*, **235**: 49-56.

Rana, T., Chandel, V., Hallan, V., Thakur, P. D., Handa, A. and Zaidi, A. A. 2007. Molecular evidence of Apple chlorotic leaf spot virus in Himachal Pradesh. *Indian J. Virol.* **18**: 70-74.

Rana, T., Negi, A., Dhir, S., Thockchom, T., Chandel, V., Walia, Y., Singh, R. M., Ram, R., Hallan, V. and Zaidi, A. A. 2011. Molecular diagnosis of apple virus and viroid pathogens from India. *Arch. Phytopathol. Plant Protect.* **44** : 505-512.

Rankovic, M. and Vuksanovic, S. 1983. The use of enzyme-linked immunosorbent assay (ELISA) in the diagnosis of viruses of apple. *Zast Bilza*, **34**: 257-264.

Rebandel, Z., Zawadzka, B. J. and Wierszyllowski, J. 1979. Effect of Apple mosaic virus on bud-take and growth of trees in the nursery. *Fruit Sci. Rep.*, **6**: 9-17.

Rodoni, B. C. and Constable, F. E. 2008. The incidence and strain variation of *Apple stem grooving* and *Apple stem pitting viruses* in Australian pome fruit. *Acta Hort.*,**781**: 167-174.

Roussel, S., Kummert, J., Dutrecq, O., Lepoivre, P. and Jijakli, M. H. 2004. Development of molecular tests for the detection of ILAR and latent viruses in fruit trees. *Comm. Agric. Appl. Biol. Sci.*, **69**: 427-432.

Roy, A., Fayad, A., Barthe, G., and Brlansky, R. H. 2005. A multiplex polymerase chain reaction method for reliable sensitive and simultaneous detection of multiple viruses in citrus trees. *J. Virol. Meth.*, **129**: 47-55.

Rybicki, E. P. 1995. The Bromoviridae. **In:** *Virus taxonomy. VIth report of the International Committee on Taxonomy of Viruses.* (Eds.): Murphy, F. A., Fauquet, C. M., Bishop, D. H. L., Ghabrial, S. A. and Jarvis, S. A. pp. 65-68

Saiki, R. K., Gelfand, D. H., Stoffel, S. and Scharf, S. J. 1988. Primer directed enzymatic amplification of DNA with a thermostable DNA polymerase. *Sci.*, **239**: 478-491.

Salem, N., Mansour, A. and Al-Musa, A. 2005. Viruses of pome fruit trees in Jordan. *J. Pl. Pathol.*, **87**: 123-126.

Sanchez-Navarro, J. A. and Pallas, V. 1997. Evolutionary relationships in the ilarviruses: nucleotide sequence of *Prunus necrotic ringspot virus* RNA 3. *Arch. Virol.*, **42**: 749-63.

Sawamura, K. and Osada, S.1980. Apple stem grooving virus isolated from apple cv. Starkrimson Delicious. *Bull. Faculty Agric. Hirosaki Univ.*, **33**: 19-27.

Seigner L. 2000. Virus detection using the Polymerase chain reaction (PCR). *Gesunde Pflanzen*, **52**: 205-211.

Semenas, S. E. and Koukharchik, N. V. 2000. Preliminary evaluation of the distribution of some pathogenic viruses in the central area of Belarus. *Poslovodstvo*, **13**: 109-111.

Serce, C. U. and Rosner, A. 2006. Molecular typing of Turkish Apple chlorotic leaf spot virus isolates based on partial coat protein gene. *Phytopathol. Mediterranea*, **45**: 117-125.

Sertkaya, G., Caglayan, K. and Ulubas, C. 2004. Detection of some viruses of stone fruits in mother plant blocks in Eastern Mediterranean Region of Turkey. *Acta Hort.*, **657**: 127-132.

Shiel, P. J. and Berger, P. H. 2000. The complete nucleotide sequence of Apple mosaic virus (ApMV) RNA 1 and RNA 2: ApMV is more closely related to alfalfa mosaic virus than to other ilarviruses. *J. General Virol.*, **81**: 273-278.

Shiel, P. J., Alrefai, R. H., Domier, L. L., Korban, S. S. and Berger, P. H. 1995. The complete nucleotide sequence of apple mosaic virus RNA-3. *Arch. Virol.* **144**: 1247–1256.

Silva, F. N., Nickel, O., Fajardo, T. V. M. and Bogo, A. 2008. Biological multiple indexing and RT-PCR detection of latent viruses in apple plants. *Tropical Pl. Pathol.*, **33**: 157-161.

Singh, B. B., Pandey, B. N., Singh, R. N. and Singh, S. P. 1979. Effect of apple mosaic virus on the growth, yield and quality of apple. *Punjab Hort. J.*, **19**: 80-82.

Smith, K. M. 1972. *A Textbook of Plant Virus Diseases.* 3rd *editon*. Longman, London, UK. pp. 541-544.

Smith, W. W. 1954. Occurrence of stem pitting and necrosis in some body stocks for apple trees. *Proc. American Soc. Hort.Sci.* **63**: 101-113.

Stouffer, R. F. 1989. Apple stem pitting virus. **In:** *Virus and Virus-like Diseases of Pome Fruits and Simulating Noninfectious Disorders.* (Ed.): Pullman W A. College of Agriculture and Home Economics Washington State University, pp. 133-144.

Sutic, D. D., Ford, R. E. and Tosic, M. T. 1999. Handbook of Plant Virus Diseases. CRC Press. London, New York, Washington, 485p.

Svoboda, J. and Polak, J. 2010. Relative concentration of apple mosaic virus coat protein in different parts of apple tree. *Hort. Sci.* **37**: 22–26.

Thokchom, T., Rana, T., Hallan, V., Ram, R. and Zaidi, A. A. 2009. Molecular characterization of the Indian strain of Apple mosaic virus isolated from apple (*Malus domestica*). *Phytoparasitica*, **37**: 375-379.

Thomsen, A. 1975. Cross protection with apple mosaic virus. *Tidsskrift für Planteavl* **79**: 57-62.

Tozzi, D. C. M., Tanaka, H., Soria, J., Barba, M. and Hadidi, A. 1995. Studies on fruit tree viruses in Uruguay. *Acta Hort.* **386**: 155-164.

Turk, B. A. 1996. Detection of apple mosaic virus, apple chlorotic leaf spot virus and apple stem grooving virus in apple trees by enzyme-linked immunosorbent assay (ELISA) using different plant tissues. *Zbornik Biotehniske Fakultete Univerze Ljubljani Kmetijstvo*, **67**: 151-157.

Tzanetakis I E and Martin R R. 2005. First report of strawberry as a natural host of *Apple mosaic virus*. *Plant Disease*, **89**: 431.

Ulubas, C. and Ertunc, F. 2005. Apple chlorotic leaf spot virus (ACLSV) status in Turkey and sensitive detection using advanced techniques. *Turkish J. Agric. Forestry*, **29**: 251-257.

Varveri, C. and Bem, F. 1995. Viruses of stone and pome fruit mother tree plantations in Greece. *Acta Hort.* **386**: 431-438.

Varveri, C., Holeva, A. and Bem , F. P. 1997. Effect of sampling time and plant part on the detection of two viruses in apricot and one in almond by ELISA. *Annals de l'Institut Phytopathologiqu Benake*, **18**: 25-33.

Voller, A., Bartlett, A., Bidwell, D. E., Clark, M. F. and Adams, A. N. 1976a. Detection of viruses by enzyme-linked immune sorbent assay (ELISA). *J. General Virol*, **33**: 165-167.

Voller, A., Bidwellm D. E. and Bartlett, A. 1976b. Enzyme Immunoassays in Diagnostic Medicine : Theory and Practice. *Bull. WHO*, **53**: 55-65.

Waterworth, H. E. and Gilmer, R. M. 1969. Dark green epinasty of *Chenopodium quinoa*, a syndrome induced by a virus latent in apple and pear. *Phytopathol.*, **59**: 334-338.

Watpade, S., Baswaraj, R., Thakur, P.D., Handa, A., Pramanick, K.K., Sharma, Y.P.,Tomar, M., 2012. Molecular detection of Latent Apple chlorotic leaf spot virus inElite Mother Plants of Apple. *Indian J. Virol.* **23**: 359–363.

Watpade, S., Baswaraj, R., Pramanick, K.K., Neeraj Sharma, Anil Handa and Usha Sharma. 2013. Simultaneous detection of Apple Chlorotic Leaf Spot Virus and Applemosaic virus in crab apples and apple rootstocks by duplex RT-PCR. *Scientia Hort.*, **164**: 88–93.

Welsh, M. F. and Uyemoto, J. K. 1980. Differentiation of syndromes caused in apple by graft transmissible xylem affecting agents. *Phytopathol.*, **70**: 349-352.

Welsh, M. F. and Van der Meer, F. A. 1989. Apple stem grooving virus. In: *Virus and virus like disease of pome fruits and simulating noninfectious disorders.* (Ed.): Pullman W A. College of Agriculture and Home Economics Washington State University. pp. 253-267.

White, R. P. 1928. An infectious chlorosis of roses. *Plant Disease Rep.* **12:** 33-34.

Wood, G. A. 1974. Elimination of latent apple viruses shows growth and yield improvements. *Orchardist of New Zealand*, **47**: 173.

Wu, Y. Q., Zhang, D. M., Chen, S. Y., Wang, X. F. and Wang, W. H. 1998. Comparison of three ELISA methods for the detection of apple chlorotic leaf spot virus and apple stem grooving virus. *Acta Hort.*, **472**: 55-59.

Yanase, H. and Yamaguchi, A. 1982. *Malus sieboldii* MO-65 as a new indicator plant for apple stem pitting virus. *Bull. Fruit Tree Res. Station, Morioka*, **9**: 69-77.

Yanase, H., Koganezawa, H. and Friedlund, P. R. 1989. Correlation of pear necrotic spot with pear vein yellows and apple stem pitting and a flexious filamentous virus associated with them. *Acta Hort.*, **235**: 157 -158.

Yanase, H., Mink, G. I. and Yamaguchi, A. 1975. Viruses causing apple top working disease in Japan. *Acta Hort.*, **44:** 221-230.

Yanase, H., Nakatani, F., Munakata, T. and Machida, K. 1986. Detection of different strains of apple chlorotic leaf spot virus and apple stem grooving virus by direct double antibody sandwich method and indirect ELISA. *Bull. Fruit Trees*, **13**: 69-81.

Yanase, H., Yamaguchi, A., Mink, G. I. and Sawamura, K. 1979. Back transmission of apple chlorotic leaf spot virus (type strain) to apple and production of apple top working disease symptoms in Maruba Kaido (*Malus prunifolia* Borkh. var. *ringo* Asami). *Annal. Phytopathol.Soc. Japan* **45**: 369-374.

Yanase, H. 1974. Studies on apple latent viruses in Japan. *Bull.Fruit Tree Res.Stat., Japan Series*, **11:** 47-109.

Yanase, H. 1983. Back-transmission of apple stem grooving virus to apple seedlings and induction of symptoms of apple top working disease in root stocks of Mitsuba Kaido (*Malus sieboldii*) and Kobano Zumi (*Malus sieboldii* var. *arborescens*). *Acta Hort.* **130:** 117-122.

Yang, C. G., Tian, Y. T. and Luo, X. F. 1997. A study on the fast detection of latent apple viruses. *J. Beijing Forestry Univ.*,**19**: 87-91.

Yardmc, N., Cevik, B. and Eryigit, H. 2008. Detection of Apple mosaic virus on apple cultivars growing in South-west Turkey by ELISA and RT-PCR methods. *Acta Hort.*, **781**: 561-565.

Ylmaz, N. D., Yanar, Y., Kadoglu, I. and Yanar, D. 2005. Study on distribution of apple mosaic virus (ApMV) in apple orchards in Tokat Province. *Ondokuz Mays Universitesi Ziraat Fakultesi Dergisi*, **20**: 12-15.

Yoshikawa, N. and Takahashi, T. 1992. Evidence for translation of apple stem grooving *capillovirus* genomic RNA. *J. General Virol.*, **73:** 1313-1315.

Yoshikawa, N., Gotoh, S., Umezawa, M., Satoh, N., Satoh, H., Takahashi, T., Ito, T. and Yoshida, K. 2000. Transgenic *Nicotiana occidentalis* plants expressing the 50 kD a protein of apple chlorotic leaf spot virus display increased susceptibility to homologous virus. *Phytopathol.*, **90**: 311-316.

Yoshikawa, N., Matsuda, H., Oda, Y., Isogai, M., Takahashi, T., Ito, T. and Yoshida, K. 2001. Genome heterogeneity of apple stem pitting virus in apple trees. *Acta Hort.*, **550**: 285-290.

Yoshikawa, N., Sasaki, E., Kato, M. and Takahashi, T. 1992. The nucleotide sequence of apple stem grooving capillovirus genome. *Virol.*, **191**: 98-105.

Yuxuan, H., Hong, N. and Wang, G. 2005. Investigation and detection of apple stem pitting virus in *Pyrus pyrifolia* by RT-PCR and TC-RT-PCR. *J. Fruit Sci.*, **22**: 343-346.

Zahn, V. 1996. Obstvirustestung in Wandel der Zeit. *Obstbau*, **21**: 547-550.

Zawadzka, B. 1983. The effect of mosaic and rubbery wood infection on the growth and yield of apple trees. *Zeszyty Problemowe Postepow Nauk Rolniczych*, **291**: 385-392.

Zhang, D.M., Chen, S. Y., Wang, X. F., Wang, W. H. and Wu, Y. Q. 1998. Comparison of three ELISA methods for the detection of apple chlorotic leaf spot virus and apple stem grooving virus. *Acta Hort.* **472**: 55-59.

Zhang, Y. Y., Wang, G. P. and Hong, N. 2005. The biological characteristics and molecular identification of some Apple stem grooving virus isolates. *Acta Phytophylogica Sinica*, **32**: 266-270.

Zheng, Y. Y,, Hong, N., Wang, G. P. and Hu , H. J. 2006. Cloning and sequence analysis for the CP gene of Apple stem grooving virus from pears. *Acta Phytopathol. Sinica*, **36**: 62-67.

Chapter 20

Disease Dynamics and their Management in Temperate Fruits and Nuts

Satish Kumar Sharma

Joint Director Research (Horticulture),
Dr Y.S. Parmar University of Horticulture and Forestry,
Solan – 173 230, H.P.
E-mail: satishsharma_27@yahoo.com

1. INTRODUCTION

Diseases are one of the limiting factors in production of temperate fruits and nuts. These attack at various plant growth stages and reduce both quality and quantity of produce. With the cultivation of temperate fruits over a wide range of elevations including lower warmer areas the biotic stresses have multiplied tremendously. Major diseases are caused by three classes or types of micro-organisms: fungi, bacteria, and viruses. Diseases may be largely responsible for the limited adaptability of temperate fruits particularly soil borne diseases. In India, premature leaf fall, apple scab, root rot, collar rot, cankers, crown gall and hairy root, bacteriosis, gummosis, leaf curl and virus diseases are recognized as common and often destructive diseases on a members of temperate fruits. Other potentially damaging diseases on these fruits are leaf spots, powdery mildew, post-harvest diseases, Stigmina blight, rust of stone fruits, frosty mildew, seedling blight *etc.* in commercial plantings. Most of these diseases have been investigated at individual level and management practices have been devised in each case.

It has also been realized that no single method of disease control can give lasting solution. In this compilation efforts have been made to gather the available information on different aspects of a various diseases. The published literature has been scanned starting from the first report of occurrence of fungi on temperate fruits (*Pyrus* spp.) in India by Berkeley in 1856 followed by observations on *Sphaerotheca pannosa* by Cooke in 1876, causing powdery mildew of almond in Harwan Kashmir and subsequently first ever record of fungi from Himalayan region in 1878 to date in India.

Important diseases of temperate fruits and nuts at global level are listed below:

Almond

Almond leaf scorch	Bacterial canker/gummosis
Bacterial leaf spots	Hairy root
Leaf spots	Brown rot
Rust	Stigmina blight

Apple

Bitter rot	Black mould rot
Blue mould rot	Brown rot
Bumpy fruit vioroid	Canker
Core rot	Crown gall
Collar/crown rot	Dapple apple viroid
Fly speck	Fire blight
Grey mould rot	Hairy root
Leaf spots	Little leaf
Leaf pucker	Marssonina blotch
Mosaic	Powdery mildew
Rust	Seedling blight
Scab	Soft rot
Sooty blotch	Sooty mould
Whisker's rot	White root rot

Apricot

Bacterial canker/gummosis	Bacterial spot
Brown rot	Canker
Crown gall	Frosty mildew
Hairy root	Rust
Stigmina blight	

Cherry

Bacterial canker/gummosis
Bacterial spot
Cherry leafspots
Crown gall
Hairy root
Stigmina blight

Peach-Nectarine

Bacterial canker/gummosis
Bacterial spot
Brown rot
Crown gall
Frosty mildew
Hairy root
Leaf curl
Powdery mildew
Rhizopus rot
Rust
Sooty mould
Stigmina blight

Pear

Bitter rot
Brown rot
Cankers
Crown gall
Crown rot
Fly speck
Hairy root
Soft rot
Sooty blotch

Plum

Bacterial canker/spot
Brown rot
Crown gall
Leaf spot
Plum line pattern
Plum pox virus
Rust

Strawberry

Leaf spot

Soil-borne diseases like white root not and collar rot lead to the death of trees from nursery stage onward and even grown up trees are killed. Seedling blight also leads to death of seedlings at nursery stage whereas crown gall and hairy root lead to stunted plant growth. Further, these are more damaging and difficult to manage as compared to foliar diseases. Every year large number of seedlings/grafted plants are destroyed resulting in huge losses to the nursery grower and upsetting the planting targets. The incipient infection of apparently healthy looking plants/roots help in the dissemination of pathogens to the planting sites in different geographical locations of temperate and sub-temperate regions. The incidence of various diseases reported by different workers from time to time has been presented in the Table 1. In most cases however, the reports are lacking and present assessment has been made on personal communication basis. However, the present compilation gives the status of research work done as well as depicts the deficiencies and gaps in the research efforts on various aspects of the important diseases.

Table 1: Incidence and Present Status of Various Diseases in India

Plant Species	*Disease*	*Incidence (per cent)*	*Reference(s)*	*Present Status/ Assessment*
Apple				
1.	Bitter rot	2.12	Kaul and Munjal (1982)	Moderate
2.	Blue mould rot	5.8-8.0	Kaul and Munjal (1982)	Moderate
3.	Brown rot	5-15.2	Sharma and Kaul (1989)	Moderate
4.	Cankers	1-25	Shandilya *et al.*, 1973	Severe
5.	Collar rot	Not quantified	–	Moderate
6.	Core Rot	10-40	Raina *et al.* (1971)	Moderate
7.	Leaf spots (*Alternaria* sp.)	40	Sharma *et al.* (2005)	Moderate
8.	*Marssonina* Blotch	65-80	Sharma (2000)	Severe
9.	Pink rot	1.81	Kaul and Munjal (1982)	Moderate
10.	Powdery mildew	Not quantified	–	Severe in nursery and moderate in orchards
11.	Scab	70	Gupta (1978)	Traces
12.	Seedling blight	40	Bhardwaj and Agarwala (1986)	Severe in warmer areas
13.	Sooty blotch and flyspeck	40-50		Severe
14.	Storage ots	10-15	Agarwala and Sharma (1968)	Moderate
15.	Whisker's rot	0.75	Kaul and Munjal (1982)	Moderate
16.	White root rot	30-35	Sharma *et al.* (2003)	Severe
Pear				
1.	Brown rot	24.6	Sharma (1989)	Moderate
2.	Grey mould	33.0	Sharma (1989)	Moderate
3.	Storage rots	11.8	Sharma (1989)	Moderate
Stone fruits/Nuts				
1.	Crown gall almond (wild peach, bitter almond, *behmi*)	10-35	Jindal and Sharma (1988)	Severe
	Peach (wild peach)	10-60	Jindal and Sharma (1988)	Severe
	Cherry	5-15	Jindal and Sharma (1988)	Severe
	(Plum or apricot)	5	Jindal and Sharma (1988)	Severe
2.	Frosty mildew	Not quantified	–	Moderate
3.	Leaf curl	Not quantified	–	Severe
4.	Rust	10.3-77.6	Thakur and Xu (2004)	Severe
5.	Stigmina blight	91.06	Sharma and Gupta (1988)	Severe

Most of these diseases are managed by soil or foliar application of chemicals, *viz.*, formaldehyde, carbendazim, thiophanate methyl, thiabendazole, mancozeb, aureofungin, copper sulphate, metalaxyl, EBI fungicides etc. However, due to high cost of chemicals, low levels of disease control, and growing environmental awareness, attention is shifting to develop an integrated approach consisting of chemical and non-chemical approaches. Various practices such as selection of site, phytosanitation, destruction of infected root system/plant debris, fallowing, deep ploughing, hot water treatment (45°C for one hour), improving poorly drained soils, applying lime in acidic soils, soil amendments such as neem cake, pine needles, *nirgundi* leaves, garlic leaves, vascular arbuscular mycorrhizae, soil solarization (10 weeks during peak summers resulted in increase in soil temperature 10-7°C at soil depths ranging from 10-30 cm), soil burning/heating (two hours) *etc.* have been developed and applied as per feasibility in the nurseries. A few promising biocontrol agents such as *Trichoderma viride, Trichoderma harzianum, Pseudomonas fluorescence, P. putida, P. pseudoalcaligenes, Bacillus subtilis, Bacillus licheniformis, Enterobacter aerogenes etc.* have been found inhibitory to the soil-borne pathogens under *in vitro*, nursery and field conditions. These practices either alone or in combination reduce the survival of propagules of soil-borne pathogens and consequently disease incidence; increase shoot length, root length, fresh and dry root weight and affect soil pH and microbial population. Hence, these should be followed for production of healthy pome and stone fruits.

It is now felt that judicious use of chemicals can be the safest means of disease control under the concept of integrated disease management which emerged in the late sixties. It is in fact, the best approach, which is environmentally safe and eco-friendly. It requires the use of a combination of different strategies such as cultural practices, resistant genes, biocontrol and need based chemicals. Important diseases in the present context and their management in temperate fruits and nuts are discussed in this chapter.

2. FOLIAR AND FRUIT DISEASES

(a) Leaf Curl

Hosts: Peach, nectarine, plum and apricot

The fungus mainly attacks the leaves, but on occasion new shoot growth and fruit may become infected. Flowers may be invaded, but they drop soon after opening showing no distinctive symptoms. The disease is widely distributed in Himachal Pradesh and cause heavy damage to peach crop in cool, humid areas.

The atmospheric temperature (15-20°C) and high relative humidity during later half of March and April are optimum for development of disease in mid hills of Himachal Pradesh (Agarwala *et al.*, 1966). The infection of *T. deformans* results in marked increase in weight of fresh and dry leaves moisture, nitrogen, phosphorus, calcium, potassium, copper, zinc, ash, protein contents, total chlorophyll, chlorophyll a, chlorophyll b, molar ratio of magnesium, manganese, iron chlorophyll a and b decreased in disease leaves (Sharma and Sharma, 1990 Singh *et al.*, 1981).

The disease in peach is initially characterized by the appearance of yellowish to reddish areas on leaf blade of young leaves in the spring. These areas progressively thicken and pucker along the midrib causing the leaf to curl. Later on, white to silvery coating of spores develop on the upper leaf surface and the affected leaves become crisp and brittle. Finally the infected leaves abscise pre-maturely or may sometimes remain attached, gradually turning dark brown on severely infected trees. Infected twigs become swollen, distorted and are killed. Sometimes gum-like substances exude from the affected twigs. The fruit infection is characterized by irregular, raised, wrinkled and reddish lesions. The infected flowers and fruits drop rapidly. The fruit set on severely affected tree is drastically curtailed. In apricot, the affected leaves in the growing tip give an appearance/effect of red flower blooming on the tree. The infected leaves later wither and finally fall by August-September. In cherry and nectarines, leaf curl stimulates excessive branching and results in the formation of witche's broom. In plum, the disease also known as 'Plum pocket', 'Bladder plum', 'Mock plum or fools' is characterized by the appearance of white spots or blisters on the fruits. The affected fruits are largely misshapen. The spongy interior of such fruits dries up leaving only exterior shell. This hollow distorted structure is characteristic, and is thus named as 'Bladder plum or plum pocket.' The shoot and leaves are often infected. The tips of infected shoots become greatly enlarged and are often twisted or curled. The disease is incited by fungus *Taphrina deformans* (Berk.) Tulansne. In all seventeen species of *Taphrina* including *deformans* are parasitic on stone fruits. *Taphrina cerasi* (Fuckel) Sadebeck causes plum pocket and *T. communis* and *T. pruni* cause cherry witches broom and leaf.

Management

Leaf curl can be economically and effectively prevented by one application of a recommended fungicide before bud swell during the spring. Prolonged periods of cold, rainy weather in the spring retard tree growth more than growth of the fungus, thus extending the infection period. A single application of fungicides, *viz.* lime sulphur (1:10), Bordeaux mixture (5.5:50), Blitox or Fytolan (0.25 per cent), Cupravit at bud development stage was highly effective in controlling the disease. Cuman sprays were also highly effective in controlling this disease. A single spray at leaf fall stage either of ferbam (0.2 per cent) or ziram (0.2 per cent) or lime sulphur (1:10) and for addition protection during spring is sufficient to control curl. Calixin 80EC (0.05 per cent) applied three times at 15 days interval controlled the disease (Sharma, 1981). A single dormant spray of Bavistin (0.05 per cent) gave complete control of the disease whereas when applied at flowering stage incidence was 5.6 per cent (Sharma *et al.*, 1988). A single application of either carbendazim at the bud swell stage or guazatine at pink bud provide best control of leaf curl disease (Thakur *et al.*, 1991). Rana and Jain (1992) obtained successful control of the disease by three sprays of Karathane (0.05 per cent) followed by Bavistin (0.1 per cent) and difolatan (0.1 per cent) at three weeks interval.

Four peach cultivars, *viz.*, Stark Early Giant, Starking Delicious, World's Earliest and Tesia Samisto are resistant to leaf curl disease (Nautiyal *et al.*, 1988). Cultivars Shan-i-Punjab, Bed Wills Early, July Alberta and World's Earliest are less susceptible/tolerant to peach leaf curl irrespective of congenial environmental conditions (Badiyala and Lakhanpal, 1994; Sharma and Badiyala, 1994).

(b) Leaf Spots

Hosts: Apple, almond, apricot, cherry, peach, pear, walnut and strawberry

Determining the causes of leaf spots is often difficult because the spotting caused by different pathogens or injuries looks very much alike. In some cases, leaf spots develop as a result of interactions among multiple factors, any one of which would not cause leaf spotting on its own. Causes of leaf spots can sometimes be deduced from disease patterns within the nursery, the time of symptom appearance, and characteristics of the individual leaf spots. Apple scab, rust and powdery mildew cause spots on leaves although they are not usually classified as "leaf-spotting fungi." Alternaria leaf spot looks very similar to frog-eye leaf spot caused by *Botryosphaeria obtusa* exemplifies the more stereotypical leaf spot disease.

The leaf spots occurring in nursery plants and orchards are almost common and their appearance depending upon the prevailing climatic conditions. All types of leaf spots occurring in temperate fruits and nuts have been discussed in the following text. Normally spots can be grouped in to two types. The less frequent one occurs on the leaf between the veins as a small, light-gray spot with a dark dot in its center. The more common leaf spots occur on the veins, midribs, and petioles (leaf stems) as long, narrow, slightly sunken, light-coloured lesions. These contain several dark dots, the fruiting structures (pycnidia) of the fungus. When petiole infections are numerous leaf drop may result. New shoot infections first appear similar to the petiole infections except they are longer and are seen more easily. They occur at the junction of the petiole with the shoot (node) or between the nodes. Once the lesion becomes established, it may continue to enlarge for 3 or 4 years, becoming noticeably larger than the diameter of the normal limb.

Various leaf spot causing fungi reported in temperate fruits and nuts include *Oothecium indicum, Linospora ochracea, Canavalis mali, Discosia artocreas, Mycosphaerella pyrina, Helicotrichum obscurum, Sphaeropsis malorum, Alternaria mali, Phyllosticta solitaria, Pestalotiopsis* Sud and Agarwala (1972) reported the occurrence of *Coniothyrium pyrinum* on *Prunus amygdalus* and *Pyrus pashia* which are new host records of the world. *Pestalotia hartigii* was recorded on living leaves of *Malus pumila*. Various other fungi recorded on different pome and stone fruits are *Sphaeropsis malorum* on *M. pumila; Cladosporium* sp. on *Prunus persica; Phoma limitata* on *P. armeniaca; Alternaria alternata* on *P. armeniaca* and *P. avium* and *Phyllosticta prunicola* on *Prunus avium* and *P. salicina* and *phyllosticta mali*. Alternaria leaf spot (*A. alternata*) has recently appeared in severe form causing 40 per cent defoliation in Kotkhai area. The disease was characterized by appearance of 1-3 mm dia reddish brown necrotic spots on young apple leaves resulting in bigger blotches. Dark brown to black elongated lesions appeared on leaf petioles and mid ribs causing yellowing and defoliation of infected leaves (Sharma *et al.*, 2005).The other fungi sucha as *Alternaria pruni* and *A. tenuis* (leaf spot), *Gloeosporium amygdalinum* (anthracnose), *Sphaerotheca pannosa* (powdery mildew), *Phoma glomerata, Sphaeropsis* sp., *Botryosphaeria dothidea* and *Microdiplodia pruni* of almond and *Phyllosticta cerasicola* on apricot, *Cercospora rubrotincta* on sweet cherry.

Management

Pre-infection application of dithane Z-78 (0.25 per cent) was effective for the control of leaf spot caused by *Pestalotia hartigii* or *Phyllosticta mali* (Bhardwaj and Agarwala, 1985; 1986). Frog eye leaf spot was favoured by high rainfall, relative humidity (›75 per cent) at 20-22°C (Singh and Agarwala, 1987). Frog eye leaf spot was best controlled with the application of Blitox whereas leaf spots caused by *Conithyrium pyrinum* and *Mycosphaerella* sp. were best controlled by application of ferbam (0.25 per cent) and aureofungin (0.004 per cent), respectively

Application of N, P fertilization + bavistin (0.1 per cent) before disease appearance and subsequently at 20 days interval controlled leaf spot caused by *Phyllosticta pyrina* (Roy *et al.*, 1985). Alternaria blight of apple (*A. mali*) can be successfully controlled by applying ziram (0.1 per cent) or ferbam, zineb or captan (0.15 per cent) or Bordeaux mixture (1 per cent) (Sud and Agarwala, 1971). Protective application of dodine, mancozeb, zineb, iprodione and mancozeb + hexaconazole exhibited effective control of Alternaria leaf spots (*A. alternata*). All the commercially grown cultivars are highly susceptible, however, pollinizers are tolerant (Sharma *et al.*, 2005).

All the pear varieties are susceptible to leaf spot (*Phyllosticta pyricola*), however, Kieffer was least susceptible (Gautam and Gupta, 1989). Four sprays of mancozeb (0.25 per cent) or carbendazim (0.05 per cent) control this disease under field conditions. Two sprays of benomyl at fortnightly interval have been found effective for management of disease (Sharma *et al.*, 1986).

Sprays of bavistin, benlate, difolatan ziram and fernacol have been found best for reducing disease incidence in strawberry.

(c) Marssonina Blotch

Host: Apple

The disease appeared in 1992 in some apple growing localities of Himachal Pradesh. The cause of the disease could not be worked out for 6 years and the various reasons such as nutritional disorders, nematodes, and soil factors were supposed to be associated with the problem. The disease resulted in leaf fall in mid season during June to August and only immature fruits are seen there on the trees. It assumed serious proportions on the lower areas and ill managed orchards where the orchard micro climate was wet with high humidity and at shady locations. Delicious cultivars were worst affected by this disease during 1996 and well managed orchards were less affected (Sharma and Gautam, 1997).

The detailed investigations on the etiology of the disease conclusively proved that Marssonina blotch is responsible for premature leaf fall (Sharma and Kaul, 1997). The disease affects fruit size, development of colour, and lesions on fruits, reducing its quality and marketability. Almost 80 per cent orchards have been affected by this disease (Sharma, 2000) covering more than 70,000 ha (Sharma *et al.*, 2004).

Disease symptoms appear both on leaves and fruits. On upper surface of mature leaves initially dark green spots becoming dark brown in colour appear on lower portion of the tree. Appearance of large number of lesions turn leaf colour to yellow, which develop abscission layer and fall of pre-maturely. Small dot like structures (acervili) are also seen in such lesions (Sharma *et al.*, 2004). Dark brown, oval, depressed spots (3-5 mm) develop on fruits which later on turn almost black dotted over with small black acevuli in these lesions.

Marssonina coronaria (*Diplocarpon mali*) was isolated and identified from infected leaves during 1997 and 1998 (Sharma, 1999).The causative fungus *Marssonina coronaria* (*Phyllachora pomigena* syn, *Marssonina mali*) has been isolated in pure culture and pathogenicity has been proved (Sharma, 2005). The fungus produces typical hyaline, bicelled guttulate conidia on short clavate conidiophores in acervuli with an average size of 20-24 x 6.5-8.5 µm. There is a wide variability in its morphology with respect to size and shape of the conidia collected from different geographical locations of the state. The fungus produced small, brown to dark brown colonies which were round, subgloboid and wrinkled. The pathogen grows best at 20°C and pH 5.0 (Sharma and Sharma, 2003). The perfect stage of pathogen *Diplocarpon mali* is found in over wintering leaves and it also perennates in leaf litter in the form of conidia, which showed upto 20 per cent viability in the next season (Sharma *et al.*, 2003 b).

The prevalence of disease is influenced by leaf aging followed by continuous average temperatures above 20°C and 70 per cent relative humidity for 5-6 days. Prolonged hot and dry weather adversely affect primary infection under favourable weather condition, the defoliation begins within 3-5 days, whereas under interrupted/ unfavourable conditions, the infected leaves are retained upto 3 months (Thakur *et al.*, 2003). Sharma (2003) reported that disease spread at a faster rate during July-August at average temperature ranging between 23.5-25.4°C and frequent rains (3-6 rainy days/week) of moderate to high intensity and is positively correlated with relative humidity and rainfall.

Management

Protective sprays of copper oxychloride (0.35 per cent), zineb (0.3 per cent), dodine (0.075 per cent), carbendazim (0.05 per cent), dithianon (0.05 per cent), mancozeb flowable (0.35 per cent), companion (0.25 per cent) and ziram (0.3 per cent) provide complete control of the disease (Sharma, *et al.*, 2003b; Sharma *et al.*, 2004). However, most of antiscab sterol inhibiting fungicides, sulfur and captan fail to control the disease. An integrated spray schedule consisting of spray at pink bud, petal fall, fruit set (pea size), fruit development-I (walnut size), fruit development-II (20-35 days) after walnut stage and pre-harvest stages with mancozeb (0.3 per cent), carbendazim or thiophanate methyl (0.05 per cent), dodine (0.075 per cent), mancozeb (0.3 per cent) carbendazim/thiophanate methyl (0.05 per cent) and mancozeb (0.3 per cent), respectively has been recommended for controlling this disease (Sharma, 2005; Verma and Sharma, 2003).

Amongst protective spray schedules, a schedule comprising three consecutive sprays starting with mancozeb (0.25 per cent) + carbendazim (0.05 per cent) at fruit development stage-I (20 days) after walnut stage followed by two more sprays, first with zineb (0.3 per cent) at fruit development stage-II (20 days after satge-I) and later ziram (0.3 per cent) at pre-harvest (20 days prior to fruit harvest) stage are quite effective (PDC = 99.4 per cent) and highly economical (C B ratio 1:5.92).

Similarly, a curative spray schedule comprising of three consecutive fungicidal sprays beginning with mancozeb (0.25 per cent) + carbendazim (0.05 per cent) at the first appearance of the disease followed by another two sprays with carbendazim (0.05 per cent) and ziram (0.3 per cent) at an interval of 14 and 21 days, respectively, was also quite effective (Sharma, 2003; Sharma and Bhardwaj, 2003). Earlier, maximum disease control was obtained with carbendazim + mancozeb followed by single application of carbendazim and mancozeb. Five sprays of carbendazim + mancozeb were superior to all other treatments in the management of the disease. All the commercial cultivars *viz.* Red Delicious, Royal Delicious, Rich-a-Red, Golden Delicious and Red Gold are highly susceptible. Cultivars TEW, Winter Delicious and Granny Smith are moderately resistant to disease under field conditions (Sharma, 2003).

(d) Powdery Mildew

Hosts: Apple, pear, stone and nut fruits

Powdery mildew occurs in all major temperate fruits and nutsapple and is especially serious in semiarid regions. Losses from the disease vary depending upon the inherent susceptibility of the cultivar, environmental conditions, and management practices. The disease can cause extensive foliar infections in dry years, following mild winters, or on seedlings and grafted plants of highly-susceptible cultivars. Mildew is usually severe on emerging saplings from the seed and young leaves and young terminal shoots remain susceptible throughout the growing season. Expanding young plant tissues are the most susceptible to pathogen attack. Further light intensity significantly affects the appearance and development of disease and it develops intensively in the shade environmental conditions (Miltich, 2012).

On the freshly infected leaves, the mildewed area appears on the lower surface as indistinct white patch. Later, it is more readily visible and may result in curling of leaves. Eventually, these patches may turn pale purple. The disease symptoms are observed immediately after bud-burst when all the freshly produced tissues are completely mildewed and resulting flowers and leaves appear as white rosette. Infected lateral buds may produce completely mildewed side shoots in the same manner. From the mildewed surfaces numerous spores are released which infect young unexpanded leaves, young buds, lateral buds on new shoots and growing points. When the terminal buds begin to grow in early spring, the fungus advances with the new succulent growth. Infected terminal shoots appear stunted and young shoots may be killed outright in the spring, or may survive the season and die in late fall or winter.

Powdery mildew is caused by fungus *Podosphaera leucotricha* (Ell. and Ev.) E.S. Salmon. The fungus consists of weft of white hyphal threads and draws nutrition through the haustoria from the infected plant parts. It produces chains of conidia at the host surface giving powdery appearance to the leaf surface. The fungus overwinters as mycelium inside infected buds. As these buds open in the spring, all of their parts become covered with a powdery coating of spores. The spores are easily windblown and they serve to infect new leaves and shoots.

Management

The management practices can be summarized as:

Winter and summer pruning of silvered twigs up to 1-2 inches before spray. All silvered areas of shoots should be removed by cutting few millimeters behind the limit of the diseased area. During this practice shaking of twigs should be avoided to prevent the spread of free mature conidia from leaf surface.

Fungicidal sprays, *e.g.*, carbendazim, dinocap, bitertanol, triforine, penconazole, difenconazole etc. at 7 days interval have been reported to control powdery mildew. Routine spraying helps to reduce secondary infections, protecting the very young buds being laid down for next year and keeping the leaves clear of mildew patches. Although sulphur based preparations may provide cheap and best control of mildew but some varieties are 'sulphur shy' and are injured when sprayed with lime sulphur. Four sprays of carbendazim (0.05 per cent), binapacryl (0.1 per cent), or Morocide (0.1 per cent) given at dormant, bud swell, petal fall and fruit development stages are effective in reducing the disease. Some EBI fungicides such as tetraconazole (1.0l/ ha), triadimenol (0.18 l/ha), hexaconazole (0.28 and 0.34 l/ha) and bitertanol (0.4 l/ ha) sprayed at 15 days interval have been reported to control powdery mildew.

Delicious group of apple varieties is highly susceptible to this disease, however, crab apple cultivars, *viz.* Renetka, Purpurnaya, Yantarka, Altarkya, Dolgoe and Saynets Karchenio and two apple cultivars Maharaja Chunth and Golden Chinese are resistant to powdery mildew.

Coccinellid *Halyzia hauseri* feeds on the mycelium of *Podosphaera leucotricha*. *Ampelomyces* a common mycoparasite of powdery mildews over-winters as resting hyphae in the dried powdery mildew mycelia covering the shoots and in the parasitized ascomata of *P. leucotricha* on the bark and the scales of the buds. Soon after the bud burst both the fungi start their life cycle but *Ampelomyces* slowly follows the spread of its myco-host on infected apple leaves (Szentivanyi and Kiss, 2003).

(e) Rust

Hosts: Apricot, plum, peach and almond

Rust occurs on almost all the stone fruits. It is particularly harmful in warmer areas. It is caused by *Tranzschelia discolor*. Rust pustules start appearing in June at temperature 17.3-32.4°C. Symptoms appear on both surfaces of leaves as pale yellowish spots, which later change into bright yellow islands prominent on upper surface only. On lower surface, cinnamon brown dusty pustules appear. Severe infection leads to defoliation. On fruits, water soaked dark green spots appear and their growth gets arrested. The spots become sunken, which are deeper green than the surrounding tissue and later turn deep yellow. Twig infection is uncommon on peach, but on plum, the bark gets infected during the early spring when buds are swelling.

The occurrence of rust in almond has been recorded in 1985 and the pathogen was identified as *Tranzschelia discolor* f. sp. *Dulcis* (Sharma *et al.*, 1987). The occurrence of Himalayan rust caused by *Gymnosporangium cunninghamianum* and its life history on *Pyrus pashia* was reported quite early. This pathogen was also reported on

Cotoneaster bacillaris from Dalhausie. Rust development is positively correlated with humidity and rainfall and negatively with temperature (Ram *et al.*, 2003). The maximum temperatures (21.6-28°C) during July-August trigger the disease severity wherein rain and relative humidity had positive and insignificant role.

Management

Following control measures have been suggested for the control of rust. Spray Bordeaux mixture (1 per cent) during February and March. Pre-symptom application of Anvil and Impact (0.05 per cent) and Dithane M-45 (0.25 per cent) at 21 days interval provided best control of rust (Bhardwaj, 1991).The disease can be controlled by spraying SAN-619 (Cyproconazole) (Ram *et al.*, 2000). Application of Delan (0.05 per cent) at bud swell and after petal fall proved promising for controlling rust, leaf spots (*Phyllosticta prunicola* and *Cercospora* spp.) and powdery mildew mildew (*Sphaerotheca pannosa*) on peach, plum, apricot and almond (Thakur *et al.*, 1991). Sprays of triadimefon (0.2 per cent), carbendazim (0.1 per cent) or bitertanol (0.2 per cent) at 15 days interval effectively control rust (Sharma and Gupta, 1993). The disease is best controlled with a single spray of mancozeb (0.3 per cent), prophylactically at 20-23°C at all locations in H.P. (Thakur and Xu, 2004).

The rootstocks St. Julian K (M), St. Julian A (G) and *Prunus multiflora* were free from rust infection whereas myrobalan B showed resistant reaction (Sharma *et al.*, 1988). Frog More Early, Kaisha and Shakarpara were free from rust whereas Palampur Special and Parmarosa rated as highly resistant while Bailadi, Kaisi and Ladakhi as moderately resistant (Sharma *et al.*, 1987, Ram *et al.*, 2000). Although most of the almond cultivars are susceptible, however, a few have been reported resistant in dry temperate zone and mid hills of Himachal Pradesh. These include California Paper Shell (Bhardwaj, 1989), Thin Shelled, Badamjor Spillo No.2, Local Kangra Selection, Local Selection-6, Afghanistan Seedling, Tree No. 106, Badamjor Spillo No. 2A (Sharma *et al.*, 1989), Bruce, Drake and Tree No. 96 (Sharma and Gupta, 1990).

(f) Scab

Hosts: Apple and pear

Apple scab was reported to be prevalent in Kashmir and Lahore in its conidial stages ,*i.e. Fusicladium dendriticum* Fuck. by Nath in 1935. None of orchards surveyed in Kashmir Valley in 1975 and 1976 were free from scab and all the cultivars were found susceptible (Gupta and Lele, 1980). Later scab was reported from Himachal Pradesh. Scab was detected in the first week of July, 1977 in Bhawani orchard of Mauholi village (ht. 2422 m amsl) of Kotkhai (Gupta, 1978). Occurrence of *Spilocaea pomi* stage of *V. inaequalis* from Nilgiri hills in the garden of a private grower at Adha Sholai, Ooty. Occurrence of scab in Uttar Pradesh, Sikkim and Arunachal Pradesh in 1983. Pear scab resembles apple scab in many respects and is caused by a closely related fungus *Venturia pirina*. It attacks limbs, fruits and succulent young twigs. The disease is severe in humid areas.

The symptoms of scab are produced on all the aerial parts like leaves, petioles, blossoms, sepals, fruits and pedicels but less frequently on bud scales. The first infections often occur on the leaves surrounding the flower buds. The appearance of dull, smoky areas on the leaf, 8 to 18 days after infection, is the first evidence of the disease. The early symptoms normally appears as small dull olive-green patches mainly on the underside on young leaves during the early spring. Later, as the disease progress, darker spots develop on both the surfaces of the leaves. Young lesions are velvety brown to olive-green and have indistinct margins. When several lesions coalesce on young leaves they become curled, dwarfed and are distorted. Small superficial lesions or large patches develop on the fruit surface. Fruit may become infected at any time in their development. Typical fruit lesions are distinct, almost circular, rough-surfaced, olive-green spots up to 3/4 inch in diameter.

The apple scab fungus overwinters in infected leaves that have fallen on the ground. Fruiting bodies are produced within the dead leaf tissue. As spring approaches these begin to mature and produce spores (ascospores) that are discharged into the air currents and carried to the developing apple buds. Three main criteria, *viz.* quality of primary inoculum (quantity and relative maturation of pseudothecia), tree phenology and the occurrence of infection periods (Mill's criteria) form the basis of apple scab forecasting in Himachal Pradesh and other parts of the world. For this purpose electronic devices such as apple scab predictor has been put into use to measure each components contributions quantitatively. Apple scab forewarning service has been provided to the orchardists through personal contacts or by blowing characteristic signaling, telephonic communication, messages flashed through 'All India Radio' to undertake immediate sprays or reschedule already recommended spray programme (Gupta and Thakur, 1992). The main source of primary inoculum of scab is ascospores being discharged from the pseudothecia embedded in the over wintered leaves. Weather plays a decisive role in the perennation of apple scab fungus in perfect stage on fallen leaves and subsequently in the events leading to initiation and build up of the disease.

Management

Scab infections may be prevented by applications of fungicides at regular intervals throughout the growing season. The objective is to provide a protective coating that will inactivate any spores landing on the fruit and foliage. The critical period for scab control is early in the season from bud emergence through flowering and fruit setting period. If scab infection can be prevented during the time of ascospores are discharged from the fruiting bodies in the fallen leaves, the disease cycle is broken and no further source of infection remains for the rest of the season. However, if control is not obtained, and leaf and fruit infection does occur, then conidia are produced on the lesions and scab will remain a constant threat all season, whenever, wet weather occurs.

Later on Gupta (1979) reported that Daconil, Cuman L., Euparan M, Syllit, Bavistin, MBC, Aureofungin, flowable sulphur, RH-2161 and Dithane M-45 when used in spring and summer sprays provided almost complete control of both leaf and fruit infection. Difolatan (0.15 per cent) gave 81 and 90 per cent control of leaf and fruit scab, respectively. Postharvest sprays of Bavistin delayed the ascospore discharge and acted as antisporulent for wood lesions. Urea spray (5 per cent) prior to leaf fall in autumn provided complete eradication of perithecial stage in comparison to other treatments. Thiophanate methyl allowed the development of more perithecia yet was able to suppress ascospore discharge. Adequate and timely sprays consisting of a single leaf fall spray of urea (5 per cent) and alternate sprays of a number of fungicides during on season were very effective in bringing down scab infection to a negligible level in Himachal Pradesh.

Timely application of 4 fungicides in 6 sprays initiated with the maturity and discharge of ascospores in relation to tree phenology was able to provide almost complete control of the disease. He has further observed the suppressive and delaying effect on ascospore discharge of the pathogen. Difolatan sprays provided excellent control of primary scab and almost very good control of secondary scab. Urea spray (5 per cent) just before leaf fall had deleterious effect on ascospore productivity and perethecial development due to decomposition of leaves. It reduces the pseudothecia and ascospore production by 85.03 and 95.84 per cent, respectively (Sharma and Sharma, 1988). Effective and economic control strategy should include single leaf fall spray of urea followed by 4-5 sprays alternatively of contact and systemic fungicides. Timely application of Bavistin, mancozeb, captafol and Cuman L in 6 sprays provided almost complete control, however, captafol caused russetting of delicious apple fruits on shady parts of the tree. Natural cleaner washing and waxing of apple fruits increased the storage life and reduced the inoculum load of the fungus (Gupta and Gupta, 1990).

Later on control strategy was devised which consists of a single spray of urea (5 per cent) in autumn and 6 spray of different fungicides *viz.* difolatan, bavistin, dithane M-45 in one season. It was also economical providing 8 times more profit from the crop. It reduced the disease 97.97 and 93.30 per cent on leaves and fruits, respectively (Sharma and Sharma, 1989). Baycor (72 h), Saprol and Venturol (48 h) and bavistin (24 h) showed curative activity applied after inoculation of fungus. However, bavistin inactivated established latent lesions reducing inoculum pressure and improving chances of better scab control with succeeding sprays (Gupta and Verma, 1985).

Protective fungicides like captafol, captan, mancozeb, zineb, folpet, dithianon and a sterol inhibiting fungicide bitertanol at higher concentration provide effective control of primary scab and secondary scab and secondary scab at reduced rate of application in four sprays (Gupta and Verma, 1988). EBI fungicides had high level of pre-symptom activity whereby lesions developed on the leaves were a typical, chlorotic or reddish brown flecks such atypical lesions have reduced viable sporulation. Curative sprays of myclobutanil (72 h) and bitertanol (96 h) provide control of primary and secondary scab on leaves and fruits. Sterol-inhibiting fungicides (flutriafol, hexaconazol, myclobutanil and fenarimol) have excellent curative activity even after 72 h of inoculation (Khosla and Gupta, 1992). Flutriafol is superior to all other EBI followed by fenarimol in inhibiting spore germination (Khosla and Gupta, 1992). EBI fungicides had also better protective activity than conventional protectants. This activity in case of clortriafol, penconazole + captan, prochloraz and fenarimol was upto 120 h which was at par with dodine and carbendazim under controlled conditions providing 90 per cent control of primary and secondary scab. Sterol inhibiting fungicides and dodine were effective in controlling established scab lesions on apple foliage which were as effective as protective and curative spray programme (Sharma and Gupta, 1995). Penconazole + captan (0.2 per cent) persisted on the foliage of Red Delicious upto 120 h and fenarimol (0.03 per cent) and hexaconazole (0.05 per cent) or (0.075 per cent) upto 96 h, however, former combination was less effective under field conditions (Thakur and Malhotra, 1995). Post symptom eradicative sprays of mycolobutanil (0.075 per cent), dodine (0.1 per cent), prochloraz (0.075 per cent), fenarimol (0.04 per cent) and bitertanol (0.1 per cent), were very effective for checking established scab lesions (Thakur *et al.*, 1995). Fenarimole was more effective when applied 48, 72 and 96 h and hexaconazole at 24, 48 and 74 h after inoculation (Thakur *et al.*, 1994), but beyond this interval their activity either decreased or not reliable (Khosla and Gupta, 1992). Myclobutanil (45 ppm) had maximum inhibitory effect on sporulation and colony growth while bitertanol (187.50 ppm) showed only a temporary inhibition of growth and sporulation (Thakur *et al.*, 1992; Gupta and Thakur, 1992; Gupta, 1990). The application of mixture of stobilurin fungicides (Kresoxim methyl and trifloxystrobin), Bellis (Pyraclostrobin and boscalid) and difenconazole reduce sporulation and active scab lesions for 3-5 weeks after treatment as compared to DMI fungicides (Pasini *et al.*, 2006).

Application of urea (5 per cent) at leaf abscission in the last week of October on affected apple leaves prevent ascospore discharge, accelerate litter (Verma and Gupta, 1992) decomposition of fallen leaves and increase microbial population. As many as 59 organisms were isolated including potential biocontrol fungi, *viz.*, *Athelia*, *Chaetomium globosum*, *C. olivaceum*, *Myrothecium roridum*, *Trichoderma harzianum*, *T. hirsutum*, *T. viride* and *Trichothecium roseum*. Significant inhibition of *V. inaequalis* was observed with *Athelia* sp., *C. globosum* and *C. olivaceum*, which also resulted in reduction in pseudothecial initials and production of asci and/ascospores (Thakur and Sharma, 1999). The apple scab has not been observed in Kullu Valley since 1999. Fifty-nine microorganisms including 31 fungi, 13 bacteria and 15 actinomycetes were isolated from urea treated apple leaf litter. The population of cellulytic and amylolytic fungi, bacteria and actinomycetes exhibited a rapid increase with enhanced leaf litter decomposition.Three fungal isolates including *Chaetomium globosum*, *C. olivaceum* and *Athelia* sp. showed antagonistic microbial association to *V. inaequalis* (Thakur and Gupta, 2000). The probable reason for zero level incidence of this disease primarily includes the early leaf fall and prevalence of adverse climatic conditions. It contributed in reducing primary source of inoculum (pseudothecia) through early leaf fall caused by severe incidence of premature defoliation disease from 1994 onwards, whereas on the adverse conditions did not allow the primary infection to manifest (Sharma, 2003). A multiscale selection approach is required for choosing cultivars for integrated, and especially for organic growing methods in order to facilitate successful apple growing in environmentally benign production systems (Holb *et al.*, 2012). The hail protection nets using black hail netting increase the intensity of apple scab than white hail netting in apple cultivars Royal Gala and Fuji (Bogo *et al.*, 2012).

(g) Shot Hole/Stigmina Blight

Hosts: Almond, apricot, plum, peach and cherry

Shot hole also called as stigmina blight, is caused by *Wilsonomyces carpophilus* (*Clasterosporium*, *Coryneum* and *Stigmina*). The disease was recorded first time in India on cherry trees in Kashmir Valley during March-May 1987 and is very serious in Kinnaur district of Himachal Pradesh where the maximum disease incidence has been recorded in Nichar. The fungus attacks the twigs, leaves, and unopened buds. On leaves, dark brown scattered lesions enlarge rapidly and loss of the dead areas results in the formation of shot holes. On twigs, small, purplish raised spots appear, which expand into necrotic cankers. On fruits small, circular, deep purple and sunken lesions developed (Puttoo and Razdan, 1991). High temperature and prolonged rain favoured the spread of the disease.

Management

The disease can be effectively controlled by spraying difolatan (0.2 per cent) followed by foltaf, blitox, dithane M-45 and dithane Z-78 (Sharma and Gupta, 1988). Spray application of captan (0.2 per cent) at leaf fall in autumn (November) and at bud swell stage in spring gave 96 per cent control of the disease. Difolatan (0.2 per cent) sprayed at 3 different stages, *i.e.* leaf fall, bud swell and half fruit development was effective in controlling the disease followed by foltaf, blitox, dithane M-45 and dithane Z-78 (Sharma and Gupta, 1988). Except cvs. Nari all other cultivars are highly susceptible to this disease. In peach, out of thirty-seven genotypes of peach Nemaguard, EC09015 and IC-34993 showed below 10 per cent Stigmina leaf spot severity (Verma *et al.*, 2006).

3. STEM AND ROOT DISEASES

(a) Canker Diseases

Hosts: Apple, apricot and pear

Localized infection on the trunk or branches of trees is usually the result of canker diseases. Cankers may appear as sunken areas or swollen areas on infected branches. Most canker diseases are caused by fungi, but there are a few bacterial canker diseases. The fungi that cause canker diseases will of ten colonize the bark of newly planted trees under water stress. If a canker disease girdles the entire branch or trunk, all growth beyond the canker will die.

The incidence of cankers range from 1-25 per cent in Himachal Pradesh.The important cankers reported from India are in the following Table 2. Amongst 14 fungi causing cankers, smoky blight canker and stem brown causes maximum damage (Gupta and Sharma, 1979). Out of these cankers *S. malorum* (smoky blight) was found most widely distributed in mid hills of Kullu valley of Himachal Pradesh. Disease incidence has been found to reduce with rise in altitude. Apple plants 2-3 years were rated as susceptible and 6-8 years age as highly susceptible irrespective of altitude. Possible infection sites are unprotected pruning wounds and exposed sun burnt surfaces of the limbs. Delicious cultivars are highly susceptible.

Nectria canker is a perennial canker. Callus tissue forms at the margin of the canker each year. As this is repeated over several years, concentric rings of callus tissue are visible at the site of infection.When the canker girdles the trunk, all the tissue above the canker dies. Nector spore play important role in survival of the fungus. Fungus also survives through the deep seated mycelium in the infected wood of apple tree (Verma, 1988). Initial entry of the fungus takes place through the lenticels in the bark. Atmospheric temperature (28°C) and relative humidity (90 per cent) favour the development of cob-web, corticium or basidial and sterile nodular stages during July and August, whereas temperature 24°C and 75 per cent RH during September and October/support the formation of nector pustules. The disease is more severe on the branches facing North-East, South East and South-West. Orchard above, 1900 m.a.s.l are free from the disease. Carotenoid pigment synthesis increases with the addition of pectin (2 per cent) or maltose or asparagines (0.5 per cent) in the medium. Isolates of *Corticium salmonicolor* from distinct geographical areas are highly variable with respect to their ability to use C, N and vitamin sources and pathogenicity on apple.

Table 2: Various Cankers Reported from India

Name of Canker	*Causal Organism*	*Reference*
European canker of apple	*Nectria galligena*	Agarwal and Gupta (1968)
Silver leaf diseases of apple	*Stereum purpureum*	Agarwal and Gupta (1968)
Apple tree canker	*Cylindrocarpon mali*	
Smoky blight of apple	*Sphaeropsis malorum*	Agarwala and Gupta (1971)
Stem black of apple	*Coniothecium chomatosporum*	
Brown canker of apple	*Botryosphaeria ribis*	
Black rot canker	*Botryosphaeria quercum*	Gupta (1977)
Nail head canker of apple tree	*Nummularia disoreta*	Gupta and Agarwala (1970)
Coniothyrium canker	*Coniothyrium fuckelii*	Gupta and Agarwala (1973); (Shandilya and Agarwala, 1974)
Cytospora canker of die back	*Cytospora ambiens* Perfect stage *Valsa ambiens*	
Pink disease	*Corticium salmonicolor*	
Fusicoccum canker	*Botryosphaeria dothidia*	
Canker and die back disease of apple	*Haplospora malorum*	Shandilya (1984)
Cryptosporiopsis canker	*Cryptosporiopsis corticola*	Sharma and Bhardwaj (2003)

Management

Spraying Blitox (0.25 per cent), Dithane Z-78 (0.25 per cent) and Captan (0.2 per cent) were helpful for controlling apple tree canker (*Cylindrocarpon mali*). Wound dresser Santar-A was most effective followed by Chaubattia paste and Bordeaux paint. Aureofungin paint was least effective against the same disease. The wound is first cleaned, disinfected with mercuric Chloride (0.1 per cent) and dressed with wound dresser in dormancy. Smoky blight canker was effectively controlled by Captan (0.2 per cent) and Cuman (0.1 per cent) applied as spray (Gupta, 1975). Cowdung paint (soil + lanolin + linseed oil and bavistin) proved to be the best for the control of these cankers (Gupta and Sharma, 1979; Sharma, 1988; Sharma and Gupta, 1977). The pink canker can be controlled by cutting back of affected branches or excising lesions beyond the extent of infection and painting the cut surfaces either with Chaubattia paste or Bordeaux paint or calixin paint.

(b) Collar Rot

Hosts: Apple and pear

Collar rot continues to be a major cause of tree death in orchards. It is often observed on 3- to 8-year-old trees. Certain rootstocks are more susceptible to the pathogen than others. The disease is often observed in low-lying areas of orchards with heavy, poorly drained soils. The importance of this disease has increased with the introduction of more dwarfing rootstocks. The first report of occurrence of disease was from Mashobra Fruit Research Station in 1960. The occurrence of collar rot and root rot causing considerable loss to bearing trees in scattered orchards in village Roga, Simble, Bhutti, Kotgarh, Thanedhar, Shimla, Kotkhai and Mashobra has been reported. Both the diseases are prevalent in orchards planted on newly cleared soils with a pH 5.4-7.0. Different isolates of the fungus vary in their virulence however, variability exists in *P. cactorum*. These isolates also differ in sensitivity to dithane M-45 and Apron SD-35, thus indicating the existence of variability in the pathogen (Garg and Gupta, 1990).

On the above ground parts, the leading edge of the lesion is irregularly mottled and attains olive green, orange or brown colour which merges into the healthy tissues.Collar rot symptoms mostly appear near the graft union of the plant, but sometimes may originate on the lower trunk or at the pruning wounds. The bark in the diseased plant parts is discoloured and turns into a moist rot in which necrotic tissues eventually turn dark brown and develop an alcoholic odour.Wood beneath the necrotic bark is stained dark brown and such staining may extend beyond the edge of the lesion. Bark above ground dries out and eventually splits away from the wood. In the subsequent years infected trees become unthrifty with pale, sparse leaves; fruits remain small and colours prematurely and there is little or no shoot growth. Death of the tree usually follows, especially where the tree has been girdled.

Seven Pythiaceous fungi, *viz., Phytophthora cactorum, P. citricola, Pythium ultimum, P. debaryanum, P. vexans, P. polytylum* and *Pythium* sp. have been isolated from the soil collected from different apple orchards in Himachal Pradesh near the base of collar rot affected trees. All these fungi are pathogenic on young seedlings. The association of *P. polytylum* was reported for the first time in the world and *P. ultimum, P. debaryanum* and *Phytophthora citricola* from India (Rana and Gupta, 1984). *Phytophthora cactorum* (Lebert and Cohn) is the main causal agent. Other pythiaceous fungi namely *Phytophthora syringae, P. megasperma, P. cryptogea, P. cinnamomi, P. drechsleri, P. cambivora, P. citricola, P. pseudosyringae* sp. *nov, Pythium ultimum* and *P. vexans* have also been found to be associated with similar symptoms.

Management

Number of preventive measures such as sanitation to keep the orchard disease free, non-piling of soil around tree trunk, bud union at higher level from soil, maintenance of proper drainage, correction of acid soils by applying lime, avoiding spread of pathogen by irrigation water has been suggested. However, curative measures include scrapping the affected portion and exercising diseased roots, disinfection of exposed surface with 50 per cent methylated spirit in water or 0.1 per cent mercuric chloride followed by application of Chaubattia paste (4:4:5).

Rootstock M-4 is hardy enough for preventing entry of the causal organism. Budding than grafting has been found useful. Rootstock budded 9 inches above the soil level escape invasion of the fungus. Bacterial antagonists such as *Bacillus subtilis* and *Enterobacter aerogenes* and their culture filtrate inhibited growth of the fungus *in vitro*. Combination of *B. subtilis* and metalaxyl was also best in inhibiting fungal growth (Gupta and Garg, 1990). Some of rootstocks such as M2, M4, MM110, MM 114, MM115 and crab apple are resistant to *P. ultimum* infection. The total and ortho dihydroxy (OD) phenols, activity of phenylalanine ammonia-lyase (PAL), tyrosine ammonia lyase (phenylalanine ammonia lyase) and Polyphenol oxidase (Catechol oxidase) (PPO) were maximum in resistant rootstock (MM115) as compared to susceptible one such as MM103, MM104 (Sharma, 2003).

Aggallol-3 has found effective in killing the fungus both in dry and wet soil upto a depth of 15 cm (Rana and Gupta, 1979) and reduced lesion size when applied after infection (Rana and Gupta, 1981). Metalaxyl, fostyl aluminium and Dithan M-45 inactivate the fungus in soil at a depth of 8-10 cm and check the infection in young seedlings (Rana and Gupta, 1983).

Thus, the management strategies can be summarized as:

Elevating susceptible scions out of the soil splash zone by high grafting at 40-70 cm above soil level.

The problem of collar rot in young trees can be avoided by planting them at properly drained sites.

Planting grasses, which help to reduce water infiltration, is useful in controlling collar rot. The use of non-host crops as green manure in apple orchards has also been suggested.

Removal of soil from the tree base has been advised to reduce infection where the scion is at or below soil level.

Sanitation practices such as removal of crop refuse and fallen fruits prevent the accumulation of *P. cactorum*.

Avoidance of injuries to stem, removal of weeds and long grass from tree basins are other practices, which are helpful in reducing inoculum.

Pre-plant dipping of seedlings in fungicides may help to reduce inoculum density on infested rootstocks.

Injection of effective fungicides such as metalaxyl and fosetyl-Al into the soil near the crown region of the tree helps to reduce crown rot.

Copper fungicides can also be applied as paint or paste on the affected parts for treating wounds. The diseased portion is exposed by removing the soil at the crown region and is scarified upto healthy tissue, disinfected with methylated spirit or mercuric chloride and after its evaporation Bordeaux paint, copper paint, metalaxyl paint or Chaubattia paste is applied.

In the infested areas rootstocks such as M2, M4, M9, M26, MM114 and MM115 are planted which are resistant to collar rot.

Enterobacter aerogenes, a potential bacterial antagonist has been isolated from orchard soil which inhibit *P. cactorum* in culture and is unaffected by the fungicides such as metalaxyl, fosetyl Al and mancozeb. This bacterium reduced root rot of apple seedlings due to *P. cactorum* and the number of viable oospores of the pathogen in sterilized field soil.

(c) Crown Gall and Hairy Root

Hosts: Peach, nectarine, apricot, plum, cherry, apple, pear and quince.

The name 'crown gall' refers to the most common site for infection, on the stem just above or below the soil line, although infections higher on the stem or on roots are also common. The crown gall bacteria infect through wounds, especially those near the soil line arising from cultivation and on roots broken at planting time. Survey of different nurseries in different agro climatic zones had revealed wide spread distribution of this disease in the state (Jindal and Sharma, 1988a, b). Jindal *et al.* (1994) reported that crown gall and hairy root are major problems of most of stone and nut fruits nurseries in Himachal Pradesh. The occurrence of crown gall on cherry peach, and plum in Himachal Pradesh was first of all recorded by Durgapal (1974). Biovar-I of the bacterium was found associated with the disease (Jindal and Sharma, 1988). *Agrobacterium tumefaciens* isolated from plum can attack *Aesculus indica, Cotoneaster microphylla, Mirabilis jalapa, Prunus cerasoides, P. cornuta* and *Pyrus pashia,* which are new additions to host range (Durgapal and Rai, 1973).

Initially crown gall symptoms appear as small, round, soft overgrowths on the stem and roots, particularly near the soil line. As the galls or tumors start increasing in size, the surface becomes convoluted having dark brown outer tissues due to the death of peripheral cells. Anatomically, a gall is a sphere of disorganized vascular and parenchymatous tissues, having soft and spongy to hard texture, depending upon the amount of vascular tissues they contain. The tumor may surround the stem or root, or may be connected to the host through a narrow neck of tissues. Tumors on pome and stone fruits become woody and hard, appearing knobby and reaching upto 30 cm in diameter. Certain tumors on herbaceous host plants and vines are spongy and may become detached from the plant. Few tumors may rot during fall and winter and reappear during spring.

Once a gall is initiated, the bacteria play no further part in its growth and development. In this respect, crown gall has been called a plant cancer. Secondary galls may develop a considerable distance from the initial infection. These too are formed in the absence of the crown gall bacteria, apparently due to a tumor-inducing substance, produced at the site of the original infection.The crown gall is caused by a gram-negative soil bacterium *Agrobacterium tumefaciens* (Smith and Townsend) Conn. belonging to Rhizobiaceae family *Agrobacterium rhizogenes* (Riker *et al.*) Conn is the causal organism of hairy root.

Management

Plants infected with crown gall cannot be cured. Cutting off galls will not cure the problem and may actually make it worse. New planting stock found to be infected should be destroyed. Crown gall bacteria survive in plant debris in the soil for several years, so planting sites with a history of crown gall are likely to continue to show problems. Nursery beds with a history of crown gall problems should be planted to non-host plants, fallowed, or fumigated. Since the crown gall bacterium has a very wide host range including not only woody plants but many vegetables and weeds, rotation crops should be carefully chosen.

Good sanitation and cultural practices are important in reducing the disease which includes discarding all symptomatic planting stock, budding rather than grafting, choosing a rootstock with low susceptibility, and adopting management practices that minimise wounding.

Susceptible nursery stock may not be planted in areas known to be infested with *A. tumefaciens*. Do not purchase plants having galls on stem or roots. Such infested fields should be planted with monocots for a few years before they are planted with nursery stock.

Root damaging insects in nurseries should also be controlled to reduce disease incidence.

Nearly cent per cent control of crown gall is achieved by dipping seeds or root system of nursery seedlings or rootstocks upto crown region in *Agrobacterium radiobacter* strain K84 suspension. Bacterial suspension can also be applied as soil drench. This biocontrol agent occupies the infection sites and produces a bacteriocin agrocin 84 which is inhibitory to the pathogen. After continuous use of strain K84, *A. tumefaciens* developed resistance to this biocontrol agent through the transfer of agrocin 84 resistant genes from K84 strain to the pathogen. At present, a genetically engineered strain K-1026 of *A. radiobacter,* which lacks the ability to transfer its agrocin 84 resistance gene to pathogenic *A. tumefaciens* strains is recommended in Australia. A non-pathogenic agrocin producing native isolate of *Agrobacterium tumefaciens* strain UHFBA-218 has been tested as a biological control agent against peach crown gall. No incidence of crown gall has been observed in the three seed dip treatments *i.e.* UHFBA-218 ½ h dip and one hour shade dry, stratified seeds dipped for ½ h in 0.1 per cent suspension of strain UHFBA-218 and K84 and then shade dried for 1h before sowing as compared to 14.76 per cent incidence in untreated plants (Aditi *et al.*, 2015).

(d) Gummosis/Bacterial Canker

Hosts: Cherry, peach, nectarine, plum, apricot and almond.

Gumming is common on stone fruit trees, whether on trunks, limbs, twigs, or fruit when injuries and fruit malformations occur. Agarwala (1971) reported the occurrence of bacterial canker/gummosis on almond, apricot, peach, and plum and reported the association of *Pseudomonas* sp. The average incidence of 65.6, 37.5, 25.0 and 85.5 per cent have been recorded in peach, plum, apricot and cherries, respectively.

The disease occurs on branches, twigs, buds, leaves, and fruit. The most conspicuous symptoms are the cankers and dying branches girdled by them. Disease symptoms depend upon plant age, host variety, pathogen strain involved, tissue invaded and other edaphic factors. The most characteristic symptom is the formation of cankers accompanied by gum exudation. The cankers on the twigs are darkened areas often at the base of leaves and flower buds, at the sites of pruning wounds including those made during removal of suckers and at the bud union. Canker tissues become slightly sunken usually in late winter or early spring and the infected areas turn dark brown. The cortical tissues of the cankered areas are bright orange to brown. Gum often exudes from cankers, especially during spring and rainy season. Cankered tissues that do not produce gum are comparatively soft, moist, and may have a sour smell. Terminal shoots or twigs of cankered trees may show dieback symptoms. When the trunk or branch is girdled by a cankerous wound, the diseased lateral branches or trunk start drying and dies within weeks.

Gummosis of stone fruits have been found in all the stone fruits in almost all the orchards located in mid hills and dry and cold zone of Himachal Pradesh. *Pseudomonas syringae* pv. *syringae* have been found associated with the gumming lesions on apricot stem. The bacterium also produced reddish purple lesions on the leaves and fruits. *Xanthomonas campestris* pv. *pruni* was found associated with gumming lesions on fruits of almond, apricot, peach and plum. *Xanthomonas campestris* pv. *pruni* was found to be the cause of the disease in more than 90 per cent fruits showing gumming lesions. Gumming was more pronounced during November, December, and March at an average temperature of 18-5-20.4°C coupled with 60-65 per cent relative humidity and cloudy weather (Verma and Singh, 1986).

Management

The management practices include:

Only healthy budwood should be used for propagation.

Susceptible varieties should be propagated only on resistant rootstocks and grafted as high as possible.

New orchards established under marginal climatic and soil conditions are more prone to this disease.

Orchards should not be located in areas where trees are subjected to frost injury, water logging and prolonged drought period. When selecting species and varieties for a particular area, it should be ensured that dormancy requirement is properly met even in mild winters.

Pruning in early summer, rather than in autumn or winter, lesser the chances of disease spread.

Cankers on trunks and large branches can be checked by cauterization with a hand held propane burner. The flame is aimed at the canker and especially its margins for 5 to 20 seconds until the underlying tissues begin to crackle and char. This treatment is carried out in early to mid spring. The treatment can be repeated if required after 2 to 3 weeks.

Partial control of the canker phase of the disease is based primarily on protective copper sprays in autumn and in the spring before flowering. The sprays of fixed copper fungicides and Bordeaux mixture protect trees from initial infection but cannot prevent the canker phase once infection has occurred.

Painting of gumming branches with lime powder results in the appearance of gummosis with in 5-7 days. The best time for painting is during the rainy days and at the initiation stage of gummosis.

(e) White Root Rot

Hosts: Apple, stone fruits, strawberry, kiwifruit *etc.*

The *Dematophora necatrix* Hartig is a soil inhabiting non-sporulating fungus attacking large number of plant species consisting of annuals, perennials, and forest and fruit trees. All the temperate fruits, *viz.* apple, apricot, cherry, kiwifruit, pear, plum, peach, pecan nut, persimmon, strawberry, walnut are attacked by this fungus. The perfect stage of the fungus, *Rosellinia necatrix* (Hart.) Berl. is known to occur only in few geographical regions of the world otherwise the serious losses are only due to imperfect stage. The pathogen is widely distributed in temperate and sub-temperate regions of the world attacking more than 160 plant species belonging to 45 families. The disease is of major importance in fruit trees where the grown up bearing trees are destroyed. The fungus was first encountered in India in Uttaranchal hills in 1929 and later on in Himachal Pradesh. The perfect stage does not occur in India. The annual losses in apple alone in India has been worked out to be about Rs. 1.3 million which are certainly much more as the disease is now reported to occur in almost all the apple growing=regions of the country. The disease is endemic at 1,800-2,400 m altitude with soil pH 6.1-6.5 (Gupta *et al.*, 1995).

The onslaught of the fungus can be easily diagnosed by observing the under and above ground parts of the plant. The earliest above ground manifestation bronzing of the leaves, diminution in their size and a stunted tree growth resulting in the progressive decline in the vigour as a whole or certain branches. In fruit trees these symptoms are usually associated with a heavy blossom and fruiting next year. In the succeeding years, few leaves emerge and much of the fruits fail to reach maturity and dying back of branches is quite evident. Infected trees often persist for 2-3 years depending upon the infection intensity in the roots but severely infected trees may succumb within a single season (Gupta, 1977). The lateral roots turn dark brown and are covered with a greenish gray or white mycelial mat having a flocculent web of whitish strands or ribbons during monsoon season. As the disease progresses, all the roots are attacked and fibrous root system almost disappears.

Management

White root of apple is best controlled by practicing preventive as well as curative measures consisting of cultural, biological, chemical methods and resistant rootstocks. Further, the management of the disease is tedious due to devouring nature of fungus and deep-seated infection in the roots; but following the different strategies help to reduce the disease severity. Various practices recommended to combat the disease have been reviewed by Sharma *et al.* (2005) and are being discussed here-under.

The cultural practices recommended for the management of white root rot of apple from time to time are hot water treatment of infected seedlings at 45°C for one hour before plantation, digging isolation trenches and removal of rotten roots followed by application of disinfectant paste (Gupta and Sharma, 1999). In order to save the trees in the early stages of attack the diseased roots should be trimmed off and destroyed. Replanting should be done in clean soil and infested soil should be fallowed with frequent cultivation to starve out the infection hyphae. Burning and heat drying of infected roots have also been reported to increase the life of the trees. Soil moisture plays a major role in affecting the intensity of the disease, hence, ill drained soils should be improved by following central drainage system and spread of the pathogen should be avoided during irrigation or other cultural operations. Poor soil aeration also affect the disease intensity and winter injury enhance the disease.The acidic soils should be corrected by applying lime for some years.

The success in controlling the soil borne disease lies in the fact that the required and effective levels of the fungicide reach the point of infection in the soil and prevailing soil temperature. With the advent of systemic fungicides, the old recommendations were replaced with fungicides such as benomyl, aureofungin, carbendazim etc. In nursery the control of root rot was achieved by three applications of carbendazim (0.1 per cent) in the root zone of the diseased plants (Gupta and Verma, 1976; Sharma and Gupta, 1997). Gupta (1977) reported that infected apple plants in pots recovered with the application of carbendazim (0.1 per cent) even when all the roots were destroyed, thereby, resulting in development of new roots

Almost all the rootstocks have shown susceptibility to the pathogen, however, in the field the degree of susceptibility varies (Gupta and Verma, 1978). Adult plant resistance beyond 20 years of age has also been reported and trees become quite hardy after 30 years of age (Agarwala and Sharma, 1966). All the wild species of *Malus* spp. occurring in northern and north eastern Himalayas of India and other related species of temperate fruits such as *Crataegus* sp., *Cydonia* sp. *Prunus* sp., *Pyrus* sp. and *Sorbus* sp. have been evaluated at Shimla and only eleven spp. including *Cydonia oblonga*, *M. baccata* (Japan), *M. baccata* (*Rohroo*), *M. baccata* (*khrot*), *M. floribunda*, *Prunus persica*, *P. cornuta*, *P. cerasoides* and *Pyrus pashia* var. *kumabni* survived for 6 years and classified as resistant (Ram, 1982). Sztejnberg and Jabareen (1985) evaluated 468 seedlings of *Diospyros virginiana* rootstocks and 235 of *D. kaki* rootstocks and reported that 5.7 and 24.5 per cent were resistant to *D. necatrix*, respectively. Sharma and Gupta (1996d) observed some resistance against the pathogen in four rootstocks namely MM102, MM 109, M 16 and MM 104, but untill and unless, a rootstock with complete resistance is developed, the root rot problem will persist. Gupta and Sing (1985) studied the effect of culture filtrate of the pathogen on the apple seedlings and reported that rootstocks M9, MM106 and MM102 were less susceptible where as Rus Pippen was resistant. All the kiwifruit varieties are susceptible root rot (Sharma *et al.*, 2003). Kishore *et al.* (2005) identified and developed a new rootstock of apple which is resistant to wooly aphid, powdery mildew and tolerant to white root rot and suitable for hight density plantings at lower elevations. The resistance sources have been presented in the Table 3.

Table 3: Resistance Sources of Temperate Fruits against *D. necatrix*

Category	*Genotype*	*References*
Resistant	*Cydonia oblongata* *Malus baccata* (Japan) *M. baccata* Rohroo *M. baccata* Khrot *M. floribunda* *Prunus persica*, *P. cornuta* *P. cerasoides*, *Pyrus pashia* var. *kumaoni* Rus Pippin	Ram (1982); Gupta and Singh (1985).
Moderately resistant	*Prunus mariana*, *P. myrobalan*, *Corylus avellana*, *Malus floribunda*, *M. toringoides*, Spe 262, Sel 69	Sousa (1985); Gupta and Singh (1985); Agarwala and Sharma (1966); Denardi and Berton (1995).
Moderately susceptible	M9, MM106, MM102, Pusa Seb Mulvrinth-1 (PSM-1)	Sousa (1985); Kishore *et al.* (2005)

Use of various antagonists in the management of disease has been explored and few universal antagonits such as *Trichoderma viride*, *T. harzianum*, *Enterobacter aerogenes* (Gupta and Jindal, 1989) have been found to protect the plants from *D. necatrix*. Bacterial antagonists like *Enterobacter aerogenes* and *Bacillus subtilis* have been found to inhibit *Dematophora necatrix* and reduce disease up to 45 days of inoculation (Gupta and Jindal, 1989). Sharma and Gupta (1996c) reported that repeated application of these antagonists enhanced their efficacy against the pathogen and effect is long lasting. Pre-inoculation application (15 days) of *T. harzianum* and *T. viride* also reduced seedling mortality in pots and naturally infested soils (Sharma and Sharma, 2001).

Soil amendment with neem cake, *Daphne* leaves and deodar needles has been reported to reduce the incidence of the disease and delayed the symptoms appearance for 30 days (Gupta and Jindal, 1989; Sharma and Gupta, 1996b). Seedlings grown in AM and residue amended pots had more VAM colonization and spore population than non-amended pots. Further, there was significant increase in height, leaf area and dry weight of apple seedlings after 45 days of pathogen inoculation.

The sensitivity of *D. necatrix* to heat depends upon the duration of exposure at particular temperature. If the temperature is above 40°C the mycelium dies with in two hours where as below 40°C four-hour duration is required under *in vitro* conditions (Sharma and Sharma, 2002). However, in temperate regions where the summer temperatures are low, soil solarization for 8-12 weeks has been found effective reducing the population of *Dematophora necatrix* and other soil fungi, bacteria and actinomycetes and controlling white root rot disease in nursery. The soil temperature under tarped was 10-7°C higher than untarped soil (Sharma *et al.*, 2003, 2005).

3. CONCLUSION

The scab, which has appeared in epidemic form dominates the in-depth studies carried out on temperate fruit disease and voluminous publications has come out. Most of the aspects such as cause, diagnosis, role of meteorological factors and their utilization for disease prediction, and protocol for its remedy have been generated.

With the rein in of the disease, it has disappeared for the last 5-6 years from the apple plantations in India. The environmental vagaries, adoption of management practices and climate change has given birth to another problem called premature leaf fall, which is causing havoc in apple industry since 1992 and of late is considered as cure for previously existing scab disease. This disease is still at infancy and warrants in depth research in future. Sometimes leaf spots caused by other fungi also ravage the temperate fruit orchards but on erratic basis. Soil-borne problems still remain to be unsolved though some temporary relief providing technologies at infancy of infection do exit. At present most of the orchardist's whose plants start dying at the age of 30-40 years however, fruits laiden on these branches are eager the solution. The solution lies in the resistant rootstocks which we are lacking. Post-harvest management of diseases is another area which has been invented thoroughly and management practices standardized, however, recurrence is there and more sophisticated solutions are required. Amongst temperate fruits apple is the only fruit which has transformed the economic status of the Himalyan orchardist's in the upper and middle heights, as a result the marginal area farmers have also developed enthusiasm for planting apple and are facing acute canker problems and degeneration of plants at young age. The problems has further been aggravated with shift in climate. So far fourteen cankers have been reported only on apple but the management practices are only of routine nature. So cankers are also required to be studied in depth for effective and lasting management. Crown gall and hairy roots are serious problems however, due to indirect losses they cause to the plants; the detailed studies are lacking and require attention. Gummosis particularly on fruits of almond and peach is of serious concern and lacks proper and detailed in depth studies. Virus diseases are going to be very serious in future. In most of the cases studies have been restricted only to occurrence and identification hence requires detailed investigations.

4. REFERENCES

Agarwala, R.K. 1971. Bacterial gummosis of stone fruits. *Himachal Hort*, **2**: 49-51.

Agarwala, R.K. and Gupta, G.K.1968. New record of pome and stone fruit tree diseases from Himachal Pradesh -1. *Indian Phytopath.*, **21**: 299-301.

Agarwala, R.K. and Gupta, G.K.1971. Canker disease complex of apple trees. Second International symposium on Plant Pathology, Indian Phytopathological Society, New Delhi, January 1971, p. 160 (Abstr.).

Agarwala, R.K. and Sharma, V.C. 1968. Storage rot diseases of apple. *Indian Phytopath.*, **21**: 294-298.

Badiyala, S.D. and Lakhanpal, S.C. 1994. Performance of some low chilling peach (*Prunus persica* Barsch) cultivars under Paonta valley conditions of H.P. *Himachal Journal Agri. Res.* **20**: 57-63.

Bhardwaj, L. N. 1991. Chemical control of almond rust in dry temperate zone of Himachal Pradesh. *Plant Dis. Res.* **6**: 89-90.

Bhardwaj, L.N. 1989. Reaction of almond cultivars to rust under dry temperature conditions of Himachal Pradesh. *Indian J. Plant Path.* **7**: 141.

Bhardwaj, L.N. and Agarwala, R.K. 1985. Evaluation of fungicides against leaf spot of apple caused by *Pestalotia hartigii* Tub. *Himachal J. Agri. Res.* **11**: 44-46.

Bhardwaj, L.N. and Agarwala, R.K. 1986. Chemical control of leaf spot of apple caused by *Phyllosticta mali* prill. and Del. *J. Ins. Agri. Ani. Sci.* **7**: 115-118.

Bhardwaj, L.N. and Agarwala, R.K. 1986. Effect of some important edaphic factors on the incidence of seedling blight of apple caused by *Sclerotium rolfsii* Sacc. In: *Advances in Research on Temperate Fruits*. Chadha, T.R., Bhutani, V.P. and Kaul, J.L. (Eds)., UHF, Solan, pp. 403-406.

Bogo, A., Casa, R.T., Agostineto, L., Goncalves, M.J. and Rufato, L. 2012. Effect of hail protection nets on apple scab in 'Royal Gala' and 'Fuji' apple cultivars. *Crop Protec.* **38**: 49-52.

Denardi, F. and Berton, O. 1995. Resistance of apple rootstock to fungus *Rosellinia necatrix* Prill. In the Midwest of Santa Catarina. *Agropecuaria Catarinense,* **8**: 25-27.

Durgapal, J.C. 1974. A preliminary note on bacterial diseases of temperate plants in India. II. Bacterial diseases of stone fruits. *Indian Phytopath.* **24**: 379-382.

Durgapal, J.C. and Rai, K. 1973. Host range of *Agrobacterium tumefaciens* isolated from plum. *Indian Phytopath.* **26**: 594-596.

Garg, R.C. and Gupta, V.K. 1990. Sensitivity of *Phytophthora cactorum* isolates to fungicides. *Indian J. Pl. Path.* **8:** 73-74.

Gupta, V. K. and Jindal, K. K. 1989. Management of white root rot of apple (*Rosellinia necatrix*) by biological and cultural methods. *Proceedings of International Horticulturlal Congress*, Firenze, Italy Vol. 2: Abstract No. 3229.

Gupta, V.K. and Verma, K.D. 1976. Research report, pome, stone and nut fruits. ICAR Tech bulletin. 9p.

Gupta, V. K. and Verma, K. D. 1978. Comparative susceptibility of apple root stocks to *Dematophora necatrix*. *Indian Phytopath.* **31**: 377-378.

Gupta, G.K. 1977. Occurrence of black rot canker (*Botryosphaeria quercum*) of apple. *Progr. Hort.* **9**: 29-30.

Gupta, G.K. 1978. Present status of apple scab (*Venturia inaequalis*) in Himachal Pradesh and strategy for its control. *Pesticides*: **12**: 13-14.

Gupta, G.K. 1979. Role of on-season, post harvest and pre-leaf fall sprays in the control of apple scab (*Venturia inaequalis*). *Indian Journal of Mycology and Plant Pathology* 9: 141-149.

Gupta, G.K. 1990. Apple scab – a review. 1990. *Indian J. Pl. Protec.* **18:** 157-165.

Gupta, G.K. and Agarwala, R.K. 1970. Nail head canker of apple trees. *FAO Plant Protection Bulletin* **18**: 145-147.

Gupta, G.K. and Agarwala, R.K. 1973. Canker diseases of apple trees in Himachal Pradesh. *Indian J. Myco. Pl. Pathol.* **3:** 189-192.

Gupta, G.K. and Gupta, S.K. 1990. Effect of neutral cleaner washing and wax coating on the storage life of scab infected apple. *Indian J. Myco. Pl. Pathol.* **20**: 51-53.

Gupta, G.K. and Lele, V.C. 1980. Prevalence, distribution and intensity of apple scab in Kashmir valley. *Indian J. Agri. Sci.* **50:** 045-050.

Gupta, G.K. and Verma, K.D. 1985. Post symptom antisporulent activity of fungicides on apple scab foliage lesions. *Indian J. Agric.Sci.***55**: 381-382.

Gupta, G.K. and Verma, K.D. 1988. Efficacy, phytotoxicity and post-symptom activity of fungicides, in the control of apple scab (*Venturia inaequalis* (Cke.) Wint.). *Pesticides,* **22**: 40-43 and 47.

Gupta, G.K., Agarwala, R.K. and Dutt, K. 1979. Stigmina blight of peach and its control in Himachal Pradesh. *Hort. Adv.* **9**: 49-51.

Gupta, V. K. 1977. Root rot of apple and its control by carbendazim. *Pesticides,* **11**: 49-52.

Gupta, V. K. 1988. Effect of soil temperature on the degradation of different fungicides. *Indian J. Mycol. Plant Pathol.,* **18**: 70-71.

Gupta, V. K. and Sharma, S. K. 1996. Evaluation of apple rhizosphere resident mycoflora against *Dematophora necatrix*, the incitant of white root rot. *IPS 48th Ann. Meeting and Naional Symosium on Management of Threatening Plant Diseases of National Importance* at PAU, Ludhiana, Feb. 14-16, 1996, pp. 28.

Gupta, V. K. and Sharma, S.K. 1999. Soil -borne diseases of apple and their management, pp. 89-104. In: *Diseases of Horticultural Crops: Fruits,* L.R. Verma and R.C. Sharma (Eds.), Indus Publishing Co., New Delhi.

Gupta, V. K., Gupta, P. K. and Gupta, S. K. 1995. Relative efficacy of some chemicals in the management of white root rot of apple. *Plant Disease Res.,* **10**: 83-85.

Gupta, V.K. and Sharma, S.K. 1979. Practical way of combating apple canker in Himachal Pradesh. *Indian Phytopathology* 32: 163 (Abstr.).

Gupta, V.K. and Singh, N. 1985. Factors affecting production of toxic compounds in culture filtrate of *Dematophora necatrix*. *Himachal J.Res.,* **11**: 16-20.

Gupta, V.K. and Thakur, V.S. 1992. Integrated management of apple scab disease. *Information Bulletin, AICRP on Apple Scab*: 23.

Holb, I. J., Dremak, P., Bitskey, K. and Gonda, I. 2012. Yield response, pest damage and fruit quality parameters of scab-resistant and scab-susceptible apple cultivars in integrated and organic production systems. *Scientia Hort.* **145**: 109-117.

Jain, S. S. 1961. Root and collar rot diseases in Himachal Pradesh. *Himachal Hort.,* **2**: 19-23.

Jindal, K.K. and Sharma, R.C. 1988. Crown gall and hairy root in fruit plant nurseries in Himachal Pradesh. *Proceedings of 5th International Congress of Plant Pathology,* Kyoto, Japan, August 20-22, 1988, p. 98 (Abstr.).

Jindal, K.K. and Sharma, R.C. 1988. Occurrence of biovar-1 of *Agrobacterium tumefaciens* on almond in India. *Indian Phytopathol.*, **41**: 614-615.

Jindal, K.K., Gupta, V.K. and Sharma, S.K. 1994. Bacterial diseases of stone and nut fruits: their management. *Indian Hort.*, **39**: 15-16.

Kaul, J.L. and Munjal, R.L. 1982. Apple losses in Himachal Pradesh due to post harvest fungal pathogens. *Indian J. Mycol. Plant Path.*, **12**: 209-210.

Khosla, K. and Gupta, G.K. 1992. Curative activity of sterol-inhibiting fungicides against leaf infection of apple by *Venturia inaequalis* (Cke.) Wint. *Indian J. Plant Protec.*, **20**: 50-57.

Kishore, D. K., Pramanick, K. K. and Sharma, S. K. 2005. Significance of crab apples in the improvement of apples. *Acta Hort.*, **696**: 39-41.

Miletich, N., Tamas, N., Vuksa, P., Pfaf-Dolovac, E. and Dolovac, N. 2012. The influence of shading on the development of *Podosphaera leucotricha* under field conditions. *Bulgarian J. Agric. Sci.* **18**: 178-184.

Nath, P. 1935. Studies on the diseases of apples in Northern India. II. A short note on apple scab due to *Fusicladium dendriticum* Fuck. *J. Indian Bot. Soc.*, **14**: 121-124.

Nautiyal, M.C., Prakash, S. and Kumar, A. 1988. Screening of peach (*Prunus persica*) cultivars for their reaction against leaf curl (*Taphrina deformans*). *Indian J. Agric. Sci.*, **58**: 575-576.

Pasini, M., Posenato, G. and Tosi, L. 2006. A new fungicide against apple scab. *Informatore Agrario* 62: 69-71.

Puttoo, B.L. and Razdan, V.K. 1991. Stigmina blight of cherry – a new record from India. *Plant Disease Res.*, **6**: 60.

Ram, R. D. 1982. Evaluation of wild germplasm of temperate fruits against root rot of apple caused by *Dematophora necatrix* Hartig. *Progr. Hort.*, **14**: 249-251.

Ram, V., Sharma, U. and Bhardwaj, L.N. 2003. Effect of weather parameters on the development of apricot leaf rust (*Tranzschelia discolor* f.sp. *armeniacae*). *Plant Dis. Res.*, **18**: 180-182.

Ram, V., Sharma, U. and Bhardwaj, L.N. 2000. Management of apricot leaf rust. *Plant Disease Res.*, **15**: 113-115.

Raina, G.L., Bedi, P.S. and Dutt, S. 1971. Occurrence of core rot of apple in nature in the Kullu valley of Himachal Pradesh, India. *Plant Disease Rep.*, **55**: 283-284.

Rana, B.S. and Jain, V.B. 1992. Screening of peach varieties against leaf curl disease and its control. *Indian Journal of Mycology and Plant Pathology*, **22**: 87 (Abstr.).

Rana, K.S. and Gupta, V.K. 1981. Cause and control of collar rot of apple in India. *Indian Phytopath.*, **34**: 17-19.

Rana, K.S. and Gupta, V.K. 1983. Changes in phenolics and phenol oxidizing enzymes in apple rootstocks in relation to *Phytophthora cactorum*. *Himachal J. Agric.Res.*,**9** : 51-56.

Roy, A.J., Seth, J.N. and Shah, Achal. 1985. Comparative effect of fertilizers and different fungicidal treatments on the incidence of leaf spot diseases of apple var. Red Delicious. *Prog. Hort.*, **17**: 241-244.

Shandilya, T.R. 1974. A new canker and die back disease of apple from Himachal Pradesh. *Indian J. Mycol.Plant Path.*, **4**: 213.

Shandilya, T.R. and Agarwala, R.K. 1974. Perfect stage of Cytospora ambiens causing canker of apple trees in India. *Indian J. Mycol. Plant Path.*, **4**: 206.

Sharma, S.K and Gupta, V.K. 1996a. Effect of post infection application of chemicals on white root rot of apple in nursery. *National Symposium on Holistic Approaches of Plant Disease Management* INSOPP, UHF-Nauni 26 (Abstr.).

Sharma, S.K. and Gupta, V.K. 1996b. Management of root rot of apple through soil amendments with plant materials. *Plant Disease Res.*, **10**: 164-167.

Sharma, S. K. and Gupta, V. K. 1996c. Potential of *Trichoderma* species as an antagonist against *Dematophora necatrix* causing white root rot of apple. *IPS, Annual meeting and National Symposium*, PAU Ludhiana, Feb. 14-16, 1996, pp. 27-28.

Sharma S.K. and Gupta V.K. 1996d. Reaction of apple root stocks to *Dematophora necatrix* causal agent for white root rot of apple. *Nat. Symp. Plant Dis. Holistic Approaches Manage INSOPP UHF*, Nauni 24 (Abstr.).

Sharma, M. and Sharma, S.K. 2002. Effect of thermal regimes on the survival of *Dematophra necatrix*, causing white root rot of apple. *Indian Phytopath.* **55**: 310-312.

Sharma, M. and Sharma, S.K. 2001. Biocontrol of *Dematophra necatrix*, causing white root rot of apple. *Plant Dis. Res.* **16**: 40-45

Sharma, A., Gupta, A.K., Khosla, K. Singh, D. and Mahajan, R. 2015. New insights for the management of crown gall of peach through native agrocin producing non-pathogenic *Agrobacterium tumefac*iens strain UHFBA-218. In: Souvenir and Abstracts (Sharma, SK., Bharat, NK., Gupta, D. Verma, S. and Gautam, HR., Eds.), 2015. *National Symposium on Holistic Approaches for Plant Health Management*, Nauni, Solan, HP, India 28-29 Sept. p. 165.

Sharma, I. M. and Bhardwaj, S.S. 2003. Cryptosporiopsis canker –a new disease of apple in Himachal Pradesh and its management. Presented in Annual Meeting (IPS-NZ) and Symposium on *"Challenging problems in horticultural and forest pathology" Department of Mycology and Plant Pathology*, Dr. Y S Parmar University of Horticulture and Forestry, Nauni, Solan on 14-15th Nov. 2003. pp. 14.

Sharma, I.M. 2003a. Influence of environmental factors on the development of pre-mature defoliation diseases caused by *Marssonina coronaria* in apple and its management. *J. Mycol.Plant Path.*, **33**: 89-95.

Sharma, I.M. 2003. Relationship between phenylpropanoid metabolism and resistance to *Pythium ultimum* Trow in apple root stocks. *J. Mycol. Plant Path.*, **33**: 14-18.

Sharma, I.M. 2003c. Probable reasons for zero level of apple scab incidence in Kullu Valley of Himachal Pradesh. *Indian J. Mycol.Plant Path.*, **33**: 96-100.

Sharma, I.M. and Badiyala, S.D. 1994. Susceptibility of peaches to *Taphrina deformans* in relation to blooming, environmental factors and genetic inheritance. *Indian Phytopath.*, **47**: 65-71.

Sharma, I.M. and Bhardwaj, S.S. 2003. Efficacy and economics of different fungicide spray schedules in controlling premature defoliation diseases in apple. *Plant Disease Res.*, **18**: 21-24.

Sharma, I.M. and Gupta, V.K. 1988. Incidence of Stigmina blight of apricot caused by *Stigmina carpophila* (Lev.) Ellis in dry temperate zone of Himachal Pradesh and its chemical control. In: *Tree Protection*, Gupta, V.K. and Sharma, N.K., (Eds.), Indian Society of Tree Scientists, Solan, pp. 444(Abstr.).

Sharma, I.M. and Gupta, V.K. 1990. Field evaluation of some almond cultivars against rust in dry temperate zone of Himachal Pradesh. *Indian J.Mycol. Plant Path.*, **20**: 167-168.

Sharma, I.M. and Gupta, V.K. 1993. Field evaluation of some fungicides against almond rust in Himachal Pradesh. *Indian J.Mycol. Plant Path.*, **23**: 203-205.

Sharma, J.N. 2000. Marssonina blotch – a new disease of apple and its control. *Indian J. Plant Prot.*, **28**: 100-101.

Sharma, J.N. 1999. Marssonina blotch – a new disease causing premature leaf fall in apple. *Indian Phytopath.*, **52**: 101-102.

Sharma, J.N. 2005. Scab and premature leaf fall diseases of apple and their management. **In**: *Challenging Problems in Horticultural and Forest Pathology*, Sharma, R.C. and Sharma, J.N. (Eds.), Indus Pub. Co. New Delhi, pp. 11-31.

Sharma, J.N. and Gautam, D.R. 1997. Studies on premature leaf fall in apple – a new problem. *Indian J. Plant Protec.*, **25**: 8-12.

Sharma, J.N., Gupta, D., Bhardwaj, L.N. and Kumar, R. 2005. Occurrence of Alternaria leaf spot (*Alternaria alternata*) on apple and its management. **In:** *Challenging Problems in Horticultural and Forest Pathology*, Sharma, R.C. and Sharma, J.N. (Eds.), Indus Pub. Co. New Delhi, pp. 25-31.

Sharma, J.N., and Kaul, J.L. 1997. Studies on the development of storage scab in apple and its management. *Indian Phytopathol.*, **50**: 396-401.

Sharma, J.N., Sharma, A. and Sharma, P. 2004. Out-break of Marssonina blotch in warmer climates causing premature leaf fall problem of apple and its management. *Acta Hort.*, **662**: 405-409.

Sharma, J.N., Sharma, P.K. and Sharma, A. 2003b. Studies on epidemiology and management of Marssonina blotch, the cause of premature leaf fall in apple. **In:** *Challenging Problems in Horticultural and Forest Pathology*, Sharma, R.C. and Sharma, J.N. (Eds.), Indus Pub. Co. New Delhi, pp. 1-7.

Sharma, P.K. and Sharma, J.N. 2003. Studies on morphology and physiology of *Marssonina coronaria* causing premature leaf fall in apple. In: *Challenging Problems in Horticultural and Forest Pathology*, Sharma, R.C. and Sharma, J.N. (Eds.), Indus Pub. Co. New Delhi, pp. 9-17.

Sharma, R.C. and Sharma, Y.P. 1990. Physico-chemical alternations in *Taphrina deformans* infected peach leaves. *Indian Phytopath.*, **43**: 382-384.

Sharma, R.C., Gupta, V.K. and Agarwala, R.K. 1986. Epidemiology and control of Phyllosticta leaf spot of pear. *National Symposium in social forestry for rural Development*, Delhi University, Delhi, January, 1986, p. 12(Abstr.).

Sharma, R.C., Gupta, V.K. and Garg, R.C. 1988. Dormant applicantion of fungicides for the management of peach leaf curl. *Indian Phytopathol.*, **41**: 259-260.

Sharma, R.C., Jindal K.K. and Kaul, J.L. 1987. Reaction of apricot cultivars to *Tranzschella discolor* rust in India. *IUFRO Workshop on Agro forestry for Rural Needs*, New Delhi, February, 1987. p. 32 (Abstr.).

Sharma, R.C., Kaul, J.L. and Sharma, R.L. 1989. Effect of post-inoculation fungicidal dips on post harvest diseases of peaches. *Int. J. Tropical Plant Dis.*, **7**: 221-224.

Sharma, R.L. 1989. Control of storage decay of china pear fruit by pre-harvest applications of fungicides. *Indian J. Mycol. Plant Path.*, **19**: 140 (Abstr.).

Sharma, R.L. and Kaul, J.L. 1989. Effect of post harvest fungicidal treatments on brown rot of apricot. *Plant Disease Res.*, **4:** 54-58.

Sharma, R.D. 1981. Calixin cures leaf curl of peach (*Taphrina deformans*). *Pesticides*, **16**: 32-33.

Sharma, S.K. 1988. Prevalence of apple cankers in Himachal Pradesh and their control. *Indian J. Plant Path.*, **6:** 125-127.

Sharma, S.K. and Gupta, V. K. 1997. Efficacy of carbendazim for the control of white root rot of apple. *J. Soil Biol. Ecol.*, **17**: 120-123.

Sharma, S.K. and Gupta, V.K. 1977. Use of cowdung paint in controlling smoky blight canker of apple. *Indian Phytopathol.*, **30**: 568-569.

Sharma, S.K. and Sharma, V.P. 1989. Efficacy of fungicide sprays against apple scab in relation to tree phenology. *Plant Disease Res.*, **4**: 199-200.

Sharma, S.K., Kishore, D.K. and Pramanick, K. K. 2003a. Effect of soil solarization on soil microflora and survival of *Dematophora necatrix* in temperate fruit nurseries. *7th International Symposium on TZFT*, Dr. Y S Parmar University of Horticulture and Forestry, Nauni, Solan, October 14-18, 2003, and pp.143.

Sharma, S.K., Kishore, D.K. and Pramanick, K.K. 2003b. The onslaught of *Dematophora necatrix* causing white root rot of kiwifruit in India. *8th International Congress of Plant Pathology " Solving problems in the real world"* held at Christchurch, New Zealand, wef 2-7th Feb. 2003, Vol. 2, Abstracts of offered papers, pp.132.

Sharma, S.K., Kishore, D.K. and Pramanick, K.K. 2005. *Dematophora necatrix*— a serious pathogen of temperate fruits. *Challenging Problems in Horticultural and Forest Pathology* (Sharma, R.C. and Sharma, J.N. (Eds.), Indus Pub. Co. New Delhi, pp. 53-71.

Sharma, S.K., Kishore, D.K. and Pramanick, K.K. 2005. Effect of soil solarization on soil microflora and survival of *Dematophora necatrix* in temperate fruit nurseries. *Acta Hort.*, **696**: 381-386.

Sharma, V.P. and Sharma, S.K. 1988. Impact of scab spray schedule on important foliar diseases of apple. *J. Tree Sci.*, **7**: 7-9.

Singh, C., Sharma, R.P. and Lakhanpal, T.N. 1981. Host parasite relationship of peach plants infected with *Taphrina deformans*. *3rd International Symposium on Plant Pathology*, Indian Society of Phytopathology, New Delhi, December 14-18, 1981, pp. 191.

Singh, D. and Agarwala, R.K. 1987. Incidence of apple frog eye leaf spot (*Sphaeropsis malorum*) and its correlation with weather conditions. *Indian J. Mycol. Plant Path.*, **17**: 169-172.

Sousa AJTD. 1985. Control of *Rosellinia necatrix* causal agent of white root rot, susceptibility of several plant species and chemical control. *Eur. J. Plant Pathol.* **15**: 323-332.

Sztejnberg A and Jabareen H. 1985. Dematophora root rot disease in persimmon and studies on resistance of rootstocks to the disease. *Alon Hanotea*, **39**: 757-762.

Szentivanyi, O. and Kiss, L. 2003. Over-wintering of Ampelomyces mycoparasites on apple trees and other plants infected with powdery mildews. *Plant Pathol.* **52**: 737-746.

Thakur, V.S. and Xu, X.M. 2004. Epidemiology and management of almond rust (Tranzschelia discolor Dulcis) in Himachal Pradesh. *Adv. Plant Sci.*, **17**): 97-107.

Thakur, V.S. and Gupta, P.K. 2000. Studies on apple leaf microflora and its microbial relationship to the perfect stage of *Venturia inaequalis* arfter urea treatment. *Adv. Plant Sci.*, **13**: 399-412.

Thakur, V.S. and Malhotra, R. 1995. The protective action of ergosterol-biosynthesis inhibiting fungicides in the management of apple scab, *Venturia inaequalis* (Cke.) Wint. *Adv. Plant Sci.*, **8**: 157-163.

Thakur, V.S. and Sharma, R.D. 1999. Effect of urea on microbial degradation of apple leaf litter and its relationship to the inhibition of pseudothecial development of *Venturia inaequalis*. *Indian J.Agric. Sci.*, **69**: 147-151.

Thakur, V.S., Gupta, G.K. and Garg, I.D. 1992. Effect of post-infection application of fungicides on the apple scab pathogen, *Venturia inaequalis* (Cke.) Wint. *J. Phytopathol.*, **135**: 160-166.

Thakur, V.S., Gupta, P.K. and Malhotra, R. 1994. Management of symptoms developed after infection of scab (*Venturia inaequalis*) in apple (*Malus domestica*). *Indian J. Agric. Sci.* **64**: 117-122.

Thakur, V.S., Gupta, P.K. and Malhotra, R. 1995. Antisporulant activity of ergosterol biosynthesis-inhibiting fungicides in the management of apple scab, *Venturia inaequalis*. *Indian Phytopathol.*, **48**: 35-39.

Thakur, V.S., Gupta, V.K. and Garg, R.C. 1991. Management of fungal diseases of stone fruits with fungicides in Himachal Pradesh. *Indian J.Agric.Sci.*, **61**: 677-681.

Thakur, V.S., Sharma, N. and Vaidya, S. 2003. Effect of weather parameters on the epidemic of apple blotch in Himachal Pradesh. **In:** *Challenging Problems in Horticultural and Forest Pathology*, Sharma, R.C. and Sharma, J.N. (Eds.), Indus Pub. Co. New Delhi, pp. 19-23.

Verma, K.D and Gupta, S.K. 1992. Effect of pre-leaf-fall sprays of urea and fungicides in suppressing the ascigerous stage of apple scab pathogen in Himachal Pradesh. *Plant Disease Res.*, **7**: 68-70.

Verma, K.D. and Gupta, G.K. 1979. Studies on leaf spot of cherry (*Prunus avium*) caused by *Cercospora circumscissa* Sacc. and its control. *Prog. Hort.*, **10**: 57-62.

Verma, K.D. and Sharma, J.N. 2003. Scab and Marssonina blotch – threatening diseases of apple and their management. *Ann. Rev.Plant Path.*, **2**: 207-220.

Verma, K.S. 1988. Role of nector stage of *Corticium salmonicolour* in initiation of pink canker disease on apple in Himachal Pradesh. *Plant Disease Research* 3: 226-228.

Verma, K.S. and Singh. A. 1986. Incidence of bacterial canker of peach in Punjab. *J.Res.Punjab Agric.Univ.*, **23**: 74-77.

Verma, V. D., Sharma, S. K., Pradheep, K. and Rana, J. C. 2006. Spotting resistance sources against shot hole a fungal disease in peach germplasm. *Indian J. Plant Genetic Resour.*, **18** : 252-254.

Chapter 21

Forecasting of Plant Diseases of Temperate Fruits

Nissar A. Qazi

Division of Plant Pathology,
Sher-e-Kashmir University of Agricultural Sciences and Technology of Kashmir
Srinagar – 190 025, J&K
E-mail: qazinissar11@gmail.com

1. INTRODUCTION

Pome and stone fruits constitute predominant component of temperate horticulture. In India, Jammu and Kashmir is the most important horticultural state with unique temperate fruit production. Amongst the fresh and dry fruits, apple and walnut are economically most important fruit crops in the state. The production of these fruits contributes significantly to regional and national economies through national and international trade. Unlike other agricultural crops, fruit crops are grown as perennials and are severely vulnerable to diseases and pests particularly under temperate environmental conditions. Various diseases and pests inflict huge losses to fruit growers in terms of reduced yield and quality, besides cost of management. Thus, cultivation of these fruit crops for optimal production is highly technical and demands improved technical support for disease management.

The success of any plant disease control programme is primarily governed by a crop production system which fulfills the essential goals of pest management. The process starts from selection of site and disease free cultivar, layout of orchard with appropriate plant density, adoption of adequate fertilizer, irrigation and drainage systems, suitable canopy management required for optimum yield and air circulation to minimize leaf wetness duration and to facilitate better pesticide coverage, and monitoring for diseases/pests at periodical intervals.

2. FORECASTING AS COMPONENT OF INTEGRATED APPROACH

Like diseases of other plants, most of the diseases of temperate fruit crops have not been completely or effectively controlled with any single management strategy. Although, chemicals remain an important tool in plant protection programmes, yet a judicious use of fungitoxicants befits the principles of modern integrated disease management, as advocated in most developed countries (Funt *et al.*, 1990; Gleason *et al.*, 1994). This approach involves utilizing various tactics *viz.*, cultural, biological, host-plant resistance, chemical, *etc.*, that provide acceptable yield and quality at economic costs with minimum damage to the environment. These activities could directly target the pathogen and/or modify the environment to make it unconducive for disease development. However, the application of any of these management strategies necessitates acquiring the knowledge about the conditions/factors under which a pathogen, when in contact with a susceptible host, can infect and become established. Disease forecasting is partly about defining such conditions.

(a) System Approaches and Methods

Plant disease forecasting provides a clue when the disease is likely to go 'critical' and have economic impact or forecasting is about predicting when a disease is likely to affect the vital parts of the plant or likely occurrence of critical periods in its development. Therefore, forecasting systems utilize their component's data on epidemiology and aetiology of the diseases and pathogens as well as requires an understanding of such aspects (Hardwick *et al.*, 2006). Complete understanding of the components of plant disease or 'disease tetrahedron' *i.e.* interaction between pathogen, host, environmental and human activity (Zadoks and Schein, 1979), becomes important for controlling diseases effectively and economically. In other words, forecasting a disease is the part of applied epidemiology. Van der Plank (1963) defined epidemiology as science of plant disease in population. Kranz (1974) defined it as the science that deals with the increase or decrease of plant disease in a host population in a specified time and space. Hence, knowledge of epidemiology is necessary for accurate forecasting. For establishing the epidemiological requirements of any plant disease, it is pre-requisite to understand the biology and potentiality of a pathogen to infect and spread in a host and the factors associated with host, pathogen and environment.

The factors of epidemic and its components should be known in advance before forecasting is done. The information required for forecasting are broadly enumerated as under:

(i) Host factors

a. Prevalence of susceptible varieties in the given locality
b. Response of host at different phonological stages to the activity of pathogen; and
c. Density and distribution of host in a given locality. Monoculture of susceptible variety facilitates quick spread of an epidemic. Growing

susceptible varieties in scattered locations and that too in a limited area are less prone to epiphytotic.

(ii) Pathogen factors

a. Amount of primary inoculum
b. Dispersal of inoculum
c. Spore germination
d. Infection
e. Incubation period
f. Sporulation on the infected host
g. Dissemination of secondary spores

(ii) Environmental factors

Meteorological factors *viz.*, the amount and duration of rainfall, length of wet periods and temperature during wetness and incubation period play key role in infection process of pathogens and disease progress, besides influencing the developmental stage of host. However, in the field, temperature is rarely constant and shifts in temperature create a situation where an infection interval is not flagged because the average temperature may be too low while the actual temperature during a portion of wetness period may be sufficient for infection. Nowadays, weather stations are well equipped with sensors to record such fluctuations for calculating the infection periods and accordingly warrants control measures for economic gain. Under temperate conditions, these factors hamper the success of effective management strategy in areas which remain frequently wet particularly during spring months.

(b) Method of Detection and Quantification of Inoculum

In epidemics of air-borne plant diseases, the number of pathogenic spores landing on plant is indicative of the number of spores present in atmosphere above the crop. Therefore, for their detection and estimation, different types of spore traps are used *viz.*, Hirst's volumetric spore trap, Rotorod sampler or rotorod spore trap, Burkard's 7 day volumetric spore trap, etc. are used for quantitative sampling of atmosphere (number of spores per unit volume of air). Cylindrical rods or microscopic glass slides, smeared with grease, are also used to recording the time of spore arrival in a locality.

Disease forecasting models assist the fruit growers in estimating the likely appearance and subsequent progress of disease in their crops as well as in selecting and timing the preventative applications Disease forecasting has become an essential tool in successful disease management and a number of forecasting models have been developed during the last few decades (Orlandini *et al.*, 2008). At present, in areas where no disease forecasting models are available, the growers adopt a spray program based on weather forecasting. For improving their accuracy, various forecast models demand further investigations to correlate the level of cultivar susceptibility and the density of inoculum with the development of disease as in case of leaf curl of peach and nectarine. Some disease models require further investigations to clarify the relationship between the predicted risk, actual disease incidence and yield in order to determine the time when fungicide sprays are economically justified. Computer-assisted Decision Support Systems (DSS's) have proven to be reliable in reducing the use of plant-protection chemicals in many countries (Murali and Secher, 1996; Tamosiunas *et al.*, 2000; Rogers and Stevenson, 2006).

(c) Use of Computer and Information Technology

Computer modeling has provided a tool for using weather data to predict disease outbreaks. Traditional plant disease models have been developed only after generating data through pioneer extensive epidemiological studies. Usually these models have used accumulated hours of wetness duration combined with temperature requirements to predict the infection process and identify infection periods of high disease risk. These types of models use recorded weather data to track when enough favorable disease hours have occurred to warrant management action. These models have been part of integrated pest management control measures. The revolution in web-based technologies has led to great strides in the development and employment of decision support systems for growers and pest management specialists.

More than six decades ago, Miller and O'Brien (1952) considered warning service as an essential tool for plant disease management and highlighted practical implications of forecasting as well, when they stated that "Forecasting involves all the activity in ascertaining and notifying the growers of a community that conditions are sufficiently favourable for certain diseases, that application of control measures will result in economic gain, or on the other hand, and just as important that the amount of disease expected is unlikely to be enough to justify the expenditure of time, energy and money for control". With the advent of the information technology and improvised weather station systems, disease forecasting has become an important tool for decision making and for restricting chemical use without hampering the economic gain. Forecasts are also available as SMS messages. A number of descriptive and mathematical as well as computer-based disease forecasting methods/ models have been successfully developed for many diseases of field crops in various countries which include potato late blight, wheat brown rust, corn bacterial wilt (*Erwinia stewartii*), tobacco blue mould, sugarbeet root rot (*Aphanomyces* sp.), sugarbeet curly top, cucurbit downy mildew, *etc.*

Forecasting models based on established epidemiological aspects have also been developed and validated for major diseases of various temperate fruits in many countries. Some of the major diseases include apple and pear scab, apple sooty blotch and flyspeck, apple fire blight, grape downy and powdery mildew, cherry leaf spot, peach leaf curl and brown rot of stone fruits, etc. (Thomidis *et al.*, 2010). However, each disease model needs to be evaluated and validated for various locations experiencing specific environmental conditions.

3. EXAMPLES OF FORECASTING SYSTEMS OF DISEASES OF TEMPERATE FRUITS CROPS

Forecasting approaches of some of the major diseases of temperate fruit with special emphasis on forecasting of apple scab are elaborated as under:

(a) Apple Scab

Apple scab, caused by *Venturia inaequalis*, is the most serious fungal disease of apple world over, including temperate regions of India. In India, the disease occurred in an epidemic form first time in Kashmir in 1973 on 'Delicious' and other cultivars of apple (Joshi *et al.*, 1975) and in Himachal Pradesh in 1983 (Gupta, 1985). In 1974, apple scab was declared as a national diaease. Since then, studies on its various aspects, including epidemiology, have been conducted. The pathogen overwinters mainly sexually as pseudothecia on infected fallen leaves producing ascospores as primary inoculum for spring foliage, and sometimes asexually as conidia or mycelium on 1-2 years old twigs (Hirst and Stedman, 1962; Hill, 1975; Qazi *et al.*, 2008).

(i) Role of weather conditions and prediction of infection periods

Maturation, production and discharge of ascospores and conidia (secondary inoculum) as well infection processes are highly influenced by weather factors such as rain, temperature, wetness duration, and wind (Hirst and Stedman, 1962; Thakur and Khosla, 1999). Pseudothecia upon wetting liberate mature ascospores and major proportion of ascospores are released during day hours (MacHardy, 1996). The prediction models based on number of hours of leaf wetness with respect to temperature required for ascosporic and conidial infection in an orchard has been developed by Mills (1944). Most of the weather-based infection risk models are more or less modified Mills' model.

(ii) Assessment of primary inoculum

The amount of initial inoculum available in the orchard during spring is predicted by assessing the leaf scab in the fall and leaf litter density and is expressed as potential ascospore dose. Various mathematical models, like Degree Days Celsius (DDC), have been used to predict the pattern of ascospore maturity. Assessment of maturity and release of ascospores (traditionally or using spore traps) is used to predict risk of scab infection. In Kashmir, ascospores start maturing in last week of March to mid-April and symptoms appear after 13 to 17 days of ascospore release (Qazi *et al.*, 2008). The peak period of ascospores discharge usually occurs in 3rd to 4th week of April (bloom to petal fall stage) and their discharge ceases by 1st week of June in most parts of the valley. In Kashmir, annual spray schedules for the management of foliar diseases of apple have been developed based on epidemiological knowledge and evaluation of fungicides.

(iii) Spread of scab infection from source of inoculum

Primary lesions formed due to ascosporic infection, start producing conidia in enormous number. These conidia are wind dispersed or rain splashed and cause repeated cycles of infection throughout the growing season, depending upon the occurrence of infection periods. Investigations at SKUAST-Kashmir, Shalimar (Qazi *et al.*, 2008) revealed a marked delay in disease appearance as well as significant reduction in disease spread and progress of infection with increase in isolated distance (90 to 120 m) from inoculum source of *V. inaequalis* (Fig. 1). The investigations clearly revealed that only after the appearance of scab symptoms on apple plantation at or nearer to source of inoculum, the disease did spread to longer distances (90 to 120 m),

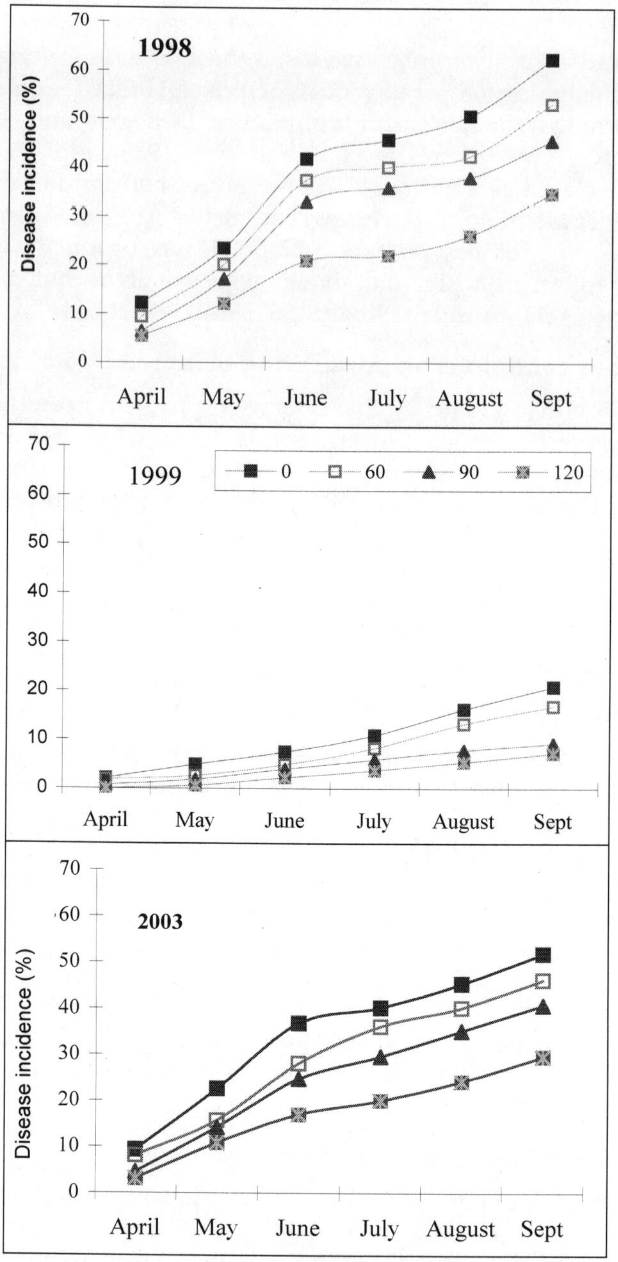

Fig. 1: Spread and Progress of *V. inaequalis* Infection at Varying Isolation Distances of Apple from Inoculum Source.

though with marked variation, in the absence of local inoculum at the start. This could be probably due to the availability and arrival of external wind borne secondary inoculum of conidia (Sutton *et al.*, 1976) as limited mobility and viability of ascospores after their discharge has been reported by Hirst and Stedman (1961). The studies indicated that sanitation provided an opportunity to delay- start the spray programme around pink bud to petal fall stages without any reduction in disease control of leaf and fruit scab and, provided more disease control as compared to initiating the spray programme in unsanitated orchard as early as silver tip or green tip stages (Qazi and Mir, 2002; Qazi *et al.*, 2008). In one of their studies on ascospore as source of primary inoculum, Hirst and Stedman (1961) considered a distance of about 91 m for differentiating local and external inoculum. Paisley and Heuberger (1972) while studying the prevention of apple infection by reducing primary inoculum in an isolated orchard considered about 183 m from the nearest apple orchard as ideal isolation distance.

(iv) Use of computer based disease forewarning systems

The computerized forewarning devices like SpecWare, RimPro, etc. have been developed for forecasting of apple scab and other diseases. The forecast systems based on host, pathogen and environmental factors, compute the infection periods by monitoring temperature, relative humidity, rainfall, duration of leaf wetness, etc. Most of the software systems developed for these devices use temperature and leaf wetness data. They help in decision making for need based chemical sprays and in turn help in avoiding unnecessary use of chemicals. In one of our study, scheduling of fungicide sprays for control of apple scab was done as per the advice provided by the METOS - a computerized forewarning system (Fig. 2). METOS – based schedule, comprised 5 to 6 sprays, proved to be equally effective as that of a fixed interval schedule (9 sprays) and a recommended spray schedule (6 sprays) in controlling

Fig. 2: 'METOS' - Microcomputer Disease Forewarning System (at SKUAST-Kashmir, Shalimar,

Furthermore, many decision support systems, developed for crop production, also possess disease management modules, such as RADAR for apple crop.

(b) Sooty Blotch and Flyspeck

Sooty blotch and flyspeck are two economically important diseases of apple usually appearing after fruit development (walnut size) stage. Sooty blotch is a disease complex caused by more than one fungal pathogen *viz.*, *Peltaster fruticola* and *Leptodontium elatius*, whereas flyspeck is caused by another fungus, *Zygiophiala jamiacacenis*. The spores of fungal pathogens of both the diseases are dispersed by rainfall and require water for their germination (Jones and Sutton, 2001). Vaious models *e.g.* SpecWare, predict the period of risk for infection based on leaf wetness after petal fall stage. The model starts prediction after 259 hours of accumulated leaf wetness and starts signaling a possible infection period only after recording a leaf wetness periods of at least 3 hours (less than 3 hours are not included). Different areas within a region need to validate the model and may require a different number of leaf wetness hours to start the model. Otherwise, the model that gives the earliest warning is preferable. Individual users of the models can run the model from the data by themselves on the basis of measured wetness hours and the accumulated wetness hours as well as any messages noting the start of an infection period.

(c) Cherry Leaf Spot

Cherry leaf spot occurs in various parts of the world where humid conditions exist. The disease is caused by ascomycetous fungus, *Blumeriella jaapi.* The fungus overwinters in diseased leaves producing ascospores (primary inoculum) in sexual fruiting bodies called apothecia. Primary inoculum spores are released into the air and prefer moderately wet conditions, with temperatures above 15.5^{0}C. Very few ascospores are ejected from apothecia, if the temperature is less than 7.7 ^{0}C. Optimal temperature for the spread of fungus ranges from 15.5 to 20 ^{0}C (Sundin and Rothwell, 2014). Once infection occurs on new leaves in the tree, visible cherry leaf spot lesions produce enormous spores that readily infect new tissue. Subsequently, more spores (secondary inoculum) develop from these lesions than ascospores discharged from leaves overwintering on the ground. In addition, spores from lesions on leaves are splash-dispersed and are much more likely to find new leaf targets within the tree than spores coming up from the ground – the distance from one leaf to another leaf is minimal, and the potential for spores to infect by moving from one leaf to an adjacent leaf is "easier" than for spores discharged from ground level. The relative ease with which the fungus spreads from infected leaves within trees can lead to significant disease spread even under marginal infection conditions. The ability of the inoculum to infect depends on temperature and leaf wetness (Jones and Sutton, 2001).

Some forecast models were originally designed to determine the risk of infection at a constant temperature for a certain period of leaf wetness. Forecast model calculates "risk factor/ rating" by assessing the temperature shifts during a wetness period. For example, the risk of infection would be 1/8, if 8 hour wetness is required at a particular temperature; and instead if it remained wet for one hour at that temperature; then the period is flagged as light, moderate and high risk for infection and when the sum of the risks reaches one or greater, there is likelihood of an infection risk.

(d) Downy Mildew of Grapes

Downy Mildew of grapes is caused by Oomycetous pathogen, *Plasmapora viticola.* The fungus overwinters in dead leaves and, sometimes, in dead berries and shoots. The initial inoculum (liberated spores) and secondary cycle of spores (resulting from initial infection), is splash- or wind-dispersed. After their landing on a leaf (or twig or grape cluster), secondary spores may germinate in a short period of time if a thin film of water is present. Depending on humidity and temperature, the infection takes 5 to 18 days to develop new inoculum. The downy mildew forecast models like SpecWare requires leaf wetness, relative humidity and temperature data to calculate the appearance of spores after the initial infection. The model estimates three levels of infection risks on a scale of 1 to 3, indicating possible or medium or high risk infections. The model starts at any point above 6.1°C temperature for estimating risk of primary infection and subsequent risk of infection from secondary inoculum. Another downy mildew forecaster namely DMCast uses hourly weather data (relative humidity, temperature and leaf wetness) to determine when infections are likely to occur. DMCast reliably predicts the date of first infection in the vast majority of seasons as well as the risk of foliar and fruit infection during the growing season. Earlier, another sophisticated computer programme "PLASMO" was developed in north Italy which stimulates critical phases in life cycle of *P. viticola* governed by temperature, relative humidity, rain fall and leaf wetness duration (Rosa *et al.*, 1995).

(e) Grape Powdery Mildew

Powdery mildew is the most common and widespread disease of grapevines in many countries, though varieties vary in susceptibility. The powdery mildew fungus (*Uncinula necator*) overwinters sexually as cleistothecia in bark, on canes, unfallen fruits and sometimes on fallen leaves. Ascospores from cleistothecia are released in spring after a rainfall of at least 2.5 mm and cause primary infection on developing plant tissue after 12-15 hours of continuous wetness at 10-15°C. Using air temperature and leaf wetness data, SpecWare and other powdery mildew models predict both infectious stages, *i.e.* primary (ascosporic) and secondary (conidial) stages (Thomas *et al.*, 1994; Weber *et al.*, 1996). A period of three consecutive days with temperature in the range of 21.1 to 29.4°C is required to initiate the conidial index. Thereafter, index increases by 20 with each day having six hours of similar temperature range. The index decreases by 10 on days with less than six hours in that temperature range, and on days with a maximum temperature greater than 35°C. The index will always be between zero and 100 with 0 to 30, 40 to 50 and 60 to 100 indicating light, medium and heavy infection risk, respectively.

(f) Bunch Rot of Grapes

Botrytis cineria, causing bunch rot of grapes, is spread through air and infects primarily post- veraison berries and flowers. The disease develops best in cool humid weather. Broome *et al.* (1995) developed a model which identifies infection periods for grape berries based on leaf wetness duration and average air temperature during wetness events. The model was programmed into an automated weather station (Envirocaster) and validated for two years on grapes. Fungicide applications based on the model allowed growers to reduce their use by approximately 50 per cent with equivalent disease control. However, SpecWare system does not calculate values for temperatures greater than 40C. There are three levels of infection risk *i.e.* light, moderate and high. Individual user can decide which risk level is acceptable for economic control.

(g) Peach Leaf Curl

Peach leaf curl, caused by the fungus *Taphrina deformans*, leads to severe early defoliation and crop losses on nearly all peach and nectarine cultivars under favourable conditions in many area throughout the world (Rossi *et al.*, 2006). The fungus overwinters on the tree and infects buds from bud-swell to bud opening under wet conditions. Thus early spring applications at or before bud break are effective in controlling this disease (Panagopoulos, 2007). Air temperatures between 10 to 21 °C are ideal for infection. About 10 to 11 hours of wetness for infection to occur and prolonged cool wet periods during bud burst can result in severe infections. The prediction model of Rossi *et al.* (2006) indicated that wetting period of at least 12.5 hours with minimum rainfall of 3 mm was required for infection and higher amounts of rainfall did not cause infection when the resulting wet period was shorter.

4. CONCLUSION

Most of the diseases of fruit crops under temperate climatic conditions start appearing during early to late spring season. In most of the years, frequent intermittent rains during spring months create conditions favourable for occurrence of repeated infection periods for many disease. Many times, prolonged wet period either hampers the application of fungicides or they are wash off. Under such conditions, any disease forecasting systems would prove less effective and overwintering inocula within the orchard or nearby orchard may become critically important for many diseases. Therefore, the impact of overwintering inoculum in the adjacent or nearby orchard has to be critically reassessed for actual and potential discharge of inoculum and long term spore dispersal, besides other interactive factors, for devising a successful forecasting model. The fluctuating climate also demands to develop models that increase the accuracy of existing forecast models through newer information on the pathogen-host behaviour. Proper validation of a plant disease forecasting system is always required for testing its accuracy and for avoiding chances of false positive predictions (forecast made for a disease, when in fact no disease was found) and false negative predictions (forecast made for a disease not to occur, when in fact the disease was found) in a location. Otherwise, there will be economical impositions involved in it, particularly the greater disadvantage associated with the latter could shatter the confidence of grower over forecasting systems. However, use of modern tools and techniques like reliable decision support systems or genetic probes for detection and identification of pathogens, etc. seems to minimize the losses associated with existing crop production systems.

5. REFERENCES

Agrios, G.N. 2005. Plant Pathology. Fifth edition. Elsevier Academic Pest. London, New York. 922 pp.

Broome, J.C., English, J.T., Marois, J.J., Latorre, B.A. and Aviles, J.C. 1995. Development of an infection model for Botrytis bunch rot of grapes based on wetness duration and temperature. *Phytopathol.*, **85:** 97-102.

Funt, R.C., Ellis, M.A., and Madden, L.V. 1990. Economic analysis of plant protection and disease-forecast based fungicide spray programme for control of apple scab and grape black rot in Ohio. *Plant Dis.*, **74:** 638-642.

Gleason, M.L., Ali, M.K., Domato, P.A., Lewis, D.R. and Dufft, M.D. 1994. Comparing integrated pest management and protectant strategies for control of apple scab and codling moth in an lowa apple orchard. *Hort. Tech.*, **4:** 136-142.

Gupta, G.K. 1985.Apple Sab (*Venturia inaequalis*). E. Merck. (India). Limited, Bombay. 36 p.

Hardwick, N.V. 2006. Disease forecasting. **In:** *The Epidemiology of Plant Diseases.* Second edition. (Eds.), B.M. Cooke and B. Kaye). Springer, pp. 239-268.

Hill, S.A. 1975. The importance of wood scab caused by *Venturia inaequalis* (Cke.) Wint. as a source of infection for apple leaves in spring. *Phytopathologische Zeitschrift*, **82**: 216-223.

Hirst, J.M. and Stedman, O.J. 1961. The epidemiology of apple scab [*Venturia inaequalis* (Cke.) Wint.]. I.Frequency of air borne spores in orchards. *Ann. Appl. Biol.*, **49**: 209-305.

Hirst, J.M. and Stedman, O.J. 1962. The epidemiology of apple scab [*Venturia inaequalis* (Cke.) Wint.]. II. Observations on the liberation of ascospores. *Ann. Appl. Biol.*, **50**: 525-550.

Jones, A.L. and Sutton. T.B. 2001. Diseases of Tree Fruits of the East. *Michigan State University Extension*, **19-20**: 57-60.

Joshi, N.C., Malik, A.G., Koul, M.L. and Anand, S.K. 1975. Some observations on the epidemic of scab disease of apple in Jammu and Kashmir during 1973. *Indian Phytopathol.*, **28**: 288-289.

Krause, R.A., Massie, L.B. and Hyre, R.A. 1975. Blitecast: a computerized forecast of potato late blight. *Plant Disease Rep.*, **59**: 95-98.

MacHardy, W.E. 1996. *Apple Scab: Biology, Epidemiology and Management*. American Phytopathological Society, St. Paul, Minnesota, USA. 545 p.

Miller, P.R. and O'Brien, M. 1952. Plant disease forecasting. *The Bot. Rev.*, **18**: 547-601.

Mills, W. D. 1944. Efficient use of sulfur dusts and sprays during rain to control apple scab. Cornell Agricultural Experiment Station Bulletin, **630:** 4pp.

Murali, N.S. and Secher, B.J.M. 1996. Integration of cultivar selection in a decisionsupport system for plant protection. *EPPO Bull.* **26**: 645-649.

Orlandini, S., Massetti, L. and Marta, A. D. 2008. An agrometeorological approach for the simulation of *Plasmopara viticola*. *Computers and Electronics in Agriculture*, **64**: 149-161.

Paisley, R.B. and Heuberger, J.W. 1972. Apple scab VI. Prevention of primary infection in an isolated apple orchard, 1970-1971. *Plant Disease Rep.*, **56**: 736-737.

Panagopoulos, C., 2007. *Diseases of the Deciduous Trees and Vineyard* (*Greek*). Stamoulis Publisher, Athens. p.195.

Qazi, N.A., Khursheed, A., Beig, M.A., Munshi, N.A. and Khan, N. A. 2008. Impact of inoculum distance on initiation and progress of apple scab [Venturia inaequalis (Cke.) Cke.) Wint.] infection. *Appl. Biol. Res.,* **10**: 40-43.

Qazi, N.A. and Mir, N.A. 2002. Comparison of different spray schedules for control of *Venturia inaequalis* infection of apple cv. "Red Delicious" in Kasahmir. *Appl. Biol. Res.,* **4:** 43-51.

Rogers, P.M. and Stevenson, W.R. 2006. Weather-based fungicide spray programs for control of two foliar diseases on carrot cultivars differing in susceptibility. *Plant Dis.* **90**: 358-364.

Rosa, M., Gozzini, B., Orlandini, S. and Seghi, L. 1995. A computer programme to improve the control of grapewine downy mildew. *Computers and Electronics in Agriculture*, 12: 311.

Rossi, V., Bolognesi, M., Languasco, L. and Giosuè, S. 2006. Influence of environmental conditions on infection of peach shoots by *Taphrina deformans*. *Phytopathol.,* **96**: 155-163.

Sundin, G. and Rothwell, N. 2014. Controlling cherry leaf spot in orchards with early infections. Michigan State University Extension, Department of Plant, Soil, and Microbial Sciences. http://msue.anr.msu.edu/news/

Sutton, T.B., Jones A.L. and Nelson, L.A. 1976. Factors affecting dispersal of conidia of the apple scab fungus. *Phytopathol.,* **66**: 1313-1317.

Tamosiunas, K., Semasskiené, R., Dabkevicius, Z. 2000. Development and implementation of cost-effective plant protection technology using decision-support systems in Lithuania. *EPPO Bulletin*. **30**: 69-75.

Thakur, V.S. and Khosla, K. 1999. Relevance of Mills infection periods to apple scab (*Venturia inaequalis*) prediction and rescheduling fungicide applications in Himachal Pradesh. *Indian J.Agric. Sci.,* **69**:152-156.

Thomas, C.S., Gubler, W. D. and Leavitt. G. 1994. Field testing of a powdery mildew disease forecast model on grapes in California. *Phytopathol.,* 84: 1070. (Abstract).

Thomidis, T., Rossi, V. and Exadaktylou, E. 2010. Evaluation of a disease forecast model for peach leaf curl in the Prefecture of Imathia, Greece. *Crop Prot.,* **29:** 1460-1465.

Van der Plank, J.E. 1963. *Plant Diseases: Epidemics and Control.* First edition. Academic Press, New York, London. 349 pp.

Zadoks, J.C. and Schein, R.D. 1979. *Epidemiology and Plant Disease Management*. Oxford University Press, New York. 427 pp.

Chapter 22

Ecosystem Services and Temperate Horticulture

N.K. Krishna Kumar

Deputy Director General (HS)
Indian Council of Agriculture Research,
KAB-II, New Delhi – 110 012
E-mail: daghort@icar.org.in; ddghort@gmail.com

1. INTRODUCTION

One of the primary moving factors in the transition of our country from an underdeveloped to a developing and progressing economy has been the edifice of Green Revolution. Over simplification of the term "Green Revolution" largely confined to wheat and rice does not justify the silent and golden revolution ushered by farmers and scientists representing various facets of Horticulture. The gap between horticulture (~ 280 million tonnes) and food grain (~260 million tonnes) production will only increase in the years to come. Unlike others, the silent, yet golden revolution is a multi-crop and multi-dimensional revolution though similarities in the use/misuse of irrigation water, fertilizer and pesticides remain. Declining soil fertility and health, depletion of water table and ground water resources, loss of bio-diversity, climate change and biotic stress shall retard the process of acceleration. We are in a way at a cross road where enhancing productivity, law of marginal returns, net economic returns and benefit/cost ratio should be weighed against sustainability and environmental safety.

What has not been fully conceptualized, understood and realized is that like every other species we are one among the animals and plants sharing the same resources and the fruits are for all. The universe has to be shared in equilibrium with due appreciation and recognition of the silent players in the complex web of ecosystem and environment. It is said that the chain is as strong as the weakest link and any weak link shall impact the fine ecological balance. The impact shall touch the *Homo sapiens* too.

The 20th Century was predominantly shaped by science-led agriculture with hybrids responding to irrigation and susceptible hybrids given an umbrella of protection against pests and diseases by powerful chemical pesticides. No doubt the increase in yield was spectacular and lives were saved. Equally important in the midst of all this high-tech science and a watershed in history was a small book *Silent Spring* by Rachel Carlson. If there was a small lamp lit to highlight the environmental concerns through indiscriminate actions, this was the moment. Unlike the 20th Century, at present environmental concerns, food safety, pesticides residues, declining water table, loss of bio-diversity, organic cultivation and organic food, clean air, resistance to development at the cost of family holding dominate the headlines. In short environment, sustainability, ecological balance has begun to shape the policy based on the concerns of today's generation. In the days and years to come environment issues are going to shape the policy within and outside.

In the midst of this one often hears the terms "Ecosystem, Agro-ecosystem, Forest ecosystem etc. Often this understanding at the grass root level is like five blind men touching and describing an elephant. Ecosystem is characterized by natural diversity, number of species, individuals within a species and above all maintenance of an equilibrium. Forest, ocean, natural soil, dessert *etc.* personify a natural ecosystem. Agro-ecosystem is a disturbed ecosystem where farming operations disturb the natural equilibrium. Horticulture ecosystem, especially of woody perennials is relatively stable compared to other agro-ecosystems. However, the purpose is not to encourage natural ecosystem at the cost of agriculture but to push horticulture to near natural ecosystem valuing the ecosystem services so that the environment is less polluted, resources are more sustainable, biodiversity less disturbed and ecosystem is well appreciated and preserved for future generations. After all let us appreciate the wisdom of Mahatma Gandhi who said that "Mother earth has enough for every one need but not enough for anyone's greed."

Ecosystem Services (ES) are the benefits that humanity derives from ecosystems. ES defines our policies on biodiversity, sustainability etc. However this concept has to take a deep scientific rooting. While knowledge and wisdom can help hardcore data based on mapping, assessing and validation should be the edifice of the decision making process. Though originally intended to highlight the importance of the ecosystem and ecological processes in human society and welfare, a few feel the policy is both welcoming and worrying. The worry is mainly derived from insufficient knowledge and half knowledge is dangerous especially in framing major policy decisions. Further it is also opined that, looking ES through the prism of economics and attaching more emphasis on monetary evaluation negates the foundation of ES.

Ecosystem services are goods and benefits people obtained as a result of natural process. The services can be in the nature of a) Provisioning services like food, water, fuel, fiber, genetic resources *etc.*, b) Regulating services like climatic balance, biological control, rainfall and water, pollination *etc.*, and c) cultural heritage, educational, spiritual etc. The biodiversity, ecosystem services and human wellbeing are closely interconnected. Biodiversity loosely representing diversity of genes, population, species, communities and ecosystem strongly impacts the supporting services which in turn support food and environmental security, basic material for good life, health and good social relations. To quote an example, the society at large has taken for granted the value of insect pollination, biological control, fresh air, pure water *etc.* These services in their own natural way are vibrant and dynamic and will impact the survival of the human race if not delicately handled. The following graph is an example of an increased awareness and papers published on ecosystem services.

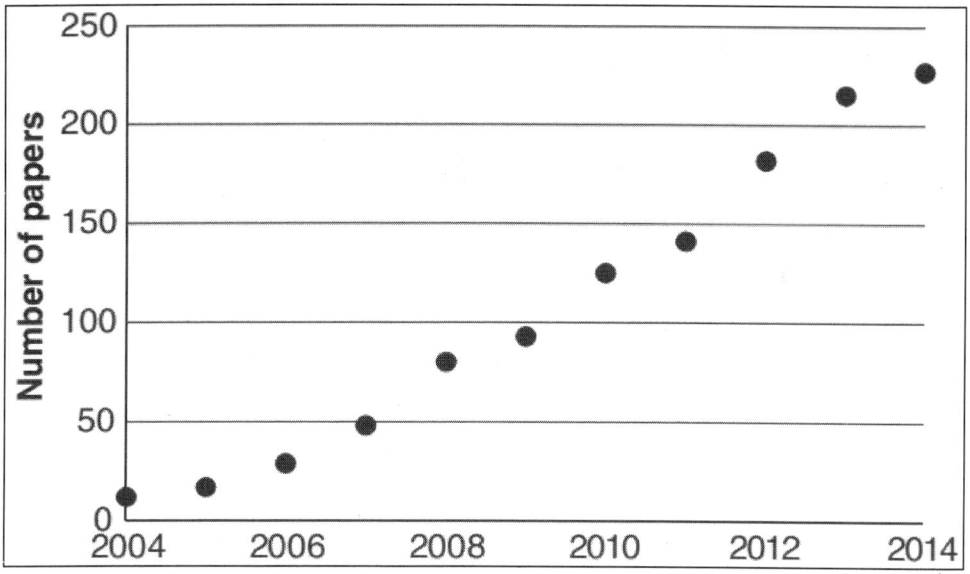

Fig. 1: No. of Papers Published on Ecosystem.

In today's world problem associated with cotton whitefly and mealy bug, papaya mealy bug, grape mealy bug, eucalyptus gall wasp *etc.,* are heard more and more. It is not to say that the problems were not there earlier, it is only the frequency and intensity that needs to be monitored. It is a common knowledge that a mono-cropping increase pest outbreak, the corollary is more the biodiversity less is the pest outbreak. We have observed but never asked why the forest trees are not planted in a straight line or symmetry or why most of them do not suffer from major pest and diseases in spite of the fact that no one sprays over them or for that matter why they are green and healthy in spite of no application of chemical fertilizer. Can the fulcrum of Horticulture move more towards a sustainable equilibrium devoid/less use of chemicals and fertilizers and while at the same time ensuring sustainable production?. The ecosystem approach is a strategy for the integrated management of land, water and using resources that promotes conservation and sustainable use in an equitable way. The approach is based on levels of biological organization, interaction among organisms and their environment.

Ecosystem services while ensuring bio-diversity are also repository of natural enemies of pests and diseases. In their own way they contribute to preserve the soil and renew the fertility, stabilize climatic aberrations. The value of ecosystem services in purifying air and water is invaluable. Ecosystem services facilitate pollination and dispersal of seeds and cycle nutrients. What is invaluable in the ecosystem services is the complex web of interactions, services delicately dependent on each other that would be very difficult to duplicate. For example over one lakh different animal species that include bats, bees, flies, moths, beetles, birds, butterflies provide free pollination services. Nearly one third of the food that comes from plants is pollinated by wild pollinators. It is estimated that the pollination from the wild pollinators alone to temperate agriculture and horticulture in US is $ 4-6 million. Horticulturists for too long have come to identify pollination predominantly by bees. It has to be appreciated that greater the diversity of pollinators better is the pollination. Only 15 per cent of the crop pollination is by domestic bees and 80 per cent by wild solitary bees. They neither charge rent nor labor nor demand a share in the profit. Their services are free and anything that is free is taken for granted. It is not alone bees but a number of other organisms such as spiders, centipedes, millipedes, earthworms, wasps, earwigs, *etc.* all play a vital role in ecological balance. How can we underestimate and not value these insects and microbes which convert the dead into decay adding to the soil nutrients?

What has to be realized is our own action; often indiscriminate driven more by greed, short term gains is impacting the ecosystem balance and services. The recent example of vehicle cess imposed at Delhi is an indicator of environmental awareness. Women and children hugging trees to prevent deforestation, the Narmada Bachao Andolan, resistance to genetically modified food, the silent valley movement are all expression of apprehensions of the common man directed at the loss of ecosystem and ecosystem services. What is required is a scientific assessment, documentation, planning that is driven more from the perspectives of sustainability and ecosystem services than direct increase in productivity and cost: benefit ratio.

The temperate ecosystem by very nature has apparently less diversity compared to tropics. To that extent may be ecologically more sensitive to the disturbances. The future horticulture revolution in our country will be greatly determined by our success in temperate horticulture. It is said that unsprayed apple trees support a large fauna of more than 2,000 arthropod species, of which 25 per cent are pests and 25 per cent are natural enemies and 50 per cent benign, through an important component of food web. The preservation of natural enemies is a crucial part in the successful IPM of especially secondary pests, like red spider mites on apple. The recent outbreak of mites, the development of pesticides resistance and the loss of predatory mites has to be more scientifically evaluated in the Indian context. Similarly, pesticides can have serious impact on the efficacy of honey bees and pollinators so vital to survival of temperate horticulture. There is very little documentation on the adverse effect of neonecotinoides such as imidacloprid, thiomethoxam that is seriously under scanner for colony collapse all over the world. The common European earwig is an important predator of codling moth, leaf curling midge, woody aphid, green aphid and psyllids on apple. No in-depth study has been carried out in India on the role of these on temperate crops especially apples.

2. CONCLUSION

For decades to come there is a need to enhance productivity under reduced factors of production- land, labour, water and capital. While appreciating the need for enhanced productivity if ecosystem services are taken into cognizance horticulture will move towards sustainability, food and environmental safety. This is more critical for temperate horticulture where the ecosystem is more delicately poised.

3. ACKNOWLEDGEMENTS

I thank Dr S. Ramani, UAS, Bengaluru for sharing his thoughts and slides on this aspect. My thanks to Dr Ranvir Singh and Dr T. Janakiram, Division of Horticultural Science, ICAR for helping me with the manuscript.

Chapter 23

Pesticide Residue Problems in Temperate Fruits

Malik Mukhtar

Research Centre for Residue and Quality Analysis
Sher-e-Kashmir University of Agricultural Sciences and Technology of Kashmir,
Srinagar – 190 025, J&K
E-mail: drmalikmukhtar@yahoo.com

1. INTRODUCTION

The main source of livelihood for mankind is plants. The plants are an important source of energy for the animals, which sustain their day-to-day needs. Apart from humans, other animals also depend on plants, that is why these plants are susceptible to about 80,000 – 1,00,000 diseases. Among the 8,00,000 insect species present in the world, 10,000 species feed on these plants and cause economic loss throughout the world. In addition to pathogens and insect pests, nematodes and rodents inflict heavy damage to economically important crops (Abelson and Phillip,1994).

Horticulture plays an important part in the Indian economy, as India earns major share of foreign exchange through exports of fresh fruits. The use of agrochemicals has predominantly increased the horticultural production during the last few decades. In order to meet the growing demand of fruits for export and to feed the inceasing population, agro chemicals particularly pesticides will continue to remain a key input in the strategy to satiate the country's needs. For sustainable production of horticulture sector and to avoid yield losses due to diseases and insect pest, pesticides are frequently used during and after production. Knutson *et al.*, 1994.

The term pesticide is a broader term meaning "killer of pests". Pesticides include insecticides, fungicides, nematicides, rodenticides, acaricides and herbicides.

Pesticides are indispensable and widely used in modern agriculture and horticulture, as they are considered economically important to reduce production losses to food security. However, their residues persist on agricultural produce for extended periods before becoming completely degraded under natural processes. These long-lived compounds are transmitted from plant and plant products to animals, air water and soil eco-system. The extensive crop production demand excessive use of pesticides on one hand and the health hazards of persistent residues on the other, raise the pertinent question that if persisting residues are inevitable, what are the prospects for their removal from foods and other critical sectors? In today's world of extensive importing and exporting of food goods, the analysis and monitoring of pesticides become more essential, but challenging. Either all are one of the classes of pesticides are used on various food crops at a particular point of time throughout the world. Pesticide use in horticulture has over casted all other control methods as a means of plant protection during the last few decades. Developed countries use 80 per cent of the produced agro chemicals whereas rest 20 per cent is being used in developing countries. India is the biggest consumer of pesticides among south Asian countries. Consumption of different classes of pesticides in India is Insecticides 76 per cent, Herbicides 10 per cent, Fungicides 13 per cent and others 1 per cent whereas the world consumption stands as Insecticides 44 per cent, Herbicides 30 per cent, Fungicides 21 per cent and others 5 per cent (Aktar 2009). Pesticide use has been a double-edged sword. On the one hand it has contributed to the stabilization of horticultural produce and on the other hand led to the development of myriad of problems. Some of the disadvantages of indiscriminate use of pesticides are, pesticide residues, insecticide resistance, insecticide resurgence, development of new pests, reduction in diversity of natural enemy and environmental pollution.

2. PESTICIDE RESIDUES

Pesticides or their breakdown products remaining in the environment following their normal use or accidental contamination is called pesticide residue. Pesticide residues end up in food supply from the usage either on crops, soil or water. The various types of pesticides residue are as follows:

- Surface or Effective Residues: The portion of the pesticides left from the initial deposit.
- Dislodgeable Residues: Readily removable, and may be used as an index for risk to farm workers
- Penetrated Residue: The surface residue becomes penetrated residue by migrating into the sub-strata.
- Terminal Residue: Breakdown products, which are stable and create as many problems as the original compound.
- Conjugated Residue: The products of secondary metabolism due to the reaction of pesticides or its metabolites with endogenous substrates.
- Bound Residue: Chemical species in soil, plant or animal tissues originating from pesticide usage that cannot be extracted by the methods commonly used in the analysis.

The knowledge of risk to public health from pesticide residues in crops and food commodities has been under the auspice of national regulatory authority for many years and since 1963 there has been international guidance from FAO/WHO joint meeting on pesticide residues (JMPR) to augment the national efforts (USDA 1994). For individual pesticides in food maximum residue limits (MRLs) are derived from supervised field trials. The logic behind setting MRLs is that if a pesticide is sprayed on crops under good agricultural practices, resulting pesticide residue in the crop should not exceed the MRL value. The MRL is a benchmark on the ground that the use of a particular pesticide will not pose risk to the health of consumers. A link between the maximum residue limit and toxicological properties of the compound is the Acceptable Daily Intake (ADI) and MRL is accepted nationally and internationally after careful observance of relevant factors. (Knutson *et al.*, 1993).

3. INSECT PESTS AND DISEASES OF TEMPERATE FRUITS

Insects are present in all types of ecosystems and they consist of more than two third of the known species of the animals of the world. Insects in a number of ways affect human beings. Some feed on crops plants, medicinal plants, forest trees and so on and so forth. Insects also feed on stored food in godowns, storage structures and cause deterioration of food quality.

Insects that cause less than 5 per cent damage are not considered as pests. Insects that cause damage between 5-10 per cent are called minor pests and those that cause damage above 10 per cent are considered as major pests. Insects that cause damage to plants are grouped into two major classes namely chewing insects and sucking insects. The former chews off plant parts and swallow them, resulting in damage to the crops, while as sucking insects pierce through the epidermis and suck the sap. Many sucking insects serve as vectors of plant diseases and also inject toxins, which are in their salivary secretions that cause irreversible damage to the crops. Table 1 deal with the dominant insect pests of temperate fruits and pesticides involved in their management.

4. PESTICIDE USE ON APPLES

Apples are the most important tree fruit of the temperate regions of the world in terms of area cropped, volume and value. It is also one of the major food crops grown in the world. The cultivated apples belong to the genus Malus, which comprises of some 25-30 species and several sub species distributed throughout the temperate part of the globe.

5. PESTICIDE RESIDUES ON APPLES

(a) UK

The most commonly occurring pesticide residues (*i.e.* found on >10 per cent of samples tested), on apples in the pesticide residue survey were chlorpyrifos, carbendazim, captan and diphenylamine. In the survey of 2004 residues of 11 pesticides were found on apple samples, which comprise of eight fungicides, one insecticide and one plant growth regulator. Around 43 per cent of samples tested

Table 1: Insect Pests and Diseases of Temperate Fruits and Commonly Used Pesticides for their Management

Sl.No.	Insect Pest/Disease	Species Name	Description	Status	Pesticide Used
1.	San Jose Scale	*Quadraspidiotus perniciosus*	Insect sucks sap resulting in weakness and loss of yield	Major	Chlorpyrifos, Dimethoate
2.	Wooly apple	*Eriosoma lanigerum*	Trees lose vigor due to drainage of sap	Major	Diazinon, Methyl oxydemeton
3.	Tent Caterpillar	*Malacosoma indicum*	Caterpillars eat leaves leaving only midribs	Minor	Dimethoate Carbaryl
4.	European red mite	*Panonychus ulmi*	Causes leaf bronzing and premature leaf fall	Major	Dicofol, Propargite, fenazaquin, fenpyroximate
5.	Codling moth	*Cydia pomonella* (Linn.)	Key pest which causes direct damage to fruit	Major	Chlorpyrifos, Methyl oxydemeton
6.	Apple root borer	*Dorysthenes hugelii*	Grubs feed on roots	Minor	Chlorpyrifos, phorate
7.	Apple stem borer	*Apriona cinerea*	Grubs make galleries in main trunk	Minor	Dichlorvos, dimethoate, methyl demeton
8.	Indian gypsy moth	*Lymantria obfuscata*	Defoliation of trees	Minor	Dimethoate, Carbaryl
9.	Bark beetle	*Scolytus nitidus*	Make galleries under the bark of the tree	Major	Dimethoate, methyl demeton
10.	Stem borer	*Aeolesthes sarta*	Bore inside the sap wood	Major	Dichlorvos, dimethoate, methyl demeton
11.	Apple scab	*Venturia inaequalis*	All parts of the tree are attacked, but scab infection of fruit is most obvious symptom.	Major	Difenaconazole, Mycobutanil, captan
12.	Apple powdery mildew	*Podosphaera leucotricha*	Causes reduction in photosynthetic efficiency and to loss of fruit size and yield and ultimately loss of leaves and blossoms	Major	Mycobutanil, Hexaconazole
13.	Apple canker	*Nectria galligena*	This fungus causes cankers on trees and also a fruit rot both in the orchard and in store.	Major	Carbendazim
14.	Crown rot and collar rot	*Phytophthora cactorum* and *Phytophthora syringae*	Collar rot is a disease of the scion and crown rot a disease of the rootstock.	Minor	Metalaxyl
15.	Blossom wilt	*Monilinia laxa f.sp. mali*	Causes loss of blossoms	Minor	Pyraclostrobin+ boscalid,

Source: Sherwani *et al.*, (2015), David, (2006).

positive for fungicides, 38 per cent for insecticides and 18 per cent for storage pesticides (David, 2006).

(b) Pakistan

Pesticide residues were determined in four fruit samples of apple, grapes, pear and peach procured from local markets. The fruit samples were contaminated with endosulfan, dimethoate, chlorpyrifos, methyl parathion, fenitrothion, cypermethrin, fenvalerate and deltamethrin. One out of eight samples was found above the MRL. Most of the fruits were contaminated with dimethoate residues (Anwar *et al.*, 2011).

(c) Czech Republic

During three year monitoring programme, 522 samples of fresh apples and six brands of temperate fruit puree were analysed for pesticide residues. Pesticide residues were detected in 59.5 per cent of the samples of fresh apples, whereas maximum residue limit (European union MRLs) were exceeded only in 1.4 per cent of samples. The level of residues were lower in fruit purees with overall residues detected in only 16 per cent of samples, with 9 per cent samples exceeding 0.01mg kg^{-1} MRL values. Multiple residues were found in 25 per cent of fresh apples and in 10 per cent fruit purees. Among the pesticides organophosphate insecticides and fungicides representing phtalimides, sulphamides, and dicarboximides were the most frequently found residues. (Stepan *et al.*, 2005).

(d) Kashmir Valley

Various pesticides used for the control of various diseases and insect pests of apple in Kashmir valley have been detected. A survey was carried out during the year 2013-2014 by the research centre for residue and quality analysis of the Sher-e-Kashmir university of Agricultural sciences and technology under the Department of science and technology sponsored Project "Monitoring of residual pesticides in Kashmir and their amelioration for sustainable agriculture". Samples were collected from the entire apple growing belts of Kashmir Valley. A total of 311 samples were collected and analysed for Pesticide residues (Table 2). In Sopore 7 per cent samples had residues of five different pesticides and 2 per cent of the samples were above MRLs. In Shopian 4 per cent of the samples had residues of five different pesticides and 3 per cent of samples violated MRLs. At the location of Tral 3 per cent of the samples contained 5 different pesticide residues and only 1 per cent were above MRLs. In Mattan 4 per cent of the samples had residues of four pesticides and 2 per cent were above MRLs. At Baramulla four different pesticides were contained in 5 per cent of the samples and 1 per cent violated MRLs. At the location of Kulgam three pesticide residues were found in 5 per cent of samples analyzed and none of the samples violated MRLs. In Ganderbal 1 per cent of the samples had residues of 3 different pesticides and none was found above the MRLs. In Bandipora residues of five different pesticides were found in 2 per cent of the samples out of which none were above MRLs. Dimethoate, Chlorpyrifos and fenazaquin were found in the apples collected from all the locations except Ganderbal. Phosalone was found only in samples collected from Kulgam (Anon., 2014).

Table 2: Amount of Pesticide Residues in Apples Collected from different Areas of Kashmir Valley

Sl.No.	*Location*	*Pesticide*	*Per cent of Samples with Pesticide Residues*	*Per cent of Samples with Pesticide Residues above MRLs*
1.	Sopore	Fenazaquin	7	2
2.		Fenpyroximate		
3.		Chlorpyrifos		
4.		Hexaconazole		
5.		Dimethoate		
6.	Shopian	Difenaconazole	4	3
7.		Fenazaquin		
8.		Carbendazim		
9.		Dimethoate		
11.		Chlorpyrifos		
12.	Tral	Dimethoate	3	1
13.		Clothianidin		
14.		Fenazaquin		
15.		Fenpyroximate		
16.		Chlorpyrifos		
17.	Mattan	Propargite	4	2
18.		Dimethoate		
19.		Chlorpyrifos		
20.		Fenazaquin		
21.	Baramulla	Dimethoate	5	1
22.		Hexythiazox		
23.		Fenazaquin		
24.		Chlorpyrifos		
25.	Kulgam	Phosalone	5	—
26.		Difenaconazole		
27.		Propargite		
28.	Ganderbal	Hexythiazox	1	—
29.		Mycobutanil		
30.		Hexaconazole		
31.	Bandipora	Bitertanol	2	—
32.		Chlorpyrifos		
33.		Propargite		
34.		Fenazaquin		
35.		Dimethoate		

Source: Anonymous (2014).

(e) Other Countries

The pesticide residue-monitoring programme in fruits was undertaken from the year 2004-2011 by the Danish veterinary and food administration. The residues were monitored for the Danish and imported fruits. The results showed that more residues were found in samples of foreign origin as compared to samples of Danish origin. Overall fruits and vegetables had higher frequencies of residues than the other groups of commodities. Fruits had higher frequencies compared to vegetables. Also samples with more than one residue were more frequently found in samples of foreign origin. Overall residues above the MRLs were found in 2.6 per cent of the samples, most frequently in fruits. Figures 1–7 shows the percentage of pesticides in different samples of fruits from various countries (Anon., 2013).

(i) Apple

The samples of apple had residues of 54 different pesticides in 80 per cent of the samples and 3 per cent contained residues above MRLs (Fig.1).

(ii) Table Grape

Table grapes contained residues in 50-75 per cent of the samples. In 49 per cent of the samples more than one residue were detected (Fig. 2).

(iii) Melons

Pesticide residues were found in 54 per cent of the samples out of which 3 per cent were above MRLs with 38 different pesticides (Fig. 3).

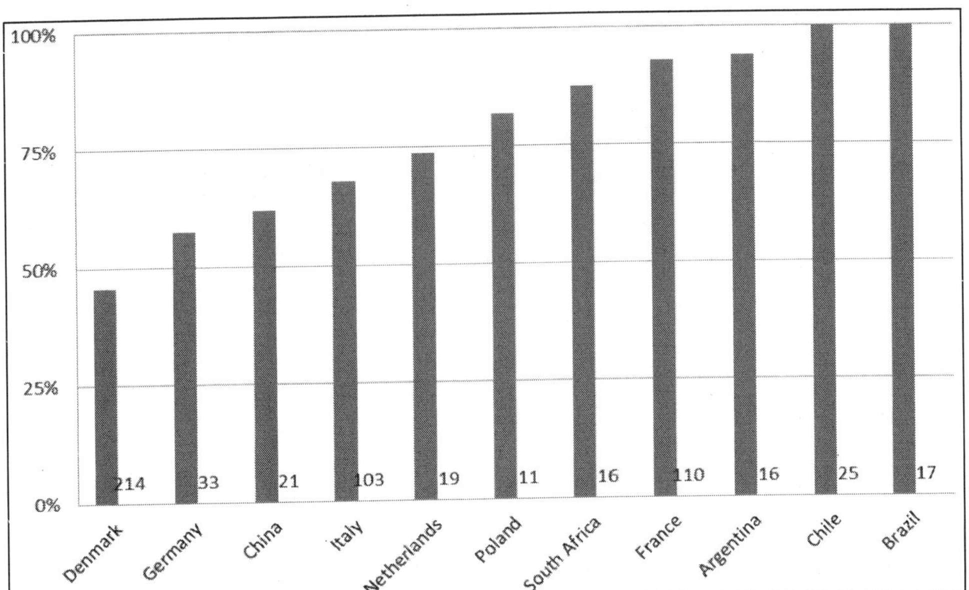

Fig. 1: Samples of Apples with Pesticide Residues from different Countries. Figures close to the bars are the number of samples analysed for the specific country.

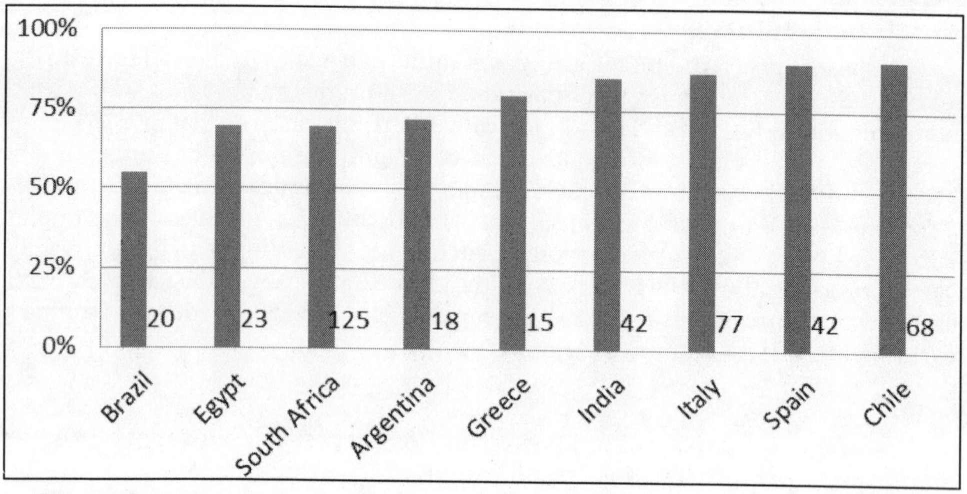

Fig. 2: Samples of Table Grapes with Pesticide Residues from different Countries. Figures close to the bars are the number of samples analysed for the specific country.

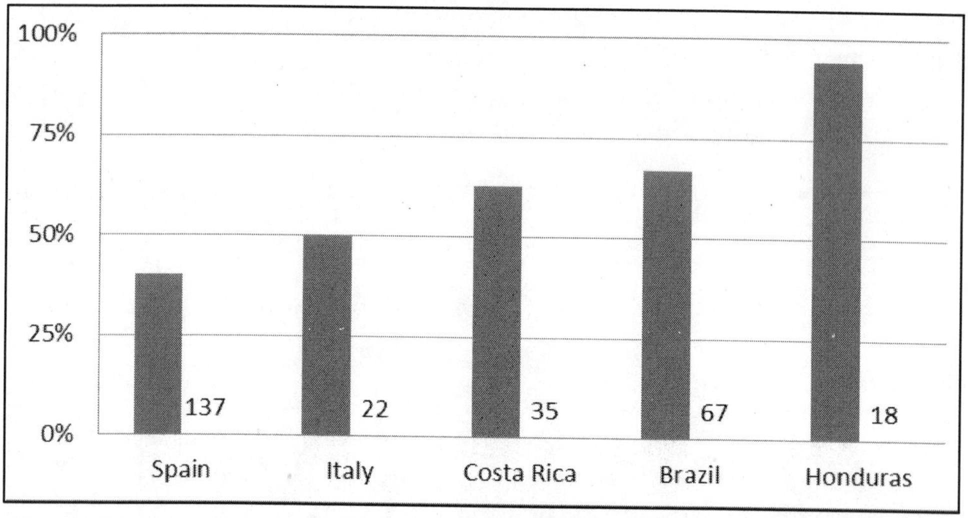

Fig. 3: Samples of Melons with Pesticide Residues from different Countries. Figures close to the bars are the number of samples analysed for the specific country.

(iv) Strawberries

In 70 per cent of the samples 45 different pesticides were found and 5 per cent were above MRL and 39 per cent of the samples contained more than one residue (Fig. 4).

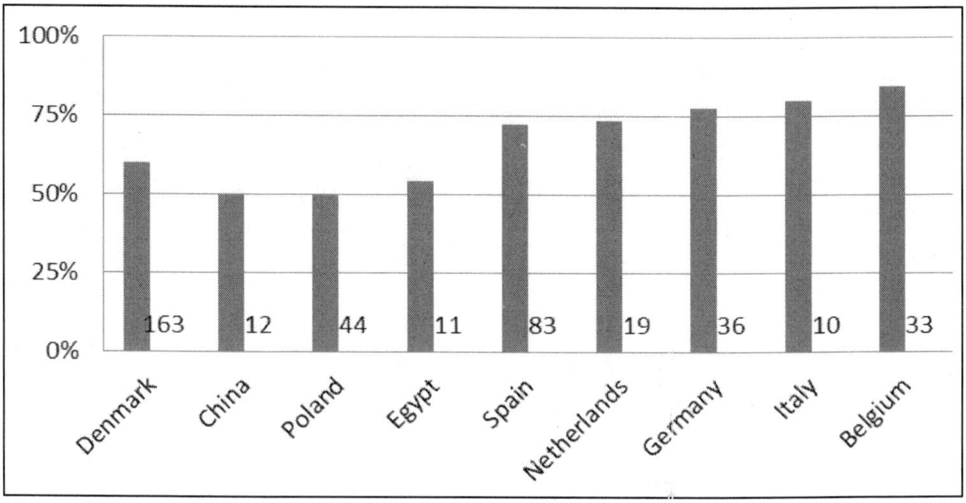

Fig. 4: Samples of Strawberries with Pesticide Residues from different Countries. Figures close to the bars are the number of samples analysed for the specific country.

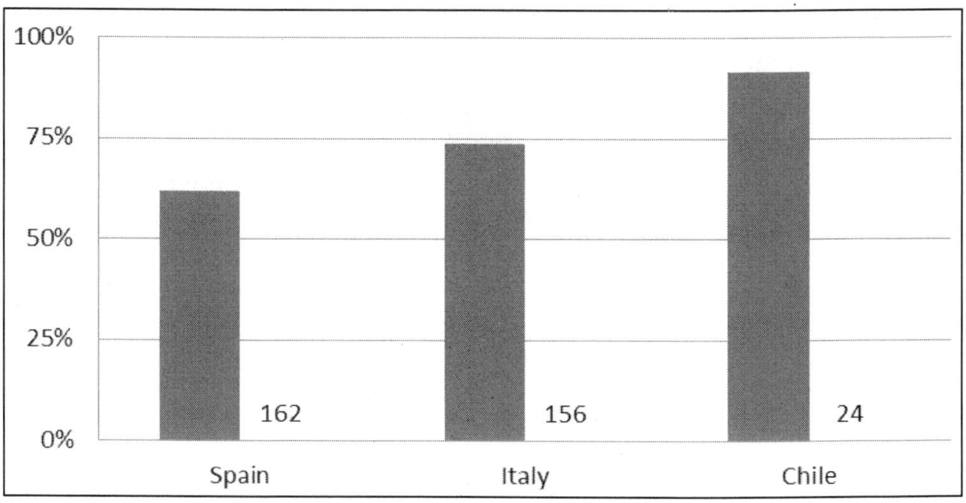

Fig. 5: Samples of Peaches and Nectarins with Pesticide Residues from different Countries. Figures close to the bars are the number of samples analysed for the specific country.

(v) Peaches and Nectarins

A total of 69 per cent of the samples contained 50 different pesticide residues out of which 1 per cent were above the MRLs and 38 per cent of the sampes contained more than one residue (Fig. 5).

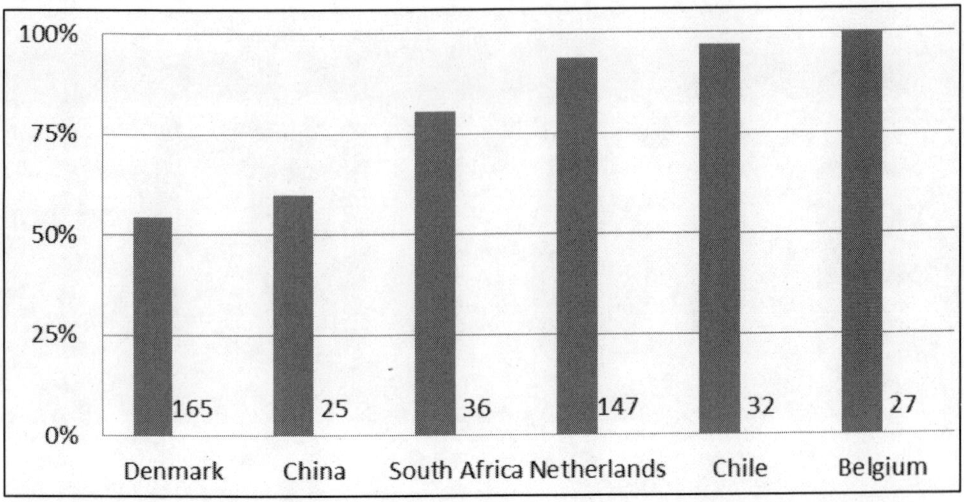

Fig. 6: Samples of Pear with Pesticide Residues from different Countries. Figures close to the bars are the number of samples analysed for the specific country.

(vi) Pear

Fourty eight different pesticides were found in 88 per cent of the samples out of which 1 per cent were above the MRL, while as 15 per cent of the samples had multiple residues (Fig. 6).

(vii) Plums

A total of 31 different pesticide residues were found in 47 per cent of the samples out of which 4 per cent were above MRLs. More than one type of residue were found to contain in 11 per cent of the samples (Fig. 7).

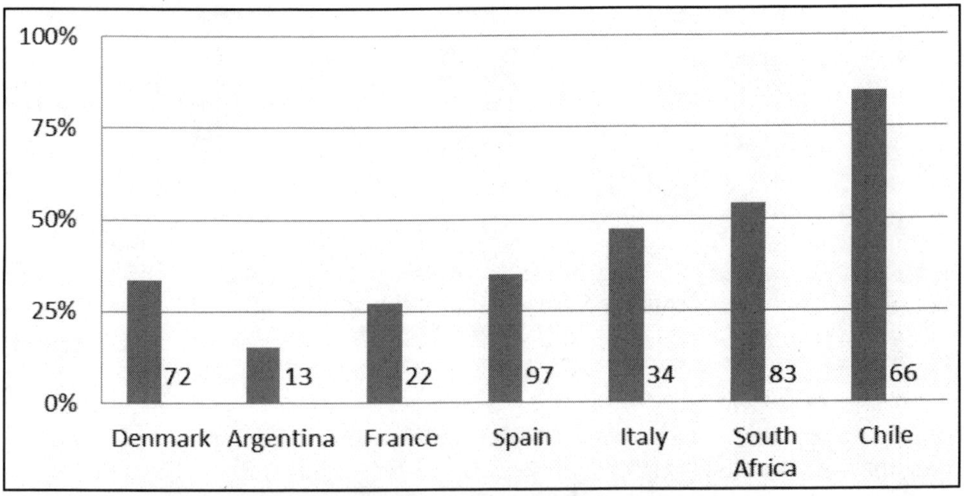

Fig. 7: Samples of Plums with Pesticide Residues from different Countries. Figures close to the bars are the number of samples analysed for the specific country.

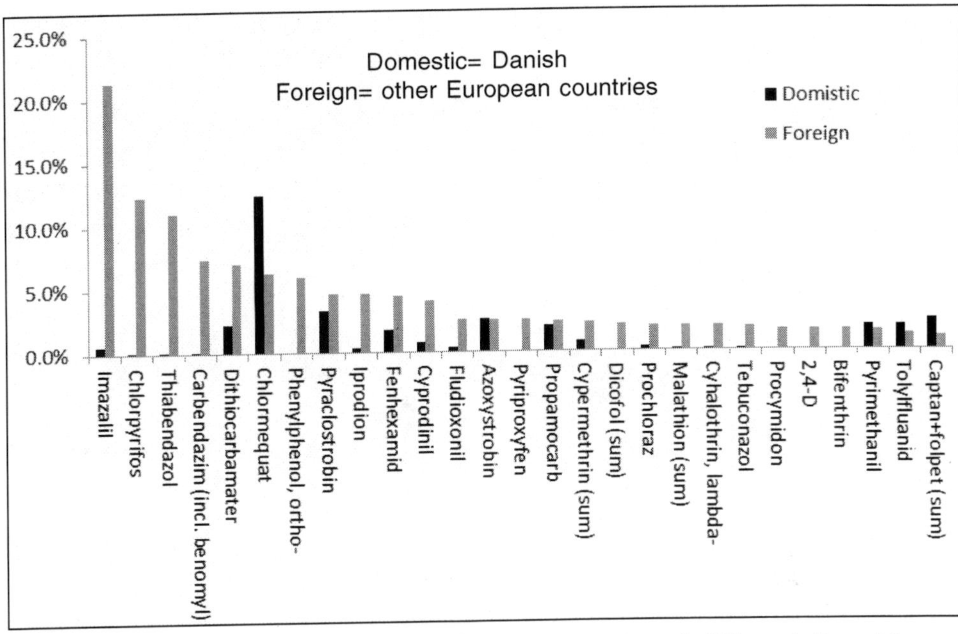

Fig. 8: Frequencies of Pesticides Found in Fruits of different Countries.

Figure 8 shows the frequency of the most often found pesticides in fruits. As can been seen from the figure imazalil was the most frequently found pesticide found in foreign samples than in the Danish produced fruits. Apart from imazalil, chlorpyrifos, thiabendazole, carbendazim, dithiocarbamates and chlormequat were found in more than 5 per cent of the foreign samples.

6. FATE OF PESTICIDES AFTER APPLICATION TO PLANTS

Distribution of a pesticide, its metabolites or derivative in a compartment, organism, system or population of concern as a result of transport, partitioning, transformation or degradation refers to as fate (OECD, 2003). After application of pesticides to the plants, they may interact with the plant surfaces, be exposed to the environmental factors such as wind and sun and may be washed off during rainfall. The pesticide may enter the plant transport system (systemic) or stay on the surface of the plant (contact) or be absorbed by the plant surface (waxy cuticle and root surfaces). While still on the surface of the plant, the pesticide can undergo chemical and microbial degradation, photolysis and volatilization. These processes are depicted in Figure 9.

All these processes can reduce the original pesticides concentration but can also introduce some metabolites in the crops. Immediately after application in the field, volatilisation of the pesticide usually occurs. The process depends on the vapour pressure of the pesticide. Those pesticides tend to volatilize rapidly into the air whose vapour pressure is high, while, as those with low vapour pressure remain longer on the surface. Volatilization rate also depends on the environmental factors such as wind speed and temperature. The faster the wind speed and the higher the temperature the more the pesticide will evaporate. Pesticide degradation through photolysis occurs

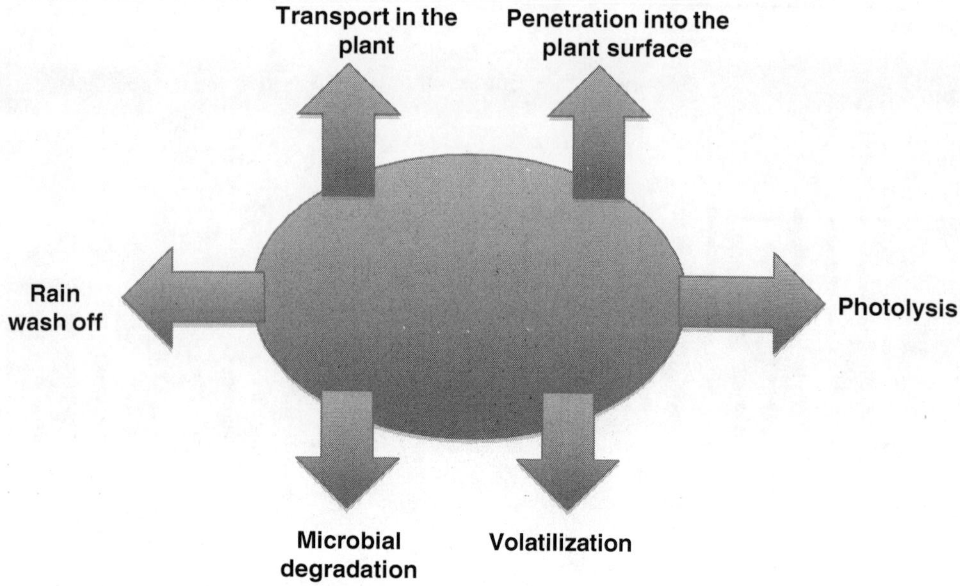

Fig. 9: Modes of Degradation of Pesticides in Environment.

when molecules absorb energy from the sunlight. Some pesticides may be degraded by microbial metabolism. Microorganisms can use pesticides as nutrients thereby breaking them into carbon dioxide and other components (Holland and Sinclair, 2004). Because of difference between naturally occurring organic chemicals and pesticide structures, the microbes cannot assimilate them but they may be altered at reactive sites. The products formed may be less or more toxic than the parent chemical. Although different environmental processes influence degradation of pesticides, Celik *et al.* (1995) concluded that under natural field conditions volatilization is the main process that affects pesticides. The degradation pattern of pesticide residues on temperate fruits follows a first order kinetics under the climatic conditions of Kashmir valley. The rate of dissipation of pesticide residues follows a perfect negative correlation (Mukhtar, 2015). Table 3 depicts the half-life and waiting period of some commonly used pesticides in Kashmir valley on apple.

Table 3: Waiting Period and Half Life of the Commonly Used Pesticides on Fruits in Kashmir

Sl.No.	*Pesticide*	*Crop*	*Half Life (days)*	*Waiting Period (days)*
1.	Chlorpyrifos	Apple	3.33	6.28
2.	Difenaconazole	Apple	6.09	13.06
		Golden delicious starkrimson	5.34	10.72
3.	Dimethoate	Apple	3.45	6.95
4.	Fenpyroximate	Apple	3.85	12.30
5.	Fenazaquin	Apple	3.62	18.55
6.	Hexaconazole	Apple	3.58	16.52

Source: Mukhtar *et al.*, (2015), Bhat *et al.*, (2015), Munazah (2014).

(a) Monitoring

The concept of pesticide residue monitoring programs is to ensure that fruits do not exceed maximum residues levels (MRLs) allowed by the government and that good agricultural practices (GAP) are maintained. Regulatory bodies also use results from these monitoring programmes for future developments in setting MRLs and risk assessment exercises for public health. In most countries, the monitoring programs are organised by a single agency designated as the competent authority. The agency designs a monitoring plan based on the previous data available from dietary consumption and risk assessment exercises or pesticide usage in the available fruits and vegetables. There is a coordinated programme in the European Union (EU) for all the member countries to follow from the European commission and the member states national programs. The European Food Safety Authority (EFSA) then yearly compiles the results as a single report.

In the case of international trade, the trading partners also influence the monitoring plan. For example, partners trading with the EU (normally referred to as third countries) have to incorporate EU standards to their food control programmes. Laboratories carrying out pesticide residues analysis should be accredited or have started accreditation procedures to some quality standard. Pesticide legislation in developing countries is generally lacking or not implemented and this also affects pesticide monitoring since it relies on legislation to be effective (Ecobichon 2001). The other problem is lack of trained personnel to enforce laws and monitor the use of pesticides and residue levels in food and the environment. However, international trade drives pesticide monitoring in some developing countries with high agricultural output. Failure to adhere to trade standards can result in a loss of revenue to the affected agricultural industry. India has also been at loss when the exports of the fruits are rejected on the grounds of high pesticide residues for which there are zero tolerance in the importing country.

(b) Maximum Residue Level

When the pesticide is used according to authorised agricultural practices, Maximum residue levels are the highest levels of residues expected to be in the food (EFSA 2010). The MRLs are always set far below levels considered to be safe for humans. It should be understood that MRLs are not safety limits; a food residue can have higher level than MRL but can still be safe for consumption. Safety limits are assessed in comparison with acceptable daily intake (ADI) for short-term exposure or acute reference dose (ARfD). MRL setting can be the responsibility of one or more authorities in a country and normally involves the health, agriculture and environmental agencies. MRL enforcement can be a responsibility of one or more agencies and may also depend on different food types. MRL setting is based on the national registered good agriculture practice (GAP) data combined with the estimated likely residue from the Supervised Trials Mean Residue (STMR), ADI and ARfD. The information is then evaluated by the risk assessment agency like Joint Meeting on Pesticide residues (JMPR) for CODEX Alimentarius.

(c) Food Processing

Fruits which are used for juice making often results in transfer factors or food processing factors (PF) of the pesticide residue in the transition from raw agriculture commodity to the processed product. These processing factors are expressed as the concentration of pesticide after processing divided by the concentration before processing. Effect of processing in fruits is influenced by the physico-chemical properties of the pesticide as well as the processing method The general processing factors for washing (0.76), peeling (0.44) and canning (0.74) for fruits have been worked out. (Holland, 1994).

(d) Risk Assessment

Risk assessment of chemicals is described as a process intended to calculate or estimate the risk to a given target organism, system or population, following exposure to a particular chemical taking into account the inherent characteristics of the chemical of concern as well as the characteristics of the specific target system (OECD, 2003). Risk assessment process includes four steps: hazard identification, hazard characterisation (dose-response assessment), exposure assessment and risk characterisation.

(e) Hazard Identification

Hazard identification is the first step in risk assessment and it involves the identification of the type and the nature of adverse effects that a chemical has as inherent capacity to cause in an organism (OECD, 2003). Recent regulations require that hazard identification be performed before a pesticide can be approved for usage in agriculture or other areas. Therefore the information on hazards posed by pesticides is readily available from the pesticide registering bodies and on their websites for public usage. Most of the information is also available from international organisations such as JMPR, OECD and EC. The hazards that have been identified concerning pesticide include reproductive and endocrine disruption, neurodevelopmental delays, immune system, cancer and respiratory distress (Gilden, 2010). Studies are carried out in test organisms (microbial, cells or animals) and the exposure level is increased until an adverse effect is produced. The highest dose of the pesticide that does not cause detectable toxic effects on the test organisms is called the no-observed-adverse-effect-level (NOAEL) and is expressed in milligrams per kilogram of body weight per day (WHO, 1997). This is important because it is used in calculation of the ADI or the ARfD.

(g) Hazard Characterisation

Hazard characterisation is the qualitative and, wherever possible, quantitative description of the inherent properties of a chemical having the potential to cause adverse effects. This should, wherever possible, include a dose response assessment and its attendant uncertainties (OECD, 2003). Hazard characterisation involves comparing the pesticide exposure concentration with the ADI or the ARfD. The ADI is the estimate of the amount of a substance in food (mg/kg body weight/day) that can be ingested daily over a lifetime without appreciable health risk to the consumer (WHO, 1997). ADI is calculated by dividing the NOAEL for animal studies with an uncertainty factor of 100 to convert to a safe level for humans. A factor 100 (10 x 10) mostly used to account for species differences and individual variability in sensitivity to the chemicals (Renwick, 2002). ARfD is the estimate of the amount of a substance in food that can be ingested over a short period of time, usually during one meal or one day, without appreciable health risk to the consumer (WHO, 1997).

(f) Exposure Assessment

Assessment of the amount of a particular chemical or concentration that reaches a target organism, or population in a specific frequency for defined duration (OECD, 2003). The potential intake or consumption of pesticide residues is divided by the body weight and compared to ADI or ARfD in exposure assessment.

Exposure = (Concentration of pesticide residue x Food consumed)/body weight

The input data used in exposure assessment comes from supervised field residue trials, national pesticide monitoring programs and food consumption surveys. The residue levels from pesticide monitoring programs might not cover the whole food supply but they are always available in most countries and they reflect samples available for consumers.

In the risk assessment of the chronic exposure, the estimated total dietary exposure is compared with the toxicological reference value, Acceptable Daily Intake (ADI). ADI is the estimated maximum amount of a substance to which an individual in a population may be exposed daily over its lifetime without appreciable health risk. Figure 10 shows the nine commodities that contribute most to the exposure to

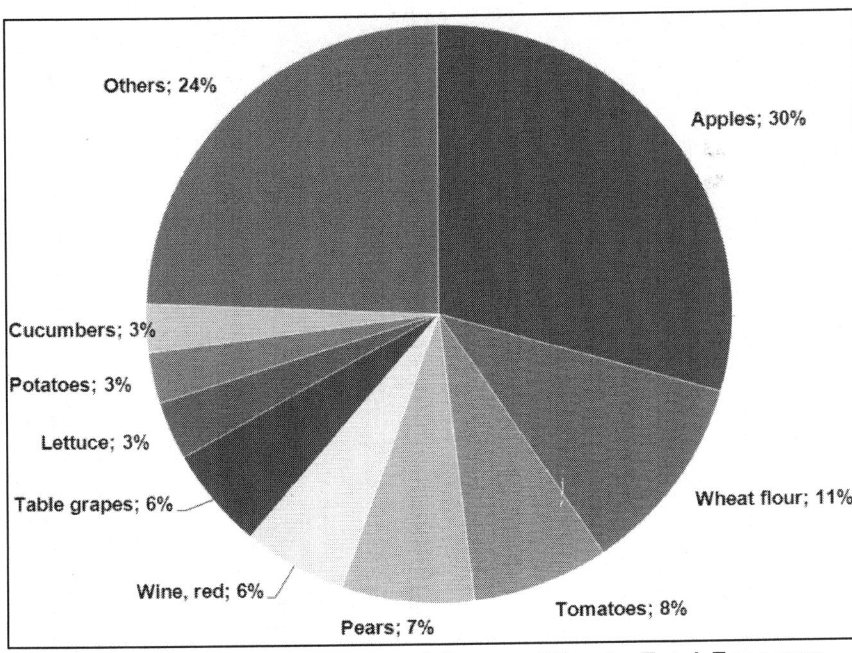

Fig.10: Relative Contribution of Commodities to Total Exposure to Pesticide Residues in the Diet.
***Source*: Anonymous 2013.**

populations together with the contributions from the rest of the commodities (Others). Apples contribute most exposure with 30 per cent (Anon., 2013).

(g) Consumption Data

Food consumption data are essential component of dietary risk assessment. The data used depend upon the type of population being assessed: children, special ethnic groups, geographical regions and estimation of the quantity of food eaten. Food consumption data may be obtained during food supply surveys, household inventories, household food use and individual food intake surveys (Hamilton *et al.*, 2004). According to EFSA guidance document on collection of food consumption data (EFSA, 2009), there are four types of dietary assessment methods, namely: diet history, food frequency, dietary records and dietary recall. In diet history, the history of the whole daily food intake of an individual and the usual meal pattern is assessed over a period of days, months and up to one year. Food frequency involves asking the consumers to estimate the usual frequency of consumption during a specified time for the foods that are listed on the questionnaire. In dietary records, the consumers weigh and record all the food including beverages before eating and also the leftovers after eating. The most appropriate source is the one that measures actual consumption instead of available food supply. Average daily consumption is the most used in exposure assessment calculations.

(h) Risk Characterisation

The qualitative and, wherever possible, quantitative determination, of occurrence of known and potential adverse effects of a chemical in a given organism or population, under defined exposure conditions is the risk characterisation (OECD 2003). The international estimate daily intake (IEDI) has been used to characterise the risk of pesticides. It is expressed as:

$$\text{IEDI} = \Sigma \text{ STMR} \times \text{E} \times \text{P} \times \text{F}$$

Where(STMR = supervised trial median residue level(E = Edible portion(P = processing factor(F = consumption of the food commodity(When the IEDI is more than the ADI the food involved is considered a risk to the concerned consumers. For the national estimated short-term intake (NESTI), the risk characterisation is compared with the ARfD.

(i) Hazard Index

In a study involving the fruits (Anonymous 2013), assessments of exposure to a mixture of pesticides were performed by using the Hazard Index (HI) method as described by US EPA (1986). The HI is the sum of Hazard Quotients (HQ) for the individual pesticides. The HI method assumes that the effects following cumulative exposure can be predicted by the mathematical model of dose-addition (OECD 2003). The Hazard Quotient (*HQ*) for a given pesticide is calculated from the total exposure from the diet to that pesticide divided by the ADI for that pesticide:

$$Hazard\ Quotient\ (HQ) = \frac{Exposure}{ADI}$$

The Hazard Index (*HI*) for a given diet is calculated by summing the Hazard Quotients (*HQ*) for each pesticide (*p*) in the diet:

$$Hazard\ Index\ (HI) = HQ_1 + HQ_2 + HQ_3 + \cdots + HQ_p$$

If the HI exceeds 1 (or 100 per cent), the pesticide has exceeded the maximum acceptable level and there might thus be a risk (OECD 2003).

Figure 11 shows the commodities that contributes most to the HI population. The seven commodities, apples, wheat flour, tomatoes, pear, table grapes, red wine and cucumber contribute most to both exposure and HI. Apples contribute most to both to the Hazard Index with 39 per cent. (Anonymous 2013).

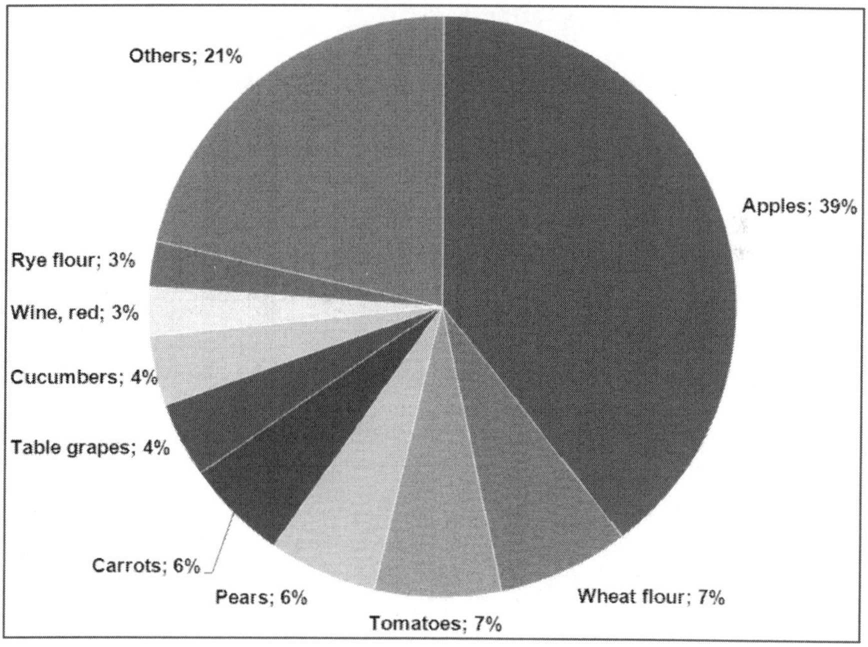

Fig. 11: Relative Contribution of Commodities to Hazard Index for Pesticide Residues in the Diet.
***Source*: Anonymous 2013.**

7. ENVIRONMENTAL EFFECTS OF PESTICIDES

(a) Pesticide Residues in Food

The majority of foods have detectable levels of pesticide residues. An analysis of several thousand samples of food, the overall assessment in 8 fruits and 12 vegetables was that 73 per cent have pesticide residues (Baker *et al.*, 2003). In five crops (apples, peaches, pears, strawberries, and celery) pesticide residues were found in 90 per cent of the crops. A study by Groth *et al.* (1999) detected 37 different pesticides in apples.

(b) Public Health Effects

(i) Acute poisonings

Human pesticide poisonings and illnesses are clearly the highest price paid for all pesticide use. Although the EPA (1992) estimated that 300,000 pesticide poisoning occurred annually, worldwide, the application of 3 million metric tons of pesticides results in more than 26 million cases of non-fatal pesticide poisonings (Richter 2002). Of all the pesticide poisonings, about 3 million cases are hospitalized and there are approximately 220,000 fatalities and about 750,000 chronic illnesses every year (Hart and Pimentel, 2002).

(ii) Cancer and other chronic effects

Ample evidence exists concerning the carcinogenic threat related to the use of pesticides. The major types of chronic health effects of pesticides include neurological effects, respiratory and reproductive effects, and cancer. There is some evidence that pesticides can cause sensory disturbances as well as cognitive effects such as memory loss, language problems, and learning impairment (Hart and Pimentel 2002). The malady, organophosphate induced delayed poly neuropathy (OPIDP), is well documented and includes irreversible neurological damage. In addition to neurological effects, pesticides can have adverse effects on the respiratory and reproductive systems. For example, 15 per cent of a group of professional pesticide applicators suffered asthma, chronic sinusitis, and chronic bronchitis (Weiner and Worth 1972). Studies have also linked pesticides with reproductive effects. For example, some pesticides have been found to cause testicular dysfunction or sterility (Colborn *et al.*, 1996).

US data indicate that 18 per cent of all insecticides and 90 per cent of all fungicides are carcinogenic. Several studies have shown that the risks of certain types of cancers are higher in some people, such as farm workers and pesticide applicators, who are often exposed to pesticides. Certain pesticides have been shown to induce tumors in laboratory animals and there is some evidence that suggest similar effects occur in humans (Colborn *et al.*, 1996).

Many pesticides are also estrogenic *i.e.* they mimic or interact with the hormone estrogen linking them to an increase in breast cancer among some women. The breast cancer rate rose from 1 in 20 in 1960 to 1 in 8 in 1995 (Colborn *et al.*, 1996). As expected, there was a significant increase in pesticide use during that time period. Pesticides that interfere with the body's endocrine–hormonal system can also have reproductive, immunological, or developmental effects (McCarthy, 1993). While endocrine-disrupting pesticides may appear less dangerous because hormonal effects rarely result in acute poisonings, their effects on reproduction and development may prove to have far reaching consequences (Colborn *et al.*, 1996).

(c) Destruction of Beneficial Natural Predators and Parasites

In both natural and agricultural ecosystems, many species, especially predators and parasites, control or help control plant feeding arthropod populations. Indeed, these natural beneficial species make it possible for ecosystems to remain green. With the parasites and predators keeping plant feeding arthropod populations at low levels resulting in low consumption of pesticides. Like pest populations beneficial natural enemies are also adversely affected by pesticides (Pimentel *et al.*, 1993).

(d) Pesticide Resistance in Insect Pests

In addition to destroying natural enemy populations, the extensive use of pesticides has often resulted in the development and evolution of pesticide resistance in insect pests and plant pathogens. An early report by the United Nations Environmental Programme suggested that pesticide resistance ranked as one of the top 4 environmental problems of the world. About 520 insect and mite species and nearly 150 plant pathogens of fruit crops are now resistant to pesticides (Stuart 2003).

Increased pesticide resistance in pest populations of fruit crops frequently results in the need for several additional applications of the commonly used pesticides to maintain fruit yields. These additional pesticide applications compound the pesticide residue and resistance problem by increasing environmental selection of pest populations for resistance. The pesticide resistance problem continues to increase despite all efforts and is spreading to other pest species.

(e) Honeybee and Wild Bee Poisonings and Reduced Pollination

Honeybees and wild bees are vital for pollination of fruits. Bees are essential to the production of about one-third of US and world crops. Their benefits to world fruit industry are estimated to be about \$120 billion year^{-1} (Pimentel *et al.*, 1997). Because most insecticides used in agriculture are toxic to bees because of harmful residues, pesticides have a major impact on both honeybee and wild bee populations. Pesticides and their residues adversely affect approximately 20 per cent of all honeybee colonies.

In addition to these direct losses caused by the damage to honeybees and honey production, many temperate fruit crops are lost because of the lack of pollination. Estimates of annual agricultural losses due to the reduction in pollination caused by pesticides and their residues may be as high as \$4 billion year $^{-1}$ (Pimentel *et al.*, 1997).

Poor pollination will not only reduce crop yields, but also equally important, it will reduce the quality of some crops, such as melons and apple fruits. In a study Mussen (1990) reported that with adequate pollination melon and apple yields increased by 10 per cent and fruit quality was raised by 25 per cent as measured by the dollar value of the fruit crop.

8. REFERENCES

Abelson, Philip H. 1994. "Adequate Supplies of Fruits and Vegetables." *Science*. Vol. 266, November 25.

Aktar, M.W. Sengupta, D and Chowdhury, A. 2009. Impact of pesticides use in agriculture: their benefits and hazards. *Interdisciplinary Toxicol.*, **2** : 1-12

Anonymous, 2013. Pesticide residues. Results from the period 2004-2011. Prepared by Division of Food Chemistry, Nutrition and Toxicology and Risk Assessment. National Food Institute, Technical University of Denmark. pp 9- 44

Anonymous, 2014. Monitoring of residual pesticides in Kashmir and their amelioration for sustainable agriculture. Department of Science and Technology, Government of India sponsored project.

Anwar, T. Ahmad, I. and Tahir, S. 2011. Determination of pesticide residues in fruits of nawabshah district, Sindh, Pakistan. *Pakistan J. Batany* **43** : 1133-1139.

Baker, B.P., Benbrook, C.M., Groth, G. and Benbrook, K.L. 2003, http: // www.consumersunion.org/food/orgnicsumm.htm (January 19, 2003).

Bhat, M. Wani, A. A. Mukhtar, M. Sherwani, A. Bhat, A. H and Showkat, A. 2015. Dissipation patterns of the fungicide difenaconazole (25 per cent EC) in apples grown in Kashmir, India. *Env. Monitor. Assess.,* **187**: 398-402.

Colborn, T., Myers, J.P. and Dumanoski, D. 1996, Our Stolen Future: How We Are Threatening Our Fertility, Intelligence, and Survival: A Scientific Detective Story, New York, Dutton.

Celik, S., Kunc, S. and Asan, T. 1995. Degradation of some pesticides in the field and effect of processing. *Analyst*, **120**: 1739 - 1743

David Pennell. 2006. *Pesticide Residue Minimisation Crop Guide for Apples*. Food Standards Agency UK. pp. 11-17

Ecobichon, D.J. 2001. Pesticide use in developing countries. *Toxicology*, **160**: 27 – 33

EFSA 2009. General principles for the collection of national food consumption data in the view of a pan-Europe dietary survey. *EFSA J.* 2009, **7**: 1435(

EFSA 2010. 2008. Annual report on pesticide residues according to article 32 of regulation (EC) No 396/2005. *EFSA J.* 2010, **8**: 1646.

EPA: 1992. *Hired Farm Workers Health and Well-being at Risk, United States General Accounting Office Report to Congressional Requesters*.

Gilden, R. C. 2010. Pesticides and Health Risks. *JOGNN*, **39**: 103 – 110.

Groth, E., Benbrook, C.M. and Lutx, K. 1999. Do You Know What You're Eating? An Analysis of US Gov- ernment Data on Pesticide Residues in Foods, http: // www/consumers union.org/food/do_you_know2.htm (January 19, 2003).

Hamilton, D., Ambrus, A., Dieterle, R., Felsot, A., Harris, C., Petersen, B., Racke, K., Wong, S., Gonzalez, R., Tanaka, K., Earl, M., Roberts, G. and Bhula, R. 2004. Pesticide residues in food – acute dietary exposure. *Pest Manag. Sci.,* **60**: 311-339.

Hart, K. and Pimentel, D. 2002, 'Public health and costs of pesticides. **In:** D. Pimentel (Eds.), *Encyclopedia of Pest Management*, New York, Marcel Dekker, pp. 677–679.

Holland, P.T. 1994. Effects of storage and processing on pesticide residues in plant products. *Pure Appl. Chem.,* **66**: 335–356.

Holland, J. and Sinclair, P. 2004. Environmental fate of pesticides and the consequences for residues in food and drinking water, **In:** *Pesticide Residues in Food and Drinking Water*: Human Exposure and Risks, Hamilton D. and Crossley S. (Ed.), 27 – 62, John Willey and Sons LTD, 0-471-48991-3, West Sussex, England.

Knutson, R. D., Hall, C. R., Smith, E.G., Comer, S. D. and Miller, J. W. 1993. Economic Impacts of Reduced Pesticide Use on Fruits and vegetables, American Farm Bureau Foundation.

Knutson, R. D., Hall, C. R., Smith, E.G., Co'mer, S. D. and Miller, J. W. 1994. "Yield and Cost Impacts of Reduced Pesticide Use on Fruits and Vegetables," *Choices, First Quarter*, 15-18.

McCarthy, S. 1993, 'Congress takes a look at estrogenic pesticides and breast cancer', *Journal of Pesticide Reform* **13** : 25.

Mukhtar, M., Sherwani, A., Wani, A. A., Ahmed, S. B., Sofi, J. A. and Bano, P. 2015. Degradation of chlorpyrifos residues in apple under temperate conditions of Kashmir Valley. *Env.Monitor.Assess.,***187**: 482-487.

Munazah, Y. 2014. Residue dynamics of some commonly used pesticides on apple in Kashmir. Ph.D thesis submitted to Sher-e-Kashmir University of Agricultural Sciences and Technology of Kashmir, pp. 49-68

Mussen, E. 1990. California crop pollination, Gleanings in bee culture **118**: 646–647.

OECD. 2003. Descriptions of selected key generic terms used in chemical hazard/ risk assessment. OECD Series on Testing and Assessment, No. 44.

Pimentel, D., Acquay, H., Biltonen, M., Rice, P., Silva, M., Nelson, J., Lipner, V., Giordana, S., Horo- witz, A. and D'Amore, M. 1993, 'Assessment of Environmental and Economic Impacts of Pesticide use.', **In**: D. Pimentel and H. Lehman (eds.), The Pesticide Question: Environment, Economics and Ethics. New York, Chapman and Hall, pp. 47–84.

Pimentel, D., Wilson, C., McCullum, C., Huang, R., Dwen, P., Flack, J., Tran, Q., Saltman, T, Cliff, B. 1997. Economic and environmental benefits of biodiversity. *Bioscience*, **47**: 747-757.

Richter, E.D. 2002. Acute human pesticide poisonings. In: Pimentel, D. (Ed.) *Encyclopedia of Pest Management*. Dekker, New York, pp. 3–6.

Renwick, A.G. 2002. Pesticide residue analysis and its relationship to hazard characterisation (ADI/ARfD) and intake estimations (NEDI/NESTI). *Pest Management Sci.*, **58**: 1073 – 1082

Sherwani, A., Mukhtar, M. and Alam, A. A. 2015. Insect pests of Apple and their Management. **In:** Pandey, A. A. and Mall, P. (Eds.), Insect Pests of Fruit Crops. Biotech Books New Delhi, pp. 295-306.

Stepan, R., Ticha, J., Hajslova, J., Kovalczuk, T. and Kocourek, V. 2005. Baby food production chain: Pesticide residues in fresh apples and products. *Food Additives and Contaminants*, **22:** 1231–1242.

Stuart, S. 2003. Development of Resistance in Pest Populations, http: //www.nd.edu/ chem191/e2.html (January 20, 2003).

U.S. EPA 1986. Office of Pesticides and Toxic Substances. Guidance for the re-registration of pesticide products containing glyphosate as the active ingredient. Washington, D.C., June.

USDA. 1994. United States Department of Agriculture, Pesticide Data Program (PDP Summary of Data, USDA AMS, April 1994.

Weiner, B.P. and Worth, R.M. 1972. 'Insecticides: household use and respiratory impairment', **In:** *Adverse Effects of Common Environmental Pollutants*, New York, MSS Information Corporation, pp. 149–151.

WHO, 1997. Guidelines for predicting dietary intake of pesticide residues. WHO, Switzerland.

Chapter 24

Insect Dynamics and their Management in Temperate Fruits and Nuts: A Global Scenario

Sheikh Bilal Ahmed and Nazir Ahmad Wani

Division of Entomology,
Sher-e-Kashmir University of Agricultural Sciences and Technology of Kashmir
Srinagar – 190 025, J&K
E-mail: sheikhbilala@gmail.com

1. INTRODUCTION

Jammu and Kashmir produces a vast range of fruit crops like apple pear, cherries, apricots, walnuts and almonds. However, insect pests have a significant effect on the fruit trees by reducing yields, lowering the quality, reducing the growth and increasing production costs. In some situations, pests completely prevent the commercial cultivation of some fruits. Consequently, every year Government of Jammu and Kashmir and fruit growers invest significant resources in combating these established and endemic pests.

The most common and damaging insect pests of apple are: codling moth (*Cydia pomonella*),green apple aphid (*Aphis pomi*), rosy apple aphid (*Dysaphis plantaginea*), red spider mite (*Panonychus ulmi*), rust mite (*Aculus schlechtendali*), Two spotted spider mites(*Tetranychus uriticae*), apple leaf miners (*Stigmella spp, Lyonetia spp*), woolly aphids (*Eriosoma lanigerum*), oriental fruit moth (*Grapholita molesta*), light brown apple moth (*Epipyas posvittana*), apple blossom weevil (*Anthonomus pomorum*), Western flower thrips (*Frankliniella occidentalis*) and San Jose Scale (*Quadraspidiotus perniciosus*) which are discussed in the chapter.

2. IMPORTANT FRUIT PESTS

(a) Codling Moth

Codling moth (*Cydia pomonella*) is one of the major and most destructive pests in apple orchards in Europe and north America. Its alternative hosts can be pears, apricots, quinces and walnuts. The control of the codling moth can prove to be difficult due to the overlapping of its generations and the limited time when spraying can be applied, which is the period between the hatching of the larvae and their entrance into the fruit. Once the larvae have entered the fruit, control is not possible. In India, Codling moth is restricted to Ladakh region of Jammu and Kashmir where it causes a huge damage to apple and apricot. In apple damage ranges from 30 -70 per cent in extreme cases. Government of Jammu and Kashmir has imposed a quarantine to prevent the export of fresh fruits from Ladakh to any other part of the country.

(b) Aphid

Aphid (*Dysaphis plantaginea*) is one of the main apple pests in Europe. Other important aphid species that cause damages to apple orchards are the apple-grass aphid (*Rhopalosiphum inserum*) and the green apple aphid(*Aphis pomi*). In conventional apple growing *Dysaphis* spp are usually tackled with an application of aphicide before flowering, followed by another after flowering or in early summer. This strategy is condemned to failure due to the appearance of resistance and the high impacts of the insecticides in the environment and for the health of farmers and consumers. Controlling this pest in spring is very difficult as the aphids are protected in tightly curled leaves. Monitoring the migration of winged adults assists in the optimal timing of autumn sprays.

(c) Mites

The stinging of mites such as the red spider mite (*Panonychus ulmi*), Two spoted spider mites(*Tetranychus uriticae*) and the rust mite (*Aculus schlechtendali*) cause a bronzing of the leaves. In the case of heavy infestations, they considerably weaken the trees, causing fruits to be smaller, less coloured and diminishing taste quality, as well as causing russeting of the fruit skin and premature leaf fall. In conventional orchards, the use of large spectrum insecticides completely eliminates the natural predators, while in IPM and organic orchards the presence of natural predatory mites and helps keep mites under control. However, in Jammu and Kashmir the application of pesticides continues to be the main tool of pest management.

(d) Wooly Aphid

Woolly Apple Aphid *Eriosoma lanigerum* is native to North America. This aphid can be found in all apple growing areas of Kashmir and in the United States and Canada. Several generations occur each year. The nymphs are produced asexually by mother aphids. Small colonies of nymphs may survive the winter on limbs, but most overwinter on roots and move onto limbs in June. Adults and immature are reddish-purple in color with a white, waxy covering. From a distance, a colony of woolly apple aphids appears as mass of cotton on the tree. This cottony mass is a result of long, white, filamentous, waxy secretions from the back of the adult aphids.

This waxy mass protects the woolly aphid from insecticides, weather, and natural enemies. Without chemical or natural controls, woolly apple aphids can increase to high numbers in apple orchards. Woolly apple aphids (WAA) overwinter as nymphs on galls formed on apple tree roots. During mild winters, some WAA may survive the winter in protected locations on above ground branches and stems.

WAA feeding on roots causes gall formation. The galls increase in size over time as aphid feeding continues. The galls inhibit root function and under severe infestations can stunt or weaken the tree. Severe root infestations can kill young trees. Heavy infestation on aerial portions of the tree can cause twig galls and leaf yellowing. Canopy infestations can lead to honeydew deposits on fruit causing the growth of the fungus that causes sooty mold, and subsequent downgrading of fruit.

(e) Western Flower Thrips

Western flower thrips *Frankliniella occidentalis* are native to western North America and are widespread throughout this region. This species feeds on hundreds of different weed and crop hosts. WFT populations can rapidly increase during fruit bloom, and are especially abundant in the warmer areas of the Mountains. These warmer areas generally have other crop and weed hosts that serve as reservoirs during the late winter and early spring, from which thrips rapidly emigrate to the more attractive and abundant resources of fruit blossoms. Heavy infestation of WFT is seen in Bandipora area of Kashmir. WFT damage the blossoms and primarily cause cosmetic damage (scarring) to fruits which lowers their quality and marketability. Nectarines and light-skinned apple fruits are most affected.

(f) San Jose Scale

San Jose Scale (*Quadraspidiotus perniciosus*) is a very serious pest in most apple growing regions in the world including Kashmir. Female Scan Jose scale gives birth to living young ones that emerge from under the edge of the scale covering. These tiny yellow crawlers wander in a random fashion until they find a suitable place to settle. Immediately upon settling, the crawlers insert their mouthparts into the host plant and begin feeding and secreting a white waxy material (white cap stage); eventually the waxy covering turns black and is known as the black cap stage. San Jose scales over winter predominantly in the black cap stage, although in mild years some adult mated females may also survive. In late January, these nymphs resume their growth. Immature male and female scales are indistinguishable until the first molt. At this time, the male scale covering begins to elongate, while the female's remains circular. Males molt a total of four times. Following the final molt, adult male scales emerge from the scale covering as tiny, yellow winged insects. They mate with the females, who remain under the scale covering. After about two months, crawlers begin to emerge from the females, usually in April; peak emergence is generally in early May. There are usually four generations a year. Summer generations overlap and crawlers are present throughout summer. Infested trees look water stressed, and wood encrusted with scale insects may die back. Scan Jose scale infested bark often cracks and dies, and heavily-infested scaffold limbs and branches die within 1 to 2 years.

3. PESTS OF NUTS CROPS

Walnut(*Juglans regia*) and almond (*Prunus amygdalus*) are the main nut fruit crops of Kashmir. They are sold at a primer price in the country and have a tremendous export potential. An array of insects that feed on their leaves and developing nuts and tunnel into buds, shoots and through the bark into the wood. However, only a few insects would be considered as pests of great significance. The insects that damage walnuts in Kashmir and else where are grouped into three categories.

(a) Insects Damaging Foliage

(i) *Chaetoprocta odata*

It is a blue butterfly which overwinters in the egg stage for a long period that ranges from 230- 260 days starting from July to ending March. The eggs hatch in the first week of April and the larvae start feeding on the young leaves. The larvae moult through 3 instars which last to 45 days before entering into the pupal stage that lasts from 20-28 days from 1st week of June to last week of June. The adults emerge in June –July, mate in 10-15 days and lay overwintering eggs in July. The adult hood lasts for 23- 27 days. The larvae are sluggish and greenish in colour and cause a serious damage to the leaves and emerging buds. The second Instar larvae are the most damaging stage of the pest. The pest is most abundant and serious at Sullar, Anatnag causing 33.90±2.088 per cent damage to foliage followed by Wussan, Ganderbal (15.83±0.731 per cent) and Kupwara (8.00±1.200 per cent).

(ii) Walnut Aphid, *Chromaphis juglandicola*

Walnut aphid can be a serious pest of walnut. Its feeding reduces tree vigor, nut size, yield, and quality. In addition to direct feeding damage, they excrete copious amounts of honey-dew that falls onto nuts, leaves and shoots. Honey-dew supports growth of the black sooty mold fungus. This fungus reduces light penetration to the leaf surface reducing its photosynthetic capacity. Being black, it also absorbs heat to predispose nuts to sunburn and subsequent kernel quality loss due to high temperatures. High populations of aphids may also cause leaf drop, exposing more nuts to sunburn. Eggs hatch as soon as leaf buds begin to open. These aphids settle on the leaflets (usually on the undersides of the leaf), mature, and reproduce without mating, giving birth to live nymphs. In autumn, the wingless females mate with smaller, winged males and they lay the over-wintering eggs.

(iii) Dusky Veined Aphid, *Callaphis juglandis*

The dusky veined aphid is a walnut pest is similar to walnut aphid. It overwinters in the egg stage on twigs. Eggs hatch as soon as leaf buds begin to open where the young aphids settle on the leaflets, and they mature into larger, yellow aphids with dusky black spots, and reproduce without mating, giving birth to live nymphs. The aphids pass through many generations a year, depending upon temperature. In autumn, wingless females mate with smaller, winged males and lay the overwinter eggs. In contrast to walnut aphid however, dusky veined aphids feed on the upper sides of leaves at the midrib.

(iv) Indian Gypsy Moth, *Lymantria obfuscate* Walker

Gypsy moth undergoes four developmental life stages; these are the egg, larva (caterpillar), pupa and adult. Gypsy moth females lay between 500 to 1,000 eggs in sheltered area such as underneath the bark of trees. The eggs are covered with a dense mass of tan or buff-coloured hairs. The eggs are the over wintering stage of the insect. Eggs are attached to trees, houses, or any outdoor objects. The eggs hatch in spring (April) into caterpillar. Where they spin a silken thread and are caught on wind currents. In early summer (June to early July), Gypsy moth caterpillars enter a pupal or transitional stage. The pupae are dark brown, shell-like cases approximately two inches long and covered with hairs. They are primarily located in sheltered areas such as tree bark crevices or leaf litter. Adult Gypsy moths emerge from the pupae in 10 to 14 days. They are present from July into August. Females have white to cream-colored wings, a tan body, and a two-inch wingspan. Female Gypsy moths cannot fly. Gypsy moth has only one generation per year. Young caterpillars primarily feed during the day whereas the older caterpillars feed at night. When present in large numbers, the older caterpillars feed day and night.

(v) Chaffer Beetles (*Holotricha* sp.)

The chaffer beetles complete life cycle in one year. Eggs hatch around mid-July and the grubs moult twice over 8 weeks. The mature grubs are well adapted to cool moist conditions and feed all through the autumn. During the winter they dig down to escape the periods of freezing conditions, but otherwise remain within 5 cm of the surface. They feed in the spring until April when they become pupae. Adults emerge in late May; fly to nearby deciduous trees to mate and feed, and subsequently females deposit up to 50 eggs each.

Most of the damage is done by the third (final) instar grubs during autumn and early spring, but damage can be masked by the abundant moisture at these times. Chaffer grubs prefer to feed on fibrous roots and can damage nursery plants of walnut by reducing their fibrous root system. The adult beetles are active at dusk and can be seen in groups in deciduous trees. Walnut foliage is one of the preferred hosts and the adult beetles feed on the foliage and cause a heavy damage to the leaves of host plants and under severe conditions they cause complete defoliation.

(vi) Walnut Blister Mite (*Eriophyes erinea*) Eriophyidae

The walnut blister mite occurs in walnut orchards. Adult mites are very small and cannot be seen without a 14 to 20 X hand lens. They have a white, slender, striated body with a few long hairs. Immature forms resemble adults but are smaller. Eggs are spherical and pearly white. This mite over winters beneath bud scales. When the weather warms up, the mites feed beneath the leaves amongst the leaf hairs. Several generations occur during the summer, which attack new foliage as soon as it unfurls.

Blister mite feeds on the lower surface of leaflets, causing characteristic blister like swellings on the upper surface of leaflets. It occurs in yellow to orange fealty masses in depressions on the underside of blistered leaves. Later in the season, these areas turn brown. Very small mite causes blister like swelling on upper leaf surface and yellowish or brown concave pocket on underside of leaf. These insects usually do not cause enough damage to warrant control.

4. INSECTS DAMAGING BARK AND TREE TRUNKS

(a) Stem Borer (*Aeolesthes sarta*) Coleoptera : Cerambycidae

Aeolesthes sarta is a polyphagous insect. Its preferred hosts include English walnut, apple, poplar and willow. However, it may also attack other fruit and forest trees. *Aeolesthes sarta* requires two years to complete a generation. Adults usually leave their pupal cells, located in the xylem, in April or the beginning of May, when the temperature averages 20°C. They are generally active in the evening and at night. During the day they hide under the bark, in larval tunnels, and in other refuges. After about 8 p.m. they leave their hiding places. The adult has an elongated, dark grey-brown body, 28-47 mm long, with the elytra being obliquely truncate at the apex and covered with short slivery hairs. The male is usually smaller than the female. The males appear first, and move about until morning on the same tree on which they have developed. Usually several generations develop on the same tree until the tree is eventually killed.

Females lay eggs in slit-like niches in the bark of the trunk and the larger branches; Egg laying begins shortly after females leave their pupal cells and continues for about two months. Usually 1 to 3 eggs are laid per niche and females may lay a total of 240 to 270eggs. The eggs are white, about 3-4mm long. The eggs hatch 9 to 17 days after being laid. Each larva makes its own tunnel, even if several eggs are deposited at the same place. The larvae are pale yellowish in colour with black mandibles and range 60-70 mm in length and are covered with golden hairs. Feeding begins in the cambium region and frass is ejected through the entrance hole. After feeding in the cambium from a period of time, the larvae enter the xylem. At the end of their first season, larvae make a long gallery that first goes upward for about 10 cm and then changes direction moving downward and resulting in a vertically gallery 15 cm long. At the bottom of this gallery, the larva over-winters protected by a double plug made from borings. The following spring, the larvae resume feeding and construct tunnels deep into the wood. At the end to July, they prepare pupation cells protected by double plugs made from borings. Pupation occurs in these cells and about two weeks later adults emerge. Adults stay in the pupation cells over winter and emerge the following spring.

(b) Pin Hole Borer or Shot Hole Borer (*Scolytoplatypus* sp.) Coleopetra : Scolytidae

Bark beetle spend majority of their lives beneath the bark of a single host tree. All bark beetles undergo four stages of development: egg, larva, pupa and adult. An adult female lays eggs in egg galleries between the inner bark and outer sapwood of the host tree. The eggs hatch to become larvae, which feed on the phloem and nutritious fluids of the tree until they are ready to pupate. After four larval stages (instars) the larva then bores into the inner bark where it forms a hardened outer casing where it becomes a pupa. Within the casing the beetle develops into the adult form. Adults are 3-8 mm long and feed on the inner bark, phloem, or both. The Adults eventually emerge from the host tree to colonize a new tree and reproduce when the surrounding air temperature is higher than about (16°C). Bark of twigs, branches and trunks of infested tree is perforated with numerous small shot holes.

5. INSECTS DAMAGING NUTS

(a) Tortrix Moth (*Tortrix* sp.) Lepidoptera : Tortricidae

The first moths appear around mid-June and are active until mid August during its one generation life cycle. Eggs are very similar in colour to the leaves and are laid in large scale like batches. Eggs are deposited in groups on the leaves. Larvae are 15mm in length, green with a red brown to black head with a brown thoracic plate. It has a medial longitudinal band which is narrow and dark with long silvery bristles which arise from small whitish warts that are disseminated over the entire body. The larvae during the 3rd instar, hibernate in a cocoon woven at the base of the leaves or at a branch axils. The adult moth has a 20 to 25mm wingspan. The female has pale brown forewings with a dark brown band and is bi-colored, the distal half is golden yellow and the basal half is grayish.

As early as the beginning of spring, they roll leaves up and devour them. Caterpillars after hatching from eggs in the month of May bore into the nut lets and feed on green hull inside, but they do not enter the fruit. The fruits often turn dark in colour and the nuts shrivel.

(b) Walnut Weevil (*Alcidodes porrectirostris*) Curculinoidae : Coleroptera

This curculios species attack walnuts and adult weevils are about ½ inch long with a snout about half the length of the body. They are brownish-gray with humps and ridges. They over winter as adults in the ground cover. The grubs are fleshy and white with brown heads. The walnut weevil has two generations per year, White creamy grubs after hatching from eggs bore into the developing fruits. Grubs of the first generation pupate inside the fruit while as the adult weevil of the second generation hibernate during winter and become active in next April.

The female insect lays eggs on the fruits which on hatching bore deeper and feed on kernels causing premature fruit drop. The adult weevils emerge from fruits and feed on petioles and tender shoots. Both the adults and the grubs damage the twigs, buds and nuts. Besides nuts, a significant damage is caused to new shoots and twigs. Under the laboratory conditions the adult weevils were found to make many holes in the walnut hull through fluids would ooze out thus making it more susceptible to secondary infection by diseases. The second generation adults would hibernate in the soil until next April. The walnut weevil is potentially important pest in Karnah area of Kashmir valley and causes varying levels of damage at different places in karnah.

(c) Walnut Husk Fly

The adult walnut husk fly (*Rhagoletis complete*) is about the size of a housefly and very colorful. A yellow spot just below the area where the wings are attached and a dark triangular band at the tip of the wings distinguishes the husk fly from other flies likely to be found in orchards. Husk flies have one generation per year. They overwinter as pupae in the soil and emerge as adults from late June until early September (in coastal areas emergence can begin as early as mid-May).Peak emergence is usually from mid-July to mid-August. The female deposits eggs in groups of about 15 below the surface of the husk. Eggs hatch into white maggots within 5 days. Older maggots are yellow with black mouth parts. After feeding on the husk for 3 to 5 weeks, mature maggots drop to the ground and burrow several inches into the soil to pupate. Most emerge as adults the following summer but some remain in the soil for 2 years or longer.

Adult female husk flies can be distinguished from males by their slightly larger size, a pointed abdomen with an ovipositor, and by the color of the first leg segment. On females, the first leg segment is straw colored, whereas on males it is brown to black. The walnut husk fly is a mid- to late season pest. It occurs in all walnut-growing areas in California. Black walnut and all cultivars of English walnut are suitable hosts for the husk fly.

The first signs of an infestation are small stings caused by females depositing eggs in the husk. After hatching, the maggots feed inside the husk, turning it very soft and black. The outer skin of the husk usually remains intact, but its fleshy parts decay and stain the nutshell. These stains cannot be removed by normal bleaching procedures, and the nut is therefore unsatisfactory for in-shell sale. A husk fly infestation early in the season (late July to mid-August) leads to shriveled and darkened kernels, increased mold growth, and lower yields.

5. PEST MANAGEMENT TACTICS

The tactics usually involves application of broad spectrum insecticides. Over reliance on the use of synthetic pesticides in crop protection programmes around the world has resulted in disturbances to the environment, pest resurgence, pest resistance to pesticides and lethal and sub lethal effects on non target organisms including humans. These side effects have raised public concern about the routine use and safety of pesticides. At the same time population increases are placing ever greater demands upon the "ecological services" Continued use of broad spectrum insecticides represents a major barrier to the implementation of biocontrol agents for many secondary pests. In this context, an IPM-approach which besides other tools, involves manipulating the behavior of the pest, provides a sound solution.

(a) Pest Monitoring

At the farm level, IPM is approached as a series of activities that culminate in a control decision made by the grower or his consultant. The first activity is gathering information about the environment, pest, potential biological control agents and the crop. This activity is termed scouting, or monitoring, and is performed frequently during the growing season to acquire periodic information about the orchard status so that timely decisions can be made and actions can be taken if necessary. Monitoring is the most fundamental, yet the most often neglected activity in an IPM program. Both the need for control and the effectiveness of any action taken are determined by monitoring pest and natural enemy populations. To decide if control is required, pest density must be related to the potential damage and balanced against how likely it is that biological control agents can maintain control below damaging levels. This information forms a sound basis for decision making – the absence of this information usually leads to the overuse of pesticides. In order for management decisions to be made in a timely fashion, information must be acquired on a frequent and regular basis, in most cases, weekly. Pheromone traps using these lures and a sticky coating are a quick and convenient way to monitor the populations of such pests. Traps are placed in the orchard before the beginning of moth emergence and are checked daily to record the first capture, or biofix, and then again at weekly intervals throughout the pest's life cycle each season. The biofix can be used to begin accumulating degree days for use in insect development models to predict future insect stage distributions for various pests throughout the season so that growers will know the optimum timing to control various pests. Insect traps alone should not be relied on for management decisions, but should be used in conjunction with direct tree inspections in the orchards.

Based on the monitoring data, a decision can be taken for the use of an appropriate control tactics.

(b) Synthetic Organic Insecticides

Synthetic organic insecticides have been the dominant pest control tactic used in orchards since their introduction soon after World War II. There are four main groups: organophosphates, organochlorines, carbamates and pyrethroids. They have been relatively inexpensive, highly effective and fast acting, often providing almost complete control. They affect a wide range of organisms and often kill natural enemies as well as pests. In addition, many pests have developed resistance to them. Many of these pesticides are highly persistent and leave their toxic residues in the food commodities at levels higher than prescribed MRL values. These powerful pest control tools should be used only as a last line of defense against pests. Natural enemies, selective insecticides and other non-disruptive control tactics should be given every opportunity to work before broad-spectrum insecticide are used.

(c) Pheromone-Based Management Programmes for Fruit Pests

Mating disruption (MD) is most effective if the application is well timed, the rate is high enough, and coverage is good. Effectiveness of MD is reduced if dispensers are applied late, since this tactic provides no control once mating has taken place. Most moths begin mating immediately upon emergence into the orchard. Dispensers should be in place prior to the predicted start of adult activity for the target pest. Codling moth (CM) and oriental fruit moth (OFM) are principal candidates for control by MD in fruit orchards. Both pests have more than one period of adult flight activity, the overwintering generation and at least one summer generation.

Experience teaches us that orchard borders require extra attention when implementing mating disruption programs. First, mated females immigrate into the border region from adjacent orchards not treated with pheromone. Second, it is suspected that pheromone concentrations are lower on the borders than the interior, thus increasing the likelihood of males locating females and mating along borders. Additional pheromone can be applied to three or four rows around the orchard perimeter to help protect orchard borders. Treating large contiguous areas with mating disruption is the best protection against border damage, as this strategy decreases the amount of border space relative to orchard interior.

Proper placement of dispensers within the tree canopy is a critical component of a MD control program. Good control of OFM can be achieved by placing dispensers in the middle of the canopy. In contrast, successful mating disruption of CM requires placing dispensers within two feet of the top of the canopy, but near foliage to protect them from UV radiation and high temperatures.

(d) Pheromone Formulations and Dispensing Systems

Three general types of MD technologies comprise the majority of commercially available products, hand-applied, sprayable and aerosol emitters. The tree fruit industry is probably most familiar with reservoir-type dispensers, such as plastic tubes, membranes, puzzle-pieces or spirals. Pheromone is dispensed from these devices at rates that are thousands of times greater than the release rate of pheromone from a female moth. Reservoir dispensers are hand-applied at a rate of 100-400 sources per acre, depending on the targeted species and other factors including pest density and the product's label.

Sprayable disruption formulations disperse up to a half million pheromone-containing microcapsules per acre. For OFM, the approach can yield a high level of disruption for about 2-3 weeks, less if hard rains dislodge the capsules from the foliage. Applying low rates at frequent intervals is the best approach for disrupting OFM with sprayable pheromone. Aerosol emitters are deployed at densities of 0.5-2 per acre, but each unit releases pheromone every 15-30 min. at a rate equivalent to that of 50-100 hand-applied reservoir type dispensers. These devices provide a stable environment for the large volume of pheromone prior to its release and control the emission of pheromone mechanically to give a constant predetermined release rate. Aerosol emitters have provided up to 97 percent disruption of male OFM orientation to pheromone-baited traps in treated plots compared with untreated controls.

(e) Codling Moth Granulovirus

Codling moth granulovirus (CpGV) has shown promise as a useful tool for codling moth control. Because CpGVs are extremely selective and have no re-entry or preharvest restrictions, they are useful as biological control agents. An additional advantage is that CpGV can be applied to apple fruit using conventional spray application methods. Codling moth larvae must ingest CpGV for infection to occur. Once inside the host organism, the virus multiplies and infects body tissues, killing the host in five to 10 days. This delay can allow continued feeding on an infected host, however, which can produce scars on the surface of the fruit. Several formulations of CpGV are commercially available. As with all insecticides, growers using CpGV should integrate it with other insecticides and cultural forms of pest control to avoid risk of resistance development. In Washington, the virus was combined with applications of the organic insecticide spinosad in pear and apple orchards. Spinosad reduced codling moth damage at harvest when compared to stand-alone use of the CpGV, and using both decreased fruit damage at harvest and reduced trap captures in subsequent seasons.

(f) Horticultural Spray Oils

Horticultural spray oils are produced by distilling and refining crude petroleum oils. They are often used in combination with other chemicals but can be used alone to suppress insects and mites. Applied during the dormant or delayed dormant period, they kill overwintering stages of pests such as scales and eggs of European red mite. They also inhibit egg laying by pear psylla females. To date, there are no documented cases of an insect developing resistance to oils. The mode of action is not clear, but several mechanisms are suspected including smothering the insect or its eggs or penetrating the insect's cuticle and interfering with nerve transmission. Oils can be phytotoxic to fruit and foliage of deciduous tree fruits. Damage varies with the type of oil (paraffin content, unsulfonated residue, distillation range, and viscosity), the tree species, application concentration, and the weather before, during or after the application. Combining oil with certain pesticides or nutrients may exacerbate phytotoxicity.

(g) Biological Control

Biological control is a pest control tactic where a pest's natural enemies (predators and parasitoids) are used to keep its density below damaging levels. Every pest has natural enemies, but whether they can provide adequate control in an orchard is difficult to predict. To make the most effective use of biological control you must know the life cycle, biology and ecology of both the pest and its natural enemies. Biological control alone will not be sufficient to control all pests attacking a tree fruit crop, but it plays an important part in an IPM program. Biological control can either occur naturally or can be encouraged by introducing natural enemies into the orchard. It also can be enhanced by protecting natural enemies that are in the orchard and providing suitable habitat near the orchard or in the cover crop to promote their survival. Natural enemies have the potential to control many secondary pests such as mites, aphids, leaf miner, pear psylla and grape mealy bug. Broad-spectrum insecticides kill most natural enemies, allowing outbreaks of pests they would normally control. Biological control is relatively safe, often permanent and economical. However, it can take time to implement a biological control program, and it may mean having to tolerate some crop injury until natural enemies build up to sufficient numbers to provide adequate control.

6. OTHER IMPORTANT ISSUES IN FRUIT PEST MANAGEMENT

(a) Pesticide Quality

Poor quality pesticides are unlikely to serve the intended purpose; provide poor value to users; encourages growers to increase the dosages or the number of sprays, trigger the development of resistance and are likely to be more harmful, directly or indirectly to humans and the environment.

(b) Adoption of GAP for Making Apples Safe for Consumption

Suggestions

Mitigation of toxic pesticide residues to allowable levels in harvested apples requires both short term and long term measures enlisted:

- Make sure that when spraying pesticides on crops, recommended doses are followed. Spraying higher doses than recommended, will lead to high levels of pesticides residues in the harvested fruits which will lead to no compliance with Codex/EU MRLs.
- Make sure that no pesticide is sprayed on the crop at least one month before harvesting of fruits is started. This is particularly very important under agro-climatic conditions of Kashmir valley, where usually a low temperature prevails, a factor conducive for high persistence of pesticides.
- State enforcement agencies like directorate of law and enforcement department, GoJK should ensure that spurious and misbranded pesticides are not pumped into the markets and all genuine pesticides recommended by university are made available to the fruit growers.
- Under the provision of section 37 of insecticide Act 1968, Government of Jammu and Kashmir should contemplate to constitute State Pesticide Regulatory Board for effective enforcement of insecticide Act 1968 and insecticide rules 1971 made there under to check the sale of spurious pesticides.
- Make sure that the safety intervals of all chemicals applied on the crop are respected. Otherwise residues from those chemicals will not get enough time to dissipate and remain in/on the fruits at levels higher than MRL's.
- Minimize the carryover of pests and diseases e.g through appropriate disposal of fallen leaves and dead/pruned twigs.
- Using alternative methods of control like cultural, mechanical, physical and biological techniques or combining these with the planned chemical crop protection programme.
- Regular testing, maintenance and calibration of sprayers needs to be given a top priority, as the spray droplet size shall affect the buildup of pesticide residues.
- Harvested apple fruits are many times brought from the orchards to the markets or to the doorsteps of consumers openly in the load carriers. Make sure that transport vehicles are completely clean when transporting harvested fruits to mandies and terminal markets, so that these fruits are not contaminated with pesticides/chemicals or other toxic substances that might have been transported before in the same vehicle.
- Avoid calendar spraying i.e pesticides are applied on a schedule without considering the actual presence or extent of pests and diseases in the crop. This practice can result in pesticides being applied when they are not needed, can increase pest and disease resistance and be a waste of time and money.

- Monitor pests and diseases regularly for pests and diseases by carefully checking plants with a magnifying glass or hand lens for the presence of pests and disease. Sticky traps can also be used to check for the presence of specific pests. Regular monitoring ensures problems are found early, making control easier and saving you time and money.
- Use pest and disease thresholds: The presence of a pest or disease does not automatically mean financial loss. In some situations, the cost of applying pesticides may be greater than the loss if no action was taken.
- Keep the farm clean: A clean farm usually has fewer pests and disease problems. Keep the farm free of weeds and carefully dispose of diseased plants and old plant matter. Ensure that any planting and propagating material is free from any disease/pest and is as clean as possible. This will reduce the number of sprays and help in mitigating pesticide residues.
- Botanical pesticides with multiple toxic, behavioral and physiological effects should find a place in IPM in order to reduce the load of synthetic pesticides.
- The pesticides should be judiciously used and extension agencies should have a role to guide the fruit growers for adoption of good agricultural practices particularly with respect to dose and application timing.

FUTURE THRUSTS

- Prudent use of pesticides is an immediate need to avoid commercial problems that affect all of us.
- Let's jointly take care of our production and commerce by applying Good Agricultural Practices.
- Large quantity of insecticides, fungicides and herbicides are pumped into the Kashmir valley annually for use on agricultural and horticultural crops. Poor efficacy of these pesticides reported at times by the growers puts a question mark on their quality and is a major headache for all stakeholders and for the environment and overall persistence of the pesticides. Government of Jammu and Kashmir needs to notify and establish more pesticide testing laboratories to check the quality of these pesticides.
- Government needs to have a policy environment conducive to IPM which would discourage the use of pesticides if the buildup of pesticide residues is to be minimized.
- Popularization of scab resistant varieties among apple growers shall surely reduce the number of sprays and minimize the buildup of residues in farm gate samples.
- Funding for IPM projects is crucial. State Government or national funding agencies should support these IPM projects to strengthen farmer's training cum awareness programmes. Adoption of IPM is a way forward to minimize the pesticide residues in ready to eat agricultural food commodities.

Chapter 25

High Density Planting of Temperate Fruits: Technology Development

M.K. Verma

Division of Fruits and Horticultural Technology,
ICAR-Indian Agricultural Research Institute,
New Delhi – 110 012
E-mail: mahenicar10@gmail.com

1. INTRODUCTION

Sustainable and maximum quality fruit production is the dream of every fruit grower. The yield and quality of the fruit are dependent factor of the number of trees per hectare, ideal tree architecture (training system) and the annual essential cultural practices like pruning, fertilization and plant protection measures. Such results are expected from high density plantings (HDPs). High density planting is the accommodation of maximum number of plants per unit area for production maximum yield of quality fruits. This technique was first commercially exploited for apple in Europe during 1960s and now majority of the apple orchards in Europe, America, Australia and New Zealand are grown under this system.

Increasing the number of trees per hectare increases yields, up to a point, in young orchards. However, as trees mature, too many trees per hectare results in crowding and shading with commensurate declines in yield and fruit quality and increases in pest and disease problems. Therefore, the ideal tree density varies greatly depending on soil type and topography, varieties, rootstocks, training systems, availability of irrigation and the grower's management capabilities. Traditionally, it was common to set an orchard "on the square," in which the distance between trees in-a-row equalled the distance between-rows. Often, the orchard floor was kept free of grasses and weeds by cultivating down the rows and across rows. High fruit quality could be attained because mutual shading was not a problem. Yields were low to moderate because a relatively high percentage of the orchard space was devoted to drive areas and, hence, not covered with fruiting wood. Planting trees close enough in the row that they almost touch at maturity and leaving an extra eight feet between rows to accommodate equipment movement results in more trees per acre, which promises potentially higher early yields plus greater yield potential in the mature orchard.

Orchard establishment is a long term process. A good orchard gives profitable returns and unplanned orchards are unprofitable venture. Therefore, it is essential for orchardists to plan their new plantings according to the availability of resources judiciously. Among the resources, land is the most limiting factor and must be utilized to the fullest of its potential. This is only possible by adoption of the closer plantings. Accommodation of maximum plants per unit area to get the maximum possible profit per unit of the tree volume without impairing the soil fertility status is called the high density planting. It is defined as planting of a density in excess of that which suffices to give the maximum crop yield of maturity if the individual trees grow to its full natural size. The underlying principle of HDP is to make the best use of vertical and horizontal space and to harness the maximum possible return per unit of inputs and natural resources. In other words, it is nothing but accommodating more number of plants per unit area than optimum through manipulation of tree size. This gives better light distribution within tree canopy increases the number of well illuminated leaves and promotes the rate of photosynthesis that leads to high yield per unit area. This system of planting is going to revolutionize the fruit industry by enhancing the production of quality fruits with minimum damage coupled with reduction in production costs. Basically, the availability of a dwarf plant is the first and foremost prerequisite for establishing high density or meadow orchard. Fertilizer dose, spacing, growth regulation by the training and pruning, use of the mechanical devices *etc.*, may also be tried either singly or coupled with other crop management practices for a successful adaptation of these concepts.

2. PRINCIPLES OF HDP

In the process of evolution in planting density several important principles have been discovered which have guided the shifting from wider planting to high density plantings. These mainly include:

- ✰ To achieve higher yields orchard canopies must intercept a high proportion of available light (70-75 per cent). Pedestrian orchards do not intercept more than 55 per cent of available light.
- ✰ Thick canopies have too much heavily shaded area with poor fruit quality. Narrow canopies or planar canopies with a depth of not more than 3 feet have better light distribution. This has led the effort to narrow the canopy of modern orchards to not more than 3 feet deep.
- ✰ High early yield to pay back the initial investment to plant the orchard through the use of use of feathered trees, planting higher tree densities,

maximizing tree growth after planting with irrigation and fertigation, minimizing pruning at planting and in the first 3 years and branch bending to induce early cropping.

- Simple and thin tree canopies are more adaptable to partial mechanization of some orchard management practices than thick, complex tree canopies.
- Orchard planting density is limited by the economic law of diminishing returns. As planting density is increased the additional benefit in yield is smaller and smaller with each additional tree. At some point, the cost of additional trees is greater than the gain in yield.

(a) Impact of Low Planting Density

When the planting density is too low, each individual plant may perform at its maximum capacity, but there are not enough plants as a whole to reach the optimum yield. Therefore, total yield of the crop becomes a limiting factor. Also, when there is too much space left between plants, weed growth is promoted, which could increase weeding costs.

(b) Impact of a High Planting Density

If the planting density is too high, plants may compete against each other, known as intra-specific competition. Under those conditions, the performance of individual plants becomes a limiting factor for maximum crop yield.

(c) Using the Right Density

As the planting density increases, the total crop yield increases and reaches a maximum, at which point further increase in planting density results in reduced yield. It is, therefore, important to use appropriate planting density, keeping in mind the fact that the density that provides the highest yield and the density that provides the highest revenue may be different. Growers should consider for both optimum yield and revenue when balancing their planting density.

(d) What is the Optimum Planting Density?

In general, it is established that the greater the planting density, the greater the investment cost to establish the orchard. However, due to higher early yield and higher cumulative yield, profitability is generally increases with increases in tree density up to a point. Nevertheless, the law of diminishing returns which results in less gain in cumulative yield as more trees are planted per ha, meant that very high tree densities were not more profitable than more moderate densities. In addition, economists suggest that risk increases with increasing level of investment, thus making the very high-density systems riskier. In a study conducted by Robinson (2013) on net present value (NPV) of the accumulated profit over 20 years showed that the per unit land area the greatest profitability was at a tree density of 1,050 trees/acre when feathered trees were used.

(e) Tree Density Guide

- Semi-dwarf orchards have densities of 123-311 trees per acre at spacings of 16' x 22' to 10' x 14'.

- ☆ Dwarf orchards are planted with spacing as tight as 2' x 10', but more commonly 4' x 14' and up to 7' x 16' for densities of 388-777 trees per acre.
- ☆ Standard orchards are no longer planted because they take too long duration to produce.
- ☆ High density means more than 500 trees to the acre and will exceed 1000+ with multi-row systems.

3. CATEGORIES OF HDP

The establishment of an orchard is a long term investment, with orchards having a life expectancy of more than 20 years. It is important that the establishment of an orchard be well planned including the site planning, the planting and the tree care. The high density plantings can require a grower investing in excess of $20,000 per acre for the establishment. The extensive planning for high density plantings will pay off in maximizing yields and returns over the productive life of the orchard. Mistakes made in the planning and planting can be very costly and are difficult to correct in later years. It is paramount that careful attention be paid to the site selection, soil preparation, tree quality, cultivar selection and early tree care. Within this range, the decision of how close to plant a new orchard should be made with consideration of the vigour of the variety, vigour of the rootstock, and soil properties. With vigorous scion cultivars, growers should use a more dwarfing clone of M.9 or B.9 rootstock and/or greater planting distances. With weak scion cultivars, a more vigorous clone of M.9 rootstock should be used and/or closer planting distances. For weak and moderate growing cultivars such as Honeycrisp, Delicious, Braeburn, Empire, Jonamac, Macoun, Idared, Gala, NY674, Golden Delicious *etc.* we suggest an in-row spacing of 3'. For vigorous varieties such as McIntosh, Spartan, Fuji, Jonagold, Mutsu *etc.* and tip bearing varieties such as Cortland, Rome, Granny Smith and Gingergold we suggest an in-row spacing of 4'. Between-row spacing should be 12'-13' on slopes and 10'-11' on level ground.

High density plantings were initially categorized in to four planting densities are re-cognized for apples, *viz.*, low HDP (< 250 trees/ha), moderate HDP (250-500 tree/ha), high HDP (500 to 1,250 trees/ha) and ultra high HDP (>1,250 trees/ha). Recently, super high density planting system has been also established in apple orchards with a plant population of 20,000 trees per ha. In some orchards, still closer planting of apple trees is followed (say 70,000 trees/ha) which is often referred as 'meadow orchards'. Corelli and Sansavini (1989) proposed 9 categories of planting densities in apple a given in Table 1.

Six type of planting systems, *viz.*, low density, moderate density, high density, ultra high density and super high density plantings or meadow orcharding. The high density plantings are the potential technologies in temperate fruit growing. The exact limits of plant density in high density planting are based on several factors and for many crops it is not yet well defined. It varies with the region, land fertility, species, variety, rootstock, economic feasibility of the farmer to adopt the system, farmer's knowledge about HDP and crop, labour and the likely return from the orchard and natural resource to support HDP. The brief about different HDPs are as under;

Table 1: Classification of High Density Orchard in Apple

Density Category	*Standard Variety*	*Spur Variety*
Very low	226/seedling	–
Low	270/M26	–
Medium low	670/M26	1000/M26
Medium	670/MM106	1000/MM106
Medium high	1000/M9	2000/M26
High	1000/M26	3000/M9
Very high	2000/M26	5000/M9
Bed system	–	10000/M9
Meadow orchard	–	40000/MM106

(a) Low Density Planting (LDP)

It is the traditional method of planting in which trees are planted at wider spacing and accommodates the minimum number of plants (100 plants/ha). There is no use of dwarfing rootstocks, minimum training and pruning, favours maximum development of canopy availing the scarce water and nutrients resource. The trees come in to commercial production potentially after 10 to 15 years of planting and the output of orchard during early years found to be very low.

(b) Medium Density Planting (MDP)

In medium density planting accommodates 250 to 500 plants/ha. Proper pruning is undertaken to restrict the tree for desired shape. It requires intensive care and labour since the returns found to be sustainable. It gives more yield and quality fruit in crops. Better water and nutrient managements are required for maintaining the orchards for a long productive life.

(d) High Density Planting (HDP)

In HDP, the tree density is more than 1,000 plants/ha depending on fruit crops. It requires heavy or rigorous training and pruning to restrict the growth. Dwarfing rootstock and chemicals are normally required maintain the optimum growth. Both yield and expenses are higher under this planting density. Establishment and maintenance of such a high density orchard could be demanding and require adequate technical backup.

(d) Ultra High Density Planting (UHDP)

Ultra High Density Planting can accommodate more than 2,000-5,000 trees per hectare and requires severe pruning and training after planting. Ultra high density in apple accommodated 1,111 plants/ha at a spacing of 3 m × 3 m (444 plants/acre). For sustainable yield and quality under ultra high density, proper canopy management, chemical assistance and nutrient management are required.

(e) Meadow Orcharding

"Meadow - grassland system" is also known as "Super High Density Planting", widely popular in temperate crops. It can accommodate more than 10,000-4,0000 or more plants per hectare in which severe top pruning is practiced similar to mowing of grass land. The tree yields after one or two years after planting. Heavy use of growth regulators as well as judicial canopy managements are required for fruiting. In apple, 10,000 plants/ha can be accommodated at a spacing of 1 × 1 m (4,000 plants/acre). For high density planting, compact varieties are used, whereas, ultra high density and super high density planting, dwarfing rootstocks are used to limit the natural growth of the plant besides the constant use of growth retardants every now and then to have a control over the growth.

4. FACTORS AFFECTING HDP

(a) Tree Quality

Tree quality is essential components must be considered on priority. Several studies have shown that the greater the number of lateral branches or feathers, the greater is the yield in the second and third year. Modern high-density orchards can have significant 2nd and 3rd year yield if highly-feathered trees are planted and if trees grow well in the first year. If growers use whips or small caliper trees these do not produce significant quantities of fruit until year 4 or 5. It is recommended that nursery trees have 10-15 well positioned feathers with a maximum length of 1 ft (30 cm) and starting at a minimum height to 30" (80 cm) on the tree.

(b) Scion Varieties

Table 2: Compact Growth Habit Type Temperate Fruit Varieties

Crop	*Genetically Dwarf Cultivar(s)*	*Desirable Features*
Apple	Spur bearing varieties Wijik McIntosh and Nugget Golden Delicious	Bear on short stems, spurs; grow to 60-70 per cent the standard cultivar in vigour and bear more spurs and yield more
Peach	Redheaven	Dwarfing and high yielding
Cherry	Sweetheart, Hedelfinger, Compact Lambert, Meteor and North Star	High yielding, self fruitful, dwarf

(c) Dwarfing Rootstocks/Interstocks

Table 3: Dwarfing Rootstocks and Interstocks Recommended for Temperate Fruits

Crop	*Dwarfing Rootstock(s)*
Apple	M-9, M-26, M27, MM-111, MM-106, Bud.9, Bud.146, P-22 and Ottawa 3
Pear	EM Quince A and C
Peach	Siberian C, St. Julien × *P. besseyi* and Rubira
Plum	Pixy
Cherry	Growth is 35 to 50 per cent of standard (Gisela 5, Edabriz, Weiroot 72) to semi-dwarfing (Gisela 12, MaxMa 14, Weiroot 158) to 90 to 110 per cent of standard (Gisela 6, M × M 60, Colt); Gisella-5, Edabriz, Ma x Ma 14, Weiroot 158, Colt, Charger and Rubira

☆ Rootstocks, dwarfer than M.9, like P.16 and P.22 are able to increase yield efficiency by 5 to 8 per cent but lose total production up to 30 per cent on a hectare basis.

(d) Tree Establishment

In high density planting, the allotted space should be filled rapidly to ensure that the ceiling production level is quickly reached. Fill planting holes with fresh, organic and soil mixtures and providing the trees with adequate water by localised irrigation. Enriching the irrigation water with nutrients add improved flower-bud formation, indispensable for a crop in the second leaf. As fertigation may have more consistent effects on growth than soil sterilisation, it might become an environment-friendly alternative to overcome replant problems in apple. The quantity of water required for a good start is depending on tree size, soil and climatic conditions. More insight in the water requirements of newly planted trees is needed to get a good balance between growth and early cropping on the one hand and to avoid waste of costly irrigation water on the other. Obviously, more water is needed for heavily feathered trees to maintain a certain water status in and around the planting hole. Although it is clear that shoot growth, fruit set, yield, and fruit weight are reduced by low quantities of water, it is not known what quantities are optimal.

(e) Growth Retardants

Various growth retardant have been used to restrict the vegetative growth of the plant. Among them the commercially adopted are AMO 1618, CCC, ancymidol, paclobutrazol, B-9 (Phosphon D) and chloramequat.

(f) Induction of Viral Infection

Though still not popular and emerged as commercially but tree size can be reduced by inducing mild viral infection, *e.g.* in apple, virus-free rootstocks series of East Malling Long Ashton (EMLA) are vigorous than their infected counterparts.

(g) Use of Incompatible Rootstock

The use of graft incompatible scion and rootstocks also induces dwarfness in the composite plants, *e.g.* in pear Bartlett and Quince.

(h) Branch Angle Manipulation

An important method of shifting the balance between vegetative growth and cropping in young trees is tying down of the scaffold branches below horizontal to induce cropping. In the vertical position, a shoot grows more vigorously than in a horizontal or pendant position and tends to remain non-fruitful. A horizontal or pendant limb grows less vigorously and then crop heavily the next year and bend under the weight of the fruit. The fruits are also strong competitors for resources and limit the growth of the branch even more. If a vertical limb is manually bent horizontal, lateral buds are released from dormancy. If the vigour of the branch is excessive, these buds may grow into vigorous upright shoots themselves and remain unfruitful. However, if the branch has more moderate vigour, the lateral buds grow into short shoots which become fruitful.

After the initial tying or weighting down of feathers at planting, new lateral branches that arise along the leader do not need to be tied down. In most climates, if moderate vigour shoots are not pruned, often cropload in the third year will bend branches down below horizontal and a natural balance between vigour and cropping will be established without additional limb positioning. Thus, with the Tall Spindle, no additional limb tying is needed after the initial tying or weighting down of the feathers at planting. However, in vigorous and/or warmer climates where winter chilling is insufficient, often limbs become too large before they set sufficient crop loads to bend the branches down. In these climates, tying down of all vigorous limbs must be done annually for the first 3-5 years until the tree settles down and begins to crop heavily. However, in most traditional apple growing areas, growers often invest too much money in limb tying, which should be limited to only the feathers at planting. Thereafter, the precocity of the rootstock induces heavy cropping and a natural balance is established.

(i) Irrigation and Fertigation

Irrigation is a crucial for viability of the production system. Large, highly-feathered trees often undergo water stress shortly after planting despite adequate soil moisture levels. This is due to the damaged root system of a transplanted tree which cannot adequately support the large top without frequent irrigation. Large, highly feathered trees produce much more leaf area shortly after planting than unfeathered trees which creates a high water demand before the root system can re-grow sufficiently to support the trees. In-addition during many years, dry weather following planting results in water stress of newly planted trees, which can limit tree growth. Frequent and early trickle irrigation can help these trees produce good growth in the first year. In humid areas growers are unaccustomed to installing irrigation immediately after planting and delay its installation until mid summer. It is recommended that growers install trickle irrigation soon after planting with high density orchards that use feathered trees to prevent water stress and maximize first year tree growth. Frequent low doses of nitrogen fertilizer delivered at least twice weekly through the trickle system (fertigation) for the first 12 weeks of the season will greatly improve tree growth during the first 2 years to speed development of the canopy. Ground fertilization with dry fertilizers is often ineffective since too much time elapses between applying the fertilizer and uptake by the tree. With high tree densities as with the Tall Spindle system and highly-feathered trees, almost no lateral tree growth is required and only vertical extension growth is needed. In this case fertigation for the first 2 years is essential both for the water to avoid water stress and for the nitrogen which is rapidly moved to the root zone and is readily available to the tree as soon as it starts growing. Adoption of fertigation considerably improves tree growth and yield potential in the 2nd and 3rd year. For moderate tree densities such as with the vertical axis, slender pyramid or Y-trellis, trees must be grown vigorously for several years to fill the allotted space with canopy and relatively high nitrogen fertilization is desirable for 2-3 years after planting. However, excessive fertilization, especially nitrogen, can cause too much growth which results in delayed flowering, reduced yields, poor fruit quality and greater pruning.

After the first few years, low nitrogen fertilization is desirable to keep the trees calm with a balance between fruiting and cropping. Many mature high density orchards receive excessive nitrogen fertilizer rates, which cause severe canopy management problems. Soil strength or fertility must be considered when calculating the amount of nitrogen to apply to mature high density orchards especially with vigorous and poor coloring varieties. In many soils excess fertility often results in excessive vegetative growth, delayed cropping and soft and poorly coloured, unmarketable fruits.

(j) Weed Control

Weed competition can drastically reduce tree growth during the first few years and can cause a failure of the orchard to fill its allotted space, which always result in diminished yield and profitability. Good weed control during the first 3-4 months of a growing season is the most critical time period of the season. In later summer months if weed control is poorer it is not detrimental to the trees. Thus, it is we recommended that growers provide excellent weed control for the first 4 months of the season for the development years of the orchard. In India, weed control in often carried out manually involves maximum production cost. Therefore, chemical weed control offers fruitful results to increase the net returns per unit area.

(k) Cropload Management

Management of cropping with high-density orchards during the first 4 years to avoid biennial bearing is critical to maintaining a proper balance between vegetative growth and cropping as the trees begin to bear. With precocious dwarfing rootstocks, young apple trees can often overset in the 2nd or 3rd year resulting in biennial bearing as early as the 4th year. These then results in increased vigour in the 4th year just when the trees have filled their allotted space and when reduced vigour is needed. Varieties differ in their biennial bearing tendency and this must be incorporated into the cropload allowed on young trees. For annual cropping varieties like Gala, it is recommended cropload of 6 fruits/cm^2 TCA (25-40 apples/tree in the second year, 40-60 apples/tree in the third year, and 100-120 apples/tree in the fourth year). For slow growing and biennial bearing varieties like Honeycrisp, we recommend cropload of 4 fruits/cm^2 TCA (15-20 apples/tree in the second year, 25-40 apples/tree in the third year, and 50-70 apples per tree in the fourth year). Within each year, the low end of the range should be used for low vigour trees and the high end of the range for high vigour trees.

(l) Pruning

During the developmental years of an orchards life the trees have a juvenile character and the balance between vegetative growth and cropping is shifted toward vegetative growth. With high density systems such as the Tall Spindle, the goal is to get the trees into cropping as soon as possible. This is the best accomplished by minimizing pruning during the first 4 years. No heading cuts should be done to the leader or lateral branches at planting or for the next 4 years since the maximum growth and earliest cropping is achieved with no pruning. For the first 4 years pruning should be limited to the complete removal of unsuitable branches such as those lateral branches that larger than 2/3 the diameter of the leader. Of much greater importance during the first 4 years is limb angle manipulation to change a young vigorous tree from a vegetative state into a reproductive state. With the Tall Spindle and Vertical Axis systems, the artificial limb bending is limited to the 1st tier of branches, while with the Solaxe even upper branches tied down. With the Tall Spindle and the Vertical Axis systems, the bending of upper branches is achieved naturally by cropload.

As the orchard reaches maturity, containment pruning of the canopy is essential to maintaining trees within the allotted space and to improve the light distribution to the lower portion of the tree. Good light distribution and good fruit quality can be maintained as trees age if the top of the tree is kept narrower than the bottom of the tree and if there is a good balance between vegetative growth and cropping. Pruning strategies based on shortening or stubbing back permanent branches that outgrow their allotted space generally are not as successful as limb renewal pruning strategies. This is partially because the most productive fruiting wood is cut off when a branch is shortened. In addition, stubbing cuts stimulate localized vigour on the affected branches which results in shading of the lower canopy. In our studies on how to manage the canopies of high-density systems, treatments where branches were shortened to maintain the conic shape to the tree, resulted in unacceptable yield reductions, a dense canopy resulting in interior canopy shading, excessive vigour compared to an unpruned control and poor fruit quality.

(m) Use Columnar Apple

Columnar growing apple trees have short internodes and axillary buds grow into spurs rather than lateral branches, hence the tree naturally grows like a sturdy cordon.

(n) Optimum Use of Light Resource

High density plantings achieve earlier cropping than wider plantings, but they show a decline in fruit quality, fruit bud formation and yield. These are mainly attributed due to its influence on light interception and distribution. Therefore, high light interception is a prerequisite for higher yield, better fruit quality, especially colour. Light interception increases as the height increases for trees of triangular in shape.

5. COMMERCIAL USE OF HDP IN DIFFERENT TEMPERATE FRUIT CROPS

High density production technology in various fruit and nut crops has already made a dent in the several countries, which showed significant results in terms of productivity and net returns. The salient details of the high density production technologies are given in the tables includes the high yielding varieties, compatible rootstock combination, tree training system, appropriate spacing, tree density per hectare and yield per hectare.

(a) Apple

In modern high-density orchards in-row spacing range from 3-5 feet and between row spacing from 10-14 ft. (780-1,450 trees/acre). Within this range, the decision of how close to plant a new orchard should be made with consideration of the vigour of the variety, vigour of the rootstock, and soil strength. With vigorous scion cultivars, growers should use a more dwarfing clone of M.9 or B.9 rootstock and/or greater planting distances. With weak scion cultivars, a more vigorous clone of M.9 rootstock should be used and/or closer planting distances. Despite some latitude in planting distances, growers should remember that to obtain high early yields high tree densities are essential. For weak and moderate growing cultivars such as 'Honeycrisp', 'Delicious', 'Braeburn', 'Empire', 'Jonamac', 'Macoun', 'Idared', 'Gala', 'NY674', 'Golden Delicious' *etc.* we suggest an in-row spacing of 3. For vigorous varieties such as 'McIntosh', 'Spartan', 'Fuji', 'Jonagold', 'Mutsu' *etc.* and tip bearing varieties such as 'Cortland', 'Rome', 'Granny Smith' and 'Gingergold' we suggest an in-row spacing of 4'. Between-row spacing should be 12'-13' on slopes and 10'-11' on level ground. In New Zealand in the 1970s, semi-intensive system of apple growing was found most profitable. The results from Lincoln University in apple grown on 'Malling 9' rootstock with tree density of 1,905 trees/ha gave higher yields as compared to grown on 'Malling-Merton-106' rootstock at 635 trees/ha.

In apple, most productive system recorded in 'Empire' grafted on dwarfing rootstock M-26 and trees trained on Y-trellis system at a closer spacing (2.1 m x 3.7 m) accommodated 1283 trees per hectare gave maximum yield (275 t/ha). High density in apple around the world entirely depends upon few varieties (Red Delicious, Empire, Granny Smith, Fuji, Gala, McIntosh, Golden Delicious, Oregon Spur and Red Chief) and rootstocks (M-9, M-26, MM-106 and MM-111). Open canopy and vertical type training systems like Y-trellis, slender spindle, central leader as well as modified leader systems are highly efficient for photosynthesis results in to greater yields. Closer tree spacing along with open canopy were found more productive. However, higher densities gave significantly more yield during initial years (Table 4, Fig. 1).

Fig. 1: High Density Orchard of Apple at CITH, Srinagar.

Table 4: Summary of High Density Orcharding Systems in Apple Assessed at various Places in India and Abroad

Variety	*Rootstock*	*Training System*	*Spacing (m)*	*Density (trees/ha)*	*Yield (t/ha)*	*Reference (Country)*
Empire	M-26	Y-trellis	2.1 x 3.7	1,283	275.0	Robinson *et al.* (1991)
Granny Smith	EM-9	V- shape	0.4 x 3.5	7,140	210.0	–
COE Red Fuji	M.26	Vertical axis	0.75 x 1.5	8,889	135.30	CITH, 2013-14
Red Delicious	M-26	Y-trellis	–	9,60	130.0	–
Mollies Delicious	M-9	Head and spread	0.75 x 1.5	8,889	126.33 (5th yr)	CITH, 2012-13
Red Delicious	M-26	Y-trellis	2.4 x 4.3	9,61	122.0	Robinson *et al.* (1991)
Mollies Delicious	M-9	Modified leader	0.75 x 1.5	8,889	108.33 (5th yr)	CITH, 2012-13 (India)
COE Red Fuji	M-9	Espalier	1.5 x 3.0	2,222	95.36 (5th yr)	CITH, 2012-13 (India)
Red Delicious	M-9	Slender spindle	–	1,495	95.0	–
COE Red Fuji	M-9	Vertical Axis	0.75 x 1.5	8,889	93.82 (5th yr)	CITH, 2012-13 (India)
Fuji	M-26	Spindle bush	2.0 x 4.0	1,250	88.00	
Royal·Delicious	MM 106	Modified leader	2.5 x 3.5	1,143	67.73 (10th yr)	CITH, 2012-13 (India)
Marshal McIntosh	M-9	Slender spindle	1.7 x 4.0	1,429	60.0	Wunsche and Lasko (2000) (USA)
Golden Delicious	M-9	Central leader	1.5 x 3.0	2,222	47.4 (8th yr)	CITH, 2010 (India)
Red Spur	Standard	Modified Leader	4.0 x 4.0		52.5	CITH, 2012-13 (India)
Vista Bella	M-9	Central leader	1.5 x 3.0	2,222	42.5(8th yr)	CITH Annual Report, 2010 (India)
Golden Delicious	MM-106	Central leader	2.5 x 3.5	1,143	39.5 (8th yr)	CITH, 2010 (India)
Mollies Delicious	M-9	Central leader	1.5 x 3.0	2,222	35.5 (8th yr)	CITH, 2010 (India)
Red Fuji	EMLA-111	Modified leader	3.0 x 3.0	1,111	16.9 (9th yr)	Bhatia and Kumar, (2009) (India)
Scarlet Gala	EMLA-111	Modified leader	3.0 x 3.0	1,111	15.6 (9th yr)	Bhatia and Kumar, (2009) (India)
Oregon Spur	MM-106	Modified leader	2.5 x 2.5	1,600	13.2 (7th yr)	Verma (2009) (unpublished) (India)
Red Chief and Silver Spur	MM-106	Modified leader	2.5 x 2.5	1,600	12.6(7th yr)	Verma (2009) (unpublished) (India)

In the management of an ultra high density (20,000 trees/ha) apple orchard, control of tree size is an important objective to obtain optimal yield of quality fruit. If horticultural practices are applied to supplement the effect of rootstock in vigour management then these practices should be economical, effective, and environmentally sound. Root pruning was effective in reducing vegetative growth in the 2nd year after treatment.

Table 5: Spacing and Planting Density for different Scion-Stock Combinations

Scion/Cultivar	*Rootstock*	*Tree Size (vigour)*	*Spacing (m × m)*	*Density (tree/ha)*
Standard	MM 109	Semi-vigorous	6.0 x 6.0	278
	MM 111	Semi-vigorous	6.0 x 6.0	278
	M 7	Semi-dwarf	4.5 x 4.5	494
	MM 106	Semi-dwarf	4.5 x 4.5	494
	M 9	Dwarf	1.5 x 1.5	4444
Spur type	Seedling	Semi-vigorous	5.0 x 5.0	400
	MM 109	Semi-dwarf	3.5 x 3.5	816
	MM 111	Semi-dwarf	3.5 x 3.5	816
	M 7	Dwarf	3.0 x 3.0	1111
	MM 106	Dwarf	3.0 x 3.0	1111

In India, Red Spur variety was found productive on PSM-1, indigenous rootstock (*Malus baccata* bio-type 'Shillong') and gave maximum yield (62.79 t/ha) at 4,444 trees/ha density with lowest leaf nutrient content. However, tree density 1,111 (3 x 3 m) was found ideal with moderate yield (25.13 t/ha) for sustainable orchards life under Himachal Pradesh conditions (Pramanick *et al.*, 2012).

The seven leading systems in the world are Tall Spindle, Super Spindle, Vertical or V-trellis, Solaxe, Bi-Axis and Fruiting Wall. All seven systems use high tree densities (900-2,200 trees/acre) and depend on early production to repay the initial investment.

- ☆ The Tall Spindle is the most preferred system with most growers using 1,000-1,300 trees/ha. It is simple to learn and requires moderate initial investments, has high early yield and is estimated to provide the highest profitability over 20 years.
- ☆ The Super Spindle orchards are planted at 2,200 trees/acre but these are limited to growers who produce their own trees. The super spindle has a simplified pruning recipe and high fruit quality.
- ☆ The Organized trellis systems (V-trellis and Vertical trellis) are the two systems which are designed for precision pruning where bud numbers are reduced through pruning to a pre-calculated number. These systems also have less sunburn to fruits.
- ☆ The Solaxe system uses extensive limb bending and manual bud extinction to achieve a balance of vegetative growth and cropping even when vigour is excessive. The bud extinction also helps controls biennial bearing.

- ☆ The bi-axis system uses a two-stem tree to achieve a very high number of leaders per acre (1,800) with only a moderate number of trees per acre (900). It also distributes tree vigour to two leaders giving more moderate growth in each leader. The system has its greatest potential when tree vigour is difficult to manage with traditional Tall Spindle training. This system is well adapted to mechanical pruning.
- ☆ The "Mur Frutiere" or Fruiting Wall system uses mechanized sidewall shearing during the early summer to reduce pruning costs and achieve high yields and good fruit quality. It is interesting that some successful apple growers manage apple trees with complete precision of bud load as with the organized trellis systems while other successful apple growers use a completely mechanized system with no precision of bud management.

(b) Pear

In pear, high density planting systems gave very high yields in varieties like Abbe Fetel, Conference and Doyenne du Comice. These all varieties were found productive on dwarfing rootstock Quince and its various clones. Higher yields were mostly recorded at very high densities (5,555 to 7,936 trees/ha), which indicate that the planting in Hedge-row pattern and trained with open centre systems like V-shape and Y-trellis facilitates maximum sunlight interception and maximum yields (Table 6).

(c) Peach

Peaches are the very important fruit crops adapted from temperate to subtropical climate. In this group, both peaches and nectarines are very productive. Genetically dwarf rootstocks are not common in stone fruits. Therefore, these fruit crops could not be planted as closer as pome fruits. Results indicated that few varieties like Royal Giant, Flavour Crest, Baby Gold-5, Stark Red Gold, Early Grande, Red Globe, Cresthaven, Glohaven and Fantasia are found highly productive under high density (900 to 1,600 trees/ha). Open canopy type training system always favours the higher productivity in peaches and nectarines as in case of pome fruits (Table 7).

In peach, the main tools for tree size control are rootstocks, summer pruning, deficit irrigation, genetic modification of tree growth habit, root restriction, growth regulators. Around the world, the most important training systems for HDP are Free and Sprint Palmette, Fusetto, delayed Vasette, Tatura trellis and related training systems. In medium or low fertile soil the Fusetto facilitates uniform distribution of leaf area, early entry into production, good light penetration and good fruit quality; maximum planting density of 1200-1,500 trees ha. In contrast, on extremely fertile soil it is necessary to adopt training systems that allow the canopy to develop as a single or double wall of limited thickness. The palmette allows maximum planting densities of 700-900 trees/ha. For orchards with planting densities greater than 1,000-1,500 trees/ha, the Tatura trellis and its variants (KAC V, perpendicular V) appear the most appropriate. Overall, the right choice of training system and orchard density will have to be made on the basis of careful economic analysis, taking into account the greater costs involved for the establishment and often also for the management of HDP.

Table 6: Summary of High Density Orcharding Systems in Pear

Variety	*Rootstock*	*Training System*	*Spacing (m)*	*Density (trees/ha)*	*Yield (t/ha)*	*Reference (Country)*
Abbe Fetel	Quince clone (MC)	V-shape	3.6 x 0.50	5555	262.0 After 7th year cumm.	Musacchi *et al.* (2005)
Abbe Fetel	Quince clone (MC)	Vertical axis	3.6 x 0.50	5555	257.0 After 7th year cumm.	Musacchi *et al.* (2005)
Conference	Quince clone (MC)	V-shape	3.6 x 0.50	5555	181.0 After 7th year cumm.	Musacchi *et al.* (2005)
Doyenné du Comice	Quince clone (MC)	Vertical axis	3.6 x 0.35	7936	132.0 After 7th year cumm.	Musacchi *et al.* (2005)
Conference	Quince-A	Y-trellis	4.0 x 1.50	1666	47.5	–

Table 7: Summary of High Density Orcharding Systems in Peach Assessed at various Places in India and Abroad

Variety	*Rootstock*	*Training System*	*Spacing (m)*	*Density (trees/ha)*	*Yield (t/ha)*	*Reference (Country)*
Royal Giant	Nema guard	V-shape	2.0 x 5.5	909	353.0	–
Flavour Crest	Nema guard	V-shape	2.0 x 5.5	909	206.0	–
Baby Gold-5	Nema guard	V-shape	2.0 x 5.5	909	169.0	–
Cresthaven	Nemaguard	Tatura trellis	2.5 x 2.5	1600	47.26 (6th Yr)	CITH, 2013-14
Glohaven	Nemguard	Tatura trellis	2.5 x 2.5	1600	42.99 (6th Yr)	CITH, 2013-14
Red Globe	Nemaguard	Tatura trellis	2.5 x 2.5	1600	41.12	CITH, 2013-14
Glowhaven	Standard	V-Shape	3.0 x 3.0	1111	40.08 (5th Yr)	CITH, 2012-13
Fantasia	Balle	Central leader	2.4 x 4.5	900	38.0	–
Stark Red Gold	GF-655-2	Free Spindle	2.0 x 4.0	1250	34.3	–
Early Grande	–	Meadow system	–	–	29.6	–
Fantasia	Standard	Modified leader	2.5 x 2.5	1600	20.6 (3rd Year)	CITH, Annual Report, 2010 (India)
Red Globe	Standard	Modified leader	3.0 x 3.0	1111	16.86 (3rd Year)	CITH, Annual Report, 2010 (India)
Red Globe	Standard	Modified leader	2.5 x 2.5	1600	15.77 (3rd Year)	CITH, Annual Report, 2010 (India)
Cresthaven	Standard	Modified leader	2.5 x 2.5	1600	13.8 (3rd Year)	CITH, Annual Report, 2010 (India)
Glohaven	Standard	Modified leader	3.0 x 3.0	1111	13.22 (3rd Year)	CITH, Annual Report, 2010 (India)
Fantasia	Standard	Modified leader	3.0 x 3.0	1111	.... (3rd Year)	CITH, Annual Report, 2010 (India)
Cresthaven	Standard	Modified leader	3.0 x 3.0	1111	10.89 (3rd Year)	CITH, Annual Report, 2010 (India)

(d) Plum

Growing of plum trees under high density is also a successful venture around the world. Only a very few varieties were reported to grow under high densities. Simca, Mallard, Opal and Elda grafted on Nemaguard and St. Julian rootstocks were found compatible as well as productive combinations (Table 8).

(e) Apricot

High density apricot cultivation is not common due to non availability of the dwarfing rootstocks. Therefore, plantations were made in medium density. In 10th leaf stage, the maximum productivity was recorded in cultivar 'CITH-Apricot-3' as compared to exotic cultivar, *viz.* Harcot and Rival (Table 9).

(f) Sweet Cherry

Production of sweet cherry in high density systems can be accomplished in a number of ways, but underlying principles exist across all systems that are critical to their success. Inherent vigour and productivity factors - scion, soil type, climate, and rootstock - must be taken into account before deciding upon a particular training strategy to impose. To develop high density sweet cherry (*Prunus avium* L.) orchards, for which growers have the choice of traditional rootstocks (Mazzard and Mahaleb seedlings) that tend to promote excessively vigorous growth or relatively new dwarfing rootstocks (Gisela, Edabriz) that, tend to promote excessive cropping. Poor management of a high density orchard system that is prone to either excessive vigour or excessive cropping can provide disastrous results for yields, fruit quality, and even tree longevity.

For sweet cherry, increased tree density translates into more manageable cropping targets (yield and fruit quality) than traditional systems. Calculation of the number of fruit per tree per tonne of yield of desired fruit size readily creates manageable production goals on a tree-by-tree basis. For example, at the relatively low density of 300 trees/ha and a target fruit size of 10 g, for each metric ton (per ha) of marketable yield, a tree must bear ~366 fruit (333 marketable fruit plus 33 additional fruit to account for a 10 per cent factor representing cullage or fruit missed at harvest). In an orchard target of 15 MT/ha, each tree must to produce 5,490 fruit. At a higher density of 900 trees/ha, each tree must only produce 1,830 fruit. In view of the production and management potential of various training systems such as Central leader, Slender spindle, Vogel spindle, Zahn spindle, HYTEC, Solaxe, Tatura trellis, Open vase, Spanish Bush, Steep Leader, Brunner, Palmette *etc.*

Sweet cherries are the choicest fruit among stone fruits. It has been successfully growing in selected pockets of Kashmir valley. Bing and Summit grafted on Gisela-6 rootstock planted at closer spacing found very productive. New Cherry clones selected by CITH, Srinagar also showed promising results under high density. Trees trained with vertical as well as open centre canopies are have positive effect on yield and fruit quality (Table 10).

(g) Almond

Almond is the most important nut crop grown only in the monsoon free areas of Kashmir valley. Domestic demand of almonds is very high in India, but productivity

Table 8: Summary of High Density Orcharding Systems in Plum Assessed at Various Places in India and Abroad.

Variety	*Rootstock*	*Training System*	*Spacing (m)*	*Density (trees/ha)*	*Yield (t/ha)*	*Reference (Country)*
Simca	Nemaguard	Central Leader	2.0 x 5.5	909	169.8	–
Mallard	St. Julian A	Y-trellis	0.5 x 4.0	5000	95.1	Meland (2005) (Norway)
Opal	St. Julian A	Y-trellis	0.5 x 4.0	5000	77.6	Meland (2005) (Norway)
Elda	St. Julian A	Y-trellis	0.5 x 4.0	5000	53.0	Meland (2005) (Norway)
Mariposa	Standard	Modified leader	5. 0 x 5.0	625	27.26 (9th leaf)	CITH, 2014-15
AU-Cherry	Standard	Modified leader	5. 0 x 5.0	625	23.99 (9th leaf)	CITH, 2014-15
Tarrol		Standard	Modified leader	5. 0 x 5.0	625	23.40 (9th leaf)CITH, 2014-15

Table 9: Summary of High Density Orcharding Systems in Plum Assessed at Various Places in India and Abroad

Variety	*Rootstock*	*Training System*	*Spacing (m)*	*Density (trees/ha)*	*Yield (t/ha)*	*Reference (Country)*
CITH-AP-3	Standard	Modified leader	5 x 5	625	23.64 (10th Yr)	CITH, 2014-15 (India)
CITH-AP-2	Standard	Modified leader	5 x 5	625	23.02 (10th Yr)	CITH, 2014-15 (India)
CITH-AP-1	Standard	Modified leader	5 x 5	625	19.06 (10th Yr)	CITH, 2014-15 (India)
Harcot	Standard	Modified leader	5 x 5	625	18.20 (10th Yr)	CITH, 2014-15 (India)
Rival	Standard	Modified leader	5 x 5	625	17.51 (7th Yr)	CITH, 2014-15 (India)

Table 10: High Density Orcharding Systems in Sweet Cherry Assessed at Various Places in India and Abroad

Variety	*Rootstock*	*Training System*	*Spacing (m)*	*Density (trees/ha)*	*Yield (t/ha)*	*Reference (Country)*
Bing	Gisela-6	Palmette	2.6 x 4.9	864	27.1 (9 yr old)	Whiting *et al.* (2005) (USA)
Bing	Gisela-6	Central leader	2.6 x 4.9	864	25.8 (9 yr old)	Whiting *et al.*(2005) (USA)
Bing	Gisela-6	Y-trellis	2.6 x 4.9	864	23.9 (9 yr old)	Whiting *et al.* (2005 (USA)
Van	Damil	Y-trellis	0.5 x 4.0	5000	14.5	Meland (1998) (Norway)
Summit	Tabel Edabriz	V-shape	0.7 x 3.8	3759	19.3 (6th year)	Radunic *et al.* (2011) (Croatia)
CITH Cherry-1	Colt	Modified centre	2.5 x 2.5	1600	7.93 (8th year)	CITH Annual Report 2010 (India)
Bigarreau Noir Grossa	Colt	Modified centre	2.5 x 2.5	1600	5.2 (8th year)	CITH Annual Report 2010 (India)
CITH -05	Colt	Modified centre	2.5 x 2.5	1600	8.3 (8th year)	CITH Annual Report 2010 (India)
CITH-05	Colt	Modified centre	3.0 x 3.0	1111	8.3 (8th year)	CITH Annual Report 2010 (India)
CITH-01	Colt	Modified centre	3.0 x 3.0	1111	8.3 (8th year)	CITH Annual Report 2010 (India)
CITH-02	Colt	Modified centre	3.0 x 3.0	1111	8.3 (8th year)	CITH Annual Report 2010 (India)

is very low. India is annually importing about more than 60 per cent of the world almond imports. This demand could be compensated by resorting to high density orcharding in almond using Non Pariel, Waris, Waris, Pranyaj and Shalimar (Table 11). Lack of dwarfing rootstocks for almond is primary hurdle in establishing high density orcharding. Therefore, high density orchards are more productive during early 5 years of tree bearing, but later they are at par with medium density orchards for nut yield.

In India, there is a lack of data from commercial temperate fruit plantings in relation to high density plantings. In apple, few experimental orchards have been established in apple growing areas of India including Jammu and Kashmir, Himachal Pradesh and Uttarakhand. The trials carried out in India have compared different rootstocks, densities and cultivars.

The effect of higher early yield and increased yield at full production with increased tree density evident in several high density experiments around the world. However, it is the labour intensive system involves the higher expenses. A study conducted at Lincoln University showed that the more labour is required per hectare basis in semi-intensive orchards than the high density orchards during first 6 years for training and pruning. Because dwarf trees (0.6-2.4 m height) require half cost than the semi-intensive vigorous trees (3.0-3.7 m height). Although establishment costs for high density plantings are higher than any other semi-intensive systems, total establishment costs are not proportional to tree numbers.

6. ADVANTAGES OF HDP

- ☆ Early production and higher returns many higher density orchards are breaking even in 6-7 years compared to 10-12 years for traditional systems.
- ☆ Efficient use of labour to harvest and prune from the ground or from a short stool.
- ☆ It has potential to produce higher quality fruit for a longer period of time by maintaining light interception in the smaller trees of higher density orchards.
- ☆ Higher pesticide application efficiency.

7. DISADVANTAGES OF HDP

- ☆ It cannot be adopted in very steep, unfertile, shallow and drought-prone areas.
- ☆ The high cost of orchard establishment.
- ☆ Require highly skilled manpower.
- ☆ Require more training and minimal pruning during the first 6 years than traditional systems, especially during the summer. A lack of attention in the early life of this type of orchard creates a very high probability that the orchard will never be profitable given the high costs of establishment.

Table 11: Summary of High Density Orcharding Systems in Almond Assessed at Various Places in India and Abroad

Variety	*Rootstock*	*Training System*	*Spacing (m)*	*Density (trees/ha)*	*Yield (t/ha)*	*Reference (Country)*
Supernova	Seedling	V-shape	2.5 x 5.0	800	24.5 (cummu. Yield 10 yr)	Monastra *et al.* (1998)
Ferragnus	Seedling	V-shape	2.5 x 5.0	800	23.5 (cummu. Yield 10 yr)	Monastra *et al.* (1998)
Waris	Seedling	Modified leader	3.5 x 3.5		3.24 (12th yr)	CITH, 2012-13
Non Pariel	Nemaguard	Standard	3.0 x 6.6	500	3.28	Duncan R., Univ. California (USA)
Non Pariel	Hensin	Standard	3.0 x 6.0	500	3.63	Duncan R., Univ. California (USA)
Pranyaj	Seedling	Modified leader	4.0 x 4.0		3.07 (12th yr)	CITH, 2012-13
Non Pariel	Price	Hedge row	2.2 x 6.7	667	2.45	Duncan R., Univ. California (USA)
Pranyaj	Seedling	Modified Leader	4.0 x 4.0	625	2.94 (10th yr)	CITH Annual Report 2010 (India)
Waris	Seedling	Modified Leader	4.0 x 4.0	625	2.78 (10th yr)	CITH Annual Report 2010 (India)
IXL	Seedling	Modified Leader	4.0 x 4.0	625	2.58 (10th yr)	CITH Annual Report 2010 (India)
Shalimar	Seedling	Modified Leader	4.0 x 4.0	625	2.37 (10th yr)	CITH Annual Report 2010 (India)

8. ORCHARDS OF THE FUTURE

Tall spindle: This system was developed to revive the New York apple industry economically. In 1993, Robinson and his cohorts at Cornell gave the birth of the high-density Tall Spindle system. "It contained concepts from Slender spindle, Vertical axe, Solaxe and Super spindle. Adoption of the tall spindle system is yielding more and showing its economic superiority. The basic elements of tall spindle are:

- ✰ *High density*- The optimum tall spindle spacing for an average vigour variety and soil is 3′ by 11′, or 1,289 trees per acre.
- ✰ *Full dwarfing rootstocks*- The most successful tall spindle orchards established to date have been on M.9 and B.9. Precocious dwarfing stocks are important since early cropping is essential.
- ✰ *Highly feathered nursery trees*- Nursery stock ideally have from 10 to 15 feathers per tree.
- ✰ *Minimal pruning at planting*- No heading.
- ✰ *Support system*- To carry the crop and support potentially brittle rootstocks.
- ✰ *Branch devigoration*- Upright scaffold branches are devigorated by bending them below the horizontal. This early investment of labour pays off in early cropping and less pruning later.
- ✰ *Limb renewal*- All scaffolds are removed as they become too large for the available space and become out of balance within the tree.
- ✰ *Early fruiting*- Fruiting in the second and third leaf is essential to keep a low tree vigour level and provide income from early fruit sales.
- ✰ This system gave cumulative production of more than 81.65 tonne in the first five years, resulting in a 40 percent higher crop value than the slender vertical axis and Sol Axis planting systems.

9. FUTURISTIC NEEDS FOR HDP

The planting of a new high density orchard is a long time (more than 20 year) commitment. Therefore, before planting a grower should consider the possible changes in orchard management that will occur in the next 20 years and try to plant the new orchard so that one can benefit from advances in orchard management that are likely to occur. The 5 basic principles or orchard design outlined above (i. high light interception, ii. good light distribution throughout the canopy, iii. high early yields, iv. simple canopies are more adaptable to partial mechanization, and v. planting density is governed by the law of diminishing returns) will be a part of any future orchard system. Based on these five principles the design of a HDP should be as given here under:

- ✰ Tall (10-11 feet) enough due to the need of intercepting 70-75 per cent of available light.
- ✰ It will have narrow simple canopies no wider than 4-6 feet due to the need to have good light distribution in all regions of the canopy.

- ☆ Use of highly branched trees for high early yields. The only exception to this rule will be those growers who choose to plant very high tree densities (> 1,500 trees/acre) where the cost of feathered trees is too high and the value of feathers is less due to very small space between trees.
- ☆ Simple canopies will offer significant benefits in mechanization with simple platforms for pruning, thinning and harvesting.
- ☆ The optimum planting density will remain close to 1,000 trees/acre.

In addition to high, early yields, a key goal is to increase the percentage of high quality fruit. That means having narrow canopies that allow good exposure to light from bottom to top. This can be achieved by narrowing canopies with mechanical hedging to form tree walls at suitable time. The next economic challenge is labour reduction. We need to grow temperate fruits with half the labour (180 h/acre) against which it is no (360 h/acre). Key ingredients are machines, platforms to thin, prune, and serve as carriers for harvest crews, but also hedgers to do much of the pruning now done by hand. Chemicals play a role as well. Chemical thinning can cut thinning labour by half.

10. CONCLUSION

The change in planting systems over the last 50 years has been dramatic. High yield and high fruit quality can be achieved with a high-density orchard when the orchard has good light distribution throughout the tree canopy and there is a balance between vegetative growth and cropping ('calm trees'). This can best be accomplished by maintaining a narrow canopy shape and through regular limb renewal pruning and the development of pendant fruiting branches. Successful growers will maintain a balance between vegetative growth and cropping by regularly renewing the fruiting wood on the tree, moderate levels of nitrogen and proper cropload management. Growers should try to obtain 50 cm of leader shoot growth in the first year, 75-100 cm of leader shoot growth in the second and third years and 50 cm of leader shoot growth in the fourth year. If this is combined with minimal pruning and a precocious rootstock, significant production should be obtained in the second-fourth years which will limit vegetative growth in future years resulting in a "calm" tree.

11. ACKNOWLEDGEMENT

The author is grateful to Central Institute of Temperate Horticulture for use of information available in public domain.

12. REFERENCES

Bhatia, H.S. and Kumar, J. 2009. Performance of new apple cultivars on different rootstocks under high density plantation. *Agri. Sci. Digest*, **29** 4: 303-305.

CITH Annual Report. 2010, 11, 12, 13, 14. Central Institute of Temperate Horticulture, Srinagar, J&K.

Corelli, L. and Sansavini, S. 1989. Light interception and photosynthesis related to planting density and canopy management in apple. *Acta Hort*. **243**: 159-174.

Kumar, Dinesh, Ahmed, N., Verma M.K. and Dar, T.A. 2013. Growth, yield, quality and leaf nutrient status as influenced by planting densities and varieties of apricot. *Indian J. Hort.*, **70** : 195-199.

Kumar, Dinesh, Ahmed, N. and Verma, M.K. 2012. Studies on high density planting in almond in Kashmir valley. *Indian J. Hort.*, **69** : 328-332.

Meland, M. 1998. Yield and fruit quality of 'Van' sweet cherry in four high density production systems over seven years. *Acta Hort.* **468**: 425-432

Meland, M. 2005. High density planting systems of European plums- the effect of growth and productivity of three cultivars after nine years. *Acta Agric. Scandinavica Section B-Soil and Plant*, **55**: 51-57.

Monastra, F., Martelli, G. and Chiarotti, A. 1998. Economic advantages of high-density planting of almond orchards. **In** : *X GREMPA p. 81-85 Meknès* (Morocco).

Musacchi, S., Ancarani, V., Gamberini, A., Gaddoni, M., Grandi, M. and Sansavini, S. 2005. Response of training system, planting density and cultivar in pear. *Acta Hort.* **671**: 463-469.

Pramanick, K.K., Kishore, D.K., Singh, R. and Kumar, J. 2012. Performance of apple (*Malus × domestica* Borkh) cv. Red Spur on a new apple rootstock in high density planting. *Scientia Hort.*, **133**: 37-39.

Raduniæ, M., Jazbec, A., Pecina, M., Èosiæ, T. and Pavièiæ, N. 2011. Growth and yield of the sweet cherry (*Prunus avium* L.) as affected by training system. *African J. Biotech.*, **10** 24: 4901-4906. DOI: 10.5897/AJB10.278.

Robinson, T.L., Lasko, A.N. and Carpenter, S.C. 1991. Canopy development, yield, and fruit quality of 'Empire' and 'Delicious' apple trees grown in four orchard production systems for ten years. *J. American Soc. Hort. Sci.* **116**: 179-187.

Robinson, T., Hoying, S., Sazo, M.M., DeMarree, A. and Dominguez, L. 2013. A Vision for Apple Orchard Systems of the Future. *New York Fruit Quart.* **21** : 11-16.

Webster, T. 2006. Control of growth and cropping of temperate fruit trees. *Chronica Hort.*, **46** : 20-26.

Whiting, M.D., Lang, G. and Ophardt, D. 2005. Rootstocks and training systems affect sweet cherry growth, yield and fruit quality. *Hort. Sci.*, **40** : 582-586.

Wunsche, J.N. and Lasko, Alan, N. 2000. The relationship between leaf area and light interception by spur and extension shoot leaves and apple orchard productivity. *Hort. Sci.*, **35** : 1202-1206.

Chapter 26

Farming Systems for Temperate Fruits: An Approach Towards Sustainability and Livelihood Security

Dileep Kachroo

Chief Scientist and Head, FSR Centre and Head,
Division of Agronomy
Sher-e-Kashmir University of Agricultural Sciences and Technology of Jammu
Jammu – 180 009, J&K
E-mail: dileepkachroo@rediffmail.com

1. INTRODUCTION

Horticulture forms an important sector in the economy of J&K State and is one of the key sector for employment and income generation because large scale industrialization suiting to ecosystem has not taken place. The sector is currently having an annual turnover of fruits worth Rs. 3,000 crores and giving employment to over 25 lakh people. Keeping this in view the state government has declared horticulture a thrust area and is taking a number of steps to boost it. Temperate zone of Jammu and Kashmir state is the fruit bowl and perhaps no state has greater ambience for horticulture; as indigenous apple, pear, grape, vine, mulberry, walnut, hazelnut, peach, apricot, raspberry, gooseberry, currant and strawberry are the potential examples. Besides saffron and *kaala zeera* the temperate crops are the other very important crops that have a history of many centuries. The state of J&K has the monopoly for producing saffron and is placed second in the world after Iran. The productivity is half in comparison to that of Iran and Spain. Commercial floriculture and cultivation of medicinal and aromatic plants have great potential which need to be promoted on a large scale due to their tremendous biodiversity and climatic advantages.

These temperate fruits grown at elevation of 1,000 to 3,000 m above sea level are important cash-fetching fruits of the state. These fruits not only supplement the diet of the people in the state and country but form an important item of our exports. From the stage of tree plantation to the point of its marketing, it has a good potential in employment creation. There is need to explore other options, that too in the field of value added agriculture under its diversified practice. However, the state is facing many problems with regard to the development of horticulture. It includes low productivity, great variability in important fruit crops like walnut and almond, higher percentage of offgrade fruits, poor connectivity with the market place and small and fragmented land holdings having marginal farmers, which has forced us to introduce high-density plantation of apples and soft fruits like strawberry and currants around cities and towns.

Therefore, horticulture sector continues to remain the important sector for socio-economic development of the people. With serious constraints on area expansion and declining scope of other sources of growth in agriculture output, the diversification of agriculture towards non-food grains and high value cash crops including fruits and vegetables, compatible with the comparative advantage of the region is suggested as a viable solution. These crops have potential of income augmentation, employment generation, poverty alleviation and export promotion. According to a study, in the state of Jammu and Kashmir, the scope to raise output through diversification is highest in the country as one per cent shift in area from food grains to non-food grains entails more than 3 per cent growth rate in crop output.

During the last several years, diversification of Agriculture in State towards high value commodities, *i.e.* fruits, vegetables and livestock products is taking place at a faster pace and is reflected in the high share of High Value Commodities (HVCs) in agricultural production in a number of districts. Sustained economic and income growth, urbanization and globalization are fuelling, rapid growth in demand for high value commodities in the state, high value horticulture has a comparative advantage in production and labour absorption over staple food items and thus is reckoned as an important activity for small orchard holders equally important, the consumption of high value commodities (HVCs) is on the rise, the evidence shows that by 2025, demand for fruits, vegetables, milk, egg, meat and fish would almost be doubled than that was in 2000. This sector is most vibrant sector for economic development in state. The per capita income of Jammu and Kashmir is 324 US $, which is more than the national average. The monthly expenditure of J&K is also higher than normal average. The urban/rural difference is lesser due to horticulture earning of the farmers. Keeping this gap in production and requirement in view there is vast scope for boosting the agriculture in the state and simultaneously going for the diversification of agriculture, which besides horticulture and crop sector includes mushroom cultivation, bee keeping, medicinal and aromatic plants and various animal products as the state has to invest more than Rs.1,400 crores in procurement of meat, poultry and animal products including fodder, feed etc.

2. OVERVIEW OF AGRICULTURE PRODUCTION SCENARIO IN J&K

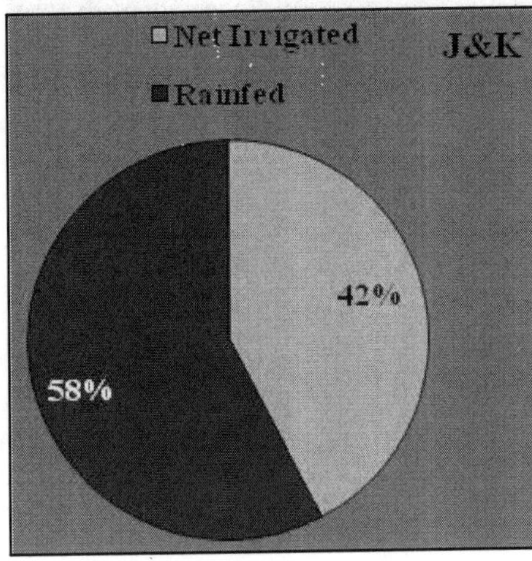

Fig. 1

In the state of Jammu and Kashmir, 58 per cent of the area under agriculture is rainfed and remaining 42 per cent is irrigated (Fig. 1). Agro-climatically the state is divided into four zones: Sub-tropical zone (2.53 lakh ha), intermediate zone (3.56 lakh ha), temperate zone (4.04 lakh ha) and cold arid zone (0.29 lakh ha). About 64, 90 and 44 per cent of the area in sub-tropical, intermediate and temperate zones respectively is rainfed (Fig. 2).

The mean annual rainfall in these agro-climatic zones is 1,069 mm (Sub-tropical zone), 1,478 mm (intermediate zone), 680 mm (temperate zone) and 83 mm (Cold arid zone). Socio-economically the biggest constraint of the agriculture in the state is small land holding size of 0.66 ha with more than 77.97 per cent farmers having less than one hectare of land against 1.33 ha holding size on national basis.

In J&K state, 11.28, 92.72, 56.99 and 96.15 per cent area under rice, maize, wheat and barley, respectively is cultivated as rainfed crop, which attributes to low productivity and production in comparison to national yields. The productivity of various crops including rice, maize, wheat, pulses and oilseeds in the State is 21.83, 20.05, 17.35, 5.10 and 8.5 q/ha, respectively compared to 22.03, 27.85, 19.21, 6.38 and 10.86 q/ha on national average.

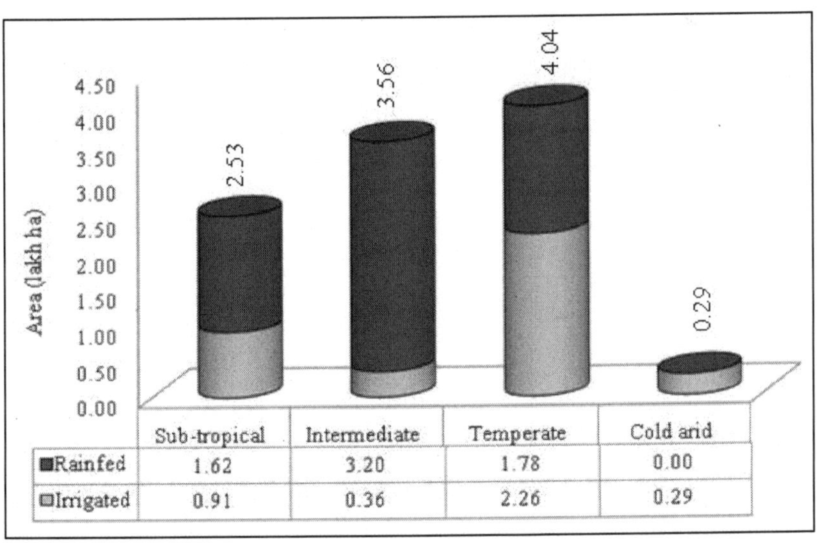

Fig. 2: Area under different Agroclimatic Regions in J&K State.

The cultivated apple (*Malus* x *domestica* Borkh) is considered as one of the most important and widely grown fruit in temperate zones of the world with regard to its acreage, production, economic returns and nutritive value and popularity. The tree is small and deciduous reaching 3 to 12 m tall with broad, often density twiggy crown blossoms are produced in spring simultaneously with the budding of the leaves. The fruit matures in autumn and is typically 5 to 9 cm in diameter enriched with nutrients; the apple is fourth widely produced fruit in the world after banana, orange and grapes. In spite of certain problems the comparative feasibility and profitability in apple production has proved to be better as compared to the other farm enterprise during the last one and a half decades or so. The area under apples has increased significantly and resulted to higher production (Table 1), which clearly indicates its potential in Integrated Farming System in temperate fruits.

Table 1: Comparative Productivity of Apple and Rice Cultivation in J&K State

Year	*Area Under Apple (000' ha)*	*Apple Production (000' Mt)*	*Productivity*	*Area Under Rice (000'ha)*	*Rice Production (000' q)*	*Productivity (g)*
2004-05	107.93	1093.33	10.12	250.04	4928	19.70
2005-06	111.88	1151.34	9.01	259.01	5574	21.52
2006-07	119.04	1222.18	9.51	252.52	5546	21.96
2007-08	127.80	1311.85	10.54	263.25	5620	21.34
2008-09	133.10	1332.81	8.99	257.63	5637	21.88
2009-10	138.19	1373.00	9.16	259.89	5011	19.28
2010-11	141.17	1852.41	10.15	261.35	5077	19.42
2011-12	154.72	1747.22	10.46	262.54	5045	19.21
2012-13	173.63	1500.25	9.5	262.56	5078	19.34

Source: Digest of Statistics Jammu Kashmir, Financial Commission (Rev), J&K .

Socio economic status of farming community is one of the most well thought issue that needs a special attention. Smaller and fragmented holding size as small and marginal farmers constitutes 94 per cent of the total land holdings in J&K state, higher population growth rate, illiteracy, unorganized marketing and distribution, poor infrastructure *etc.* are the socio-economic constraints of temperate zone in particular and other zones of the state in general. As per the 8th Agriculture Census (2005-06) of the State, the total number of operational holders has been worked out to be 13,77,808 and average size of operational holding was found out to be 0.67 ha. About 94 per cent of the operational holders fall in the category of marginal and small farmers. About 5 per cent of the operational holders fall in semi-medium category holding, only 1 per cent of the operational holders fall in the medium category holding and only 0.04 per cent of the operational holders fall in the large category holding. Adoption of Integrated Farming System is the only solution for uplifting the socio-economic status of the farming community of the state, which is having fragmented small and medium land holdings.

3. PARADIGM SHIFT TOWORDS APPLE CULTIVATION

The state has the largest potential for production of quality temperate horticulture crops. It has created niche production of apple, pears and dry fruits, among temperate fruits apple ranks first in terms of production and productivity. The annual production of apple in the state is about 9.09 lakhs tonne at an average yield of 10.09 tonne per hectare. However, the production and productivity of apple crop has been fluctuating during last two decades this is due to drought or some other conditions. In spite of above problems the apple production in the state has increased from just 6,000 metric tonne in 1950-51 to more than 9 lakh tonnes, area increased from 7,000 ha to more than 90,000 ha and yield per hectare in Jammu and Kashmir state is highest in the country, which is much higher than national average of 6.86 tonne/ha. It is also compared well with the world average of 10.82 tonne/ha or with China (9.93 tonne/ha) which is world's largest producer of apple.

(a) Need for Farming Systems

Adoption of horticulture based Integrated Farming System in the state is having vast potential for taping the natural resources and increases the productivity, profitability and sustainability. Exploding population, urbanization and industrialization, are leading to decline in per capita availability of vital agricultural resources and also fragmentation of farm holdings, making them operationally uneconomic. The process of marginalization of land holdings is likely to continue due to various demographic reasons. As discussed above, it is evident that per capita arable land is continuously decreasing and is expected to shrink further. Such farm families have certain basic needs including food (cereal, pulse, oilseed, milk, fruit, honey, fish, meat, egg *etc.*); feed, fodder, fibre, employment etc. To fulfill these, farmers have been doing their own farming system for a long time.

At SKUAST-Jammu lot of efforts have been made, aiming at increasing the productivity of different components of farming system like crop, dairy, livestock, poultry, apiculture, vegetable production, fruit cultivation, mushroom cultivation, sericulture *etc*. However, the national scenario is clearly indicative of the fact that normally, a farming family maintains multi-enterprise systems, depending upon his/her family requirements, knowledge base, socio-economic setup, agro-climatic conditions and available farm resources. However, to have a systematic integration of multi-enterprise systems in a scientific manner, components need to be chosen in such a manner that product or by-product of one component becomes the input for other, becoming complementary and are organically well interlinked to each other without much wastage. For example, by-product and waste material generated from cultivation of many crops is used as raw material for many industrial uses or as fuel or as feed/fodder to the cattle of the farm. The animal dung may also be used as fuel or a mix of dung, urine and crop residues may be used for production of compost/farmyard manure that is applied to crops and fishpond. The silt deposited at bottom of a fishpond can again be utilized as manure to crops. The FYM can substitute about 25 per cent of recommended NPK for crops. Crops like mustard serve as a good source of nectar for honeybees, besides providing edible oil for human and cake for animal feed. The crop residues can substitute 12 per cent of recommended NPK. If fishpond embankment is used for cultivating fodder/flower/fruit/vegetable crops, it provides extra output either for human or animal.

'Farming systems approach of agricultural research is one venture to harness the complementarities and synergies among different enterprises and augment the total productivity, profitability and gainful rural employment.' Unlike commodity research, where the focus is on increasing productivity/profitability of individual crop/animal, under farming system approach it is the farm household that remains the focus of the research programme.

The university has developed a model of Integrated Farming System, which includes multiple enterprises mainly Crop + Horticulture + Animal unit + Fish-cum-Poultry + Apiculture + Vermi-composting + Biogas + Mushroom unit coupled with boundary plantation targeting small and marginal farmers. The model is having the potential to revolutionize the farming practices and the approach will not only boost the vertical productivity and sustainability but also will improve the livelihood security of the farmer many-folds. Therefore, the state govt. has to emphasize on launching of mega project on Integrated Farming System on the lines of the SKUAST-Jammu model with the intervention of including potential temperate crops.

4. POTENTIAL OF FARMING SYSTEM APPROACH TOWARDS SUSTAINABLE AGRICULTURE

Integrated Farming System leads towards sustainable agriculture (with crops, trees, livestock *etc.*), which is based on locally adapted agro-diverse cropping patterns and use of local resources (natural resources and natural processes), based on local knowledge, skills and innovations. The integrated farming system approach also highlights the potential of sustainable agriculture, in terms of mitigation of GHG emissions as well as adaptation to climate change. Further, it holds great potential for meeting global and national food security requirements even as it leads to improvement of farmers' livelihoods through enhancing their net incomes and improving the productivity of their resources in the long run.

5. FOOD SECURITY

A question that is often posed with regard to sustainable agriculture is whether it will be able to feed the growing population. We respond by saying that sustainable agriculture does not imply lowered yields, as experience of successful farmers' bears out on the ground. This is reinforced by an FAO report (2007), which says that "Conversion of Global Agriculture to Organic management, without converting wild lands to agriculture and without using N-fertilisers would result in a global agricultural supply of 2,640 to 4,380 Kcal/person/day". Sustainable intensification in developing countries through organic practices would increase production by 56 per cent. A meta-analysis of 133 scientific papers concluded that organics was particularly competitive under lower yield environments, a feature that is common in developing countries. Organic yields on an average are comparable to conventional yields although yields do decline initially when converting from high-input systems and almost double when converting from low-input systems. In India, it should be remembered that a majority of land is rainfed and continues to be low-input by default.

6. IMPROVEMENT IN RURAL LIVELIHOODS

While macro-level food production and availability levels are likely to increase through sustainable agriculture, at the individual and community level too, there are bound to be improvements. It has been felt that access to food will increase by livelihood improvement both for farmers and agricultural workers through integrating the various components in a farming system mode. Recycling of farm waste through various composting methods will improve food access by increasing productivity, diversity and conservation of natural resources, by raising incomes, improving employment and by reducing risks. It has been recorded that shift to sustainable agriculture practices can reduce the outward migration from rural areas.

7. REDUCTION IN GHG EMISSIONS

Changes in farming models and practices towards sustainable agriculture offer a significant opportunity at reducing GHG emissions. Under Integrated Farming System there is a great potential to reduce its fossil fuel dependence in many ways. For instance, for soil productivity management, internal inputs and practices are used, which in the longer run minimizes the use of chemical fertilizers– for instance, creating the micro-climate required for increased soil (beneficial) microbial activity. This is done by returning bio-mass to the soil. Legume production, crop rotation, mixed cropping *etc.*, are other ways of achieving this.

Shifting to organic agriculture instantly under Indian conditions is a debatable issue but adopting the Integrated Farming System can lead us to move towards organics and as per the estimates it prevalent that the IFS approach is having the potential for avoidance of methane emission to considerable limits through the promotion of aerobic micro-organisms and high biological activity in soils, oxidation of methane can be increased. Through practices like System of Rice Intensification, which is mostly based on principles of ecological farming, flooding in rice can be reduced and thereby, methane emissions.

Nitrous oxide, result of overdoses and losses on nitrogen, can be effectively minimized through integrated farming practices. While production of chemical fertilizers is an energy-intensive process that emits carbondioxide and nitrous oxide, application of nitrogen fertilizers makes the soil emit nitrous oxide. These can be avoided through by using recycled inputs in the system, *i.e.* interdependence of waste output of one component to be used as input for the other component.

Integrated Farming System also increases the Soil Organic Carbon (SOC) by incorporating organic materials into the soil. Soil can be a major source of storage of carbon, about twice as much carbon as in the atmosphere. Minimizing fertiliser use replaces soil organic matter in intensive systems, which reduces potential sequestration. Crop, tree and livestock integration with a systematic recycling of organic wastes is an integral part, which can lead to sustainability. Long term studies have shown that compost application and cover crops in rotation were particularly adept at increasing soil organic matter even in comparison to no-tillage techniques. While conservation tillage is promoted elsewhere as a way of sequestration of carbon dioxide, this is often done by the use of chemical herbicides and GMOs, which have their own ecological implications. In sustainable agriculture however, mitigation of climate change can be addressed both carbon sequestration and minimized emissions of GHGs. Agro-forestry as a boundary plantation is also a desired principle of integrated farming systems, which further adds to the potential towards sustainability in carbon sequestration.

8. ADOPTION OF INTEGRATED FARMING SYSTEM

Integrated Farming System (ecological farming/sustainable agriculture/LEISA *etc.*) approaches are now acknowledged for the wide set of ecological and economic benefits that accrue to the practitioners as well as consumers of agricultural products. These approaches, which are based on low external inputs are also low energy intensive and less polluting, hence mitigate and help in adapting to the climate change. However, the promotion of integrated farming system on a large scale is often confronted about its potential as well as its practical limitations.

These successful experiences had three elements in common. First, all have made use of locally adapted resource conserving technologies. Second, in all there has been coordinated action by groups or communities at local level. Third, there have been supportive external (or non-local) government and/or non-governmental institutions working in partnership with farmers. Almost every one of the successes has been achieved despite existing policy environments which still strongly favor 'modern and established' approaches (technology and support systems) to agricultural development. Now the challenge is to create a policy environment to scale up across the temperate region of the J&K in particular. This needs a newer approach in terms of capacity building, horizontal learning, newer institutional systems and newer forms of financial support to be put in place.

The programmatic support to agriculture today favour only high external input based agriculture. As a result, none of the mainstream programmes provide any support for promotion of these models. This needs the recasting of programme guidelines or initiating newer programme to provide support to more sustainable models in agriculture, which can be easily accessible to small and marginal farmers. The Integrated Farming System can initiate a programmatic support to scale up sustainable agriculture and livelihood security with the objectives to:

- ☆ Characterize existing farming systems to know the productivity, viability and constraints.
- ☆ Develop and validate region-specific Integrated Farming System Models for enhanced system productivity, profitability and sustainability and reduce the risks and vulnerabilities with uncertain weather conditions and degraded and limited natural resources in these regions, by adopting suitable cropping patterns and production practices, which can build livelihood security systems to cope up with the natural disasters like drought, floods and other climate uncertainties.
- ☆ Assess relative efficiencies of the IFS model in terms of economics, resource use and energy and diversify the assets and income sources to sustain the livelihoods by integrating livestock and horticulture into agriculture and promoting on-farm and off-farm employment opportunities.

- ☆ Optimize individual components of IFS in regional perspective. Conserve and efficiently use the available natural resources like soil and water, and promote biomass generation.
- ☆ Capacity building of stakeholders in Integrated Farming Systems through appropriate trainings and organize farmers into institutions, which can help them to have better planning, greater control over their production, help to access resources and support, improve food security and move up in the value chain.

9. APPLICABILITY OF INTEGRATED FARMING SYSTEM AND ITS BASIC GUIDELINES

(a) Sustainable Agriculture

Sustainable Agriculture can be defined as an integrated farming system, which is based on locally adapted cropping patterns and local resource (natural resources and natural processes) use based on local knowledge, skills and innovations. The capacity of a farming system to adapt to changing climate and weather conditions is based on its natural resource endowment and associated economic, social, cultural and conditions. The viability of these elements also constitutes the basis for sustainable agriculture, understood as agricultural production that: ensures adequacy of food production; does not harm the resource base; is economically viable; and enhances quality of life. Many climate and weather risk management strategies fit squarely into sustainable agriculture practices and can, therefore, be promoted with several of the programs and policies targeting environmentally responsible production.

Strategies to be Adopted

- ☆ Changes cropping patterns and cropping systems to suit the local resource and weather conditions. Multiple/mixed cropping, intercropping systems with legume components etc.
- ☆ Ecological farming practices which can maximise the local resource use. Many of these practices are based on indigenous knowledge and focus on building soil biological productivity. Less dependence on chemical Pesticidal Management, Organic Soil Management, Community Seed Banks, System of Rice Intensification, soil moisture management etc have already proven to be useful.
- ☆ Locally adopted crop varieties specially in saline and flood prone areas, drought prone areas, making suitable selections adopting Participatory Plant Breeding and Participatory Varietal Selection.
- ☆ Developing suitable farming systems integrating agriculture, horticulture and livestock.

(b) Farmers' Institutions

Organized communities have proven to be more effective in planning and managing their resources and livelihoods, lobbying for a policy change and securing their entitlements. Appropriate institutional systems for each of the purposes need be established.

Strategy to be Adopted

- ☆ Integrated farming research model needs to be extended to dry land agriculture as the total area under the rainfed situation of the state is about 77 per cent and their will be the real breakthrough for food security and livelihood for the poor resource farmers.
- ☆ Capacity building programmes for farmers shall made mandatory in which the provision may be made for making and arranging the need based trainings and visits of the farmers to the Integrated farming system model developed at SKUAST-Jammu. It has been observed that the farmers of the temperate region in particular and J&K region in general are lacking in technical efficiency, which has been estimated about 45 per cent.
- ☆ Demonstration of IFS models on the farmers field near the University vicinity/research institutions shall be under taken, so that more and more farmers shall adopt.
- ☆ There is a great potential of value addition and marketing aspects shall be accorded priority under the IFS components.
- ☆ The farmers would be organised into common interest groups federated into producer collectives. Existing institutions like Women SHGs etc. would be used to initially anchoring the program.
- ☆ These institutions would take the roles of planning, mobilising resources, organising production, and take up post harvest management and marketing activities.
- ☆ The producer collectives will improve the collective bargaining power of the farmers, will internalize market activities like bulking, primary and secondary processing which improve the village economy.

(c) Food and Livelihood Security

Shift to Integrated Farming system is often seen as a compromise on food security. This is mainly because food is understood as only wheat and rice, few pulses, oilseeds and vegetables. The food basket can be increased if we can expand the scope to include millets, coarse cereals, dryland fruits, uncultivated greens *etc.*, which can also bring in nutrition security. Data from National Centre for Organic Farming (NCOF), ICRISAT and CMSA have proven that crop productivity can also be maintained with organic/ecological farming. Going beyond the current food security systems like PDS and mid-day meal schemes, systems need to be established to improve livelihood security in terms of sustaining food production in the village, improving income generation opportunities to the small farmers and agriculture labor is important in the rural areas especially in temperate regions. The frequent monsoon failures results in droughts and support systems needs to be build to help the farm families and livestock to tide over.

a. Building household food security systems by adopting suitable cropping patterns
b. Village level management systems for alternative models like grain banks would be appropriate and attempted.
c. Suitable off-farm and non-farm employment opportunities would be identified and promoted.

(d) Financial Support Systems

Currently, all the financial support systems to agriculture are given only for external inputs. We need to create proper support systems for farm internalized inputs, community based infrastructure, knowledge and skill building and sharing *etc.*

- ☆ Direct subsidies to farmers rather than input subsidies.
- ☆ Integrating NREGA with integrated farming system so that each farmer gets 100 labour days for farming can provide ample scope in this direction.
- ☆ Explore tools like Direct Income Support which exist in many developed/developing countries need to be explored as decent living income cannot be explored only through pricing mechanism.

(e) Partnerships

At the district/village/block level we need build partnerships between various governmental and non governmental agencies to implement the program. At the state level we need to build an alliance of public sector research organisations, extension agencies, departments dealing with rural livelihoods, SKUAST-Jammu and NGOs, which are working on sustainable agriculture/organic/natural/ecological farming.

10. BROADER GUIDELINES FOR ADOPTION OF INTEGRATED FARMING SYSTEM

(a) Selection of Area

Select a contiguous piece of land measuring 1.0-1.5 ha, to represent small and marginal farmers of the region. The yardstick for defining small and marginal farmers may vary among the stake holders and accordingly area selected for the IFS may also vary.

(b) Selection of Enterprise (Module)

There is no thumb rule for selecting a specific number of enterprises and allocation of area to different enterprises. It would largely depend upon the common logic of household requirements of an average farm-family of the region, especially in respect of food, feed, fodder, fuel, fibre and reasonable cash. In farming system approach we have to ensure that output from the individual farm unit (IFS Project Area in our case) fulfils aforementioned requirements for a normal size family (size to be specified) along with its farm animals/fish *etc.* Moreover, general agro-ecological situations, infrastructure availability, government policies and social acceptability have to be duly considered. As far as possible, select the farming system enterprises, which may fit in a modular format, *i.e.*, they may be developed as self-contained components of the farming system as a whole, with well-defined interface(s) so that they can easily be connected to or disconnected from the system. It should be further diversified with additional components with well-defined objectives and principles of farming systems approach.

(c) Selection of Crops/Cropping Systems/Animals

Within each component, identify crops and their most suited varieties to be grown, cropping systems to be adopted, type and number of animals to be raised, species/breeds of livestock/fish/honeybees to be raised. Allocate optimum area under each unit. This should also be taken care in accordance with abovementioned criteria. Since under temperate conditions the important cash crop are the fruits, hence, Horticulture-based cropping system will have much more relevance.

(d) Technologies to be Adopted

All the latest available production technologies for different production/processing units have to be adopted with optimum levels and should be specified in technical programme.

11. AN OVERVIEW OF INTEGRATED FARMING SYSTEM MODEL DEVELOPED AT SKUAST-JAMMU

The SKUAST-Jammu after characterizing the existing farming of the region including the temperate areas has developed the model of 1ha for small and marginal farmers on its farm to study the synergistic of various components of farming. The farming system module developed comprises of Horticulture (Fruit, Vegetable and Flowers)+ Crop + Animal + Fish cum poultry + Vermicomposting cum farm waste recycling + biogas + mushroom unit + Small poly house + Boundry tree plantation. The area distribution with regard to 05 family members for individual enterprise has been standardized as under (Table 2).

Table 2: Component-wise Summary

Component	*Year of Start*	*Net Area (ha)*	*Gross Return*	*Total Cost (Rs./ha)*	*Net Returns (Rs./ha)*
Crop Unit					
Cropping systems	2011	0.3802	63650	38068	25582
Horticulture unit					
Fruit crops	2011	0.3	13328	17887	9094
Vegetable (intercrop)	2011	–	29649	4234	11762
Floriculture	2011	0.07	7800	2890	4910
Agro-forestry	2011	0.1	1315	597	718

Contd...

Table 2–*Contd...*

Component	*Year of Start*	*Net Area (ha)*	*Gross Return*	*Total Cost (Rs./ha)*	*Net Returns (Rs./ha)*
Livestock Unit					
Dairy animals (including Vermicompost and biogas unit)	2011	0.08	318212	174599	143613
Fish cum Poultry Unit					
Poultry/Ducks	2011	0.001	76578	31989	44589
Fisheries	2011	0.1	42958	12258	30700
Apiary unit (03 boxes)	2011	0.001	4500	3300	1200
Green fodder (Fodder-Fodder Cropping System and on bunds planted perinial Fodder crops (Hybrid napier)	2011	0.07	14543	1763	12780
Mushroom unit (6 qts wheat straw)	2014	0.02	31034	9780	21254
Grand total	–	**1.0**	**603567**	**297365**	**306202**

The profit is quit comparable to existing Farming system, which works out to be 0.80- 0.90 lakhs/ha.

It is evident from the above study of integrated farming system that their is having the potential not only to increase the profit to Rs 3.062 lakhs/ha as compared to existing farming system were the farmer is realizes only Rs 0.80 - 0.90 lakhs but also increases the sustainability issues. About 150 kg N is being realized through farm wast composting.

12. CONCLUSION

Present day commercial Agriculture can be profitable and sustainable only when it is carried in the Farming Systems Mode. Temperate Horticulture is the system based enterprise and this can help the orchardists to meet the different requirements by adopting different farming activities. such models need to be developed and popularised among the growers for meeting different needs. Success stories also need to be replicated for their upliftment and better livelihood opportunities in the region.

Chapter 27

Good Horticultural Practices and Farming Systems in Temperate Fruits

Jayant Kumar and Shashi K. Sharma

Regional Horticultural Research and Training Station,
Dr Y.S. Parmar University of Horticulture and Forestry,
Bajaura, Kullu – 175 125, Himachal Pradesh
E-mail: jayantkumar_in@yahoo.com

1. INTRODUCTION

Good horticultural practices are the practices that address environmental, economic and social sustainability for on-farm processes and result in safe and quality production of fruits and other horticultural commodities under a given set of ecological conditions. In simple terms the concept of GHP stands on four pillars (economic viability, environmental sustainability, social acceptability and food safety and quality). In recent years, the concept of GHP has evolved to address the concerns of different stakeholders about quality production and the environmental sustainability of horticulture. These stakeholders include governments, farm produce retailing industries, farmers and consumers who seek to meet specific objectives of food safety, food production, production efficiency, livelihood and environmental benefits.

Appropriate promotion and adoption of GHP in temperate fruit production is supposed to help in improving the safety and quality of fruits and their products while safe guarding the highly fragile Himalayan ecosystem. In addition, producers and consumers will be benefit from expansion of horticultural horizons across the globe. Access to global markets will improve their livelihoods and the national economy as a whole. It will help to promote sustainable horticulture and contribute in meeting national and international environmental and social development objectives. Adherence to food quality and safety will protect people's health – an important factor in national development (ASQ, 2000). Today, the National agricultural development programs aim at creating an enabling environment for improving horticultural productivity and profitability thereby improving farm income and contributing to socio-economic upliftment of growers leading to poverty reduction. To achieve the target of economically viable and ecologically safer temperate fruit production system awareness is needed (both at producer and consumer level) regarding 'Good Horticultural Practices', which will lead not only to improvements in terms of environment and health safety but also for the appreciable increase in yield and productivity of the horticultural crops (Mushobozi, 2010).

Presently, the country needs a proper policy and adequate resources to facilitate improvement of the fruit production especially in the fragile Himalayan ecosystem. To respond to the present challenges and opportunities of temperate fruit production; various technical, economical, social, infrastructural and institutional challenges need to be overcome in order to ensure fruit quality and safety to protect people's health and participate in the global economy. Furthermore, there is a need to ensure participation of small-scale producers, processors, extension staff and retailers who are in the horticultural supply chain. Empowerment of these stakeholders within the framework of GHP can enhance basic quality standards that foster the health and economy of both producer and consumer as well (WHO, 1999).

2. GHP IN SELECTION OF SITE, CROP/VARIETY AND PLANTING OF ORCHARD

(a) Planning for New Plantings

Under the changing climatic scenario many apple growers are looking hard at making the investment in new orchards. With the large investment of money and management required to be successful, it is important to plan well and consider all the details. Any mistake made in the beginning about site selection, choice of rootstock or cultivars, pollination needs, soil preparation, training system, etc. may have long term detrimental effects on the orchard's performance and profitability. This means planning for a new orchard should begin well in advance of the actual planting. There is much talk about new technologies of high density plantings, designing orchards for mechanization, and the latest cultivars and rootstocks. But it is really the horticultural basics that will make or break a new orchard. Here are GHP that needs to be followed for establishment of a new orchard:

- ✰ Choose the site – consider slope, elevation, wind movement, soil type and condition, drainage.
- ✰ Get the land history.
- ✰ Is altitude is sufficient enough to fulfill the basic chilling requirements of the crop you are planning?
- ✰ Is the crop to be planted is fit for the aspect of the slope on which orchard is to be planted?

- ☆ Get expert advice on expected inter-annual climatic variation at the site.
- ☆ Get information on market accessibility.

As climate change is the major issue in orchard planning so a case study of Kullu district is discussed hereunder for having elaborations on the issue.

Temperate fruit production is highly vulnerable to climate change slight change meteorological variables has caused a significant shift in cropping zone. Kullu valley was known for good apple production but now-a-days growers are diversifying their crop production with stone fruits, kiwifruit and pomegranate which have quite lower chilling requirement and can tolerate high summer temperatures than apple. It may be attributed to the change in temperature variables in the recent past. From early sixties there has been increase of 1.8°C in maximum temperature, 3.1°C increase in Minimum temperature. Overall increase of 2°C has been observed in the mean temperature. Increase of 2°C over the past fifty years has made the region marginal far apple production (Kuniyal, 2014). Under such a changing climate, GHP is selection of varieties which are having lower requirements of chilling hours and better production and quality under changing climate. Alternatively, growers have started diversifying the horticulture in the Kullu valley with other temperate fruit crops which are having lesser chilling requirements.

Effect of climate change on maximum and minimum temperature from 1962 to 2010 at Katrain in Kullu valley

Decade	Maximum Temperature (°C)	Minimum Temperature (°C)	Mean Temperature (°C)
1962-70	20.6	8.8	14.7
1971-80	20.7	9.3	15
1981-90	20.3	9.2	14.8
1991-00	21.4	9.7	15.6
2001-10	21.4	11.9	16.7
Difference 1963-2010	1.8	3.1	2.0

Availability of Fruits Round the Year in the Mid Himalayan Region of Himachal Pradesh

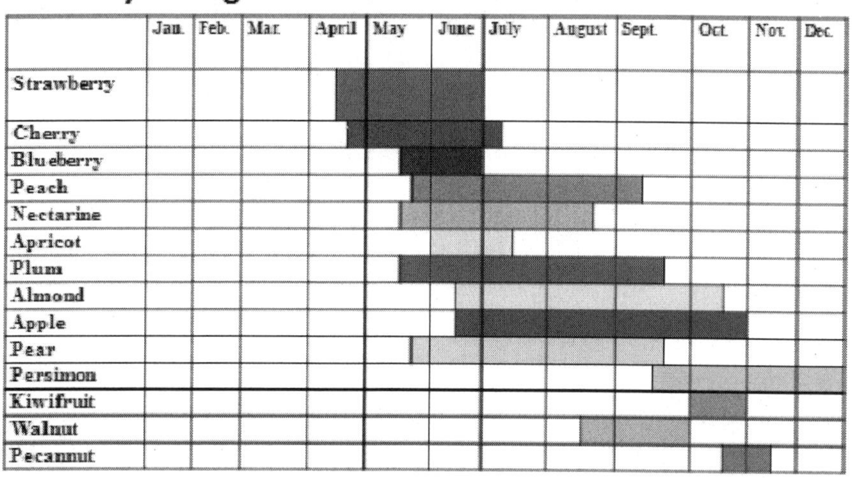

*Shaded area shows the availability of the fresh fruits

Therefore, under this changing climatic scenario is having knowledge on availability of fruit crops round the year in the target growing region can be considered as GHP. In mid Himalayan region of Himachal Pradesh, there is wide variety of fruit crops which have potential for diversification. Fruits become available right from the month of March to end of December. In Kullu valley Pomegranate, Kiwifruit, Pear, Peaches, Plum has made major dent to acreage of apple in the region. There is a wide choice of varieties in all the temperate fruits which can be adopted under changing climatic scenario.

3. GHP IN CHOOSING CULTIVAR

Choosing the correct cultivars for the markets and your climate has the pivotal role to the profitability of your new orchard. Remember that you are planting for the consumers of tomorrow, so consider the trends in buying patterns *i.e.* younger people, ethnic diversity and where populations are increasing.

In recent years, the most popular cultivars in new plantings include Gala, Super Chief, Red Velox, Jeromine, and Ambrosia, in apple and in pear Carmine, Concord, Packham's Triumph because demand for these cultivars is growing. Growers with retail markets often include cultivars for their niche markets like Royal Delicious, Vance Delicious, Red chief apples and Red Bartlett and William Bartlett pears etc. that require delicate handling (Kumar and Thakur, 2015). Each grower needs to evaluate cultivars suitable for markets over the next 20-30 years.

Following factors should be considered at priority when choosing cultivars to plant:

1. **Climatic suitability:** Will this cultivar grow well at my site? Winter hardiness, temperatures during the growing season, length of growing season, and temperature during ripening are important to consider. Royal Delicious, Vance Delicious, Red Fuji, Red Chief etc. prefer cooler growing areas, while cultivars like Michael, Schlomit, Dorset Golden, Anna require higher heat accumulations and frost-free season.
2. **Markets:** What cultivars do my buyers want? Your buyers may include packers, farmers markets, retail stands and local stores. Your orchard will be more successful if you grow the apples your markets demand. Otherwise, the low-priced juice market may be your only alternative.
3. **Disease susceptibility:** One should be ready to manage disease problems in each cultivar? Most commercial cultivars are susceptible to scab (which requires a strict management strategy), Cultivars of Delicious group are susceptible to scab, but this problem is worse when grown in humidity persisting niche areas (Rosenberger, 2003). Therefore, intensive management of serious diseases and pests should always be kept in mind.
4. **Pollination:** Will my cultivar mix provide the cross-pollination required for high yields and quality fruit? To ensure an adequate pollen supply when blooms are open, check if bloom dates overlap adequately. In areas with high pollination problem combination of two to three pollinizers along with 2-3 bee hives per hectare as pollinator (Jindal and Gautam, 2002).

Some of the new pollinizer varieties producing good quality fruit are: Gold spur, Gale Gala, Red Loom Gala, Granny Smith. Red Gold and Commercial. Besides, Crab apple varieties like Snow Drift, Golden Hornet, *Malus floribunda*, Crimson G are preferred for areas with pollination problem. In plum new varieties like Red Beaute, Frontier, Mariposa, Angeleno and Black Amber require Santa Rosa as pollinizer.

Choosing cultivars for your new plantings is a complicated process, but pre-evaluating all of these factors will help you plan better for the cultivar mix that is right for your operation. Photographs of some of the potential crops and their varieties are presented on the next two pages for having quick look on some of the important temperate fruit crops which can be considered as GHP under changing climatic conditions.

4. GHP IN CHOOSING ROOTSTOCKS

Rootstocks influence temperate fruit trees in many ways: their growth rate, final size, early cropping, fruit size, yield potential and response to nutrition. Rootstocks vary in their susceptibility to insects and diseases, tolerance of soil conditions, and compatibility with various scions. Rootstocks also vary in their ability to withstand cold winter temperatures, as well as freeze—thaw cycles during the dormant season. So it is important to match the correct dwarfing rootstock to your soil, site and cultivar so your trees grow the appropriate size and produce high early yields of large, quality fruit. Sharma and Karkara (2003) conducted a long term trial with EMLA 26, EMLA 7a, EMLA 106 clonal rootstocks for Well Spur, Starkrimson Delicious, Silver Spur, Red Chief, Oregon Spur –II and Brighten Early varieties of apple. It was found that yield of different varieties was recorded highest with EMLA 7a followed by EMLA 106. Similarly in another experiment on inducing dwarfing in different cultivars of apple using M9 as dwarfing rootstock, it was found that highest yield was recorded for Hardeman followed by Red Spur. Whereas, yield efficiency

Influence of rootstock and cultivars on yield (MT/ha)

Cultivars	EMLA 26	EMLA 7a	EMLA 106	Mean
Well Spur	58.16	64.97	61.45	61.52
Starkrimson Delicious	55.20	63.91	70.52	63.21
Silver Spur	58.95	89.41	66.06	71.47
Red Chief (Campbell)	58.58	64.62	67.91	63.70
Oregon Spur II	56.80	91.17	64.91	70.96
Bright- N- Early	60.69	68.88	63.57	64.38
Mean	58.06	73.82	64.38	

Growth and yield of different apple cultivars under high density planting on M9 rootstock

Varieties	Girth (cm)	Height (cm)	Spread (cm)	Extension Growth (cm)	Yield (kg/tree)	Yield efficiency (g/cm²)
Royal Delicious	35.93	392.4	318.4	34.81	29.57	269.0
Top Red	39.24	379.3	362.7	36.20	38.71	338.1
Vance Delicious	41.29	412.9	388.1	27.74	38.71	293.3
Skyline Supreme	37.64	415.0	362.0	34.51	33.29	332.4
Hardeman	51.69	467.9	455.9	36.83	52.29	264.4
Red Spur	38.26	378.3	395.6	52.51	51.71	439.3
$CD_{0.05}$	7.24	54.74	61.99	10.63	NS	NS

Jeromine

- Highly productive
- 5-6 days early than Red Delicious

Apple - Oregon Spur

- Ripening 3-4 days before Royal Delicious
- Spur type and very productive
- Suitable for higher altitude

Scarlet Spur II

Superchief

Early maturing variety. Develop good taste and colour early

Apple - Scarlet Gala

- Early maturing apple
- Scarlet red in colour
- Sweet in taste with slight acidic blend, firm fleshed

Red Chief

- Ripening 10-12 days before Royal Delicious
- Spur type and very productive

Pear-Red Bartlett

- Maturing in last week of June to mid July
- Sweet in taste
- Fruit large blushed with red colour

Pear-Starkrimson

- Ripening one week late to Red Bartlet
- Sweet, fetches good price in market

Kiwifruit

Varieties

Hayward and

Allison

Male: Tomuri

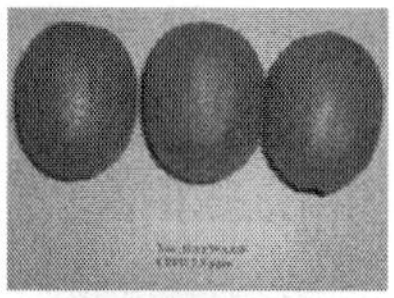

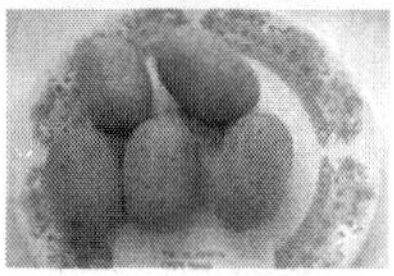

Walnut

CITH Walnut-1

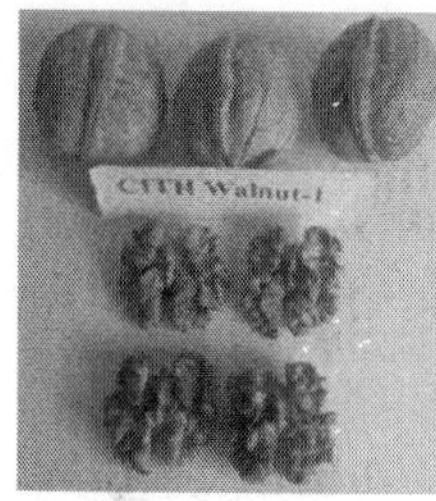

Peaches

Red Haven

- Ripening during mid June
- Fruit dark red coloured
- with sweet taste

Sun Crest

- Ripening in the last week of July
- Fruit orange red, attaractive
  Sweet taste

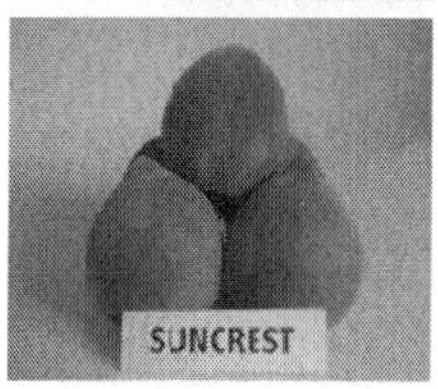

Chandler

Pecan nut

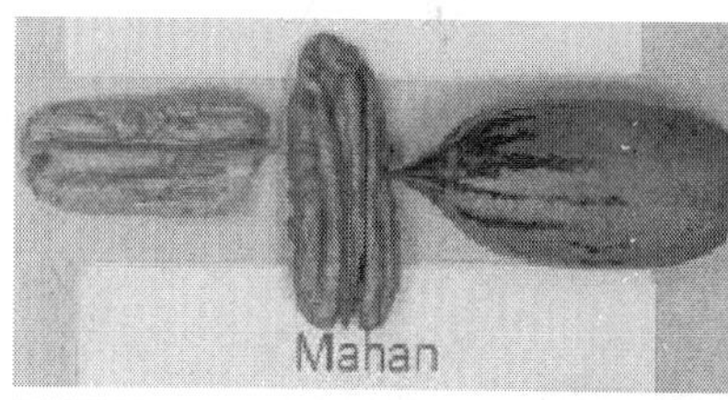

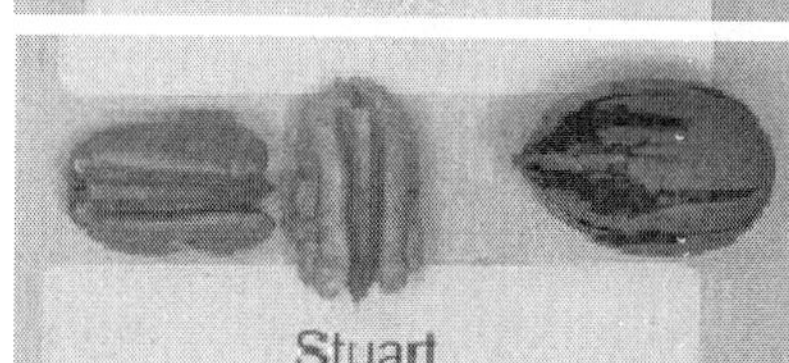

Nectarine

Silver King

Fruit yield: 70 to 80 Kg/ tree

Fruit weight: 80g

TSS: 7-8°

Ripening mid-May. Fruits medium in size. Ground colour greenish white, 3/4 fruit covered with dark red blush, Fruit very attractive, Taste is sweet.

Plum

was recorded highest for Red Spur followed by Top Red.Overall vegetative and reproductive performance was recorded best for Top Red and Vance Delicious apple cultivars (Kumar and Thakur, 2015). In a series of other experiments it has been concluded that MM 106 is susceptible to collar rot hence should not be tried under clay soil conditions. M9 has been recommended for different standard apple cultivars as a dwarfing rootstock. For spur type varieties M7 yields best performance under varied growing conditions.

5. GHP IN CHOICE OF NURSERY PLANTS FOR NEW ORCHARD

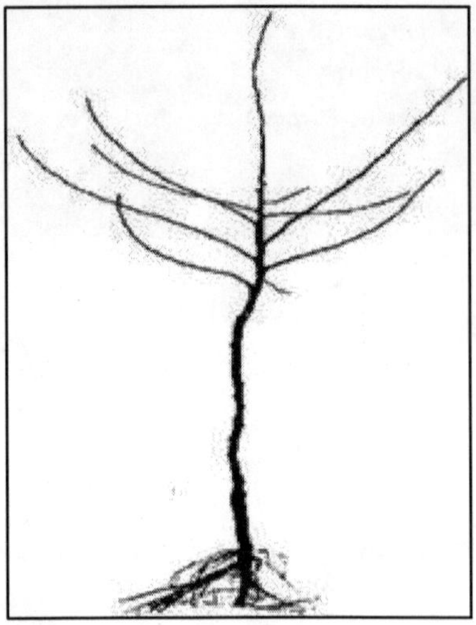

Starting an orchard with high quality nursery plants is the first key to a successful planting. High quality plants will quickly establish, grow to the desired height, and fill their space. The right trees will be ready to produce fruit in the 2nd year, and will reach full production a few years later (Omafra, 2012). Planting the right tree means early cropping, which is the key to profitability of your new orchard. Trees of lesser quality can eventually fill their space and produce good yields, but the early returns are lost, so profitability over the life of the planting will be less.

Under GHP, the ideal nursery plants should have following qualities:

- ☆ Well feathered nursery plant.
- ☆ At least 1.7 m tall, preferably 2-2.2 m.
- ☆ An abundance of healthy roots.
- ☆ A dominant straight leader.
- ☆ 6-10 "feathers" that are 12-25 cm long.
- ☆ The bottom feather no lower than 24" above the soil when the tree is in place in the orchard.
- ☆ Feathers distributed along the leader at regular intervals.
- ☆ Delivered unbroken, with moist roots and no disease.

6. GHP IN PLANTING DENSITY

We've been talking about the concept of high density apples in India for more than 20 years. In fact, and a number of cost of production studies have shown that high density was the most profitable. Hence, high density orchards are not new, but many growers have difficulty accepting that it is the way to go – mainly due to the high investment, as well as the many changes in management required. However, there is a basic concept that is most important to understand why trees should be planted closer.

High density works because fruit produced in the early years is much more valuable that fruit produced inlater years. Think about this: If someone offered you one lakh rupee now or one lakh rupee in one year, which would you take? Of course, you'd take the Rs.1,00,000 now. Similarly if your orchard produced 100 boxes this year or 100 boxes next year, which would you, take? Again, this year's fruit is more valuable. This concept is called Net Present Value (NPV), and is a basic economic concept. Think of NPV as the opposite of interest paid. NPV says that money received in the future will be at a discounted value compared to its value today. Choosing to plant at a high density ensures that early yields will be higher. And the higher the density, the higher the early yields will be. Each tree produces only a certain yield (depending on the quality of the nursery tree). So, a newly planted orchard planted at 1,000 trees/acre will produce double the yield of 500 tree/acre planting in the early years.

There are number of planting system which can be adopted for high density planting depending upon the growing conditions, cultivar and rootstock etc. Mostly practiced system are either single row planting or double row planting or triple row planting in different apple growing region of the world Kumar (1995). As a GHP presented here are the recommendations on optimum spacing of different rootstock and variety combinations under single row or double row or triple row of planting with the respective per hectare number of plants that can be accommodated for establishing a high density orchard of apple.

Planting Distance According to Variety, Rootstock and Planting System

Cultivar/Rootstock	Spacing/m	Trees/ha
Single Row Red Delicious		
Standard/M9	3.4-3.5x1.3-1.4	2040-1840
Spur type/M6	3.0-3.3x0.9-10	3330-2730
Spur type/M7	3.3-3.5x1.0-1.2	2730-2140
Spur type/MM111	3.5-3.7x1.5-1.7	1700-1430
Golden Delicious/M9	3.2-3.5x1.0-1.2	2810-2140
Golden Delicious/MM106	3.5-3.8x1.4-1.5	1840-1560
Double Row		
Golden Del./M9	3.3-3.5x1.0-1.5	2750-2670
Red Del. Spur type/M26	3.0-3.3x1.3-1.5	3640-2930
Triple Row		
Red Del. Stock/M9	3.2-3.4+2x0.9x1.75	3090-2570
Spur type/M26	2.8-3.0+2x0.0x1.4	4820-4550

7. GHP IN ORCHARD FLOOR MANAGEMENT AND NUTRITION

The best orchard floor system will provide a level, solid working area for labour and equipment, in all types of weather including rain, wind and heat. It will minimize problems with weeds, pests, nematodes and dust, and will protect and maintain the soil and soil structure as a fertile resource for growing large crops of fruit.

In a long term experiment conducted on Starking Delicious apple grown under different orchard soil management practices Kumar *et al.* (1999) recommended herbicide plus mulching with hay as GHP for apple grown in the mid Himalayan region.

Orchard Management Practice	*Trunk Girth (cm)*	*Shoot Length (cm)*	*Fruit Yield (kg)*	*Fruit Weight (g)*	*TSS (°Brix)*
Herbicide + Hay mulching	55.8	46.6	89.4	260	14.3
Herbicide (Simazine 6 kg/ha)	53.2	43	85.7	249	14.9
Hay mulching	54.4	38.3	81.7	255	14.4
Mulching with white netted polythene	53.2	36.7	78.3	246	14.5
Clean cultivation	52.4	32.8	66.3	232	15.1
CD (0.05)	**NS**	**4.9**	**7.8**	**7.8**	**0.7**

In another field trial Thakur *et al.* (1993, 1997) recommended Black polythene mulch as an effective orchard floor management technique in Red Delicious apple. Most orchard floors combine a grassy sod in the row middles (or alleyways) and a weed-free strip under the trees. This system is a compromise between three factors:

i. Tree growth and early cropping (no competition from vegetation)
ii. The soil (protection from erosion and compaction)
iii. Orchard operations (stable areas for moving equipment and labour efficiency)

It is critical to prevent weeds from growing under the trees, especially in the first 3 months after planting (known as the Critical Weed-free Period). Usually the final tillage after the trees are planted will level the ground around the tree and remove early weeds. The weed-free strip or clean basin should be wider than the drip line of the tree (new orchards should have a 3 to 4 foot wide strip or basin). Floor management in combination with mulch should take into consideration the time of emergence of weeds mulch is to be applied you are planning to use a mulch (which will suppress weeds as well as preserve soil moisture), apply it before the next weeds emerge.

Long term experiments on apple and pears have revealed that the high yield and growth can be obtained with application of 700 g N, 350 g P and 700 g K_2O in apple (Kumar and Sharmal, 2006) and pear (Kumar and Chandel,2002). Pre-bloom sprays of urea (0.5 per cent) plus boric acid (0.1 per cent) increased growth and fruit yield of apple (Kumar *etal.*, 2002). Now we are recommending urea + N:P:K (19:19:19) + Multiplex® + boric acid at pink bud stage to increase fruit set and fruit yield.

8. YIELD FORECASTING

Yield is a function of complex interacting factors which are governed by environmental conditions and intrinsic potential of the plant. It shows a cause and effect relationship which can be measured in numeric terms. The important factors governing yield are flower density, per cent of flowers which yield mature fruit and rate of fruit growth. Most of the available information does not distinguish between these components. Keeping these factors constant for a given variety, the crop estimates are also influenced by other factors like temperature, rainfall, humidity during flowering and fruit set. Other morphological factors such as trunk girth, tree volume and number of fruiting spurs per unit tree volume are also potential explanatory variables for yield variation under a given set of growing conditions.

Kumar and Awasthi (1998), and Awasthi and Sharma (2000, 2001, 2005) presented various mathematical models for predicting yield of apple of different age groups grown at different altitudes. Some of these models predicting yield of fully grown standard apple tree under the mid hill conditions, on the basis of different parameters are presented below:

i) Meteorological parameters

$$Y= -161.1 + 0.11CU + 2.34Nh^{14-20} +2.22RH^{40-60}$$

ii) Morphological parameters

Y= -102.8 + 0.38C + 1.28V + 0.62PS + 0.77SS - 0.27S1 + 2.23S2 + 0.64S3 + 2.80S4 - 0.01S5

iii) Nutritional parameters

Y= - 73.87 + 27.60N + 284.5P + 56.14K + 24.53Ca -13.08Mg

iv) Bud Differentiation

Y= -172.3 + 0.12R1 + 7.36R2 + 4.75R3 + 10.11R4 - 0.26R5

Where, Y=Yield in Kg per tree, CU= Cumulative chill units, Nh^{14-20} = Number of hours with temperature between 14 to 20°C, RH^{40-60}= Number of hours with relative humidity between 40-60 per cent, G = trunk girth, V= tree volume, PS = No. of primary scaffolds, SS = No. of secondary scaffolds, S1= Buds on shoot terminals, S2=buds on branched spurs, S3 = buds on spurs which bore fruit during previous year, S4 = buds on solitary swollen spurs, S5 = buds on week spurs, R1, R2, R3, R4 and R5 are proportion of reproductive buds determined through histological studies respectively on S1, S2, S3, S4 and S5.

9. GHP IN USE OF GROWTH HORMONES

Growth hormones have become very important in regulating crops in apple and other fruits. Especially in the years of heavy fruit set, fruit thinning agents NAA have to be used to reduce the crop and get good size fruit and also the return bloom.If not thinned the there may be no return bloom. Promoline has been found very effective in improving shape of apple fruits if sprayed at full bloom in apple (Kumar *et al*,. 2003). Similarly in case of Cherries, 10 ppm spray of GA is very effective in improving the fruit size. In kiwifruit 10 ppm Sitofex (CPPU) is commercially used to improve fruit size.

10. GHP IN AUTOMATION, MECHANIZATION AND ROBOTICS

Rapid increase in area under fruit crops especially the temperate fruits (>1.5 lakh hectare in Himachal Pradesh, (DESHP, 2012) and significant decrease in labour availability is raising the concern of automation, mechanization and use of robotics in temperate fruit production. In a study conducted it was found that cost of manual basin preparation is more than double than the cost involved in doing the same operation with the use of machines. Similarly, 4 to 6 times cost can be saved in grass cutting and pit making with the use of machines than undertaking the same operations manually. Tremendous saving in time is also observed in different orchard operations.

In the time to come the pressure will continue for growers to reduce costs of production, harvesting and packing. Traders will continue to manipulate prices received for their benefit and not the producers, growers will be forced to sell the produce at cheaper rates. Labour costs will continue to increase. Today, per day living wages have exceeded Rs. 300/- or even Rs. 500/- per day for partially skilled workers. Hence, in order to reduce the production cost more sophisticated, accurate and faster sensors based machines are required to be developed. In some developed

Time and labour cost of basin preparation

Size of basin	Hand /Machine	Time taken to prepare basins	Cost of basin preparation
One meter radius	Hand hoeing	20 minutes	Rs 7.50/-
	Machine hoeing	2 minutes	Rs2.00/-
Two meter radius	Hand hoeing	42minutes	Rs 13.12/-
	Machine hoeing	6 minutes	Rs 6.00/-

Time and labour cost cutting grass

Type of grass	Hand cutting/ Machine cutting	Time taken to cut the grass	Cost of cutting
Grass	Hand cutting	30 hrs/acre	Rs. 562.50
	Machine cutting	6.66 hrs/acre	Rs 164.00
Hard grass	Hand cutting	66.8 hrs/acre	Rs. 1248.35
	Machine cutting	10.8 hrs/acre	Rs. 242.50
Soft grasses during rainy season	Hand cutting	259.2 hrs/acre	Rs.4728.75
	Machine cutting	96.2 hrs/acre	Rs. 1503.75

Time and labour cost of digging pits

Type of pit	Hand/Machine	Time taken to dig the pit	Cost
Single bore	Hand	30 minute/pit	Rs. 9.37/-
	Machine	1.25minute/pit	Rs. 2.39/-
Bigger pit with	Hand	1.40 minutes/pit	Rs. 31.25/-
	Machine	5 minutes/pit	Rs. 9.56/-

countries functional robots are just around the corner for fruit production systems.But, due to having poor access for orchard mechanization either due poor availability of machines or due to existence of standard plantations which are not much compatible with the orchard mechanization equipment. It is very crucial that multidisciplinary research, involving plant scientists, engineers, food scientists, economists and marketing expertise should be focused on creating new production systems. Innovation will come through intimate understanding of all parts of the system. Growing systems will be designed for specific crops and then dedicated machines will fit. Some of the machines as a part of GHP have demonstrated by RHRTS Bajaura, Kullu are shown here for having more elaboration on the issue of mechanization.

Grass Cutter
BCS Ferrari Make Mini Power Tiller Model 710
Walking/ Hand Tractor/ Orchard Tractor
Motorized Earth Auger
Other Horticultural Machines in use by Developed Countries
Greenhouse robot sprayers
Worker safety and precision application
Crop monitoring
Octocopter
Carries sensors and cameras for aerial surveillance
Crop loads, stress conditions – including water, nutrient, harvest maturity, pest and disease load
Instrumented glove for fruit quality
Developed as harvesting aid by French scientists
Equipped with non-destructive miniature sensors and artificial intelligence
Measures color, sugars, firmness.
Programmed to specific attributes and enables picker to make objective decisions.
Not commercialised
Harvesting aids
Pneumatic branch shaker
Used for cherries and apples
Fruit falls onto collecting surface
Physical damage an issue – fruit hits branches, other fruit or surface of catching frame

10. MANAGING WEATHER RISKS

Weather has influence on many aspects of the business of growing tree fruit, as growers are well aware. In a study conducted on peaches by Sharma *et al.* (2014) reported that performance of many peach varieties is greatly influenced by the inter-annual variability of weather parameters. Therefore, one should fore see the expected weather variations and should prepare accordingly because more economic losses are caused by these factors than any other production related parameters. One should consider following weather risks as potential threat to his temperate fruit orchard:

(a) Heavy Rainfall

Mid Himalayan weather patterns usually create this problem in the spring and fall only. However, sometimes many locations suffer with excessive rain throughout the entire year. Apples, peach and pears do not like "*wet feet*", although some newer rootstocks are more tolerant of imperfect drainage and subsequent Phytophthora infections. Tree roots standing in water-logged soils during the growing season are at much greater risk than when trees are dormant. Wet soils or potholes also make it challenging for spray or harvest equipment access. Rain close to harvest can cause cracking in sweet cherries and russetting in apples.

Potential GHP

- ☆ Installation of systematic tile drainage before planting the orchard.
- ☆ Choice of rootstocks tolerant of phytophthora.
- ☆ Use of cover crops before and during plantings to improve soil structure and drainage.
- ☆ Avoidance of imperfectly-drained sites, especially clay soils.
- ☆ Use of sub-surface drainage.

(b) Drought

Rainfall is often insufficient during the growing season to maintain tree health and to produce adequate yields and quality fruit. Water needs of fruit trees are increased under intense sunlight, and incoming solar radiation has been gradually increasing in recent decades. Dwarfing rootstocks have shallow root systems so are more susceptible to lack of rain.

Potential GHP

- ☆ Supplemental irrigation, preferably a permanent trickle system.
- ☆ Use of under-tree mulch to conserve soil moisture.
- ☆ Addition of soil organic matter before planting with cover crops or other organic amendments.
- ☆ Installation of subsurface tile drains which encourage deeper growth of roots, so this helps sustain trees through periods of drought.

(c) Hail

The devastation from a short episode of hail has been felt by most tree fruit growers at some time. Hail can either bruise or cut fruit, starting at calyx until harvest.

Hail cuts can also allow disease infections in leaves and bark (pears and apples) that can lead serious losses to tree health and productivity.

Potential GHP

- ☆ Hail netting, most useful for high value cultivars.
- ☆ Production insurance to cushion the economic impact.
- ☆ Multiple locations of orchards to spread the risk.

(d) Winter Injury

Although pears can tolerate colder air temperatures than most tender fruits, they can be damaged by sudden air temperature drops in the fall or very deep cold in the winter. The trunks and limb crotches are the last tissue to harden off in the fall, so may be vulnerable to air temperature changes in November and December. Extreme cold air temperatures, especially after January when trees start to de-acclimate, may result in fruit bud damage or dead shoot tips. Infections from powdery mildew also make shoots most vulnerable to cold injury. Peaches and other stone fruits can be very vulnerable to temperature drops.

Potential GHP

- ☆ Avoid cold sites.
- ☆ Apply nitrogen only early in the season.
- ☆ Grow sod or cover crops between rows after July to encourage trees to harden off.
- ☆ Do not prune in the fall, or early winter.
- ☆ Use reflective white paint on young tree trunks to avoid southwest injury.

11. INSECT PEST AND DISEASE CONTROL

Healthy plants yield more fruit than either less vigorous or more vigorous ones. A reasonable supply of mineral nutrients provided by the application of fertilizers and/or organic matter helps assure a harvest.

Potential GHP

- ☆ Scab and leaf spot diseases overwinter on leaf debris that remains on orchard floors. Therefore complete decomposition of leaf debris from all rosaceae family plants should be taken at priority.
- ☆ Crops that are new to an area are likely to be initially free of pest and disease attack.
- ☆ The threat will build up quickly with continual cultivation initial phyto-sanitary measures are important, therefore.
- ☆ Prevent build-up of pests by crop rotation and biological or integrated methodsof control should be preferred, where feasible, to chemical control. Risks of pesticide residue are particularly critical with food products. If chemicalapplications are unavoidable and cannot be reliably controlled

SPRAY SCHEDULE FOR THE CONTROL OF DISEASES OF APPLE

S. No.	Tree Stage	Name of Chemical	Quantity of Chemical for 200 L of Water	Control of Diseases	Remarks
1.	2.	3.	4.	5.	6.
1.	Green tip	Captan	600g	Scab	
		or			
		Dodine	200g		
		or			
		Ziram	600 ml		
2.	Pink bud	Mancozeb	600 g	Scab	
		or			
		Difenoconazole	30 ml		
3.	Petal fall/ Pea Stage	Carbendazim	100 g	Scab	
		or			
		Thiophanate Methyl	100 g		
		or			
		Hexaconazole	100 ml		
		or			
		Flusilazole	50 ml		
4.	Fruit development (Walnut size)	Myclobutanil	80 g	Scab	
5.	Fruit development (20 days after 4th spray)	Dodine	150 g	Scab	Do not spray Dodine if temperature is above 30°C
		or			
		Propineb	600 g		
		or			
		Zineb	600 g	Scab, Sooty-blotch, Fly speck	
6.	Pre harvest (20-25 days before harvest)	Captan	600 g	Scab	
		or			
		Ziram	600 ml		
		or			
		Mancozeb Flowable	700 ml	Scab, Sooty-blotch, Fly speck	

SPRAY SCHEDULE FOR THE CONTROL OF INSECT & MITE PESTS OF APPLE

S. No.	Tree stage	Name of Chemical	Quantity of Chemical for 200 L of Water	Control of Pests
1.	2.	3.	4.	5.
1.	Pink bud	Thiacloprid	100 ml	Thrips
		Chlorpyriphos	400 ml	Aphid
2.	Petal fall	Spiromesifen	60 ml	Mites
		Malathion	200 ml	Aphid
3.	Fruit development (Walnut size)	Hexythiazox	200 ml	Mites
4.	Fruit development (20 days after 3rd spray)	Fenazaquin or Propargite or Hexythiazox	50 ml 200 ml 200 ml	Mites
5.	Fruit development (20 days after 4th spray)	Fenazaquin or Hexythiazox	50 ml 200 ml	Mites
6.	Pre-harvest (20-25 days before harvest)	Malathion	200 ml	Aphid
7.	Post harvest	Chlorpyriphos	400 ml	Aphid

Note:- Avoid repetition of Acaricides for the control of phytophagous mites in subsequent sprays.

and traced, the areais likely to be unsuitable for production of domestic and export horticulture.

☆ It is important, however, not to apply too much nitrogen to fruit plantings because fruit set is reduced and succulent growth is more susceptible to diseases and pests. Onlyrecommended manure and fertilizer application schedule should be followed.

☆ Always the recommendations of local research and development agencies should be followed for effective management of pests and diseases as these agencies keep a track of surveillance of the pests and diseases infestations in the area. The spray schedule recommended by Dr YS Parmar University of Horticulture and Forestry, Solan and Department of Horticulture Himachal Pradesh for managing various diseases and pests in apple is presented herewith for elaborations on disease and pest management in Himachal Pradesh (UHF, 2014).

12. CROPPING SYSTEM RESEARCH IN TEMPERATE FRUITS

Sustainable and optimized return assuring cropping system is the priority these days. It envisages a combination of traditional cropping systems and the knowledge of how ecosystems exist, maintain their productivity, stability and resilience. A long term study have been conducted by Kumar and Chauhan (2003), Kumar *et al.* (2001) reports the performance of different crop combinations with plum as main crop. Wheat, maize, pea, tomato, lentil, black gram and their combinations were tried as intercrops in plum orchard. It was concluded that intercrops had significant effect on net returns from plum orchard. The highest net return was obtained when combination

Effect of Intercrops on Fruit Yield and Physico-chemical Characteristics of Plum

Intercrop	Yield (Kg/tree)	Increase in yield over control (%)	Fruit Wt(g)	TSS (^{0}B)
Wheat-Maize	68.2	4.9	66.6	12.2
Pea-Tomato	86.4	32.9	67.7	12.9
Lentil-Blackgram	79.8	22.7	63.2	13.0
Lentil-Maize	68.4	5.2	61.4	12.4
Wheat-Blackgram	73.7	11.8	63.3	13.4
Clover	69.3	6.6	59.4	12.4
Setaria	52.4	-19.3	49.3	13.0
Control (Clean basin and natural sward)	65.0		54.4	13.8
CD(0.05)	2.4		4.1	0.4

of pea and tomato was taken as intercrop in plum orchard. However, the net returns from intercrops declined as the plum trees advanced toward full canopy. Hence it was concluded that intercrops are beneficial cropping system for the early age of the orchard; as the orchard attain full canopy intercropping should be avoided. Only solid plantations of temperate fruits yield sustained and high returns in practice.One of the finnest example of intercrop croping system is depicted here in the photograph showing cabbage as intercrop in Super Chief apple orchard during initial years of orchard establishment. Likewise peas, beans and lentils have been proved good intercrops for the first ten years of standard apple plantation. It is not adviseable to go for intercropping in high density orchards.

13. REFERENCES

ASQ.2000. ISO 9000 Standards for quality. Amer. Soc. Qual. Available via the Internet at http: //www.iso-9000-2000.com

Awasthi, R.P. and Sharma, Shashi K. 2000. Studies on apple yield forecasting on the basis of bud examination. *Acta Hort*. **519**:103-11.

Awasthi, R.P. and Sharma, Shashi K. 2001. Studies on apple yield forecasting on the basis of flowering,fruitset and tree biometrical parameters. *Indian J. Hort.* **58**: 309-13.

Awasthi, R.P. and Sharma, Shashi K. 2002. Yield forecasting. **In:** *The Apple-Improvement, Production, Management*. Malhotra Publishing House, N. Delhi, pp. 278-286.

Jindal, K.K., Gautam, D.R.2002.Apple cultivation in India-problems, prospects and future strategies for enhancing its productivity.p.83-101. **In:** K.K. Jindal and D.R. Gautam (Eds.), *Enhancement of Temperate Fruit production*, Dr Y. S. Parmar University of Horticulture and Forestry, Nauni, Solan (H.P.), India.

Kumar, J., Rehalia, A.S., Rana, S.S. and Chandel, J.S. 2002. Effect of pre and post bloom sprays of urea and boric acid on growth, fruit set, yield and fruit quality of apple cv. Starking Delicious. *Prog. Hort.*, **34**: 22-26.

Kumar, J., Rana, S.S., Verma, H.S. and Jindal, K.K.2003. Effect of Promalin on fruit shape, set, yield and quality of apple cv. Starking Delicious. *Indian J. Hort.*, **60** : 143-146.

Kumar, J. and Chandel, J.S. 2004. Effect of different levels of N, P and K on growth and yield of pear cv. Red Bartlett. *Prog. Hort.* **36**: 202-206.

Kumar, J. and Sharma, V.P. 2006.Effect of different levels of N,P and K on growth and yield of apple cv. Starking Delicious. *Prog. Hort.*, **38**: 172-177.

Kumar J. and Thakur, D. 2015.Long term performance of some spur and coloured strains of apple (*M. domestica* L.) under low altitude conditions of Kullu valley of Himachal Pradesh. *Indian J. Hort.* Communicated for publication (Ref. No. 157/ 8/11).

Kumar J., Rana, S. S., Verma, H. S. and Parmar, D. K. 2001. Long-term effect of intercrops on growth, yield and fruit qualityof plum. *Indian J. Agric. Sci.* **71**: 687-690.

Kumar, J. and Chauhan, J. S. 2003. Effect of orchard soil management practices at different levels of nitrogen on nutrient availability and uptake by plum. *Indian J. Agric. Sci.* **73:** 307-310.

Kumar, J., Rehalia, A.S., Rana, S. S. and Verma, H. S. 1999.Long-term effect of orchard soil management practices on growth, yield and fruit quality of apple. *Indian J. Agric. Sci.* **69**: 355-358.

Kumar, J., Parmar, D.K. and Chauhan, J. S. 2003. The effect of orchard soil management practices at different levels of nitrogen on growth, yield and quality of plum. *Hort. J.* **16**: 1-7.

Kuniyal, C.P. 2014. Climate change is affecting apple cultivation in Himachal. www.current science.ac.in/Volumes/106/04/0498.pdf

Mushobozi, W. L. 2010. Good Agricultural Practices (GAP) on horticultural production for extension staff in Tanzania. FAO GAP Working Paper Series: 13. 186 p.

Omafra. 2012. Nursery Trees for the New Orchard. www.omafra.gov.on.ca/ neworchard/english/apples/4nursery.html

Rosenberger, D.A. (2003). Susceptibility of new apple cultivars to common apple diseases. *New York Fruit Quart.*, **11**: 17-22.

Sharma, R.P. and Karkara, B.K.2003.Performance of some cultivars of Delicious apple in low altitude warmer valley areas of Shimla. *Acta Hort.* **662**: 175-177.

Sharma, Shashi K. and Verma, K.S. 2014.Adaptation potential of low chill peach varieties to inter-annual climatic variability in the lower Shiwalik Himalayas. *International J. Agric. Env. Biotech.* **7**: 157-164.

Shashi Kumar and Awasthi, R.P. 1998. Yield forecasting in apple based on meteorological parameters. *Indian J. Hort.* **55**: 190-95.

Thakur, G. C., Chadha, T. R., Verma, H. S. and Kumar, J. 1993. Effect of soil management systems on bearing and fruit quality of apple cv. Red Delicious. *Indian J. Hort.* **50**: 10-13.

Thakur, G. C., Chadha, T. R., Verma, H. S. and Kumar, J. 1997. Effect of clean cultivation, mulching and sod culture on mineral nutrition and root growth of apple cv. Red Delicious. *Indian J. Hort.* **54**: 53-57.

UHF. 2014. Spray schedule of diseases and pests of apple. Extension Bulletin by: Dr YS Parmar University of Horticulture and Forestry, Nauni, Solan and Department of Horticulture Himachal Pradesh.

WHO. 1999. *Basic Food Safety for Health Workers*, WHO/SDE/PHE/FOS/99.1

Chapter 28

Mechanization in Temperate Fruit Production System in Kashmir Valley

Rohitashw Kumar, Jagvir Dixit and J.N. Khan

Division of Agricultural Engineering,
Sher-e-Kashmir University of Agricultural Sciences and Technology of Kashmir
Srinagar – 190 025, J&K
E-mail: rohituhf@rediffmail.com

1. INTRODUCTION

Jammu and Kashmir, being a hilly state, is blessed with naturally occurring micro agro-climatic regions suitable for cultivation of a wide range of horticultural crops with a great potential for development. Jammu and Kashmir is India's northernmost state, lying between six mountain ranges and covering an area of 2,22,236 sq. kilometers. It is located between 32°17' and 36°58' North latitude, and between 37°26' and 80°30' East longitude. Kashmir valley is covered by forested mountains, lakes, waterways and terraced fields. The State falls under the high Altitude temperate Sub-zone under the agro-climatic zone- I, *i.e.* Western Himalayan Region. Different districts of Jammu and Kashmir is shown in Fig. 1.

2. STATUS OF FRUIT CROPS

Jammu and Kashmir state is well known for its horticultural produce both in India and abroad. The state offers good scope for cultivation of horticultural crops, covering a variety of temperate fruits like apple, pear, peach, plum, apricot, almond and cherry. Fruit crops have witnessed most significant increase in area contributing

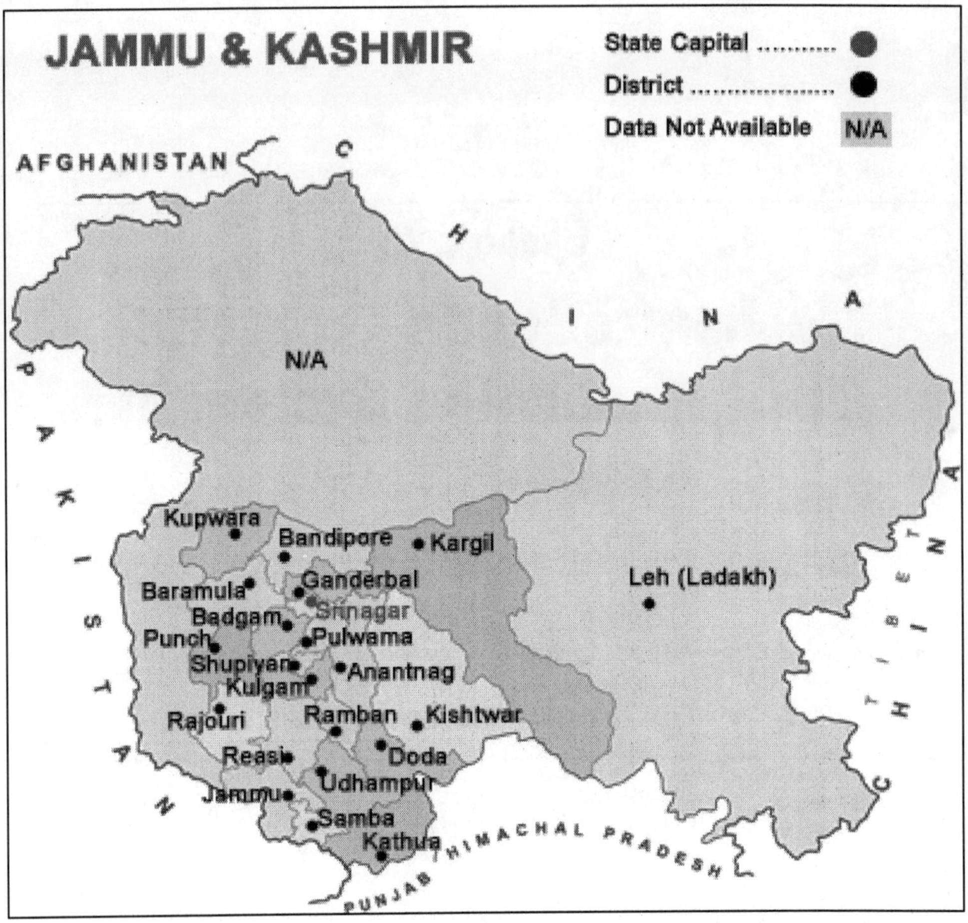

Fig. 1: Jammu and Kashmir Map Showing Different Districts of State.

to the shifts at a very high pace with bright prospects towards horticulture sector. Important reason being handsome returnsto the farmers. Horticulture, in the state has developed as an industry and more and more land is apportioned to this sector each year. Horticulture is gaining momentum in the state and its contribution to GSDP remains around 7-8 percent over the past few years. Presently, a total of 3,47,223 ha of land are under fruit cultivation in Jammu and Kashmir with 2,30,187 ha in Kashmir and 1,17,036 ha in Jammu region. Of this 1.76 lakh hectares are under apple cultivation alone (Table 1).

Global trends demand that appropriate mechanization in horticultural cultivation should be taken in order to increase the production and reduce the drudgery involved in traditional cultivation. Although, there has been a general increase in the average yield of horticultural crops during the preceding years, but it is relatively far less than the requirements of vehemently increasing population. Also the potential yield achieved in experimental fields does not realize on farmer's field.

Table 1: Trends of Area Under Cultivation and Production of Temperate Fruits and Nuts in J&K
(Area: 0000' ha and production: 0000' M Tonnes)

Crop	Year 1990-91		Year 1999-2000		Year 2004-05		Per cent Increase (+) or Decrease (-) 1990-91 to 1999-00		Per cent Increase (+) or Decrease (-) 1999-00 to 2004-05	
	Area	Production	Area	Production	Area	Production	Area (ha)	Production (tonnes)	Area (ha)	Production (tonnes)
Almond	1.92	0.22	1.88	0.97	1.54	1.35	–2.1	+340.9	–18.1	+39.2
Apple	6.87	65.82	8.67	92.90	10.79	109.33	+ 26.2	+ 41.0	+24.5	+17.7
Apricot	–	–	0.39	0.66	0.49	1.20	–	–	+25.6	+81.8
Cherry	0.14	0.42	0.23	0.53	0.26	0.74	+64.3	+26.2	+13.0	+39.6
Peach	–	–	0.17	0.14	0.19	0.21	–	–	(+)11.8	+50.0
Pear	0.81	1.67	0.91	2.92	1.05	4.03	+12.3	+74.9	+15.4	+37.7
Plum	–	–	0.29	0.38	0.34	0.37	–	–	+17.2	–2.6
Walnut	4.09	3.87	5.94	7.49	7.49	10.06	+48.5	+93.5	+26.1	+34.3

Directorate of Horticulture, J&K, Govt. of J&K.

Thus a large gap in potential yield and actual yield of fruits in the state is witnessed. Timeliness and efficient energy utilization in various farm operations along with other inputs are key factors in improving horticultural productivity. There is a positive relationship between farm power and productivity. Still most of the horticultural operation like nursery raising, pit digging for planting fruit saplings, spraying, inter-culture and harvesting are not mechanized in the state and country also. Horticulture mechanization is not a luxury but a necessity. Highly sophisticated costly machines may not be affordable to the fruit growers, as majority of the fruit growers are small and marginal farmers. Some of the machines may have to be made available on custom hiring basis. The human labour will still be required for large number of operations for propagation (budding and grafting), pruning and harvesting aids of fruit crops.

The availability and quantity of sufficient labour for hand harvesting is a major concern to many growers (Ibanez *et al.*,1997) Fruit harvest is a very seasonal activity and the uncertainty of stable, skilled harvest labour at the right time is a major problem. The shortage, stability and cost of labour for hand harvesting, declining fruit prices and increasing competition from lower cost world producers. This is driving force behind development in mechanized harvesting. The 'shake and catch' bulk-harvesting systems developed in the 1980's and 1990's were unsuitable for fresh market fruits because of excessive fruit damage and lack of uniformity in fruit maturity on the trees. Nearly 75 per cent of the country's temperate fruits, mainly apples, are grown in the state.

3. SCOPE OF FARM MECHANIZATION

The farm mechanization is badly hampered by stepped, small and irregular fields, undulating topography, lack of skilled man-power, poor facilities of repair, maintenance and manufacture of implements and high cost solar gadgets.Hand-harvesting fruit crops is labour-intensive, and the supply of dependable, skilled labour is a concern of the fruit industry. Only a small portion of all fruit crops is harvested mechanically, primarily for processing. Immediate attention of the state government and other funding agencies is required to strengthen the agricultural engineering wing in Jammu and Kashmir. A strong cell of agricultural engineering should be created to handle the farm mechanization problems.

The strategies need to be adopted to achieve appropriate farm mechanization in the hills can be listed as follows:

- ☆ Design and development of low-cost improved hand tools/equipment especially for marginal and small farm groups.
- ☆ Introduction and development of specific power source and equipment which is light in weight, small in size, low in turning radius, easy to handle/carry from one plot to the other.
- ☆ Introduction of crop-specific package of tools and equipment for different soil conditions.
- ☆ Systematic testing of the available tools/equipment particularly on ergonomical aspects to reduce drudgery of farmers.

- ☆ Introduction and popularization of power tillers to reduce the drudgery of the farmers and also to encourage the educated youth to go for opening custom hiring centres.
- ☆ Awareness camps and demonstrations need to be arranged for promoting and disseminating improved farm technologies on a larger scale at different locations in the state.
- ☆ Development of proper infrastructure and trained technical man-power for conducting demonstrations and trainings on improved tools/ equipment in all parts of the state.
- ☆ Establishment of small-scale units for/equipment at different locations in the hilly region.
- ☆ Strengthening of a network of repair and service facilities for the machines/ tools in the region.
- ☆ Formulation of adequate farm mechanisation policies and programmes.

4. MECHANIZATION OF HORTICULTURE CROPS

The following different aspects should be considered for mechanization in horticulture:

- ☆ Fruit crop mechanization equipment for pit making, transplanting of samplings, pruning, spraying in tall crops, harvesting of fruits etc. need to be identified/adopted/developed and popularized.
- ☆ Crop mechanization equipment for seed-bed preparation, planting, transplanting of seedlings, inter-culture, irrigation, spraying harvesting etc., need to be identified/designed and introduced.
- ☆ Modern manual and power operated garden tools and equipment need to be introduced and popularized for landscape and horticulture crops.
- ☆ Plasticulture has wide scope in hill agriculture demanding intensive R&D and extension effort especially in green houses, low and high tunnels, plastic mulches, micro-irrigation and mist formation.
- ☆ Post harvest equipment and technology are needed for cleaning, grading, drying, cooling, evaporative cooling, storage, cold storage and handling of farm produce to improve their quality and shelf-life. Cold chains for transport of perishable products of fruits with minimum losses.
- ☆ Agro-processing activities should be promoted in the production catchments to reduce losses.
- ☆ Expansion and upgrading of horticultural activities requires appropriate short and medium term storage facility.

5. STRATEGY FOR MECHANIZATION

- ☆ Mechanization is essential for timeliness in field operations and precision in placement of costly inputs to increase productivity, reduce unit cost of production and drudgery in farm operations as well as conservation of natural resources.

- For intensification of agriculture farm power availability should be increased from the present level of about 0.60 kW/ha to about 2.0 kW/ha in Jammu and Kashmir, by 2020 as well as other states of Western Himalaya.
- For stationary operations like water lifting, harvesting, cleaning, grading and other agro-processing and value addition activities, adequate electrical energy is essential and its hould be provided.
- Equipment and power units suitable for hill agriculture, agro-processing and rural living should be selected, tested, adapted and introduced in the region.
- Improved Agricultural Implements and Machinery for crop production.
- For efficiency and economy in tillage and sowing/planting/transplanting operations, large scale adoption of planters and trans-planters should be given high priorities.
- For timely and efficient plant protection, Aero-blast sprayers, orchard sprayers and electro-statics praying equipment are required to be introduced.
- Improved harvesting equipment like serrated sickles, fruit pluckers and fruit harvester should be introduced.
- To overcome drudgery in harvesting.
- For increasing production and productivity snowmelt and rain water harvesting, conservation,storage, efficient conveyance and application using plastic pipes, laser land levellers, sprinklers and drip irrigation systems should be intensively promoted.
- Power operated trenchers, angle dozers, drudgers, buck scrapers and other earth moving machinery will be required to be introduced in the region for making farm ponds, bunds and terraces, irrigation channels and waterways.

6. MECHANICAL EQUIPMENT'S/TECHNOLOGY FOR DIFFERENT CROPS

(a) Fruit Harvesting

Harvesting of apple is done manually by using a traditional orchard ladder in the state. Fruits are generally damaged due to lack of a suitable harvesting device and operation is prolonged and done by climbing on a tree or using a ladder. After plucking some fruits, ladder has to be placed at new position, which is not economically viable. About 30 man-days are required to pluck fruits from 1 ha of land. These labour days are required in the month of August to November when the farmers are busy in paddy harvesting. Study conducted in USA indicated that the amount of labour needed for harvesting was often well over 50 percent of the total annual labour requirement (USDA, 1969). Some fruit harvester working on the principle of impact cum shear, push and cut type and hold and twist type are developed and it is shown in Figure 2.

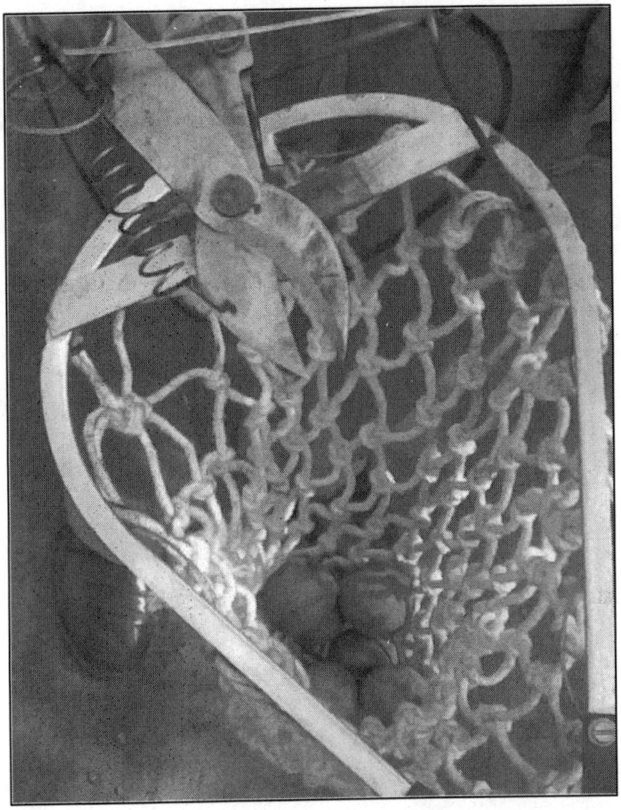

Fig. 2: Clutch Lever Operated Fruit Picker for Apple and Pear.

(i) Mechanical and robotic harvesting technologies

New mechanical and robotic harvesting technologies now a day is used to increase fruit production efficiency (Peterson and Millar, 1999; Robinson *et al.*, 1990)). The demand for fresh fruit is increasing in world markets as the standard of living increases. Harvesting is the most labour intensive operation in fruit production accounting for up to 60 per cent of the total labourr equipment. The latest developments and availability of mechanical and robotic harvesting technology and has identified potential opportunities and benefits for the use of this technology used for deciduous fruit industries. The technology for harvesting options for the tree of fruit crops such as walnut, apricot and berries and new developments in sensor technology and for non-destructive fruit quality assessment. The key features of technology are:

- ☆ New developments in mechanical harvesting aids provide the most immediate opportunities to improve harvest efficiency in quality fresh fruit production. The Dutch designed Pluk-O-Trak was particularly impressive because of its high harvest rates, extremely low fruit bruising and damage, versatility for different orchard and operations, low running costs, quietness of operation, ease of maintenance and suitability to pickers.

- ☆ Robotic bulk harvesting is a medium term proposition which needs further improvement for use on apples but has immediate application for cherries and possibly pears and plums. The main limitations for fresh apples are the detachment of stalks in some varieties, non-uniformity of fruit maturity on the tree and the potential for bruising in delicate varieties. The system is suited to narrow inclined canopies.
- ☆ Robotic selective harvesting of individual fruits is and exciting longer term proposition. The advances in orchard production systems, mechanical, artificial intelligence and sensor technology since the first prototypes were developed in France in the 1990's now make robotic harvesters an achievable and commercially viable proposition.

There were significant improvements in fruit removal rates and harvest efficiency (Bourley, Per. Comm., 2001).In Japan, researchers at Iwate University are conducting work on the robotic harvesting of Fuji äpples (Kataoka *et al.*, 1999). Robotic harvesting hand for apple harvesting has been developed. The Pluk-O-Trak system was developed in Holland for apples and is now used on a range of other fruit including stone fruit, pears and oranges (Sanders, Pers. comm., 2001).

(ii) Pluk-O-Trak

The most impressive harvesting aid in Europe was the 'Pluk-O-Trak'. The Pluk-O-Trak is manufactures in the Netherlands by Munckhof. Several hundred machines

Fig. 3: 'Pluk-O-Trak' Harvesting Golden Delicious apples in France. (Photographs Courtesy of Munckhof).

Fig. 4: Glove.
(Photograph courtesy of CEMAGREF).

Fig. 5: Ladder for Fruit Plucking.

are now being used throughout Europe and other parts of the world. The unique feature of the Pluk-O-Trak is the 6 to 8 conveyor chutes, which can be positioned close to the picker and fruit in canopy for ease of picking. The picking conveyors can be adjusted in all directions to minimize the distance to the picker and the crop. Harvest rates are high, 250 kg/hr/person. The machine can reach tops of trees from 3.25 to 4.25 m high and be accommodated in very narrow rows from 3 to 6 m wide. Many of the mechanical adjustments can be converted to hydraulic control. The Plük-O-Trak is also very versatile and can be used for pruning and hand thinning fruit operations. Pluk-O-Trak' harvesting Morganduft apples for processing is shown in Fig. 3.

Technology, particularly in sensors for use in non-destructive fruit quality assessment and fruit grading is advancing rapidly around the world. *The instrumented 'GLOVE'* is given in Fig. 4. The different field operations and man-days/ha required for an apple orchard is summarized in Table 2. The apple shaker (Pneumatic) harvester is shown in Fig. 6.

Post hole digger play an important role for trans-planting of fruit crops. Different type of post hole digger, rotary dibbler for sowing seed in pot and leaf mulcher are shown in Figs. 8-10.

(b) Walnut Cultivation in Jammu and Kashmir

Walnut is a major dry fruit crop grown in Jammu and Kashmir state. The production of this crop has touched 175.09 thousand M. Tonnes in Kashmir Division and 49.50 thousand M. Tonnes in Jammu Division. Latest techniques of walnut budding/grafting have been introduced which have helped in reducing gestation

Table 2: Schedule of Field Operations and Man-Days/ha Required for an Apple Orchard

Operation	Type	Month												Total
		Jan	Feb	March	April	May	June	July	Aug	Sep	Oct	Nov	Dec	
Land cleaning and development	Bullock										10	10		20
Digging of pits	Manual												31	31
Filling of pits	Manual	10												10
FYM and fertilizer application	Manual			4	4									8
Basin preparation	Manual		10		10									20
Plant protection	Manual			2	2	2	2	2	2					12
Fruit Harvesting	Manual								10	10	10			30
Grading	Manual									15	15			30
Packing	manual	15									15	15	15	60
Pruning	Manual	25											25	50
Total		**50**	**10**	**6**	**16**	**2**	**2**	**2**	**12**	**25**	**50**	**25**	**71**	**271**

Fig. 6: Apple Shaker (Pneumatic) Harvester (American Society of Agricultural Engineers, USA).

Fig. 7: Automatic Mechanical Man Positioning System for Fruit Harvesting.

period of this crop. Polyhouse technology has been used for raising budded/grafted walnuts. Walnut cultivation plays a significant role in the economic profile of the farmers living in hilly and backward areas, where economic condition of the people is extremely fragile.The walnut grafted plants inside polyhouse is illustrate in Fig. 11.

Fig. 8: Post Hole Digger for Mechanized.

Fig. 9: Rotary Dibbler for Sowing Seed in Pot.

Fig. 10: Leaf Mulcher.

Fig. 11: Walnut Planting Material grown under Polyhouse.

(c) Almond and Walnut-Dehulling

Almond and walnut dehulling are the labour, money, energy and time consuming operations. Keeping in view de-hulling machines save the time and energy.

(i) Manual Walnut Dehuller

The manual dehuller is very useful tools for dehulling walnut for small and marginal farmers. Its capacity is about 90-100 kg/h with an efficiency of 90-92 per cent. The broken percentage is about 2-3 per cent. The manual walnut dehuller is shown in Fig. 12.

Fig. 12: Manual Walnut Dehuller.

(ii) Motorized Walnut Dehuller

The capacity of a motorized walnut Dehuller is 335-350 kg/h, as compared to 200-250 kg/day/person by conventional method. The de-hulling Efficiency is 93-94 per cent with a broken percentage of 2 per cent. The overall dimensions of the developed machine are (L × B × H) 1800 × 890 × 1630 mm. The operation and working of motorized de-huller is presented in Fig. 13.

Almond de-hulling hand operated machine is shown in Fig. 14.

Lack of cold storage in valley forces growers to sell produce at cheap rates. The main reason of mismatch in supply and demand is the lack of cold and conditional atmospheric storages in the valley.

7. APPLICATION OF PLASTICULTURE IN MECHANIZATION

The term plasticulture refers to the practice of using plastic materials in agricultural applications. Plastic has many valuable applications in high-tech horticulture includes drip irrigation, plastic mulches, row covers, low tunnels, high tunnels, silage bags, hay bale wraps, plastic trays and pots used in transplant and bedding plant production packaging and storage for superior quality of produce and in postharvest management. Use of plastic in horticulture crop production has increased dramatically in the last ten years. The use of plasticulture in the production of horticultural crops (small fruits and tree fruits) helps to mitigate the sometime

Fig. 13: Motorized Walnut Dehuller.

Fig. 14: Almond De-hulling Hand Operated Machine.

extreme fluctuations in weather, especially temperature, rainfall and wind, which occurs in many parts of country. A variety of plastics materials and end products are deployed in plasticulture applications - for water conservation, irrigation efficiency, crop and environment protection, as well as end product storage and transportation. Plastics can play a major role in energy conservation. They require minimum energy in production and conversion to finished products. Plastics have definite advantages over conventional materials.

Plasticulture applications are as follows:

(a) Water Management

- ☆ Lining of canals, ponds and reservoirs with plastics film.
- ☆ Drip and sprinkler irrigation.
- ☆ PVC and HDPE pipes used for water conveyance.
- ☆ Sub-surface drainage.

(b) Nursery Management

- ☆ Low tunnel, nursery bags, pro-trays, plastic plugs, coco-pits, hanging baskets, trays *etc.*

(c) Surface Cover Cultivation

- ☆ Protected cultivation.
- ☆ Soil solarisation.
- ☆ Plastics mulching.

8. USE OF PLASTICS IN HORTICULTURE

Following are some of the uses; i) Green house ii), Drip irrigation, iii) Mulching, iv) Plastic for raising planting material, v) Low Tunnel polyhouse, vi) Shading net house and vii) Packaging.

(a) Apricot Harvester

See the Fig. 16.

Fig. 15: Drip Irrigation in High Density Apple.

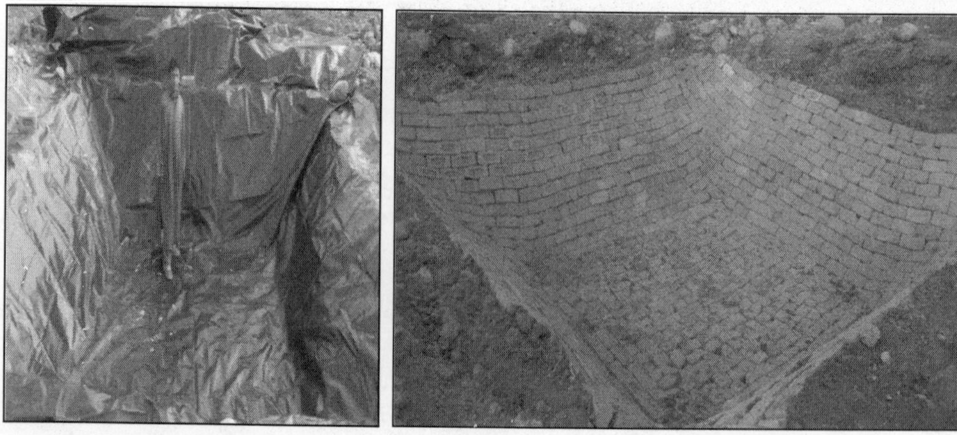

Fig. 16: LDPE Farm Pond for Water Storage.

Fig. 17: Apricot Harvesting.

Fig. 18: Horticultural Tool Kit.

Fig. 19: Weeder for Inter-cultural Operation.

Fig. 20: Brush Cutter and Sprayer.

(b) Poly Tunnel Type Solar Dryer

Drying of vegetable and fruits is a common practice for offseason use in Kashmir and Ladakh region. Farmers normally practice open sun drying for vegetable/fruits, which is a time consuming process and also results into deterioration of qualities of some products. The poly tunnel type solar dryer developed by Division of Agricultural Engineering, SKUAST-K could be successfully used for drying vegetable and fruits. The use of equipment results in saing of time duration of drying by 60-70 per cent without deterioration of quality of produce. The small size solar dryer is shown in Figure 21.

Figure 21: Solar Dryer.

9. FUTURE THRUST

- ✰ Design and Development of ergonomically suitable tools and equipments for hill agriculture.
- ✰ Need to develop women friendly tools and equipments for hills.
- ✰ Need of ergonomically suitable light weight power tiller for high hills.
- ✰ Introduction of appropriate and suitable horticultural tools
- ✰ Strategies for Hill Farm Mechanization

(a) Infrastructural Improvements

- ✰ Testing facilities for agricultural machinery and agro-products for quality control should be created in the region
- ✰ For creating awareness amongst the farmers, extension workers and entrepreneurs of the region and display centres of improved agricultural machinery.

(b) Policy Issues

- ✰ To encourage agricultural mechanization on sound footing there should be a state policy for agricultural mechanization.
- ✰ For efficient use of stationary farm power units and equipment, the electricity to the farmers, should be provided at subsidized rate.

10. CONCLUSION

With shrinking labour base in agriculture, it is imperative that mechanization has to take the centre stage in modern fruit cultivation. Machinery developed elsewhere cannot be adopted as such. Effort are to be made to develop prototype for small tools and equipments for hill and mountainous areas of temperate regions of the county. These equipment would definitely ease drudgery of the fruit grower and would make every operation easier and efficient thereby enhancing productivity and quality besides better price realization.

11. ACKNOWLEDGEMENTS

The author is highly thankful to the Division of Agricultural Engineering for providing necessary information and the Principal Investigator of HTM (3.27) project.

12. REFERENCES

Bourely, A., Rabatel, G., Grand d'Esnon, A. and Sivila, F. 1990. Fruit harvest robotization: 10 years of CEMAGREF experience on apple, grape and orange. In *Proc. AGENG '90 Conference*, 178-179, Berlin, Germany.

Brown, G.K. 2001. Challenges and benefits for mechanical harvesting of fruit crops. *6th International Symposium on Fruit, Nut, and Vegetable Production* 11-14 September, 2001, Potsdam Germany, Abstracts: 23.

Dixit, J and Sharma, M. K. 2009. *Status and Prospects of Mechanization in Apple Orchards.* Kalyani Publisher, New Delhi, pp. 297-307.

Ibanez, M., Hetz, E. and Vevegas, A. 1997. Evaluation of two walnut and apple harvesting systems. *Agro-Ciencia* **13**: 325-329.

Kataoka, T., BulAnon, D.M., Hiroma, T., Ota, Y. 1999. Performance of a robotic hand for apple harvesting. *Transactions of the Amer.Soc.Agri. En. Paper* No. 993003. St. Joseph, Mich.: ASAE.

Peterson, D.L., Miller, S.S. and Kornecki, T.S. 1998. Over-the row harvester for apples. *Transactions of the Amer. Soc. Agri. En.*, **28**: 1393-1397.

Robinson, T.L., Millier, W.F., Throop, J.A., Carpenter, S.G., Lakso, A.N. 1990. Mechanical harvest ability of Y-shaped and pyramid-shaped 'Empire' and 'Delicious' apple trees. *J. Amer. Soc. Hort. Sci.* **115** : 368-374.

Chapter 29

Emerging Opportunities in Processing of Temperate Fruits in India

Devina Vaidya

Department of Food Science and Technology,
Dr Y.S. Parmar University of Horticulture and Forestry,
Solan – 173 230, H.P.
E-mail: devinavaidya@yahoo.com

1. INTRODUCTION

India is the second largest producer of fruits and vegetables in the world with several types of tropical, subtropical and temperate fruits and vegetables. The fruits and vegetables are highly perishable commodities and the ambient high temperature obtained in the tropical country like ours makes them more susceptible for rapid development of senescence, decay and rotting. Both respiratory and transpiratory rates are proportional to temperature, increases and so that the produce quickly dries, wilts and spoils unless properly preserved. It has been estimated that 20 to 30 per cent of the horticultural produce is lost before consumption which accounts for Rs. 5000 crores because of poor harvesting, handling, storage, transportation and marketing practices. Two approaches are possible for solving this problem. One is the creation/expansion of cold storage facilities in the fruit and vegetable producing regions. Another approach is to process the fruits and vegetables into various products which could be preserved for a long time and add to the value of the product.

2. GROWTH OF PROCESSING INDUSTRY

The growth of food processing will bring immense benefits to the economy, raising horticulture yields, enhancing productivity, creating employment and raising standard of life of people across the country, especially in rural areas. This industry ranks fifth in the country and employs around 13 million people directly and 35 million people indirectly. It accounts for 14 per cent of total industrial output of the GDP. India's food processing sector covers fruit and vegetables; spices; meat and poultry; milk and milk products, alcoholic beverages, fisheries, plantation, grain processing and other consumer product groups like confectionery, chocolates and cocoa products, soya-based products, mineral water, high protein foods etc. India's exports of Processed Food was Rs. 31563.43 Crores in 2014-15, which including the share of products like Mango Pulp (Rs. 841.39 Crores), Dried and Preserved Vegetable (Rs. 847.11 Crores), Other Processed Fruit and Vegetable (Rs. 2,569.93 Crores).

The food processing industry in production of fruits and vegetable is fragmented into three sectors such as organized, unorganized and small scale industries. Total production of the horticulture crops such as fruits and vegetable *i.e.* 81285 MT and 162187 MT respectively. But a marginal quantity of 1.0 to 2.0 per cent of the produce is processed and packaged in contrast with developed and developing countries, *i.e.*, 70 to 80 per cent. Export of fruits and vegetables from India In terms of global trade, India's share in agricultural export is insignificant. While India contributes 8.56 per cent and 13.5 per cent respectively to world's fruits and vegetables production, its share in global exports of these products is less than 1.0 per cent. The food processing sector is critical to India's development. It establishes vital linkages and synergy between industry and horticulture, the two pillars of India's economy. Food processing is the set of methods and techniques used to transform raw ingredients into food or to transform food into other forms for consumption by humans or animals either at home or by the food processing industry.

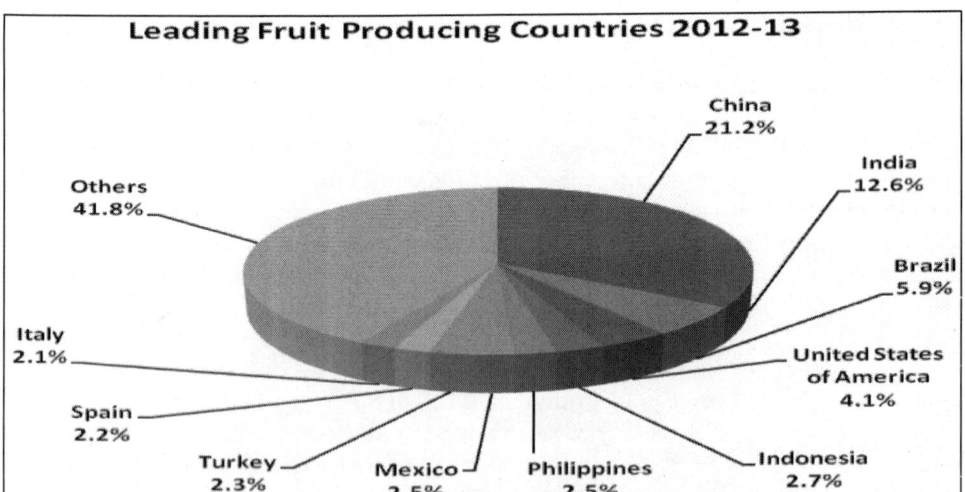

Fig. 1: Fruit Producing Countries 2012-13.
(*Source*: Indian Horticulture Database, 2014).

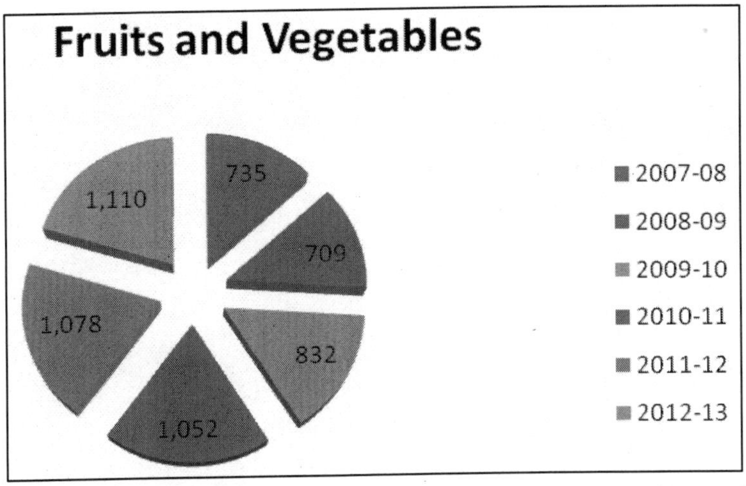

Fig. 2: Registered Food Processing Units of Fruits and Vegetables from 2007-08 to 2012-13. (*Source*: Indian Horticulture Database, 2014).

With increasing urbanization rise in middle class purchasing power, change in food habits and the dyeing out of the practice of making preserves in individual homes, there is increasing demands for factory made jams, jellies, fruit beverages, dehydrated foods, pickles etc. in the domestic market. Moreover, there is considerable demand for some of these products in foreign markets *e.g.* Cherry both fresh and canned, fruit juices, salted cashew are good foreign exchange earners. The production of fruit and vegetable products in India are canned, bottled fruits and vegetables, jams, jellies, marmalades, fruit juices, fruit pulps, squashes, crashes, cordials, fruit syrups, fruit nectars, RTS fruit beverages, fruit juice concentrates, fermented beverages and pickels, chutneys, pickles, candied and crystallized fruits and peels, dehydrated fruits and vegetables, frozen fruits and vegetables, tomato products, sauces, soups etc. In India there are 4000 processing industries in functioning.

A number of technologies have been developed to satisfy demands of the processing sector. These include: dehydration technologies for the production of apple rings, cubes and chunks; juicing technologies for the production of concentrates from apples, plums, apricots and kinnow; technologies for the debittering of kinnow juice and for the fermentation of cider, wine, vermouth, sparkling wine, medicinal wine and brandy, as well as technologies for the production of health foods, fruit based carbonated juices, osmo-canning of apples, osmotic dehydration of fruits, flavour extraction, and for the production of pectin from waste.

The main objective of preservation is to increase the shelf life by reducing the microbial load and also the water activity. Both can be achieved by either traditional method of preservation (thermal processes) or by non thermal processes. It can be achieved by various ways like addition of salt, sugars, preservatives, antioxidants, naturally occurring antimicrobial substances and also by the processes like drying, freezing, refrigerated storage and hurdle technology as well as by novel technologies such as microwave heating, Pulsed Electric Field (PEF) Technology, High Pressure Processing (HPP), Pulsed Light Technology, Ohmic Heating, Ultra sonics, Pulsed X-Rays.

3. TYPES OF TECHNOLOGIES

The selection of particular preservation method for the particular food product is based on the following criteria like cost of production, scale of production, type of product.

(a) Thermal Processing

The objectives of thermal processing are

- ☆ Render foods free of pathogenic and spoilage organisms
- ☆ Impart changes in texture, colour, flavour
- ☆ Improve digestibility
- ☆ Improve shelf-life

(i) Blanching

It is carried out before sterilization, dehydration,and freezing to deactivate enzymes and reduce surface load of microbes in vegetables and some fruits.

(ii) Pasteurization

It is carried out for low acid foods (pH > 4.5) to kill pathogenic microorganisms. (milk: 15 s at 71.5 °C) and for acidic foods (pH < 4.5) to extend shelf-life by killing spoilage microorganisms (molds, yeasts).

(iii) Heat sterilization

It is carried out to kill the enzymes and microbes (high temp > 100 °C). canned food and aseptically packaged food.

(iv) Solar energy

Solar food processing is an emerging technology that provides the good quality food at low or no additional fuel cost. A number of solar dryer, collector and concentrator are currently being used for various steps in food processing and value addition. Solar cabinet dryer with forced circulation has been used for dehydration and value added products from locally grown fruits and vegetables.

(v) Canning

Canning is the process of sterilizing and sealing foods in airtight containers to preserve them. Canning process retains nutrients and optimum quality of fruits and vegetables.

(vi) Drying, dehydration and concentration

Removal of moisture by applying heat is called "Drying". Dehydration is drying by artificial heating under controlled temperature, humidity and air flow. The basic drying processes are classified as sun drying, atmospheric dehydration and vacuum dehydration. Sun drying is still in use. These methods are accompanied by removing water. Moisture may be removed from foods by any of the methods.

Dehydrating (below 5 per cent) moisture helps to preserve nutritive value of the product although heat-labile vitamin losses are more. Low moisture preserved material can be fortified with vitamins under proper storage conditions. The dried products have more shelf life in properly packed and stored.

Retention of nutrients and overall quality is better in modern methods of dehydration such as freeze drying, vacuum drying, osmotic or solar dehydration as there are little changes in physio-chemical composition.

Inherent Disadvantages of Thermal Processing

- ✰ Loss of original flavour,
- ✰ Taste, appearance,
- ✰ Colour
- ✰ Nutritional quality
- ✰ Nutraceutical value

Now a days, consumer choose more and more for minimally processed foods with characteristics closer to those of fresh foods using minimal processing or reducing chemical preservative. Main objective of non thermal techniques are, to retain the nutrients, sensory properties and to increase the self life without any adverse effect on the quality of the food products.

(b) Non-Thermal Processing

The objectives of this processing are;

- ✰ Render foods free of pathogenic and spoilage organisms
- ✰ Retain colour, flavour
- ✰ Improve shelf-life
- ✰ Improve texture

(i) Microwave heating

Microwave heating refers to the use of electromagnetic waves of certain frequencies to generate heat in material (Fig. 3). When a microwavable container with food is placed in a microwave oven and then a oven is activated, the food at the edge of the container heats faster and a temperature gradient develops between the centre and the edges. For food applications most commonly used microwave frequencies are 2450 and 915MHz.

Mechanism

Dipolar Interaction: Once microwave energy is absorbed, polar molecules such as water molecules inside the food will rotate according to the alternating electromagnetic field. The water molecule is a "dipole" with one positively charged end and one negatively charged end. Similar to the action of magnet, these "dipoles"

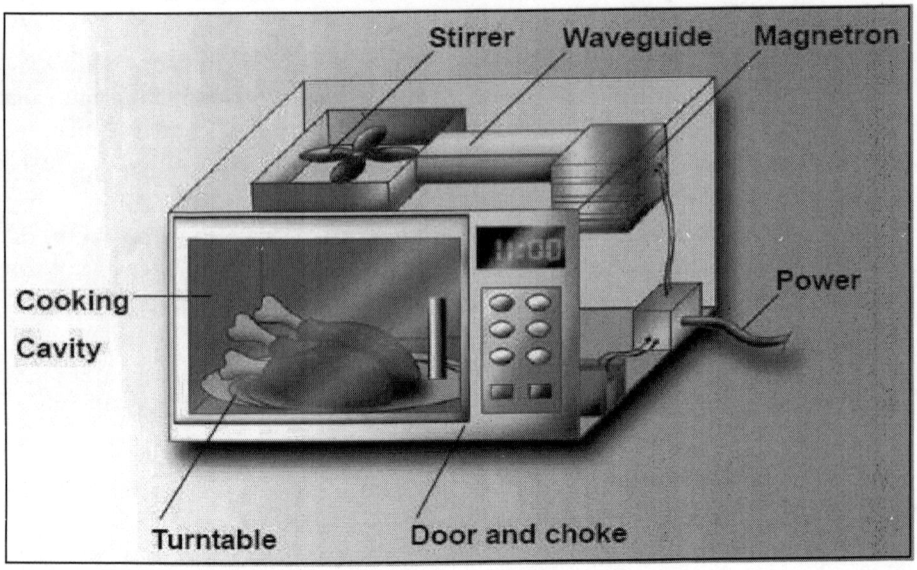

Fig. 3: Basic Structure of Microwave Oven.

will orient themselves when they are subject to electromagnetic field. The rotation of water molecules would generate heat for cooking

Ionic Interaction: In addition to the dipole water molecules, ionic compounds (*i.e.* dissolved salts) in food can also be accelerated by the electromagnetic field and collided with other molecules to produce heat.

Application

- ☆ Microwave heating offers distinct benefits in dehydration because of its penetration depth, and the uniform heating results in water vaporizing from throughout the product. This induces an inner pressure that maintains puffed character of the dried product and preserves colour, flavour, and nutritional value.
- ☆ Microwaves for pasteurization or sterilization improve digestibility of the products.
- ☆ Concentration and thawing of fruits.
- ☆ Blanching of fruits and vegetables.

(ii) Pulsed electric field (PEF) technology

PEF is a non-thermal food preservation technology that involves the discharge of high voltage electric pulses (up to 70 kV/cm) into the food product, which is placed between two electrodes for a few microseconds (Fig. 4). It is the emerging technologies for the replacement of traditional thermal pasteurization among non-thermal processes, apart from microfiltration eventually combined with moderate heat treatment. An external electric field is used to exceed a critical transmembrane potential of one volt.

This result in a rapid electric breakdown and conformational changes of cell membranes, which leads to the release of intracellular liquid, and cell death However, treatment temperature should be kept as low as possible in order to avoid heat damage to the treated product and to prevent off-flavours.

(iii) Mechanisms of microbial inactivation

Electrical breakdown

The membrane can be considered as a capacitor filled with a dielectric. The normal resisting potential difference across the membrane is 10 mV and leads to the build-up of a membrane potential difference due to charge separation across the membrane. Potential difference is proportional to the field strength and radius of the cell. The increase in the membrane potential leads to reduction in the cell membrane thickness.

Electroporation

It is the phenomenon in which a cell exposed to high voltage electric field pulses temporarily destabilizes the lipid bilayer and proteins of cell membranes. The plasma membranes of cells become permeable to small molecules after being exposed to an electric field, and permeation then causes swelling and eventual rupture of the cell membrane. The main effect of an electric field on a microorganism cell is to increase membrane permeability due to membrane compression and poration.

PEF treatment at 30 kV/cm for 172 μs at less than 35°C reduced the activity of microorganism 5 log (*E.coli*).In case of cherry juice, peach and apricot nectar there was 100 per cent reduction *Penicillium expansum* spore germination by different level of treatments. It is eco friendly method with no wastage generated. PEF is more energy saving as compare to the thermal processing. Colours, flavours and nutrients are also preserved by this method. There is no evidence of toxicity and also required short treatment time.

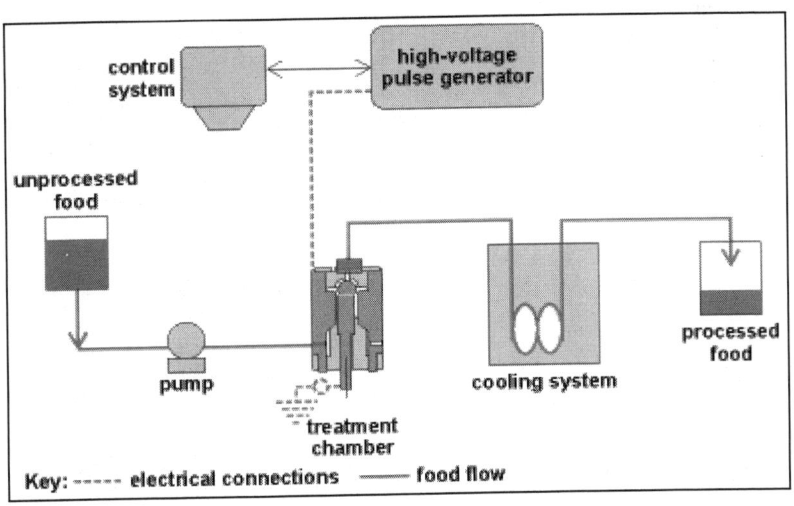

Fig. 4: Pulsed Electric Field (PEF) Setup.

Application

PEF is used for inactivation of enzyme, microorganism and disintegration of biological material, *e.g.* in juice processing, a similar juice yield with fresh-like quality and a higher concentration of functional components was found.

(iv) High pressure processing (HPP)

High hydrostatic pressure (HHP) processing uses pressures up to 1000 MPa, with or without heat, to inactivate harmful microorganisms in food products (Ramaswamy *et al.*, 2005). The application of HHP in food area started from 1900's when Hite and other researchers applied HHP on the preservation of milk, fruits and vegetables. However, it takes a long time for the commercial products to emerge in the market. In 1990, the first HHP processed fruit jams were sold in the Japanese market. Typically a pressure of 350 MPa applied for 30 min or 400 MPa for 5 min will cause a tenfold reduction in vegetative cells of bacteria, yeasts or moulds.

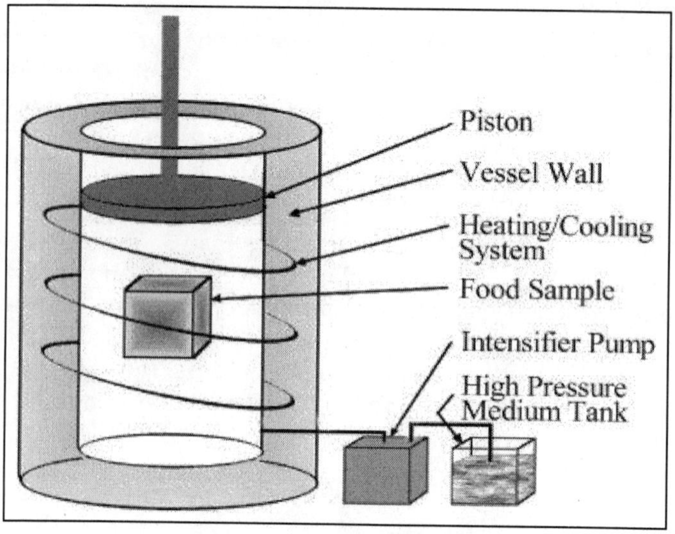

Fig. 5: HPP Equipment.

Mechanism of Action

Two principles that govern the behaviour of foods under pressure: the Le Chatelier-Braun principle and the Isostatic principle. The Le Chatelier-Braun principle indicates that any phenomenon (such as phase transition, change in molecular configuration, chemical reaction, *etc.*) accompanied by a decrease in volume is enhanced by the increase in applied pressure. The isostatic principle means that the distribution of pressure into the sample is uniform and instantaneous. Thus, the process time is independent of sample size and shape (Ramaswamy *et al.*, 2005).

Application

HPP used for pasteurization and sterilization of fruits and fruits products, sauce and pickles.

(v) Pulsed light (OR) high intensity light technology

High intensity light is also described as pulsed broad spectrum white light is a decontamination or sterilization technology that can be used for the rapid inactivation of micro organisms on food surfaces, equipments and food packaging materials. Surface decontamination of food products using pulsed high intensity light has many potential benefits to the food industry. It is a non thermal preservation intervention, with the ability to minimize the deleterious effects of thermal processing and chemical treatments on quality and sensory attributes.

High intensity light technique is used for decontamination of vegetables, dairy products. It is also used in microbial inactivation of water, sanitation of packaging materials and disinfection of equipment surfaces. The electromagnetic radiations are emitted and propagated by means of waves which differ in wavelength, frequency and energy (E). The term light is generally used to mean radiations having wavelength ranging about from 180 to 1100 nm, which includes ultra-violet rays (UV, 180–400 nm) roughly subdivided into:

- ☆ UV-A, 315–400 nm
- ☆ UV-B,280–315 nm
- ☆ UV-C, 180–280 nm
- ☆ visible light (400–700 nm)
- ☆ infrared rays, IR, (700–1100 nm)

UV-C 180-280 having the germicidal effects*

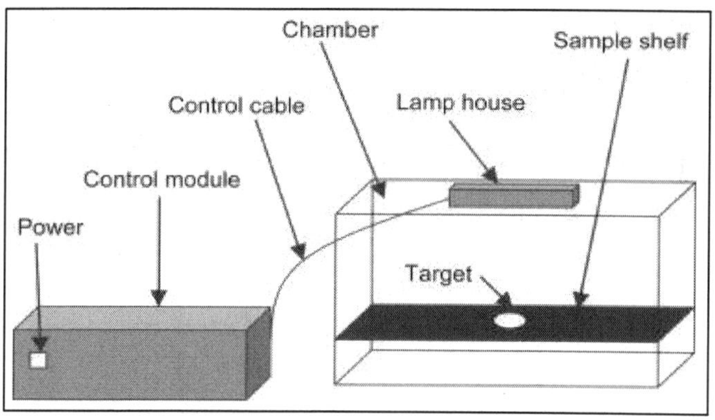

Figure 6: Pulse Light Set-up.

Mechanism

It include short term thin layer temperature effect, photochemical effect nand DNA damage.

Application

Surface sterilization of food.

(vi) Ohmic heating

Ohmic heating is based on the principle of passing an electric current through an electrically conducting product, as with microwave energy, electric energy is transformed into heat (Fig. 7). Ohmic heating is an efficient way of processing foods containing large solid particulates unlike conventional method of processes such as canning and aseptic processing. Ohmic heating cannot be used for all the food products. It is depending upon the product's electrical conductivity and on whether the product is an insulator or a conductor. In this process simultaneous and uniform heating of solid and liquid phases can be achieved, thus reducing the danger of under processing as well as nutritional loss. The critical parameters affecting ohmic heating include the electrical conductivity of the food, temperature dependence of electrical conductivity, the design of heating device, the residence time, distribution time, thermo physical properties of foods and electric field strength.

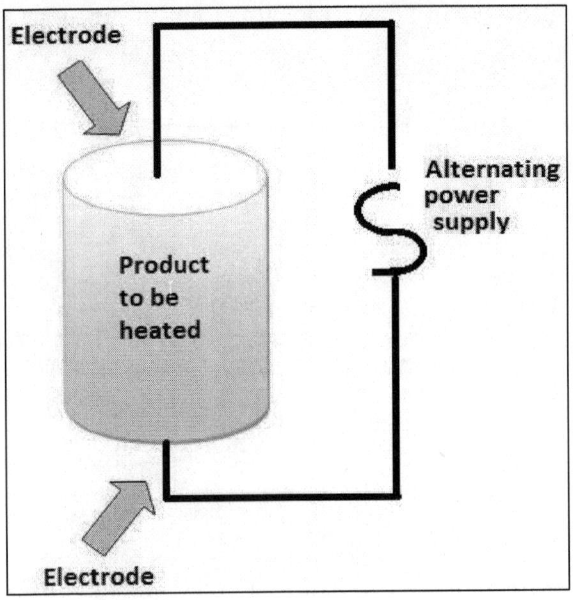

Fig. 7: Ohmic Heating of Food.

Applications of Ohmic Heating

Used in blanching, evaporation, dehydration, fermentation, extraction and value added process. Literature shows that fruits were analyzed for Electrical conductivity during Ohmic heating. At 25°C the electrical conductivity of pineapple was very low and significantly different than apples and pear. Electrical conductivity of peach and strawberry was high. At higher temperatures (40-140°C), apples and pineapple had low conductivity. Strawberry and peach had higher conductivity and significantly different compared to other fruits. The gap in the electrical conductivity between strawberry and peach, and other fruits increased with the temperature.

(vii) Ultrasonics

The frequency of sound waves audible to human ear ranges from 20Hz to 20 kHz. The sound waves having frequencies greater than 20 kHz are called "Ultrasonics" or "Supersonics" (Fig. 8). The term supersonics is used for sound waves having velocities greater than that of sound. High power ultrasound is known to damage or disrupt biological cell walls which will result in the destruction of living cells.

Mechanism

Firstly, the ultrasonic transducers convert electrical energy to sound energy. Secondly, when the ultrasonic waves propagate in liquid, small bubbles will be formed and collapsed thousands of times per second. This rapid collapse of the bubbles (cavitation) results in high localized temperatures and pressure, causing breakdown of cell walls, disruption of cell membranes and damage of DNA.

Applications

It is Used in pasteurization at mild heat, extraction, enzyme inactivation, emulsification, crystallization, viscosity alteration and degassing, spraying or coating, anti fouling and de-foaming.

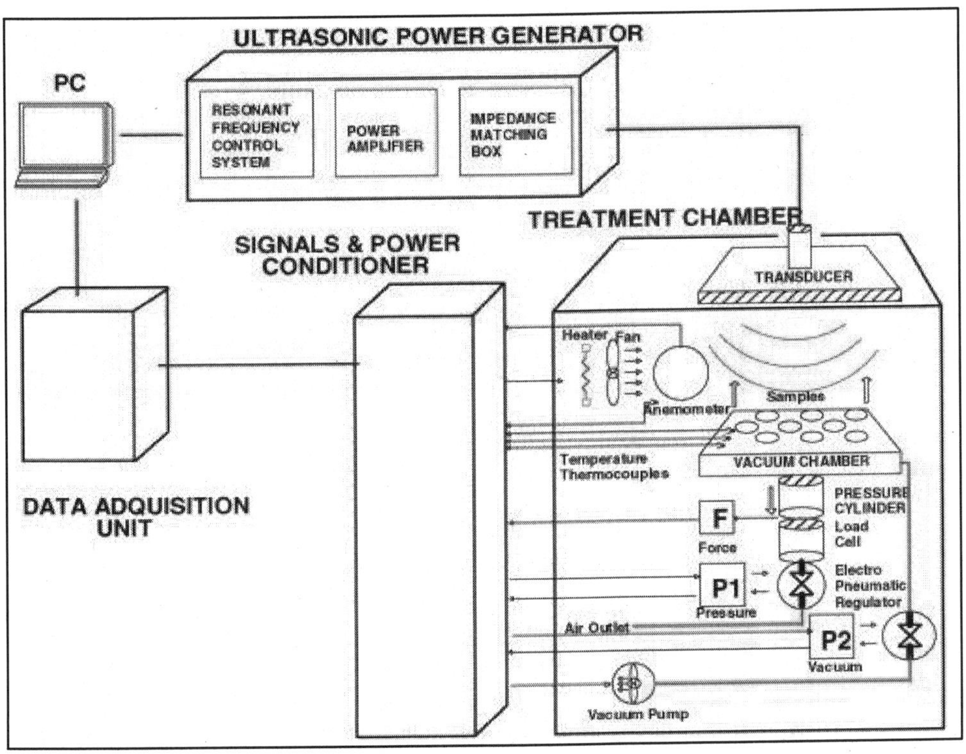

Fig. 8: Ultrasonic in Food.

(viii) Pulsed X-rays

Pulsed X-ray is a technology that utilizes a solid state opening switch to generate electron beam x-ray pulses of high intensity. Electrons have limited penetration depth of about 5 cm in food while X-ray have significantly higher penetration depths (60-400 cm) depending upon the energy used.

Applications

Used in examination of packaged food for tramp materials, inactivation of e-coli in meat also used to eliminate *Salmonella* from foods.

(ix) Irradiation

Irradiated foods are those that have been treated with ionizing irradiation. This process is a physical treatment that consists of exposing foods to the direct action of electronic, electromagnetic rays to assure the innocuity of foods and to prolong the shelflife (Fig. 9). The higher the wavelength, the smaller is the associated energy. Radio waves, with considerably long wavelengths of 30 cm to 3 km have very limited energy associated with them. The irradiation sources that are permitted for use in food processing are gammarays produced from the radioisotopes cobalt-60 (1.17 and 1.33 MeV) and caesium- 137 (0.662 MeV). Gamma-rays and X-rays transfer energy in a number of ways, each involving the liberation of fast electrons that then lose energy in electronic interactions. The main source of gamma irradiation is cobalt-60 which is a radioactive isotope produced from cobalt-59.

Doses

The treatment received by the food product is characterized by the irradiation dose, which is the quantity of energy absorbed by the food while it is exposed to the

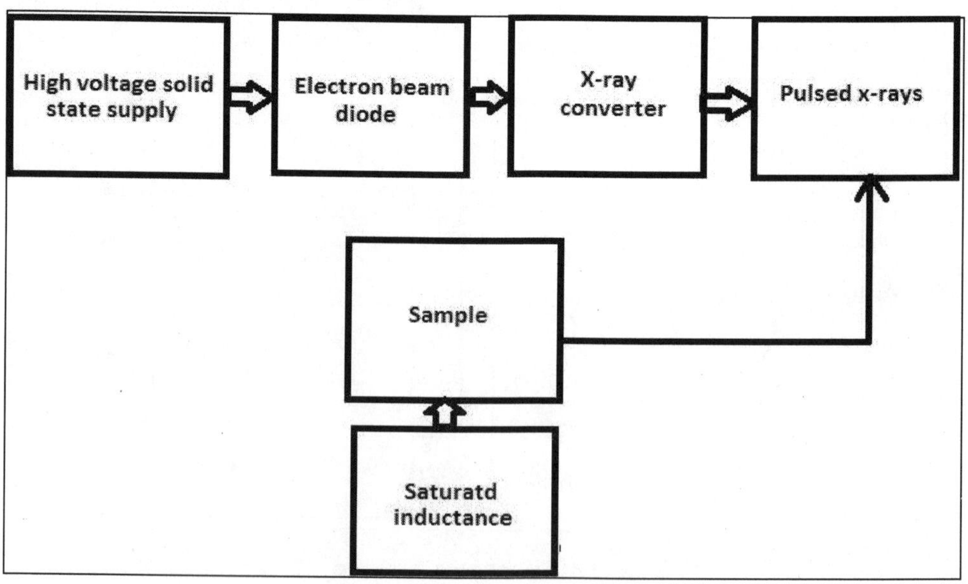

Fig. 9: Pulsed X-ray Setup.

irradiation field. The international unit of measurement is the Gray (Gy). One Gray represents one joule of energy absorbed per kilogram of irradiated product. One Gy is equivalent to 100 rad (radiation absorbed dose). Identification of the irradiated food product or food ingredient, using the international symbol, is required if it makes up more than 10 per cent of the product content. A written statement that the food has been irradiated must also be either written on the package or displayed next to the irradiated food.

Applications

Irradiation dose (2–5 kGy) could control post-harvest spoilage and diseases that affect fruit and vegetables without impairing their sensory qualities tissue or skin damage, changes in flavour or texture or by affecting the appearance, normal ripening of the fruit and disinfestations and shelf-life extension.

(x) Nanotechnology

It is an emerging technique which is applied in various fields such as computer electronics, communication, energy production, medicine and food industry. It is commonly used in food industry as in two forms: food additive (nano inside) and food packaging (nano outside). Nano scale food additive used to influence product shelf-life, texture, flavour, nutrient composition or even detect food pathogen and provide function as food quality indicators. In context of food packaging, nanotechnology are mainly consider to be of used to increase the product shelf life, indicate the spoilt ingredients, or generally increase the product quality, *e.g.*, by preventing the gas flow across product packaging.

4. CONCLUSION

The main problem with the thermal processing of food is loss of volatile compounds, nutrients and flavour. To overcome these problems non thermal methods came into food industries to increase the production rate and profit. The non thermal processing is used for all foods for its better quality, acceptance and for its shelf life. Non-thermal processing is a promising and useful approach for fruit juice and beverage preservation. The products based on these techniques show many advantages such as the retention of sensorial qualities and nutritional values over traditional thermal processing. However, among these non-thermal techniques, only high pressure processing has been adopted by the food industry so far. Additional pilot-scale testing may require for these non-thermal preservation methods to become a real alternative for thermal processing.

5. REFERENCES

Ravichandran, R. 2010. Nanotechnology application in food and food processing: innovative green approaches, opportunities and uncertainties for global market. *Int. J. Green Nanotech.: Physics and Chemistry*.**1**: 72- 96.

Hussain, P.R., Meena, R.S., Dar, M.A. and Wani, A.M. 2011. Radiation Processing of temperate fruits of Kashmir valley. *BARC Newslett*. **321**: 1-10.

Clouston, J.G. and Wills, P.A. 1969 Initiation of germination and inactivation of *Bacillus pumil- lus* spores by hydrostatic pressure. *J. Bacteriology*, **97**: 684–690.

Heinz, V., Alvarez, I., Angersbach, A. and Knorr, D. 2002. Preservation of liquid foods by high intensity pulsed electric fields – basic concepts for process design. *Trends Food Sci Tech.*, **12**: 103– 111.

Jayasooriya, S.D., Bhandari, B.R., Torley, P.D. andArcy, B.R. 2004. Effect of high power ultrasound waves on properties of meat: a review. *Int. Jo. Food Properties*, **7**: 3019-319.

Ozer, N.P. and Demirci, A. 2006. Inactivation of *Escherichia coli* O157: H7 and *Listeria monocytogenes* inoculated on raw salmon filets by pulsed UV-light treatment. *Int. J. Food Sci. Tech.*, **41**: 354-360.

Singh, S.P., Tegegne, F. and Ekenem, E. 2012. The food processing industry in India: Challenges and opportunities. *J. Food Distribution Res.*, **43**: 81-89

Chapter 30

Temperate Fruit Procssing and Marketing in India: Status and Opportunities

G.H. Shah

Member,
J&K Muslim Wakaf Board, Srinagar – 190 010, J&K
E-mail: drshahgh7@yahoo.co.in

1. INTRODUCTION

India is the second largest producer of fruits and vegetables in the world. Out of total world fruit production of 654.449 million metric tonnes, India produces about 88.971 million metric tonnes of fruit (NHB website), thereby contributing about 14 percent to the world fruit production.The country is front runner in the production of mango, papaya and banana and contributes 44.1, 42.6 and 25.6 percent, respectively, in the world production of these fruits (MoFPI, Investors' Portal).

India has diverse agro climatic conditions, ranging from cold desert temperate of below – 35 degree Celsius in Ladakh region to the hot summer temperatures of over 45 degree Celsius in tropical regions. Such diversity offers a great potential for growing different kinds of fruits and vegetables. Though most of the area falls in tropical or sub tropical zone, but a substantial area of the country falls in temperate zone as well. States like Jammu and Kashmir, Himachal Pradesh, Uttrakhand, Arunachal Pradesh and some other states have temperate zones within their geographical areas. Fruit like apple. Pear, cherry, peach, plum, apricot, walnuts, almonds and other such fruits which require low chilling temperatures are grown in the temperate zones. Out of total annual fruit production of about 88.971 million MT, temperate fruit production is about 3.5 to 4.0 million MT, which constitutes about 4 percent of country's fruit production. Among temperate fruits apple is the major fruit crop with an annual production of about 2.497 million MT (NHB website).

2. STATUS OF TEMPERATE FRUIT PROCESSING

Temperate regions of India are mostly hilly areas with many geographical barriers. Processing of fruits and vegetables in these areas was earlier confined to some traditional products like murabhas, sundried fruits, jams etc. Unorganized small and cottage scale units were set up in these states by some processors in 60's and 70's of twentieth century. Some of these units were subsequently closed for one of the other reason but some continue to be in operation. Processing and post harvest technology sector in temperate regions of the country started getting focused in the late 70's onwards. Fruit Processing plants and post harvest handling and storage units were set up. In J&K, the Government set up 8 cold stores each with a capacity of 1000 MTs, 18 packing and grading centers for apples each with a capacity of 6,000 MTs, 8 walnut Hulling and Drying units each with a capacity of 1,500 MTs and one Walnut Export centre with a capacity of 4,000 MTs, somewhere between 1979 to 1985. But unfortunately none of the units could be put to commercial operation for one or the other reason, though test operation of some units was conducted even during a full fruit season. Most of these infrastructures got damaged during the years of disturbances and presently none of these units is even fit for revival. In Himachal Pradesh also Government set up 6 packing houses, 4 grading centers, 5 cold stores, one apples processing plant somewhere around 1974. A large scale apple processing plant was inaugurated by M/s Cadbury India Ltd. in 1979, in J&K at Doabgah, Sopore. However, the company closed the plant just after three years of its operation. The same plant was taken over by J&K Government during 1985 and put to operation. It has been utilizing substantial quantities of apple for processing into apple juice concentrate. Successful operation of this plant attracted private investment also and country's largest apple processing plant was set up at Srinagar by a local entrepreneur. In Himachal Pradesh both Public and Private Sector investment was also made in fruit processing sector. Such investments made visible impacts in the horticulture sector of these states. Besides adding a new outlet for marketing/utilization of substantial quantities of unmarketable fruit, these processing plants indirectly helped to improve orchard sanitation. Moreover, prior to installation of these units the local street fruit traders used to offer much lower rates to the growers for such inferior grade fruits. In view of the option of selling their fruit to the processing units, a significant hike in the selling rates of such inferior grade fruit was noticed.

The country witnessed one additional revolution in processing sector when packing of fruit juices/beverages in Tetra packs was introduced and temperate fruit processing was also positively affected by this change. The consumption of fruit juices in traditional glass bottles started declining and the new concept of Tetra Pak started getting popularized. Some leading fruit processing companies like Parle Agro packed apple beverage (reconstituted from apple juice concentrate) in Tetra pak.The apple juice concentrate for this purpose used to be procured from Jammu and Kashmir or from Himachal Pradesh. With the changing scenario the two public sector undertakings of Jammu and Kashmir and Himachal Pradesh also ventured into packing of fruit juice/beverages into Tetra Packs. This too happened in late 80's of the previous century. Since none of these states had its own Tetra pak Plant, so both these organization were getting the juices packed from private Tetra Pak Plant. M/s SAI Foods, Gaziabad and Godrej Food Processing Plant, Mandideep, Bhopal were used by Jammu and Kashmir Horticultural produce Marketing and Processing Corporation for packing apple juice/beverage in Tetra Pak.

As a sequence of developments which took place in shaping and exploiting the processing sector of temperate fruits was use of apple juice concentrate in baby foods by one of the reputed Multinational Companies. A recipe was developed for one of the baby foods having apple juice concentrate as one of the important constituents. The annual demand of the concentrate for this particular baby food was almost equal to the annual production of apple juice concentrate by one of the states. This development also took place in the late 80's of previous century. The recipe developed by the said Multinational Company is being adopted even today and apple juice concentrate continues to be an important constituent of that baby food.

Though temperate fruits grown in the temperate regions of the country include apple, pear, peach, plum, strawberry, cherry, apricot, walnut and almond but in view of annual production of fruits, apple, pear and walnut assume a greater significance. Production of minor fruits like cherry, peach, plum, apricots, strawberry is not comparable to production of apple,walnut or pear. These fruits are generally sold as fresh fruits. However, some quantities of these fruits are processed into fruit juices, jams, and squashes in small scale processing units. Sun drying of apricot especially in Ladakh Division is very common and is a traditional practice. In the recent years some solar Dryers have been installed in the area for fruit and vegetable drying. Some small scale processing units have also been set up for jams, jellies and squash making. A cooperative unit for fruit processing has been set up in the area which is also engaged in processing of apricots into products like jams, juices, dried apricots. Small/Medium processing plants have also been set up in other temperate fruit growing areas of the country including Himachal Pradesh and Uttarakhand.

Processing of walnuts has received a focus from the private sector. Primary Processing of walnuts is generally taken up by the farmers at their respective places. The primary processing of walnuts includes de-hulling, washing/drying and grading. Though a substantial quantity of walnuts is sold as such in the market, but entrepreneurs have set up walnut processing units, which are equipped with grading, packing, kernel extraction and vacuum packing. The value addition is helping for better returns.

Primary processing of almond is also conducted by almond growers at their orchards/farms. This process also includes de-hulling, cleaning, drying and packing or at times kernels are extracted from dry almonds and packed as such or under vacuum, in packs of different sizes. Bulk packing of walnuts and almonds for sale in wholesale markets is a common practice with large fruit growers/traders and middlemen. A large quantity of these nuts is handles in this manner. However, primary processing like de-hulling, washing, drying, grading is assured before bulk packing of nuts.

Ministry of Food Processing industries, Government of India have reported 702 registered Food Processing Industries in three major temperate states, out of total number of 37,175such registered factories in the entire country. Statewise breakup of these factories (MOFI, website) is reflected in the Table 1.

Table 1: Registered Food Processing Industries in Major Temperate States

State	*No of Registered Units*
Jammu and Kashmir	156
Himachal Pradesh	163
Uttrakhand	383
Total	702

The registered units depicted above reflect the total food processing units which include all types of food industries registered in these states, including fruit and vegetable processing units as well. Like most processing plants of the country, the plants in these states are also underutilized, in respect of their capacity utilization.

Quantities of fruit being processed annually in temperate regions vary from year to year, depending on number of factors. However, apple is the major fruit crop being processed into apple juice concentrate. Other fruit products like jams, juices, squashes, jellies etc are also being processed from temperate fruits. Demand of apple jam and apple juice in the country was estimated to be of the order of 0.207 million tonnes and 0.234 million tonnes in the year 2003. Similarly, total consumption of apples as fresh and processed fruit in India was estimated to be of the order of 1.535 million tonnes during 2003 (Pandey *et al.*, 2009). If current estimates of such consumption are drawn proportionately taking into population factor alone into consideration, the estimates will work about 1.826 million tonnes. However, when the factors like changing food habits, lifestyle, circumstantial compulsions, health awareness etc are taken into consideration, the requirement is likely to be much higher.

India exported 6,726.36 MT of walnuts during 2013-14 worth Rs. 32,453.49 lacs, whereas during the year 2014-15 a quantity of 2,665.87 MT valuing Rs. 13,645.23 lacs were exported (APEDA). J&K has major share in the export of walnuts and the fruit is being exported to Vietnam, Egypt, Netherland, UK, Spain, United States, Germany, France, Thailand, Australia, and some other countries.

In the recent years it has been reported that walnut is also being imported to India. During 2013-14, a quantity of 178 MT of walnuts worth Rs. 431 lacs has been imported to India (APEDA website). Import of walnut could be either due to high demand/consumption of walnuts in India or it could be the requirement of processor/ exporters. Since many exporters extract kernels and then export the vacuum packed kernels, so the exporters/processers may be attempting to utilize maximum for their processing capacity and minimize underutilization.

3. STATUS OF MARKETING

Marketing of horticultural produce takes place through numerous marketing channels, out of which the most dominant as reported in Global Agriculture Information Network (GIAN Report 2013) is as under:

1. Grower → Pre-harvest contractor → Commission agent → Wholesaler → Retailer → Consumer.
2. Grower → Forwarding agent → Commission agent → Wholesaler → Retailer → Consumer.
3. Grower → Commission agent → Wholesaler → Retailer → Consumer.
4. Grower → Private CA store → Distributor → Retailer → Consumer.

Sale of domestic horticultural produce including the temperate fruits is generally completed through the Agricultural Markets established and regulated under the State Agricultural Produce Marketing Committee (APMC) Acts. Each state has established fruit mandies in the respective states. Most of the temperate fruits of the country are traded through local fruit mandies and Azadpur Fruit and Vegetable Market, which is ranked as Asia's largest fruit market. Fruit is traded directly from production areas to local fruit Mandies or mandies of other states like Delhi, Mumbai, Bangalore, Bhopal, Amritsar, Lucknow, Kolkata *etc*. Even fruit from Azadpur fruit Mandi is supplied to other terminal market of the country. In the recent past the private players have entered into fruit trade in a different manner. These private players have set up controlled atmospheric (CA) stores in the production areas and with an objective to utilize the capacity of these stores are either purchasing the produce and storing the same in these CA stores,or these private players have tied up arrangements with growers for storing their produce in these stores. Private players like Adani Agri Fresh Limited, Devbhumi Cold Chain Limited, Harshana Cold Store *etc*. are some such players. Many of these private players have started buying the produce directly from the farmers and then selling the same through a network of dealers/market outlets. Fruitlike cherry, peach, plum, apricots *etc*. being grown in temperate regions have less production compared to apple, and as such the major focus of the growers/fruit traders in these regions remains on apple. This fruit crop decides fate of the orchardists and fruit traders in the temperate areas. A report on apple production, consumption, exports etc has been drawn for three consecutive years (GAIN Report, 2013) taking July as beginning of market year which is reflected here under to understand the scenario of this most important temperate fruit of our country.

Table 2: Details of of Apple Production, Availability and Distribution

Apple	*2010-11*	*2011-12*	*2012-13*	*Average Lakh MT*
Production, MT	2890000	2203000	1850000	23.243
Imports, MT	135000	18100	202000	1.183
Total Supply, MT	3025000	2391100	2052000	24.893
Fresh Domestic, MT	2016000	1898000	1800000	19.046
Exports, MT	47000	30000	20000	0.323
For Processing, MT	231000	66100	46000	1.143
Withdrawals From market, MT	731000	397000	186000	4.38
Total Distribution	3025000	2391100	2052000	24.893

The above details reveal that apple is in demand in the domestic market. As against average export of 0.323 lakh MT, the country has imported 1.183 lakh MT of the fruit. This amply justifies the market potential of apple. According to industry sources demand is growing in both urban and rural India. India consumers nearly all domestic production of apples and consumption exceeds domestic production, which reflects the demand for this fruit. Apple imports to India have grown dramatically. Despite the fact that India's high basic tariff on import of fresh apples is one of the highest in the world, the produce in witnessing double digit growth every passing year. The major attraction for imported apples for a specific cliental group is the uniformity in size, colour and above all the retention of their crispness/ fresh attributes. These attributes are being maintained by the respective countries because of infrastructural facilities like grading/packing lines, CA stores, cold chain *etc.* Such arrangements also facilitate the exporting countries to push their produce in Indian market at a time when the domestic produce availability in the country has drastically reduced. Countries like United States, China are major exporters of apple with their share of about 41 and 31 percent, respectively, of the total apples imported in India (GAIN, 2013). Others countries like Chile, Newzeland, Italy and some other countries also export apples to India.

Apple from orchards reaches to consumers through a multitude of channels, each of which differs in the number of intermediaries and as such has varying degrees of complexities. Although these channel intermediaries perform specific role in the value chains but they exploit farmers to a great extent. More often than not, each intermediary adds his margin arbitrarily with little adherence to the market acts. This ultimately raises the retail price with adverse consequences to both the growers and consumers. Plethora of intermediaries jack up the final cost without significant value addition. They also exert a negative pressure on the margin of the farmer. Moreover, due to multiple handling of fruit by different stakeholders, the fruit quality is also affected. The studies (Pandey *et al.*, 2009) have revealed that the share of grower in consumer rupee is just 35 percent, whereas intermediaries' margins and marketing costs work out 44 and 21 percents of consumer rupee respectively. In countries like USA the farm realization is about 60 to 65 percent.

In the recent years a positive development has been observed in the temperate fruit regions. Setting up of post harvest infrastructures by private investors and procurement of the fruit by the same investors is gaining popularity. This arrangement is showing very positive impact on the share of grower's realization.

Table 3: Average Retail Price of Apple from J&K at Delhi for the Year 2014

Month	*Average Retail Price, (Rs.)*	*Month*	*Average Retail Price (Rs.)*
Jan	7154	Aug	0
Feb	7604	Sept	9000
March	8429	Oct	8522
April	11364	Nov	8250
May	14000	Dec	8288
June	14500		
July	0		

The average retail price reflected in the Table indicates that retail price is maximum from April to June. If infrastructural facilities likes CA stores/Cold chain are available, the quality of produce can be maintained and supplies to distant markets can be regulated which will ultimately benefit the grower.

For maximizing the growers share in the consumers' rupee, following measures can play a positive role:

- ☆ Reduce the number of intermediaries in the marketing channels.
- ☆ Select a channel with least intermediaries.
- ☆ Do not create a glut like situation in the fruit mandies.
- ☆ Tie up storage arrangements with Entrepreneurs who have facilities of CA and cold chain arrangements.
- ☆ Efforts should be made to tie up marketing with such agencies that have CA stores and marketing network as well.
- ☆ Ensure proper grading and packing.
- ☆ Regulate fruit supply to different mandies/states in such a manner that the dispatches do not result in a glut like situation.
- ☆ The Market Act needs to be implemented in letter and spirit.
- ☆ No extra charges/commissions are allowed at any stage while adopting any market channel.

4. OPPORTUNITIES FOR PROCESSING

The structure of house holds' consumption expenditure has changed as a result of rapid economic development. In general there is a positive correlation between food consumption and income. Further the life style of individuals, their exposure to other communities/countries and their nature of job has a great influence on the food habits of individuals and families. The changing scenario in food consumption is showing a positive attitude of individuals towards consumption of fruits, vegetables and their products. The survey conducted (National Accounts Statistics) revealed following facts about per capita expenditure on various food items.

Table 4: Per Capita Consumption Expenditure for the Year 1999-2000

Sl.No.	*Name of the Food Item*	*Expenditure*
1.	Cereals	894.34
2..	Pulses	116.95
3.	Sugar and gur	268.79
4.	Oils and oil seeds	268.20
5.	Fruits and vegetables	861.69
6.	Potato and tubers	75.97
7.	Milk and milk products	752.62
8.	Meat, egg and fish	291.91
	Total	***3530.47***

The tabulated data depicts that fruits and vegetables are one of the major components of our food related expenditures. Out of total per capita expenditure of Rs. 3530.47, an amount of Rs. 861.69 is incurred on fruits and vegetables, which constitute about 24.4 percent of food expenditure.

According to World Health Organization (WHO), at least 400 g of fruits and vegetables must be consumed per day per capita which is about 146 kg per capita per year. WHO studies 21 countries to find out countries whose fruit and vegetable consumption status was as per the recommended values. It was observed that out of 21 countries only 3 (Israel, Spain and Italy) had acceptable natural average intake of at least 400 g per day or more (Banwat *et al.*, 2012).

Inadequate consumption of fruits/vegetables by many countries could be due to many reasons and the low availability/production could be one of them. Temperate region of the country offer a tremendous potential for achieving much higher goals of production and productivity. Since all temperate regions are located in hilly areas and the required infrastructures matching to the production levels have not come up in the post harvest handling, transportation and processing sectors in adequate numbers. It is only in the recent years that post harvest handling and storage infrastructures are being established in these regions to cater the needs of increased levels of production.

Though fruits like cherry, peach, plum, apricot are grown and processed in the temperate region of the country, but their processing is confined to the traditional fruit products like juices, jams, squashes *etc.* With the changing trends there is a need to develop new value added products, which could bring more returns to the growers. Products like baby foods, intermediate moisture foods, toffees, fruit bars, fruit balls, pulps etc. could be developed from these fruits. Similarly fruits like apple, pear being grown in the regions also offer a scope for processing into products like baby foods, fruit bars, pulps *etc.* The traditional products may not find much demand in coming future, so product diversification may help to catch the market. Temperate fruits can be very successfully converted to wide range of processed products. However, the geographical location of each fruit producing area and the quantity of production will decide the type and processing infrastructure required for the particular fruit/ area. Fruit pulps could be made from all temperate fruits and the same could be subsequently converted to different products. Processing of fruits into pulps in the production areas, its transportation to a central place with adequate arrangements for its conversion to different products could be one of the options to avoid huge investments in small production areas.

There is a general consensus that about 25 percent of the fruit is wasted after harvesting. If such fruit is processed into various products, a huge quantity of processed products can be obtained. The following quantities of products can be processed from such estimated loses.

- ☆ Annual Apple Production = 2.497 million MT
- ☆ Assuming 25 percent inferior grade fruit/post harvest loss = 0.624 million MT

- ☆ Estimated quantity of apple juice concentrate = 0.062 Million MT
- ☆ Estimated quantity of apple pulp = 0.530 million MT (5,30,400 MT).
- ☆ Estimated quantity of fruit Jam and jelly = 0.811 million MT (8,11,200 MT)
- ☆ Natural Apple Juice = 0.4992 million MT (4,99,200 KL)

Such products from other temperate fruits like cherry peach, plum, apricots, etc could also be developed, however, the estimated production/quantity will be of much lesser magnitude. If the nutritional value of these products is worked out, one can realize that these products can make a major contribution for improving nutritional status, besides generating huge revenue for various agencies involved in fruit production/processing and marketing. None of the states has till date the required full infrastructures to face the growing potential of processing sector. Moreover, selection of product lines has not been well planned. Blending of different fruits to develop products of higher market preference has not been commercially adopted in temperate regions. Design and layout of processing plant has to be made in such a manner that during lien season some activities are performed to ensure better utilization of human resources and machinery.

Primary processing of walnuts offers a tremendous scope in walnut growing areas. Cottage/small scale units for de-hulling, washing, drying and grading of walnuts in almost all villages where walnut are grown is required to save the nuts from quality deterioration of kernels. Since walnuts are generally of seedling origin so size uniformity is missing. Proper grading is highly essential for better marketing of the produce.

Sun drying of many fruits and vegetables has been a tradition. This is one of the established practices for drying of fruits like apricots. With changing technologies one should not discard this concept, but improve the techniques of sun drying. Maintaining sanitation of the drying yard/product, uniformity in fruit slices to improve the product uniformity, protection from dirt/dust and other techniques can help to maintain product standards. The concept of solar driers could be made more popular even at large scale.

5. OPPORTUNITIES FOR MARKETING

Any arrangement/channel which eliminates or minimizes the number of intermediaries between producer and consumer can emerge a better option for marketing of horticulture produce. Despite Marketing Act, the fruit growers are being exploited by different intermediaries. Though some of the intermediaries perform commendable jobs for the fruit industry but the element of exploitation exists which nullifies their positive contributions.

The new concepts of marketing like online marketing have now been initiated by some fruit traders, but this cannot be adopted yet by common growers at their level. Marketing trends and strategies may keep on changing with the change in time, but as on date it is felt that the recent concept of tying up marketing arrangements with such agencies who have set up their CA store/grading lines in the production zones and have a market network in different parts of the country/world is becoming a better option for the growers. These business houses are offering options to the growers for storing their produce against a predefined storage charges. Growers must cooperate with these Agencies and store their fruit in such CA stores and sell the fruit either to the same agency or some other agencies, as and when it suits them. A mutual trust and confidence between the grower and the CA store owner is beneficial for overall benefit of the industry.

The Marketing Act in different mandies needs to be implemented in letter and spirit. The exorbitant and unjustified charges being charged at different levels by intermediaries need to be checked. The techniques of bidding in fruit mandies continues to be hidden. There is a need to install visible electronic bidding system in all the fruit mandies of the country and the decision on selling or retaining the fruit against a particular offered rate must be with the owner.

Adequate infrastructure for storage (CA and cold storage) need to be set up near all fruit mandies so that the farmers do not face a situation of distress sale. With changing times it is being felt to change the pack size of horticulture produce. Attractive and convenient fruit packs must be developed suiting to family packs for retails markets.

For better marketing it is the responsibility of producers/growers to take extra care at the harvesting and post harvesting stage while picking, packing and grading. Faulty techniques of fruit handling results in post harvest losses and fruit decay, which ultimately results in great loss either to the grower, or to the traders or to the retailer. It also causes dissatisfaction to the buyer.

6. CONCLUSION

Production of temperate fruits in India has received a great attention from growers and Government Agencies from time to time. As such the country has made significant achievements in production of temperate fruits. But post harvest handling and processing sector has not developed with the same pace. Hilly and difficult terrains of temperate regions, reluctance of the investors and some other factors have been major reason for poor investment in these sectors. But with the changing scenario the situation is now changing. Government of India and State Governments are focusing on post harvest segments throughout the country including temperate regions. The Private sector is also now investing in this sector. The temperate region of the country offer a great potential for processing fruits into products like juices, concentrates, fruit bars, baby foods, pulps *etc.* Products having demand in the market need to be processed from temperate fruits.

Marketing of horticulture produce is presently facing great exploitation from intermediaries. A large percentage of consumer's rupee is taken by middlemen and intermediaries. In the recent years the trend has changed and private investors have invested in infrastructure like CA stores, grading lines *etc.* The same Agencies have started procuring fruit from the growers. Such arrangements are showing a positive impact and are seemingly grower friendly. The best way to market the horticulture produce will be to minimize the intermediaries in the marketing channel. The awareness among growers, new concepts of online marketing, involvement of educated youth with horticulture sector and such like factors are likely to influence the marketing of horticulture produce and in future years it is expected that besides better quality retention of the produce, the producer will get higher share of consumer's rupee.

7. REFERENCES

Agriculture Products Export Development Authority (APEDA) website. (www.apeda.gov.in)

Banwat, M.E., Lar, L.A., Daboer, J., Audu, S. and Lassa, S. 2012. Knowledge and Intake of Fruit and vegetable consumption among adults in an urban community in North Central Nigeria. *Nigerian Health J.* **12** .

GAIN Report 2013.Global Agriculture Information Network Report No. IN 3069 by USDA Foreign Agricultural Service.

Ministry of Food Processing Industries website(www.mofpi.nic.in)

National Accounts Statistics. Transition in Food Consumption Patterns (http// mospi.nic.in/mospi cso rept pubn.htm)

National Horticulture Board website (www.nhb.gov.in)

Pandey M., Sikka, B.K. and Panthari, S. 2009. ICT Systems for Increasing Efficiency of apple- value chain. *National Seminar on ICT for Agriculture and Rural Development* held at Arunachal Pradesh.

Chapter 31

Marketing, Market Integration and Price Forecasting of Apple in India

S.A. Wani, S.H. Baba, Muzaffar M. Mir and Haris

Division of Agricultural Economics and Marketing,
Sher-e-Kashmir University of Agricultural Sciences and Technology of Kashmir
Srinagar – 190 025, J&K
E-mail: dr.shabirwani@rediffmail.com

1. INTRODUCTION

Apple is commercially the most important and widely cultivated temperate fruit and is fourth among the most widely produced fruits in the world after banana, orange and grape. There are more than 7,500 known cultivars of apples, presenting a range of desired characteristics; various tastes and uses. Commercial apples are generally propagated by grafting and the crop is prone to a number of fungal, bacterial and pest problems. Production, value addition and marketing of apple is capital intensive and has been a major employment provider among horticultural crops.

About 76 million tonnes of apples are produced worldwide in 2012 and China produces almost half of this total. The United States is the second-leading producer; producing more than 5 per cent of world production. Other important apple producers across the globe are Turkey (3.8 per cent), Poland (3.8 per cent), and India (2.9 per cent). Regarding world trade, the largest exporters of apples in 2009 were China, the U.S., Turkey, Poland, Italy, Iran, and India, while the biggest importers were Russia, Germany, the UK and the Netherlands. Other countries with a significant production are Brazil, Argentina, Ukraine, Germany and South Africa (Fig. 1).

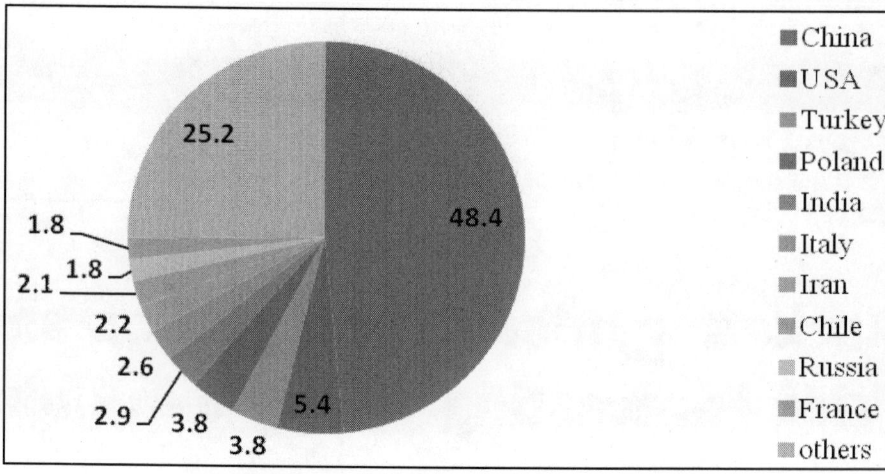

Fig. 1: Top Ten Apple Producing Countries.

Growth performance of apple in India compared to its trend in the Asia and World has been presented in Table 1. It could be seen that between 1973-74 to 2011-12, this fruit in India has shown a significant growth in production (3.5 per cent) which is higher compared to the world aggregate growth but lower than its growth in Asia. The area expansion under apple in India also exhibited same behavior. However, growth of the yield levels in India is yet lower compared to Asia and World. Average yield of commercially important apple cultivars per unit area is the highest in the country ranging between 10-12 tonnes/ha, but it compares poorly to the yields of 20-30 tonnes/ha in horticultural advanced countries of the world. Alternate bearing, defective pruning and training, use of seedling rootstock of unknown performance, lack of proper nutrients and water management, deficiency of suitable pollinizers and ineffective control of pests and diseases are the main causes of low productivity. The paralyzed growth of apple productivity in India needs to be comprehensively analyzed for sustaining this growth.

Table 1: Compound Growth in Area, Production and Productivity of Apple in Various Regions during 1973-74 to 2011-12

Regions	*Area*	*Production*	*Yield*
India	2.4** (0.044)	3.5* (0.090)	1.1 (0.088)
Asia	3.9* (0.181)	6.3* (0.093)	2.4* (0.133)
World	1.2 (0.088)	2.4* (0.044)	1.2 (0.088)

Source: FAO, 2013.

Note:* Significant at 1 per cent level of significance; ** Significant at 5 per cent level of significance.

2. NATIONAL SCENARIO OF APPLE

The trend in area and production in India from 1991-92 to 2010-11 is presented in (Fig. 2). While the area under apple has increased over the year, the growth in case of production is more prominent.

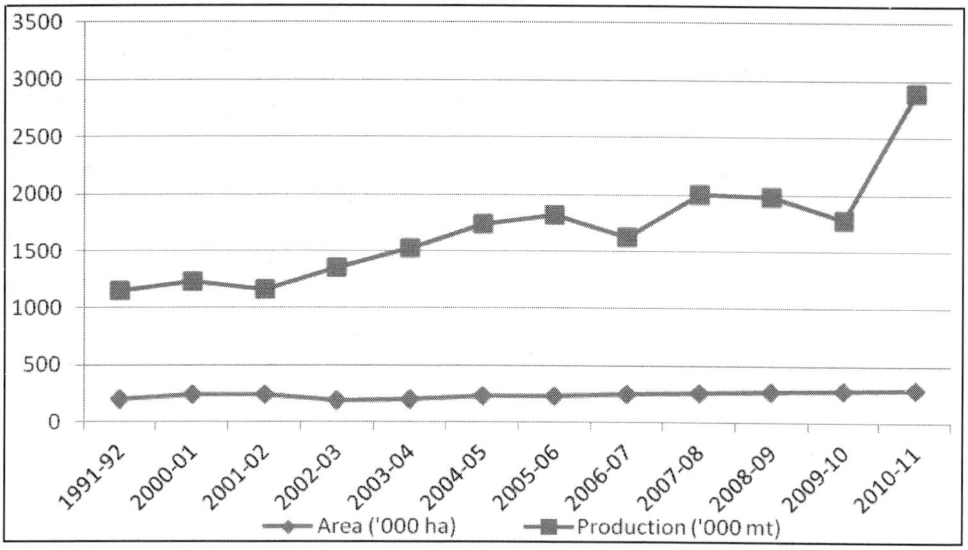

Fig. 2: Area, Production and Productivity of Apple in India.
***Source*: UN Food and Agriculture Organization (FAO), Rome.**

Jammu and Kashmir, Himachal Pradesh, Uttrakhand and Arunachal Pradesh are the major apple producing states of India. Within India, J&K and Himachal Pradesh together accounts for 92 per cent of the total production and about 85 per cent of the total area under apple in India. As far as productivity of apple is concerned, J&K has the highest productivity of apple followed by Himachal Pradesh (3.9 tonnes/ha) and Uttarakhand (2.16 tonnes/ha) (Fig. 3).

Inspite of tremendous progress in production and productivity of apple in the state, there are various issues pertaining to marketing which needed to take care of. The state does not have specialized markets, and these fruits are traded in distant markets such as the Delhi, Ahmedabad, Bengaluru, Mumbai and Kolkata markets. This leads to a rise in post-harvest costs, including marketing cost, forcing the growers to opt for pre-harvest disposal to meet their financial obligations and avoid climatic risks. Apples are characterized by strong seasonality and perishability. This induces competition and affects their prices and quantities supplied. In this background it is imperative to analyze existing marketing system and price behavior of apple in different major markets in India.

3. METHODOLOGY

Both the primary and secondary data has been used in the study. Primary data were collected from sample respondents in major fruit belts of the valley which include Sopore, Shopian, Pulwama and Baramullah. Secondary data with respect to area, production and daily data from 2005 to 2014 on wholesale prices of different markets were collected from Directorate of Horticulture, Horticulture Planning and Marketing, Directoreate of Economics and Statistics, Government of J&K,

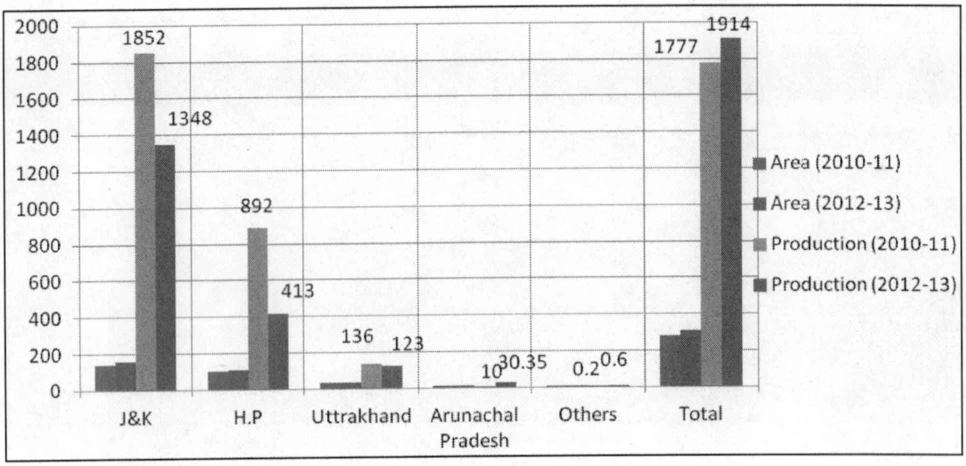

Fig. 3: State-wise Area and Production of Apple during 2010-11 and 2012-13. *Source*: Indian Horticulture Database.

It is pertinent to note here that the apple comes from every corner of the state and are disposed to different primary/secondary wholesale markets. Apple is marketed in about ten major primary wholesale markets in the country however, based on the highest volume of the apple receipts; eight markets (Delhi, Mumbai, Kolkata, Bengaluru, Ahmedabad, Amritsar, Sopore and Parimpora) were selected. Further three prominent apple varieties were considered for the study *i.e.* Delicious, American and Maharaji. This data was later averaged to obtain weekly wholesale prices set from September to December every year for each identified market.

(a) Stationarity in Data Set

The time series data has to be checked for stationarity before conducting any statistical analysis. A time series is said to be stationarity, if its properties are unaffected by a change of time origin, that is if the joint probability distribution associated with k observation Z_{t1}, Z_{t2}...... Z_{tk}, made at any set of time t_1, t_2 ....t_k respectively. Z_{t1+k}, Z_{t2+k},.......Z_{tk+m} made at any set of time t_{1+m}, t_{2+m}....t_{k+m}. Thus for a discrete time series to be strictly the joint distribution of any set of observations must be unaffected by shifting all the time of observation forward or backward by any integer amount m. Stationarity implies that the series remain at a fairly constant level overtime. If a trend exists in the time series then series is not stationary. The time series should show constant variation in its fluctuations over a period of time.A time series must satisfy the stationarity conditions in order to be eligible for the application of this methodology.

(b) Augmented Dickey–Fullers (ADF) Test

The stationarity of a series can be tested using a unit root test, the most widely used being Augmented Dickey-Fuller unit root test. It would test the null hypothesis that the series has a unit root, *i.e.* non stationary. The test is applied by running the regression of the form given in Equation (2):

Yt = β1 + δYt-1 + αi Σmi=1 ΔYt-i + εt (2)

where, åt is a pure white noise error-term and

ΔYt-i = (Yt-1 – Yt-2).

(c) Error Correction Model

The co-integration analysis reflects the long-run movement of price indices, although in the short run they may drift apart. Johansen's (1988) multivariate Co integration approach was used to examine Co integration between two price indices. Before conducting co integration test, it is mandatory to perform stationarity test. Augmented Dickey-Fuller (ADF) unit root test (Dickey and Fuller, 1979) was performed in this study to check stationarity for both the series.

A co integrated system can be written as:

$$\Delta y_t = \sum_{i=1}^{k} \Gamma_i \Delta y_{t-i} + \alpha\beta' y_{t-k} + \varepsilon_t \quad (1)$$

Where yt is the price series, Δy_t is the first difference i.e., ($\Delta y_t = y_t - y_{t-1}$), and the matrix $\alpha\beta'$ is n x n with rank ($0 \leq r < n$), which is the rank of linear independent co integration relations in the vector space of matrix. The Johansen's method of co integrated system is a restricted maximum likelihood method with rank restriction on matrix $\Pi = \alpha\beta'$. The rank of Π can be obtained by using λ_{trace} statistic. The test statistic can be given as:

$$\lambda_{trace} = -T \sum_{i=r+1}^{n} \ln\left(1 - \hat{\lambda}_i\right) \forall\, r = 0, 1, \ldots, n-1 \quad (2)$$

Where, T is the total number of observations, $\hat{\lambda}_i$'s are estimates of the Eigen values representing the strength of the correlation between the first difference part and the error-correction part. Now the null and alternative hypothesis for testing co integration rank, are, $H_0: rank\ of\ \Pi = r$ and $H_1: rank\ of\ \Pi > r$ respectively. Where, r is the number of co-integrating vectors. This test is carried under the condition that the co integrating equation has only intercept (no trend) and the original price series follows a trend since the mean and variance are non-constant over a period of time (non-stationary).

(d) Auto-Regressive Integrated Moving Average

As noted, the BJ methodology is based on the assumption that the underlying time series is stationary or can be made stationary by differencing it one or more times. This is known as the ARIMA (p,d,q) model, where d denotes the number of times a time series has to be differenced to make it stationary. In most applications d = 1 – that is, we take only the first differences of the time series. Of course, if a time series is already stationary, then an ARIMA (p,d,q) becomes an ARMA (p,q) model. The practical question is to determine the appropriate model in a given situation. From the fitted model is white noise; if they are, we can accept the chosen model, but if they are not, we will have to start afresh. That is why the BJ methodology is an iterative process. The ultimate test of a successful ARIMA model lies in its forecasting performance, within the sample period as well as outside the sample period.

4. RESULTS AND DISCUSSION

(a) Growth in Area, Production and Productivity of Apple in J&K

Apple is the principle fruit crop of Jammu and Kashmir accounting for 45.33per cent of total area (3.47 lakh ha) under all fruits grown in this state. The annual apple production in the state is 13.48 lakh M tonnes (2012). Climatic and other agro-ecological factors of Kashmir region of the state are ideally suited to the cultivation of many apple varieties. The scenario of apple fruit in the state has been presented in the Figure 4. The analysis revealed that area increased by 2.8 percent from 1980-981 to 2010-11, while as production by 4 percent during the same period. Inspite of manifold increase in area and production; apple yield has remained almost stagnant at during the past two decades. However due to the concerted efforts of institutions and farmers, yield has picked up to more than 13 metric tonnes per hectare in the year 2010. Later its yield level has again declined to about 9 Mt/ha. Area under apple has witnessed a continuous increase since 1980s. During 1980-81 area under this crop was 60,286 hectares, which increased to 1,57,280 hectares in 2012-13 (Fig. 4). Production of this crop has also shown the same pattern and has increased by more than three times. As far as productivity is concerned, after exhibiting fluctuating pattern, it has remained around 9 MT/ha from 1980-81 to 2012-13.

Table 2 documents period-wise exponential growth rate of area, production and yield of apple in Jammu and Kashmir from 1981-82 to 2010-11. Since 1980-81 there has been a significant expansion in the area under apple. Between 1980-2012 there has a commendable progress in apple production. The area, production and yield witnessed growth momentum of 2.8, 4 per cent and 1.2 per cent, respectively. It was observed that the area, production and productivity of apple has been increasing at increasing pace over the years.

Table 2: Exponential Growth Rates of Area, Production and Yield of Apple in Jammu and Kashmir

Period	*Area*	*Production*	*Yield*
1980-82 to 1990-91	1.5*	2.7***	1.2***
1991-92 to 2000-01	2.8*	3.0*	0.2***
2001-02 to 2010-11	5.2*	6.5*	1.3***
1981-82 to 2010-11	2.8*	4.0*	1.2*

*Significant at 5 per cent; **Significant at 1 per cent; ***Significant at 0.1 per cent.

(b) Marketing System

Apple is produced by large number of small farmers scattered around the valley whereas, the consumers are located throughout the country. Small produce, lack of knowledge of marketing system and less liquidity potential etc., prevent them to undertake direct marketing of apple. Accordingly, majority of farmers are compelled

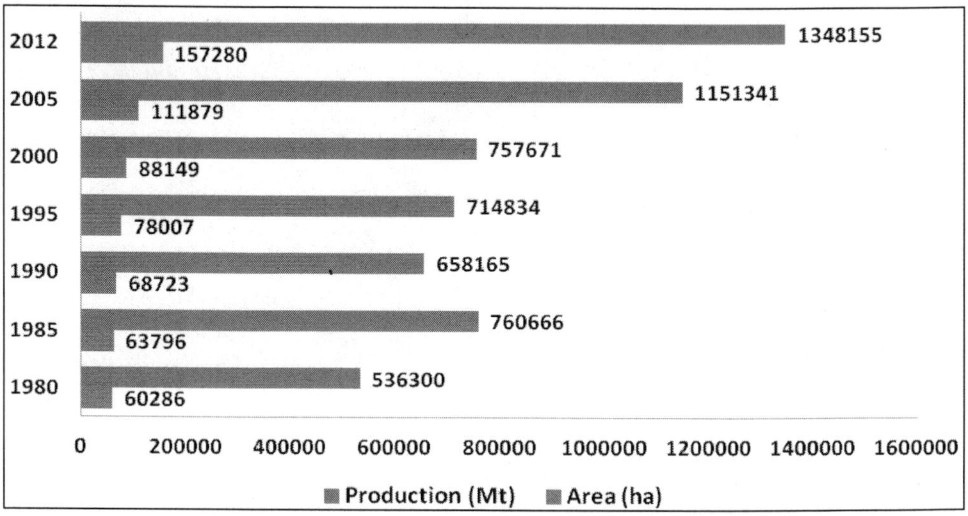

Figure 4: Trends in Area and Production of Apple in J&K.
***Source*: Directorate of Horticulture, Govt. of J&K.**

to dispose of their produce through different functionaries (Figure 5). Among different functionaries pre-harvest contractors are the major players and who purchased standing crop at blooming or fruiting stage at relatively lower prices. Only 14 per cent of the farmers were found selling produce directly to wholesellers.

The marketing system for apple is highly complex and is composed of different marketing channels for distribution of apple in different markets (Table 3). In each channel varying number of functionaries are involved and numerous specialized business activities called marketing functions are to be performed by them. Following are the commonly encountered and well identified channels in the valley.

Table 3: Marketing Arrangements of Apple

Channel-I	Producer-Wholeseller/Commission agent-Retailer-Consumer
Channel-II	Producer-Pre-harvest contractor- Wholeseller/Commission agent-Retailer-Consumer
Channel-III	Producer-Commission agent- Wholeseller-Retailer-Consumer
Channel-IV	Producer-Pre-harvest contractor- Commission agent- Wholeseller-Retailer-Consumer*
Channel-V	Producer-Post-harvest contractor/potential growers-Commission agent- Wholeseller-Retailer-Consumer*

* With respect to Delhi market.

(c) Marketing Cost and Price Spread in Apple Marketing

Since Delhi market receives 65 per cent of total apple produce exported to other states from Kashmir and in transportation cost of apple has been estimated with respect to Delhi market. The transportation cost per box of apple was found to be

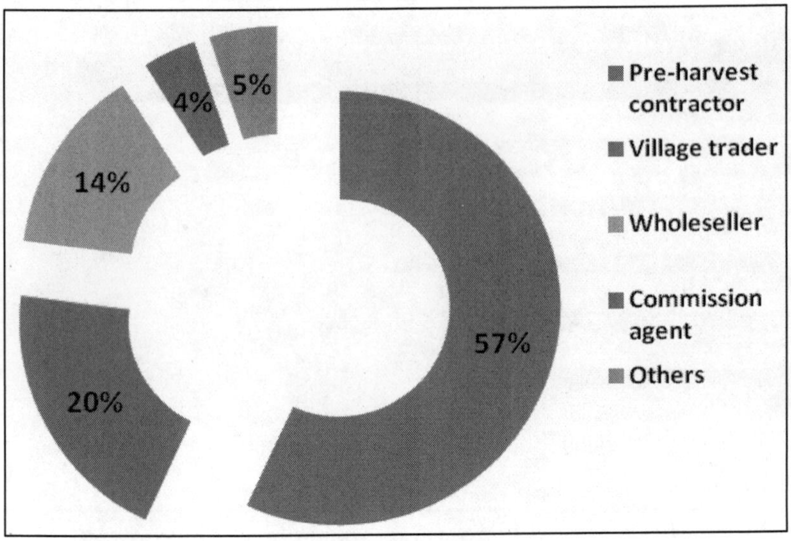

Fig. 5: Sale of Apple through different Functionaries.

56.54 per cent of total marketing cost. Pre-packing cost includes cost over picking, assembling and grading. The apples are picked manually by skilled labour and assembled at a plain place mated with paddy straw and then grade it again by skilled labours. The Table 4 revealed that the total pre-packing cost accounts for 8.73 per cent of total marketing cost. Packing cost includes cost on packing material and labour charges. The cost on packing box alone accounts for 24.8 per cent of total marketing cost.

The marketing efficiency of different channels as presented in the Table 5 revealed that the channel-III (0.73) turns out to be economically more efficient, followed by channel -I (0.68) and least efficient is channel-V (0.29). It was observed that producer got maximum share of consumers rupee in the channel where, produce was directly marketed to whole seller.The contractor in turn trade their produce to wholesaler at higher prices than producer, because of higher bargaining power. An orchardist could earn maximum share of consumers' price in the channel where he sells his produce directly to the whole seller. However, lack of liquidity potential, ignorance of market demand etc capitalizes into distress sale. Liberal cheap credit facility along with other incentives to apple growers would definitely increase their bargaining power.

(d) Marketing Facilities Generated

In order to improve marketing efficiency of apple marketing and to provide marketing facility for agriculture/horticulture produce at the doorsteps of the farmers/ growers, the various programmes have been initiated by way of establishing three Terminal Fruit and Vegetable markets *viz.*, Parimpore (Srinagar), Sopore (Baramulla) and Narwal (Jammu), which are functional and the trade is carried on in these markets.

In addition a new Terminal Fruit and Vegetable Market is being established at Jablipora Anantnag for which 441 Kanals of land has been acquired.

Table 4: Marketing Cost of Apple

Sl.No.	Cost Components	Percentage of Total
1.	**Pre-packing cost**	
	Picking charges	5.35
	Assembling charges	1.13
	Grading charges	2.25
	Total	**8.73**
2.	**Packing cost**	
	Cost of Packing box	24.80
	Cost of wrapping paper	1.69
	Cost of Paddy straw	0.90
	Cost of nails	0.56
	Cost of packing	1.13
	Closing and assembling of boxes	0.56
	Labeling and stenciling	0.56
	Total	**30.21**
3.	**Transportation cost**	
	Carnage of go down, loading and unloading charges	1.63
	Orchard to road head	5.63
	Forwarding charges	5.63
	Freight to Delhi	32.6
	State toll tax	8.51
	Loading at road head	0.79
	Unloading at destination	1.01
	Octroi at destination	0.56
	Communication etc.	0.05
	Total	**56.5**
4.	**Miscellaneous costs**	**4.50**
	Grand total**	**(100.00)**

Source: Directorate of Horticulture, Govt. of J&K; *Standard box contains 18 kgs. of apple.

**This does not include commission and market fee paid @ of 10 per cent of the total bill

The Department of Horticulture is also developing 21 Satellite F and V markets in the state out of which seven are completed and functional, six are likely to be completed and made functional, five are in progress and land acquisition process for three F and V markets is in progress. Eleven Apni Mandis are also under the process of development out of which three are completed, six are in progress and land acquisition for two Apni Mandis is under process (Table 6).

Table 5: Marketing Margins and Efficiency of different Channels in Apple (per cent of consumer price)

Particulars	Marketing Channels				
	I	II	III	IV	V
Expenses incurred by producer	22.24	–	35.92	11.61	–
Net price received by producder	45.93	43.35	37.19	27.85	28.92
Commission agents margin	–	–	7.31	7.31	7.31
Expenses incurred by pre-harvest contractor	–	22.24	–	–	35.92
Pre-harvest contractors margin	–	4.09	–	–	9.02
Expenses incurred by post-harvest contractor	–	–	–	17.02	–
Post-harvest contractors margin	–	–	–	10.29	–
Expense incurred by whole seller	0.48	0.48	3.06	3.06	3.06
Whole sellers margin	3.88	3.88	2.75	1.80	2.12
Expenses incurred by retailer	12.08	12.08	12.57	12.57	12.57
Retailer margin	15.38	15.38	8.51	8.51	8.51
Marketing efficiency	0.68	0.43	0.73	0.39	0.29

Despite government intervention, apple produce is still marketed indiscriminatly to demand into different primary wholesale markets of the country which demand strengthening of marketing intelligence. In this direction the examination of prices and marketing integration is very essential.

Table 6: Fruit and Vegetable Markets in J&K State

Sl.No.	Market Category	Market Location	Market (No.)
1.	Terminal Markets	Narwal[*] Parimpora[*] Sopore[*] Jablipora[1]	4
2.	Satellite Markets	Shopain[*] Pulwama[*] Kulgam[*] Baramulla[*] Handwara[*] Udhampur[*] Chari Sharief[*] Kupwara[1] Kathua[1] Rajouri[1] Bisnah[1] Akhnoor[1], Batingoo[1] Zazna[1] Leh[1] Kargil[1] Pryote Doda[1] Poonch[1] Samba[1] Sunderbani[1] Aglar (shopain)[1]	21
3.	Apni Mandi	Pachhar[*] Nunmai Kulgam[1] Mandi Poonch[1] Pouria, Reasi[1] Gharian (Udh)[1] Tapyal[1] Raya Bagala[1] Dyala Chak[1] (Chadwal) Bhaderwah Bani[1]	11
	Total		37

Source: Directorate of Horticulture, Govt. of J&K, [*]functional markets, [1]markets under process to be made functional.

(e) Price Behaviour and Market Integrations

The study analyses the interdependence of price changes of apple in seven different country level markets in India. The criteria for selecting the markets was the geographical distribution like, local, northern, western, southern and eastern markets and the volume of transactions as well. In addition, availability of data too was important for selecting the markets (Wani *et al.*, 2015) Price variability is the major component of market risk for both producers and consumers (Schumpeter, 1999). Government of India at the national level plays an important role in administering agriculture prices in India through various market intervention mechanisms, however, these strategies are mostly taken up in agricultural commodities especially some food grains only which too have reduced overtime (Jha and Srinivasan, 1999; Gulati *et al.*, 2000; Ramaswamy, 2002; Chand, 2003), most of which are not perishable as those of the fruits and vegetables. Usually under this particular commodity group hill and mountainous states are mostly the sufferers, despite the significant potential of pushing growth in agriculture beyond predicted values. It is noteworthy to mention here that more than 90 per cent of market surplus in fruits and more than 60 per cent of vegetables are sold in open market arrangements in these states. Under such situation the discovery of price behaviour under various market situations becomes important for risk management. The paper looks into the issue by checking for the presence of co integration among the markets. It also, examines the possibility of causal linkages among the different markets.

(i) Market efficiency

The market efficiency evaluation under co-integration analysis recognizes that the time series prices for various markets are usually non-stationery variables (Shen and Wang, 1990; Fortenbery and Zapata, 1993; Wang and Ke, 2005) and if these series are found to be non-stationery then it becomes necessary to test for co-integration, as a pre condition for market efficiency and un-biasedness (Kallar *et al.*, 1999) and also finding of no-integration of markets is normally interpreted to imply market in-efficiency. ADF test was applied at level and first difference to check the stationarity of this series. Table 7, presents the results of unit root test for three commercial varieties with two grades in selected markets. The results reveal that the null hypothesis of the unit root cannot be rejected for all the price series; therefore, we could conclude that all the prices in selected markets are non-stationery. Apple prices in different major Indian markets become stationery at first difference.

(ii) Co-integration among markets

The trace test results (Table 8) determine the number of co-integrating vectors. The figures documented in the table revealed that there are four co-integrated vectors at 5 per cent level of significance. The discussion is suggestive of the fact that even if there is geographical dispersion of markets, which are spatially segmented, the prices are linked together indicating that all the market locations are in the same economic market system. The flow of apple produce from one market to another market may affect the price levels in destiny market and accordingly the supply of the produce has to taken care of so that each market receives optimum quantity as per the quantum of demand.

(f) Causality in Various Markets/Varieties and Grades

The co-integration tests performed indicated the existence of long run relationship among the prices across the 7 markets only. Since the direction of the relationship among price series and market is equally important, therefore, Granger Causality Tests were also performed. The results presented in Table 9 revealed that the uni-

Table 7: Augmented and Dickey-Fuller Unit Root Test on Market Prices

Market		*Test Statistics*	*Mackinnon p-value*	*Critical Value*	
				1 per cent	*5 per cent*
Parimpora	Level	–2.838	0.1835	–4.178	–3.512
	Ist difference	–5.853	0.000	–4.187	–3.516
Sopore	Level	–2.784	29.2	–4.178	–3.512
	Ist difference	0.4	0.000	–4.187	–3.516
Delhi	Level	–2.624	0.2688	–4.178	–3.512
	Ist difference	–6.352	0.000	–4.187	–3.516
Mumbai	Level	–3.273	0.0709	–4.178	–3.512
	Ist difference	–5.8	0.0000	–4.187	–3.516
Bengaluru	Level	–3.766	0.0184	–4.178	–3.512
	Ist difference	–6.728	0.0000	–4.187	–3.516
Amritsar	Level	–3.139	0.0972	–4.178	–3.512
	Ist difference	–7.700	0.0000	–4.187	–3.516
Ahmedabad	Level	–2.419	0.3698	–4.178	–3.512
	Ist difference	–6.458	0.0000	–4.187	–3.516
Kolkata	Level	–3.436	0.0468	–4.178	–3.512
	Ist difference	–6.820	0.000	–4.187	–3.516

Table 8: Johansen's Co-integration Test for Selected Apple Varieties/Grades/Markets

Maximum Rank	*Eigen Value*	*Trace Statistics*	*Critical Value (5 per cent)*	*Max. Statistics*	*Critical Value (5 per cent)*
0	–	430.41	156.0	163.51	51.42
1	0.97567	266.90	124.24	88.76	45.28
2	0.86698	178.14	94.15	78.58	39.37
3	0.83236	99.56	68.52	57.68	33.46
4	0.73041	41.88	47.21	22.84	27.07
5	0.40495	19.04	29.68	13.89	20.97
6	0.27067	5.15	15.41	4.97	14.07

Trace test indicates 4 co-integrating equations at the 0.05 level.

directional causality in the Delhi market, which affect to the prices of apple in Kolkata, Sopore, Parimpora, Amritsar and Ahmedabad. These uni-directional relationships where prices of one market affects to the prices of other market without having a reciprocal impact on the prices would imply that the market for such varieties/ grades is not very efficient in terms of influencing the prices of the other markets and also would increase prices in such markets. Similarly, the bi-directional causation

Table 9: Granger Causality Test Statistics for Selected Apple/Markets

Causality	χ^2	Prob.	Relationship
SOPORE does not Cause PARIMPORA	6.949	0.1390	P → S
PARIMPORA Cause SOPORE	45.702	0.0000	
DELHI Cause PARIMPORA	25.506	.0000	D → P
PARIMPORA does not Cause DELHI	3.3248	0.505	
AHMADABAD Cause PARIMPORA	3.010	0.0000	A ↔ P
PARIMPORA Cause AHMADABAD	58.601	0.0000	
KOLKATTA Cause PARIMPORA	17.877	0.001	P ↔ K
PARIMPORA Cause KOLKATTA	22.894	0.000	
MUMBAI Cause PARIMPORA	15.102	0.004	M → P
PARIMPORA does not Cause MUMBAI	1.194	0.879	
AMRITSAR Cause PARIMPORA	4.1478	0.386	P → A
PARIMPORA does not Cause AMRITSAR	64.224	0.000	
DELHI Cause SOPORE	19.768	0.001	D → S
SOPORE does not Cause Delhi	2.5196	0.641	
AHMADABAD cause SOPORE	30.478	0.000	Ah ↔ S
SOPORE cause AHMADABAD	26.037	0.000	
KOLKATTA cause SOPORE	44.189	0.000	K ↔ S
SOPORE cause KOLKATTA	12.97.1	0.011	
MUMBAI cause SOPORE	28.411	0.000	M → S
SOPORE does not cause MUMBAI	2.415	0.660	
AMRITSAR cause SOPORE	14.736	0.005	A « S
SOPORE cause AMIRTSAR	51.335	0.000	
AHMADABAD does not cause DELHI	3.084	0.544	D → AH
DELHI cause AHMADABAD	15.927	0.003	
KOLKATTA cause DELHI	6.3923	0.172	D → K
DELHI does not cause KOLKATTA	23.834	0.000	
MUMBAI does not cause DELHI	8.7023	0.069	M → D
DELHI cause MUMBAI	7.5057	0.111	
AMRITSAR cause DELHI	5.605	0.231	D → A
DELHI does not cause AMRITSAR	37.578	0.000	
KOLKATTA cause AHMADABAD	21.393	0.000	AH ↔ K
AHMADABAD cause KOLKATTA	69.235	0.000	
MUMBAI cause AHMADABAD	30.381	0.000	M → AH
AHMADABAD does not cause MUMBAI	1.8136	0.770	
AHMADABAD cause AMRITSAR	55.358	0.000	AH ↔ A
AMRITSAR cause AHMADABAD	12.093	0.017	
MUMBAI cause KOLKATTA	23.315	0.000	M → K
KOLKATTA does not cause MUMBAI	4.9629	0.291	
AMRITSAR cause KOLKATTA	2.1441	9.7	K → A
KOLKATTA cause AMRITSAR	24.276	0.000	
AMRITSAR does not cause MUMBAI	7.1078	0.130	M → A
MUMBAI cause AMRITSAR	35.37	0.000	

was observed in Parimpora and Kolkata, Ahmedabad and Sopore, Kolkata and Sopore and between Ahmedabad and Kolkata. It should be noted here that the Granger causality results may vary for different number of lags or time horizon included in the models. Granger causality always has to be tested in the context of some model. In the specific case of the test function, the model has p past values of each of the two variables in the bi-variate test. A conventional way to choose p for this model would be to try this regression with various values of p and use keep track of the AIC or BIC for each lag length. Then run the test again using the value of p which had the lowest Information criteria in the regressions. In the present investigation, it is found that the AIC values are minimum at 3 to 6 lags. Accordingly specific lag length was used for testing the causality with the help of VAR model.

This implies that the markets by and large have enough ability to predict subsequent prices among them. The results of the study are therefore quite useful to various stakeholders like producers, traders, commission agents and policy makers. The results will be helpful in framing well oriented policy on market intervention schemes for an open commodity market in apple.

(g) Price Forecast

Daily price data of apple for seven important markets of Delhi, Mumbai, Amritsar, Kolkata, Ahmedabad, Parimpora and Sopore were tabulated and finally data was arranged fortnightly in order to carry out forecasting for the benefit of producers so that they can choose mandie of where their produce will fetch good prices. Further it will help them to diversify their market option and to some extent their harvesting time schedule for better returns. In order to carry out the price forecast daily price data of apple of different markets from the year 2005 to 2014 was taken into account and subsequently forecasting was done using VAR and VECM model and present inn Table 10.

5. CONCLUSION

An attempt was made in this study to examine the marketing of apple and marketing integration of apple across different major markets in India employing both primary and secondary data. The results revealed that the state has made a good progress on front of area expansion and increasing production. The marketing system of apple was unorganized characterized by number of marketing channels and existence of number of intermediaries which deprive farmers of real benefits. Further the major markets of apple have either one way or two way causations. Forecast of prices made on available data revealed its steady increase towards upcoming season. The findings of the study put forth few policy implications as:

The study emphasized upon scientific management to improve productivity of apple orchards across different regions of the state.

Since farming community of the state has a predominance of small/marginal category, they lack resource to be spent on marketing and production. Their resource position has to be supplemented as credit from external sources especially through government institutions wherein group lending should be given priority.

Table 10: Price Forecast of Apple for the Year 2015-16

Date	Delicious (Rs./FC)		American (Rs./FC)		Maharaji (Rs./FC)	
	Grade I (Min-Max)	Grade II (Min-Max)	Grade I (Min-Max)	Grade II (Min-Max)	Grade I (Min-Max)	Grade II (Min-Max)
Parimpora (Mandi)						
16th to 30th sep.	701-803	424-535	595-693	316-387	309-388	169-255
Ist to 15th Oct.	680-841	406-560	584-707	311-392	308-395	154-274
16th to 31th Oct.	664-869	393-580	579-719	306-397	306-400	142-288
Ist to 15th Nov.	652-895	383-598	575-730	302-402	304-404	131-300
16th to 30th Nov.	642-919	374-614	572-741	298-406	302-407	122-310
Ist to 15th Dec.	666-941	367-629	570-752	294-411	300-411	113-319
16th to 31th Dec.	658-960	360-643	568-762	291-414	298-415	111-312
1st to 15th jan	652-980	354-657	567-771	288-418	297-418	103-319
Sopore (Mandi)						
16th to 30th sep.	688-786	521-637	650-812	351-471	–	–
Ist to 15th Oct.	665-820	506-661	623-858	328-500	–	–
16th to 31th Oct.	649-847	495-682	602-895	311-523	–	–
Ist to 15th Nov.	636-871	486-699	585-927	297-542	–	–
16th to 30th Nov.	625-892	479-716	570-955	285-560	–	–
Ist to 15th Dec.	647-912	472-731	558-981	275-576	–	–
16th to 31th Dec.	638-931	467-745	548-1006	265-591	–	–
1st to 15th jan	631-949	462-759	538-1029	257-605	–	–
Delhi (Mandi)						
16th to 30th sep.	834-990	479-634	733-857	597-721	442-526	317-389
Ist to 15th Oct.	802-1043	450-669	719-888	579-751	428-547	311-407

Contd...

Table 10–*Contd...*

Date	Delicious (Rs./FC)		American (Rs./FC)		Maharaji (Rs./FC)	
	Grade I (Min-Max)	Grade II (Min-Max)	Grade I (Min-Max)	Grade II (Min-Max)	Grade I (Min-Max)	Grade II (Min-Max)
16th to 31th Oct.	777-1082	429-697	709-913	567-776	417-564	306-419
Ist to 15th Nov.	757-1116	412-721	700-935	557-798	409-578	301-430
16th to 30th Nov.	741-1148	397-743	693-955	550-818	402-591	297-440
Ist to 15th Dec.	763-1177	384-763	688-974	544-837	395-603	293-448
16th to 31th Dec.	751-1205	372-781	683-992	538-855	390-614	290-457
1st to 15th jan	739-1231	361-799	679-1009	534-873	385-625	287-465
Bengaluru (Mandi)						
16th to 30th sep.	1074-1198	796-952	764-931	520-692	614-801	441-587
Ist to 15th Oct.	1047-1244	773-992	741-975	490-729	582-846	420-594
16th to 31th Oct.	1029-1283	757-1024	720-1010	468-758	558-882	409-604
Ist to 15th Nov.	1016-1317	745-1053	703-1040	449-783	539-913	402-616
16th to 30th Nov.	1005-1348	735-1079	689-1067	433-806	523-941	397-627
Ist to 15th Dec.	1047-1378	726-1103	676-1091	419-827	509-967	392-639
16th to 31th Dec.	1040-1405	719-1126	665-1115	406-846	497-992	388-650
1st to 15th jan	1034-1433	714-1148	655-1137	394-865	486-1015	384-660
Kolkata (Mandi)						
16th to 30th sep.	1046-1165	488-598	666-797	531-643	–	–
Ist to 15th Oct.	1021-1216	472-623	643-826	512-670	–	–
16th to 31th Oct.	1004-1257	459-643	628-851	498-692	–	–
Ist to 15th Nov.	992-1293	449-661	615-874	487-711	–	–
16th to 30th Nov.	984-1301	441-676	605-894	478-729	–	–

Contd...

Table 10–*Contd...*

Date	Delicious (Rs./FC)		American (Rs./FC)		Maharaji (Rs./FC)	
	Grade I (Min-Max)	Grade II (Min-Max)	Grade I (Min-Max)	Grade II (Min-Max)	Grade I (Min-Max)	Grade II (Min-Max)
Ist to 15th Dec.	1026-1358	434-691	597-913	471-745	–	–
16th to 31th Dec.	1021-1388	428-705	589-931	464-760	–	–
1st to 15th Jan	1016-1416	422-718	583-947	458-774	–	–
Mumbai (Mandi)						
16th to 30th Sep.	809-939	625-825	–	–	–	–
Ist to 15th Oct.	801-991	592-869	–	–	–	–
16th to 31th Oct.	784-1018	569-905	–	–	–	–
Ist to 15th Nov.	774-1046	551-937	–	–	–	–
16th to 30th Nov.	764-1071	535-966	–	–	–	–
Ist to 15th Dec.	795-1094	521-993	–	–	–	–
16th to 31th Dec.	789-1116	509-1018	–	–	–	–
1st to 15th Jan	783-1138	499-1042	–	–	–	–
Ahmadabad (Mandi)						
16th to 30th Sep.	933-1049	622-742	–	–	–	–
Ist to 15th Oct.	913-1099	605-775	–	–	–	–
16th to 31th Oct.	897-1137	594-802	–	–	–	–
Ist to 15th Nov.	886-1172	585-826	–	–	–	–
16th to 30th Nov.	878-1204	579-848	–	–	–	–
Ist to 15th Dec.	916-1233	573-868	–	–	–	–
16th to 31th Dec.	911-1262	569-888	–	–	–	–
1st to 15th Jan	908-1289	566-906	–	–	–	–

Contd...

Table 10–*Contd...*

Date	*Delicious (Rs./FC)*		*American (Rs./FC)*		*Maharaji (Rs./FC)*	
	Grade I (Min-Max)	*Grade II (Min-Max)*	*Grade I (Min-Max)*	*Grade II (Min-Max)*	*Grade I (Min-Max)*	*Grade II (Min-Max)*
			Amritsar (Mandi)			
16th to 30th Sep.	523-626	341-444	–	–	–	–
Ist to 15th Oct.	496-652	321-470	–	–	–	–
16th to 31th Oct.	480-676	306-489	–	–	–	–
Ist to 15th Nov.	467-696	293-506	–	–	–	–
16th to 30th Nov.	455-715	282-521	–	–	–	–
Ist to 15th Dec.	467-732	273-534	–	–	–	–
16th to 31th Dec.	458-749	264-547	–	–	–	–
1st to 15th Jan	449-764	256-559	–	–	–	–

Cooperative production and marketing is need of the era and according they are to be encouraged in the state. There is a need to strengthen existing cooperatives through effective leadership.

There is need to strengthen market intelligence to help farmers and trading community to find remunerative markets. Although this initiative has already been taken by the Division of Agricultural Economics and Marketing, SKUAST-K under, "Network Project on Market Intelligence" with financial assistance from Indian Council of Agricultural Research (ICAR) but it has to be launched as a regular feature. The price forecasts should be disseminated through variety of media and at the same time further R&D efforts are required to assess optimum supply that should enter each wholesale market to contain price level forecasted for each market.

6. REFERENCES

Chand, R. 2003. Domestic reforms for trade liberalization: Analysis and approaches, Paper presented at the Workshop on "*Analysis of Trade Liberalization for Poverty Alleviation*", April 21-25, Colombo, Sri Lanka.

Fortenbery, T. R. and Zapata, H. O. 1993. "An examination of cointegration relations between futures and local grain markets." *J. Futures Markets*, **13**: 921-932.

Golettie, F., Raisuddin, A. and Farid, N. 1995. "Structural Determinants of Market Integration: The Case of Rice Market in Bangladesh," *Developing Economics*, Vol. 33 (2): 185-202

Jha, Shikha and Srinivasan, P.V. 1999. Grain price stabilization in India: Evaluation of policy alternatives. *Agricultural Econ.*, **21**: 93-108.

Johansen, S. 1988. A statistical analysis of cointegration vectors. *J. Eco. Dyn. Control*, **12**: 231-54.

Kellard, N., Newbold, P., Rayner, T. and Ennew, C. 1999. The relative efficiency of commodity futures markets, *J. Future Markets*, **4**: 413-432.

Ramaswami, Bharat. 2002. Efficiency and equity of food market interventions, *Econ. Political Weekly (Mumbai)*, **23**: 1129-1135.

Schumpeter, J.A. 1999. The creative response in economic history, *J. Eco. History*, **2**: 149-59.

Shen, C. and Wang, L. 1990. Examining the validity of a test of futures market efficiency: a comment. *J. Futures Markets*, **10**: 195-196.

Wang and Ke, B. 2005. Efficiency tests of agricultural commodity futures markets in China, *Australian J. Agric. Res. Eco.*, **49**: 125-141.

Wani *et al.* 2015. Market integration and price forecasting of apple in India, *Indian J. Agri. Econ.***70**: p. 169-181

2500 ppm IBA

Sand: Cocopeat : vermicompost media p.274

Soil media (Contol)

Plate 2: Chip Budded Plants in the Nursery (92 per cent success). p.274

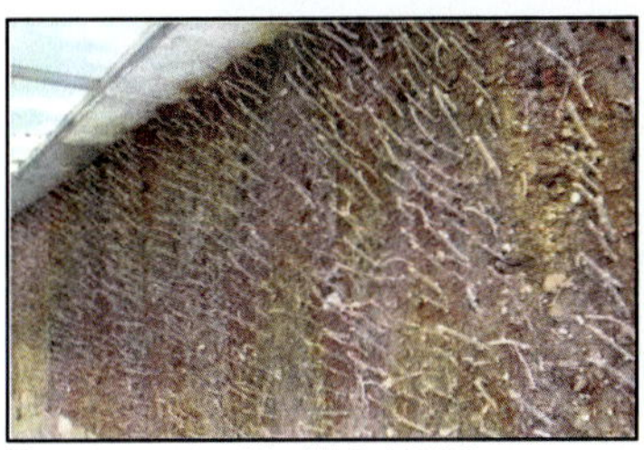

Treatment with IBA 5000 ppm and planting in growing medium of sand : forest leaf compost cocopeat (1:1:1) p.279

Plate 3: Rooting in Hardwood Cuttings of Kiwifruit.

Sowing of stratified seeds (in mid February). Chip budding (in July) Saleable plants (in December)

Plate 4: Production of Apple Nursery in a Year-under Polyhouse Conditions. **P. 280**

Fig. 1: *In vitro* Regeneration of Plantlets of *Juglans regia* and *Prunus persica* from Cryobanked Embryonic Axes. **p. 305**

Fig. 2: *In vitro* Regeneration of *P. dulcis* Plantlets in Culture Medium Raised from Cryopreserved Embryonic Axes.

A: Shoot growth initiation after 15 days of culturing; B: Well grown rooted plantlet after 35 days of culture in medium and C: Plantlet after one week of transfer to soilrite. *P. 305*

Fig. 2: 'METOS' - Microcomputer Disease Forewarning System (at SKUAST-Kashmir, Shalimar, J&K). p. 379

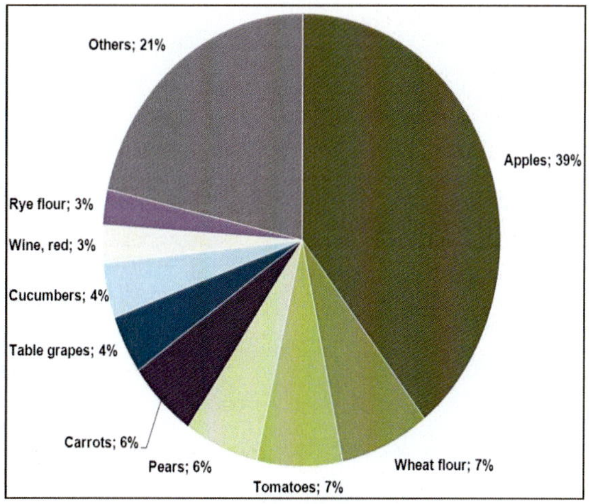

Fig. 11: Relative Contribution of Commodities to Hazard Index for Pesticide Residues in the Diet. *Source*: Anonymous 2013. p.

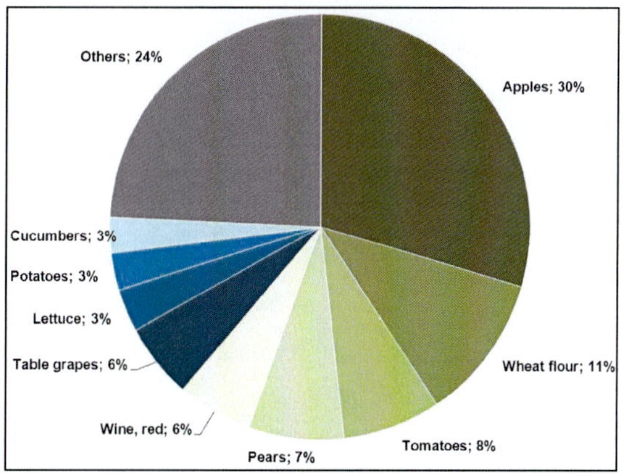

Fig.10: Relative Contribution of Commodities to Total Exposure to Pesticide Residues in the Diet. *Source*: Anonymous 2013. p. 407

Fig. 1: High Density Orchard of Apple at CITH, Srinagar. p. 437

Fig. 1: High Density Orchard of Apple at CITH, Srinagar. p. 463

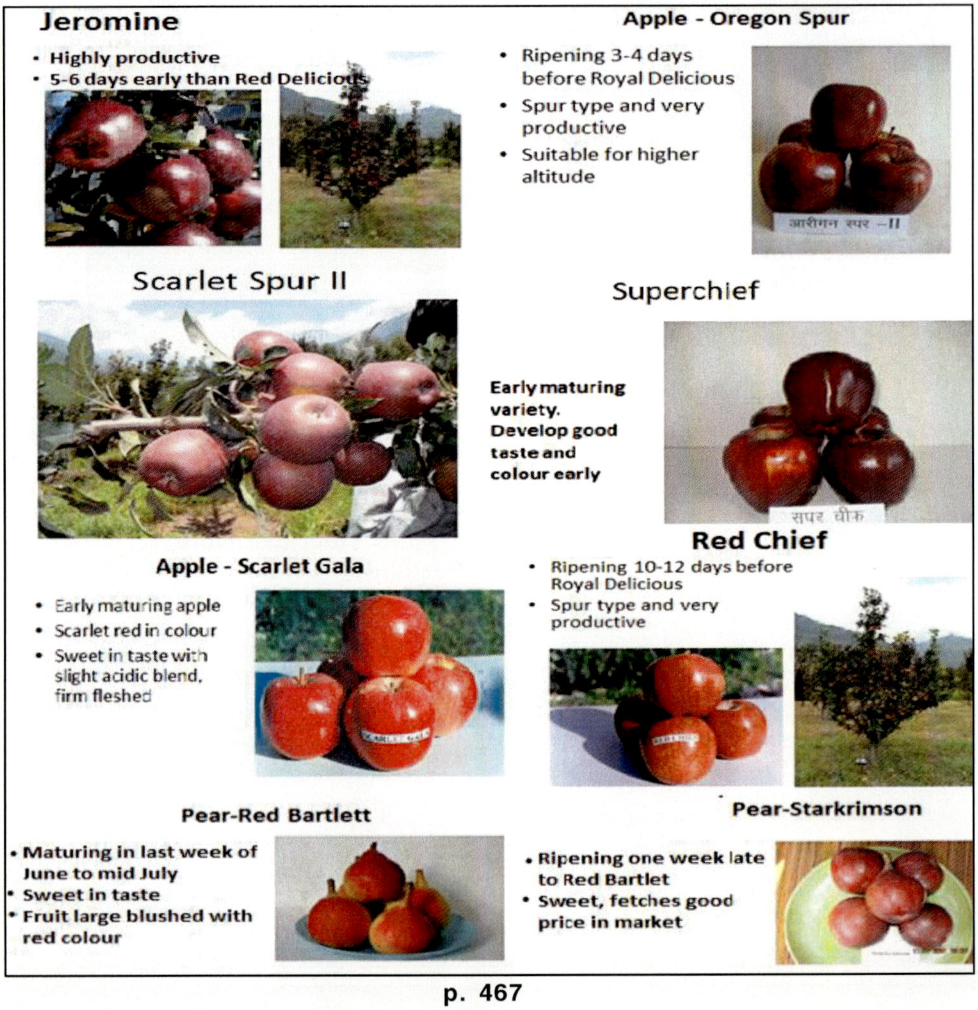

p. 467

Kiwifruit

Varieties

Hayward and

Allison

Male: Tomuri

Peaches

Red Haven

- Ripening during mid June
- Fruit dark red coloured
- with sweet taste

Sun Crest

- Ripening in the last week of July
- Fruit orange red, attaractive

Sweet taste

Nectarines

Silver King

Fruit yield: 70 to 80 Kg/ tree

Fruit weight: 80g

TSS: 7-8°

Ripening mid-May. Fruits **medium in size. Ground** colour greenish white, 3/4 fruit covered with dark red blush, Fruit very attractive, Taste is sweet.

Plum

Walnut

CITH Walnut-1

Chandler

Pecan nut

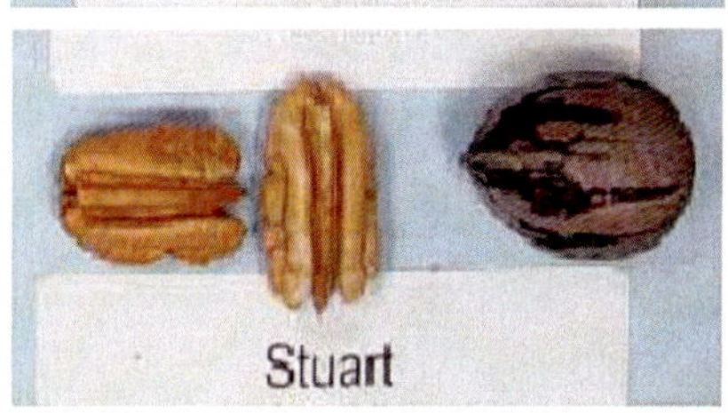

p. 467

p. 473

p. 483

p. 478

Fig. 6: Apple Shaker (Pneumatic) Harvester (American Society of Agricultural Engineers, USA). p.497

Fig. 7: Automatic Mechanical Man Positioning System for Fruit Harvesting. p.497

Fig. 8: Post Hole Digger for Mechanized. p.498

Fig. 9: Rotary Dibbler for Sowing Seed in Pot. p.498

Fig. 10: Leaf Mulcher.p.498

Fig. 11: Walnut Planting Material grown under Polyhouse. p.498

Fig. 12: Manual Walnut Dehuller. p. 499

Fig. 13: Motorized Walnut Dehuller. p. 500

Fig. 14: Almond De-hulling Hand Operated Machine.p. 500

Fig. 15: Drip Irrigation in High Density Apple. p. 501

Fig. 16: LDPE Farm Pond for Water Storage. P.502

Fig. 17: Apricot Harvesting. p. 502

Fig. 18: Horticultural Tool Kit. p.502

Fig. 19: Weeder for Inter-cultural Operation. p.503

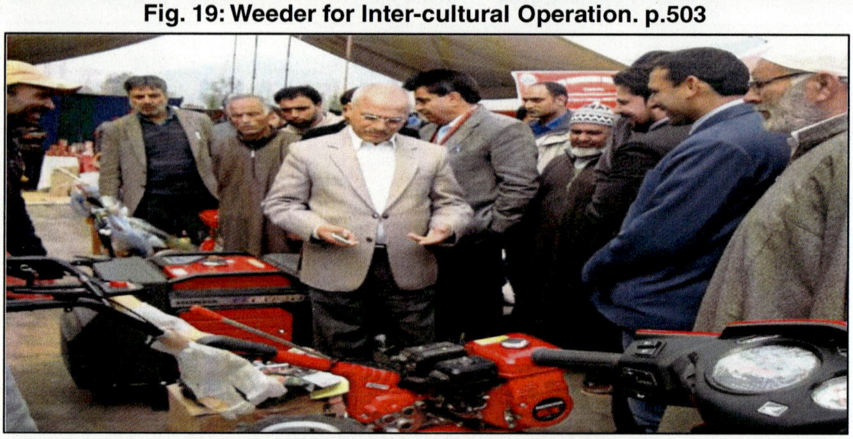

Fig. 20: Brush Cutter and Sprayer. p. 503

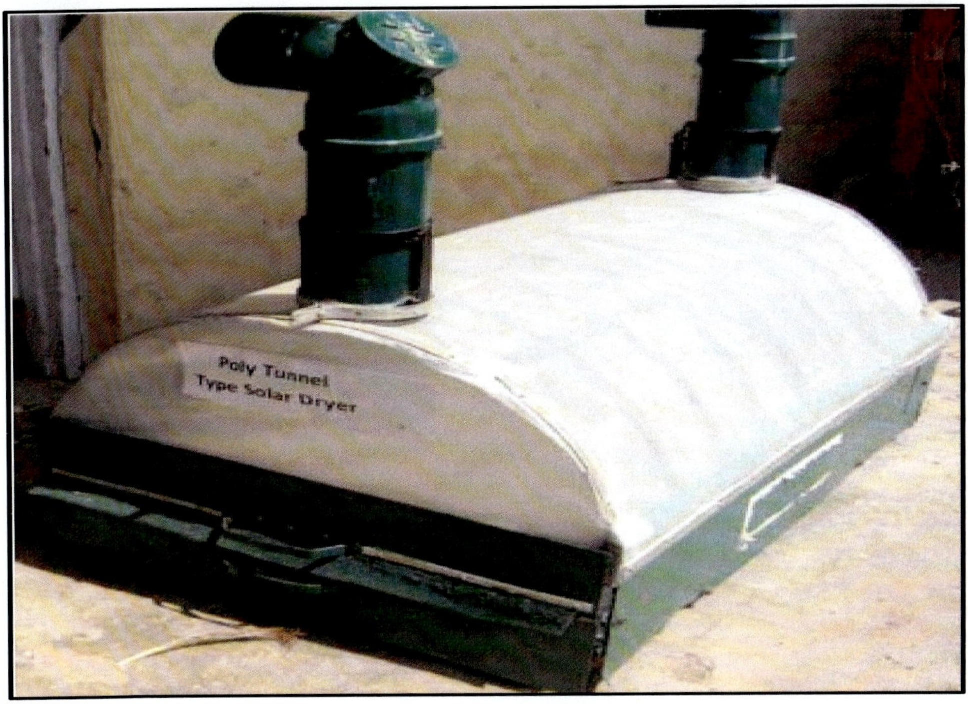

Figure 21: Solar Dryer. p. 503

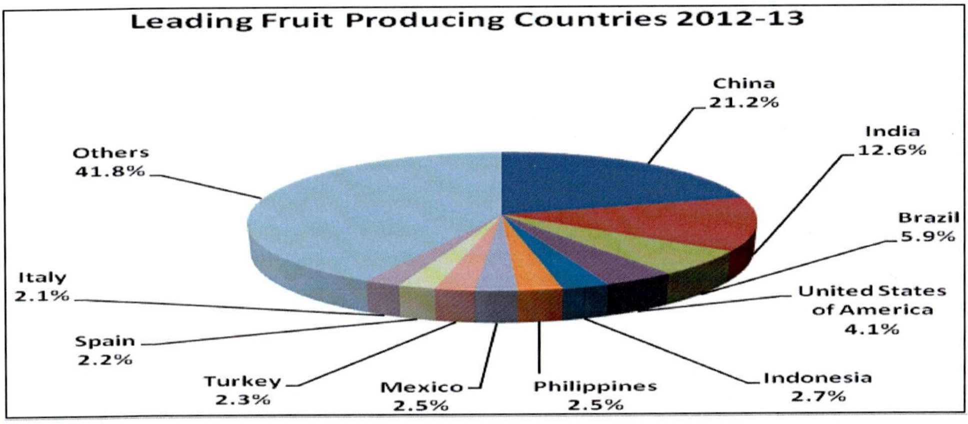

Fig. 1: Fruit Producing Countries 2012-13.
(*Source*: Indian Horticulture Database, 2014). p. 508

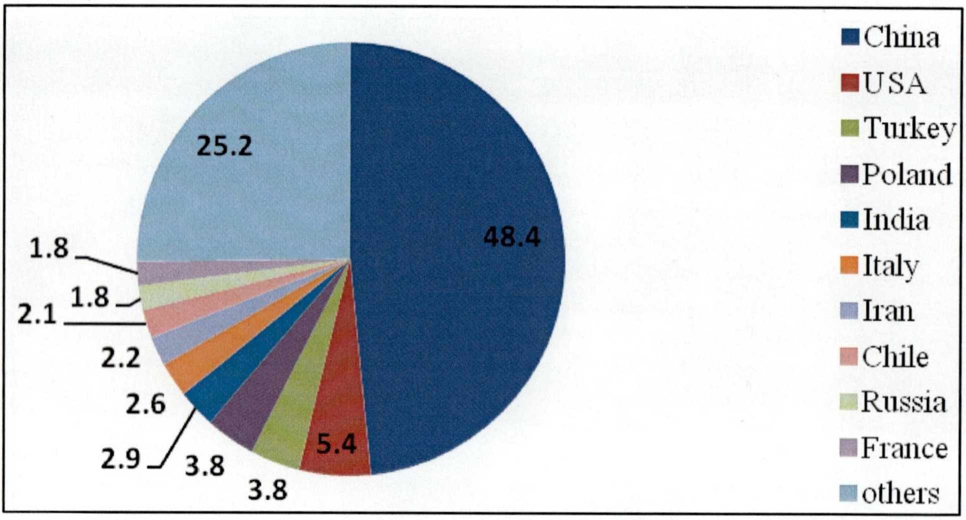

Fig. 1: Top Ten Apple Producing Countries.

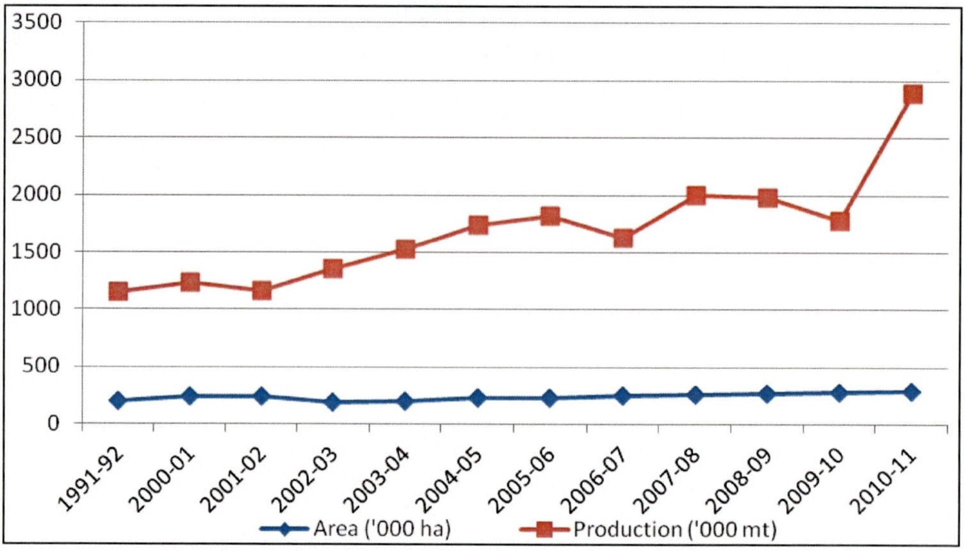

Fig. 2: Area, Production and Productivity of Apple in India.
***Source*: UN Food and Agriculture Organization (FAO), Rome.**